BEAM DYNAMICS ISSUES OF HIGH-LUMINOSITY ASYMMETRIC COLLIDER RINGS

AIP CONFERENCE PROCEEDINGS 214

BEAM DYNAMICS ISSUES OF HIGH-LUMINOSITY ASYMMETRIC COLLIDER RINGS
BERKELEY, CA 1990

EDITOR:
ANDREW M. SESSLER
LAWRENCE BERKELEY LABORATORY

American Institute of Physics   New York

Authorization to photocopy items for internal or personal use, beyond the free copying permitted under the 1978 US Copyright Law (see statement below), is granted by the American Insitute of Physics for users registered with the Copyright Clearance Center (CCC) Transactional Reporting Service, provided that the base fee of $2.00 per copy is paid directly to CCC, 27 Congress St., Salem, MA 01970. For those organizations that have been granted a photocopy license by CCC, a separate system of payment has been arranged. The fee code for users of the Transactional Reporting Service is: 0094-243X/87 $2.00.

© 1990 American Institute of Physics.

Individual readers of this volume and non-profit libraries, acting for them, are permitted to make fair use of the material in it, such as copying an article for use in teaching or research. Permission is granted to quote from this volume in scientific work with the customary acknowledgment of the source. To reprint a figure, table or other excerpt requires the consent of one of the original authors and notification to AIP. Republication or systematic or multiple reproduction of any material in this volume is permitted only under license from AIP. Address inquiries to Series Editor, AIP Conference Proceedings, AIP, 335 E. 45th St., New York, NY 10017.

L.C. Catalog Card No. 90-55857
ISBN 0-88318-767-1
DOE CONF 9002128

Printed in the United States of America.

This book was put into production on 13 August 1990 and was published on 26 October 1990.

Contents

Preface ix

PART I. SUMMARIES

Workshop Summary 2
    A. Chao, S. Chattopadhyay, E. Courant, A. Hutton, E. Keil,
    S. Kurokawa, G. Lambertson, F. Pedersen, J. Rees, J. Seeman,
    A. Sessler, M. Tigner, F. Willeke, A. Zholents, and M. Zisman

Working Groups Summaries

  Summary of the Working Group on Lattice Design 6
    P. Baigley, M. Donald, A. Garren, C. Geyer, B. Holzer, A. Hutton,
    P. Krajcik, Yu. Orlov, F. Porter, T. Risselada, R. Schmidt, T. Tanabe,
    L. Teng, N. Walker, F. Willeke, E. Wilson, and A. Zholents

  Summary of the Working Group on Instabilities 19
    P. Morton

  Report of the Working Group on Beam-Beam Effects 25
    Eberhard Keil

  Report of the Working Group on Technology 35
    Ferd Voelker and Glen Lambertson

PART II. INVITED TALKS

The Physics of BB Factories 42
    W. Schmidt-Parzefall

B-Factory Lattice Considerations 43
    A. Garren

Interaction Region Considerations 59
    H. DeStaebler

Influence of Collective Effects on the Performance of High-Luminosity Colliders 81
    Michael S. Zisman

The Beam-Beam Limit in Asymmetric Colliders: Optimization of the B-Factory Parameter Base 130
    J. L. Tennyson

The Beam-Beam Interaction: Coherent Effects 175
    Kohji Hirata

Beam-Beam Interaction: Experimental 219
    David H. Rice

RF Systems 235
    D. L. Rubin

Feedback Systems 246
    F. Pedersen

Comparison of Accelerator Physics Issues for Symmetric and Asymmetric B-Factory Rings 270
    Maury Tigner

PART III. CONTRIBUTED PAPERS

A Comparison of Flat Beams with Round 278
    S. Krishnagopal and R. H. Siemann
The Advantages of Flat Beams in Head-on Collisions 284
    Andrew Hutton
Comparison of Luminosities for Round and Flat Beams......................... 289
    V. A. Lebedev
Round Beams Generated by Vertical Dispersion 298
    P. Bagley
Beam Separation in an Asymmetric Energy Collider Using a Tilted
Experimental Solenoid... 309
    T. Risselada
A High Luminosity e^+e^- Collider with a Crossing Angle...................... 314
    Rüdiger Schmidt
B Factory with a Finite Crossing Angle 320
    B. Autin, Y. Baconnier, K. Hirata, A. Hoffmann, J. Jowett, H. Lengeler,
    D. Möhl, H. Moshammer, T. Risselada, H. H. Umstätter, T. Taylor,
    A. Verdier, and T. Wang
Dispersive Crab Crossing: An Alternative Crossing Angle Scheme....... 327
    G. Jackson
Bunch Length Compression Using Crab Cavities..................................... 336
    Y. Orlov
Monochromatic Design for the Apiary B-Factory 347
    A. Dubrovin, A. Garren, and A. Zholents
Isochronous Storage Rings and High Luminosity Electron-Positron Colliders 364
    Claudio Pellegrini and David Robin
R&D of Damped Cavities at KEK.. 381
    E. Kikutani
Reduction of Beam-Beam Synchrobetatron Resonances Using Compensating
Interaction Regions 386
    John T. Seeman
Synchro-Beam Interaction... 389
    K. Hirata, H. Moshammer, F. Ruggiero, and M. Bassetti
Determination of the High Frequency Behavior of the Impedance from
Low-Frequency Data 405
    S. A. Heifets
Curvature Effects to Beam Dynamics—Applied to the Asymmetric B Factory...... 411
    King-Yuen Ng
Trapped Ion Effects in BFI... 419
    R. Cappi
Symmetrization of the Beam-Beam Interaction in an Asymmetric Collider 424
    Yong Ho Chin
Beam-Beam Simulation with Non-Gaussian Beams.................................. 434
    E. Kikutani
The Long-Range Beam-Beam Force: Effects on the B Factory 441
    Kohji Hirata

Incoherent Beam-Beam Effect—The Relationship Between Tune-Shift, Bunch Length, and Dynamic Aperture 454
    C. D. Johnson and L. Wood

The Beam-Beam Interaction as a Discrete Lie-Poisson Dynamical System 464
    Paul J. Channell

Comments on a Linac Based Beauty Factory 484
    S. A. Heifets, G. A. Krafft, C. McDowell, and M. Fripp

Superconducting Cavities for a B-factory—Interim Progress Report 508
    H. Padamsee, W. Hartung, J. Kirchgesner, D. Moffat, D. Rubin,
    D. Saraniti, Y. Samed, J. Sears, and Q. S. Shu

PART IV. B-FACTORY PROJECTS

B-Meson Factory in the CERN-ISR Tunnel 536
    L. Rivkin

B-Factory Plans at Cornell University 561
    Maury Tigner

Study for an Asymmetric B-Factory 565
    K. Balewski, C. Geyer, B. Holzer, E. Jaeschke, D. Krämer, H. Nesemann,
    D. Proch, J. Sekutowics, F. Willeke, and S. G. Wipf

Asymmetric B-Factory Project at KEK 575
    F. Funakoshi, M. Anami, A. Asami, S. Enomoto, K. Hanaoka,
    T. Kageyama, K. Kanazawa, E. Kikutani, J. Kobayashi, H. Koiso,
    S. Kurokawa, T. Ohsawa, K. Oide, S. Sakanaka, K. Satoh, T. Shidara,
    M. Suetake, F. Takasaki, and Y. Yamazaki

Novosibirsk B-Factory 592
    A. A. Zholents

An Asymmetric B Factory in the PEP Tunnel 594
    A. Hutton and M. S. Zisman

A Linac-on-Ring Collider B-Factory Study 602
    P. Grosse Wiesmann, C. Johnson, D. Möhl, R. Schmidt,
    W. Weingarten, L. Wood, and G. Coignet

List of Participants 618

Appendix: Beam Dynamics Issues of High-Luminosity Asymmetric Collider Rings 622
    Andrew M. Sessler

Author Index 639

Preface

Machines for use in high-energy physics are advancing along two frontiers. First, there is the frontier of energy, currently being pressed by the Fermilab collider ($p\bar{p}$), and SLC and LEP (e^+e^-), and in the near future by HERA (ep), the LHC, and the SSC (pp). Second, there is the frontier of intensity, currently being pressed by a variety of low-energy machines and, at higher energies, by various linacs such as those at KEK, Fermilab, GSI, and LAMPF (p) and CEBAF (e^-). In the future there should be, along this frontier, various "factories" such as those for Kaons at TRIUMF, and those proposed for φ mesons, τ-charm particles, and B mesons. It is with the intensity frontier that these proceedings are concerned.

The elementary particle motivation to study the nonconservation of PC in the B-$\bar{B}$ system (which topic is not covered in these Proceedings, but is treated extensively in the literature) has motivated the study of very high intensity asymmetric collider rings. It was for this purpose that a Workshop on Beam Dynamics Issues of High-Luminosity Asymmetric Collider Rings was held, in Berkeley, during February 12–16, 1990.

A general introduction to the subject has been given in an article which is reprinted here as an Appendix. The nonexpert may wish to start there. The volume consists of four parts. The first part consists of Summaries; first an overall summary of the Workshop and then, second, more detailed summaries from each of the working groups. The second part consists of the Invited Talks at the workshop. The third part contains various Contributed Papers, most of which represent work that came out of the workshop. Finally, there are, in the fourth part, brief Summaries of the Various Proposed B-Factory Projects in the world.

The Workshop was supported by the U.S. Department of Energy, Office of Energy Research, Division of High Energy and Nuclear Physics. I am most grateful for this support, as well as that of the Lawrence Berkeley Laboratory.

It is a pleasure to thank the many who made the Workshop possible. They include, especially, Mollie Field, Joy Kono, Joyce Lockhart, Cheryl McFate, Louise Millard, Izetti Perry, Betty Strausbaugh, and Gladys Ureta. Special thanks goes to my colleague, Swapan Chattopadyhay, who compiled a most useful reprint volume for the participants. Finally, I want to thank Darlene Moretti for overall organization of the Workshop, for general helpfulness to all, and for preparation of these Proceedings.

Andrew M. Sessler
Lawrence Berkeley Laboratory
Berkeley, CA 94720

August 1, 1990

Part I

SUMMARIES

WORKSHOP SUMMARY

A. Chao,[1] S. Chattopadhyay,[2] E. Courant,[3] A. Hutton,[4] E. Keil,[5]
S. Kurokawa,[6] G. Lambertson,[2] F. Pedersen,[5] J. Rees,[4] J. Seeman,[4]
A. Sessler,[2] M. Tigner,[7] F. Willeke,[8] A. Zholents,[9] M. Zisman[2]

March 19, 1990

General Conclusions

1. An asymmetric B-Factory is here defined to be an e^+e^- storage ring collider capable of 10^{33}-10^{34} cm^{-2} s^{-1} luminosity in the center of mass energy range 10-11 GeV with beam energy ratios of up to 4 to 1. Based on studies of designs for such machines at eight laboratories around the world, there is no known reason to expect that such a facility cannot be built. No completely satisfactory conceptual design for such a facility exists at this time, however. Technical issues requiring further study and resolution are discussed in this report.

2. There is no question that e^+e^- collisions with luminosities in excess of 10^{32} cm^{-2}s^{-1} can be achieved in the 10-11 GeV center of mass energy regime. The success of a B-Factory hangs upon achieving 30 to 100 times this luminosity. Due to uncertainties in scaling of detector backgrounds, the beam-beam tune shift limits, or multibunch instabilities, and because the requisite extrapolation in luminosity is large, the facility designs need to be sufficiently conservative that they can be easily adjusted to accommodate the possible need for larger currents or modified collision geometry, beam energy ratio and emittances.

---

1. SSC Laboratory, USA
2. Lawrence Berkeley Laboratory, USA
3. Brookhaven National Laboratory, USA
4. SLAC, USA
5. CERN, Switzerland
6. KEK, Japan
7. Cornell University, USA
8. DESY, Germany
9. INP, Novosibirsk, USSR

Lattices

Lattices for an asymmetric collider in which beams of different energies are brought into collision can be designed with existing technology. The optics problems do not take us into unexplored ground and colliding single bunches of the necessary intensity imposes no new requirements beyond those which are already achieved in existing accelerators.

The most important new issues are the small bunch spacing and the very high total currents, which produce a considerable amount of synchrotron radiation and lost particles from which the detector must be protected.

Two alternative solutions are proposed, neither of which is totally satisfactory:

a) If the solution is based on head-on collisions, synchrotron radiation generated in the beam separation and focusing system causes the most difficult problem. The nature of this problem is well understood and we are confident that the radiation background is calculable and predictable given the beam tail distribution. An accurate model of the beam tail distribution is one of the R&D items to be resolved by more computer simulations as well as experiments.

b) If the solution is based on collision with a crossing angle, then the synchrotron radiation problem is greatly reduced. The problem, in this case, is shifted towards finding a feasible solution for a technique of controlling beam instabilities (synchro-betatron motion) that arise from crossing angles; for example, the technique of using resonators ("crab cavities"). We need to be assured of the technical feasibility of the components required for this solution, and this will probably involve some experimental work.

A final machine design requires more theoretical studies on beam dynamics issues, masking of detectors, and crossing-angle geometry and, in particular, a specific design of an interaction region lattice.

Beam-Beam Effects

The magnitude of the beam-beam Coulomb effect is normally expressed in terms of the beam-beam strength parameter ξ which has been in the range $0.03 < \xi < 0.06$ in all existing e^+e^- colliders for the last ten years. We expect that the same will hold true for asymmetric B-Factories, and their design should include the flexibility needed to cover this range of ξ with minimal loss of performance.

"Energy transparency," i.e., making the relevant combinations of the beam and lattice parameters in the two rings identical, is one way of reducing the beam-beam problems in asymmetric colliders to those of symmetric colliders, and therefore may be used as a specification for a design. Parameter combinations which are less restrictive and less complicated, but achieve the same performance are under active study by analytic work, simulation, and experiment.

4   Workshop Summary

Studies of beam-beam coherent effects indicate that although it is not required, it is preferable to build a B-Factory with equal circumference rings with equal number of bunches in each ring.

The design and performance estimates rely heavily on simulations. Our confidence in such simulations should be improved by comparing simulation codes amongst themselves and with observations on existing machines. Focused and highly controlled beam-beam experiments at operating electron-positron colliders should, despite their difficulty, be given the highest priority in order to shed light on this crucial issue, particularly in the aspects of round versus flat beams and ξ as a function of beam energy in a single machine.

Instabilities

Considerable experience has been gained over the last three decades on the subject of instabilities in storage rings. The relevant scaling parameters are the absolute intensity of the beam and the individual bunch length. The regime of bunch lengths encountered in B-Factories is in a domain where most of the major instability issues are well understood, both theoretically and experimentally. No new physical mechanism of coherent motion is expected. The intensity levels are, however, an order of magnitude larger than existing levels, due to the high luminosity (large number of bunches) required.

Because many beam bunches are utilized, it can be stated with reasonable confidence that no single bunch instabilities are expected to limit the collider performance, with proper control of the beam electromagnetic environment (impedance). The large average current, however, is certainly expected to lead to strong coupled bunch instabilities which will have to be addressed by special RF cavity designs and effective implementation of feedback systems. This can be accomplished, we believe, with careful design resulting from a well-defined R&D program.

One area that will require special attention is the phenomenon of ion trapping in electron rings. Many problems arising from trapped ions are understood and solved by well-known techniques. There remain, however, examples of ion behavior that are not understood. It should be noted that a number of operating storage rings (such as the VUV Ring (BNL), the Photon Factory (KEK), and the EPA (CERN)) utilize many bunches. Furthermore, a number of similar machines are being constructed, today, as photon factories and accumulator rings. Thus we should continue to gain much more insight into this topic in the immediate future and recommend that an R&D program be addressed to this subject.

Technology

As a basis for the final design, studies and R&D should be carried out in the following areas:

a) Design of damped RF cavities for acceleration and "crab crossing" cavities having adequate proper precision.

b) Feedback systems to control (damp) multibunch instabilities. The systems must have adequate bandwidth, gain, and power to be capable of linear operation in the presence of injection engendered oscillations.

c) The interaction region, with particular attention to the masking geometry and its effect on the beam.

d) Special magnets near the interaction region.

e) Vacuum chamber design and desorption effects.

Summary of the Working Group on Lattice Design

P. Baigley	CESR	F. Porter	Caltech
M. Donald	SLAC	T. Risselada	CERN
A. Garren	LBL	R. Schmidt	CERN
C. Geyer	DESY/MPI Heidelberg	T. Tanabe	Columbia
B. Holzer	DESY/MPI Heidelberg	L. Teng	Argonne
A. Hutton	SLAC	N. Walker	SLAC
P. Krajcik	CERN	F. Willeke	DESY
Yu. Orlov	Cornell	E. Wilson	CERN
		A. Zholents	INP Novosibirsk

reported by F. Willeke

1 Introduction

In an asymmetric B-meson factory, intense e^+ and e^- bunches with differing energies will collide with a high frequency providing luminosities in the range of $(10^{33} - 10^{34})$ $cm^{-2}s^{-1}$. The design of such a facility is a challenge for lattice designers.

The goal of the working group on lattice design was to compare the various schemes and concepts in the B-factory designs which are studied and investigated in various accelerator laboratories [1,2,3,4,5,6,7,8]. The crucial aspect of the lattice design is the layout of the interaction region (IR).

In the case of head-on collisions, the central problem is to achieve small beam sizes for high current beams which have to be separated quickly after collision. Closely related and interconnected to this are the problems arising from synchrotron radiation produced in the low-β quadrupoles and beam separator magnets, the generation of chromaticity in the IR and the required large aperture of the low-β quadrupoles.

Figure 1 summarizes the partially contradicting requirements on the lattice design. The luminosity formula

$$L = \frac{I_1 \gamma_1 \Delta \nu_1 (1 + \kappa)}{2 r_0 e \, \beta_z^*}$$

derived under the conditions of equal beam cross-sections at the interaction point indicates that a large total beam current I_1 and a small value of the

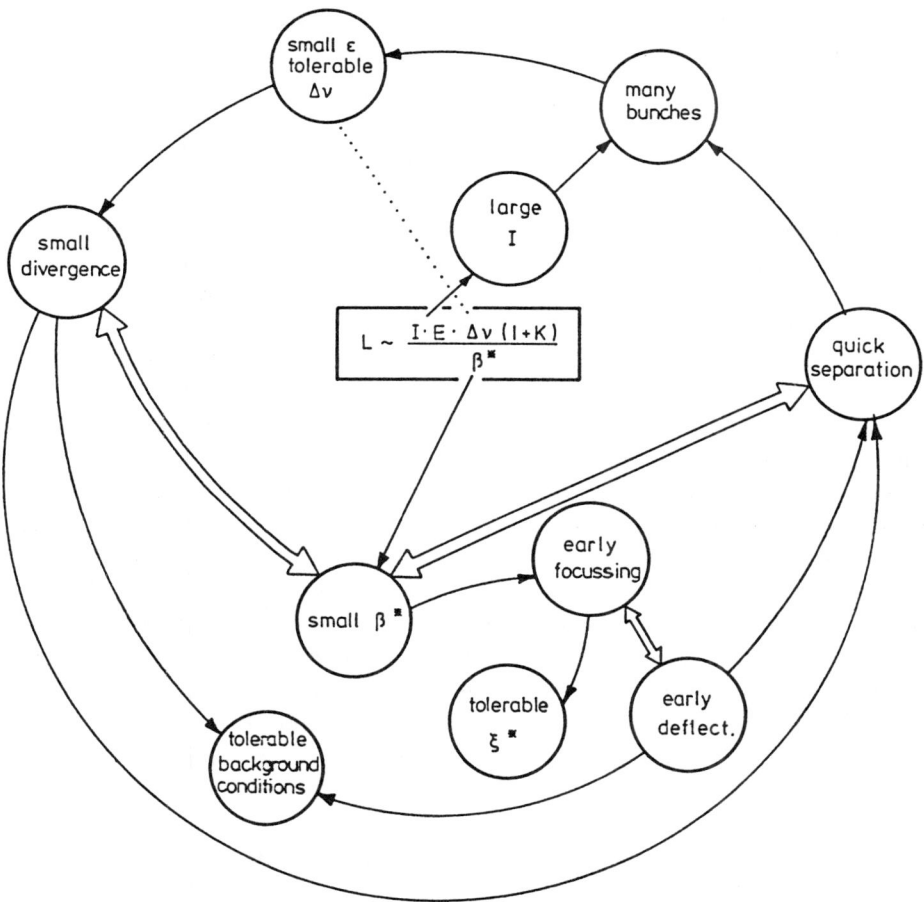

Figure 1: Lattice Design Constraints of an Asymmetric Collider

vertical β-function at the interaction point (β_y^*) are needed for high luminosity at a given beam energy E_1 and a given maximum beam-beam tune shift $\Delta\nu_1$ (κ is the aspect ratio of the beams at the interaction point (IP), e is the elementary charge, r_0 is the classical electron radius). The large total beam currents have to be distributed over as many bunches (N_b) as possible in order to allow for the smallest possible beam emittance ε_x since the transverse beam density is restricted by the tune shift limit

$$\Delta\nu_1 = \frac{r_0 N_2}{2\pi\gamma_1} \frac{1}{\varepsilon_{x1}(1+\kappa)} = \frac{r_0 I_2}{2\pi\, e\, N_b f_0 \gamma_1} \frac{1}{\varepsilon_{x1}(1+\kappa)}. \tag{1}$$

Small beam emittances mean small beam divergences $\sqrt{\varepsilon/\beta^*}$. Small divergences allow the beams to be separated quickly. This, in turn, is necessary to allow many bunches to be stored without the occurrence of parasitic beam-beam crossings with incomplete separation so that the total tune shift limit is reduced.

Small beam divergences also mean small amplitudes of the beam in the low-β quadrupole magnets. This keeps the amount of the synchrotron radiation generated very close to the interaction point at a tolerable level. The need for small values of β^* has also to be included in the picture. A small value of β^* means a large beam divergence $\sqrt{\varepsilon/\beta^*}$ which cancels the positive effect which a small emittance has on synchrotron generation background. It also prevents a quick beam separation. Moreover, the corresponding large β-values in the low-β quadrupoles ($\beta(s) = (\beta^{*2} + s^2)/\beta^*$) create a large additional contribution to the chromaticity (momentum dependence of the betatron tunes $\xi = \frac{1}{4\pi} \oint ds \beta k$, $k = $ focussing strength). Compensation of this additional chromaticity ξ^* requires additional compensation by stronger sextupole fields. There is, however, a limit to the strength of these nonlinear fields since they reduce the dynamic aperture of the machine. This problem is the more reduced, the closer the low-β quadrupoles can be placed to the IP. This, however, is in conflict with the need to place separator magnets as close as possible to the IP to achieve quick beam separation.

Various concepts have been developed in order to reduce this dilemma. They will be reviewed in section 2.

Another approach to reduce the lattice design problems is to abandon the concept of head-on collisions and to consider collisions at a certain crossing angle. Recently, a new scheme has been proposed [9] which suppresses the synchrobetatron resonances that are excited by collisions under a crossing angle [10]. Transverse rf-fields are applied to the beam which are produced in so-called crab cavities. These must be placed at locations with 90° (($2k+1$) × 90°) betatron phase advance (in the plane of the crossing angle) from the IP. But there are also scenarios in which consider a small tolerable crossing angle is considered and which would nevertheless produce sufficient beam separation. This will be discussed in section 3.

Among the subjects which have been discussed is the question of round beams. Round beams, which means equal emittances and equal β^* for horizontal and vertical betatron oscillation modes ($\varepsilon_x = \varepsilon_y, \beta_x^* = \beta_y^*$), are considered as a possibility to raise the beam-beam tune shift limit. From the point of view of lattice design, round beams would have consequences for the layout of the IR design. Furthermore, since the beams will be naturally flat, equal beam emittances require modifications of the layout of the arcs. These questions will be discussed in section 4.

Reduction of the center of mass energy spread is desirable for machine operation at narrow quarkonium resonances. This can be achieved by introducing dispersion at the interaction point such that particles with lower than nominal momentum collide with particles having a larger momentum than nominal. This situation leads to new design parameters for the whole machine and that also allows a different approach for the IR layout. This option is discussed in section 5.

2 Head-On-Collision Beam Focussing and Beam Separation Schemes

The contradictary requirements on the layout of a head-on IR are compromised in the various designs under study. Although none of the solutions proposed so far is completely satisfactory, there are many promising ideas and there is hope that several of these ideas can be combined into a good lattice design.

The S-bend solution, see fig. 2a, is an antisymmetric beam separation scheme. It has the advantage that the synchrotron radiation masks for the two beams are radially on opposite sides of the beam pipe for the two different beams. This is advantageous for the masking of the primary photon beam and may give an overall improvement in the background conditions if the synchrotron radiation from both beams is equally important. But if the synchrotron radiation background is dominated by the radiation from the high-energy beam so that secondary photons from one beam cause as many problems as primary photons from the other beam, an S-bend solution may even be slightly disadvantageous. Thus from the point of view of synchrotron radiation background, an S-bend scheme is potentially more preferable for low-asymmetry scenarios.

The largest benefit from an S-bend geometry is that it includes the option to change the head-on geometry to a crossing angle geometry with only minor changes of the lattice. S-bend schemes are part of the Cornell and SLAC design studies.

The combined function scheme (fig. 2b) resolves the problem of how to focus and to separate the beams as closely as possible to the IP. The beams begin to be focussed and deflected simultaneously in the low-β quadrupoles. In the

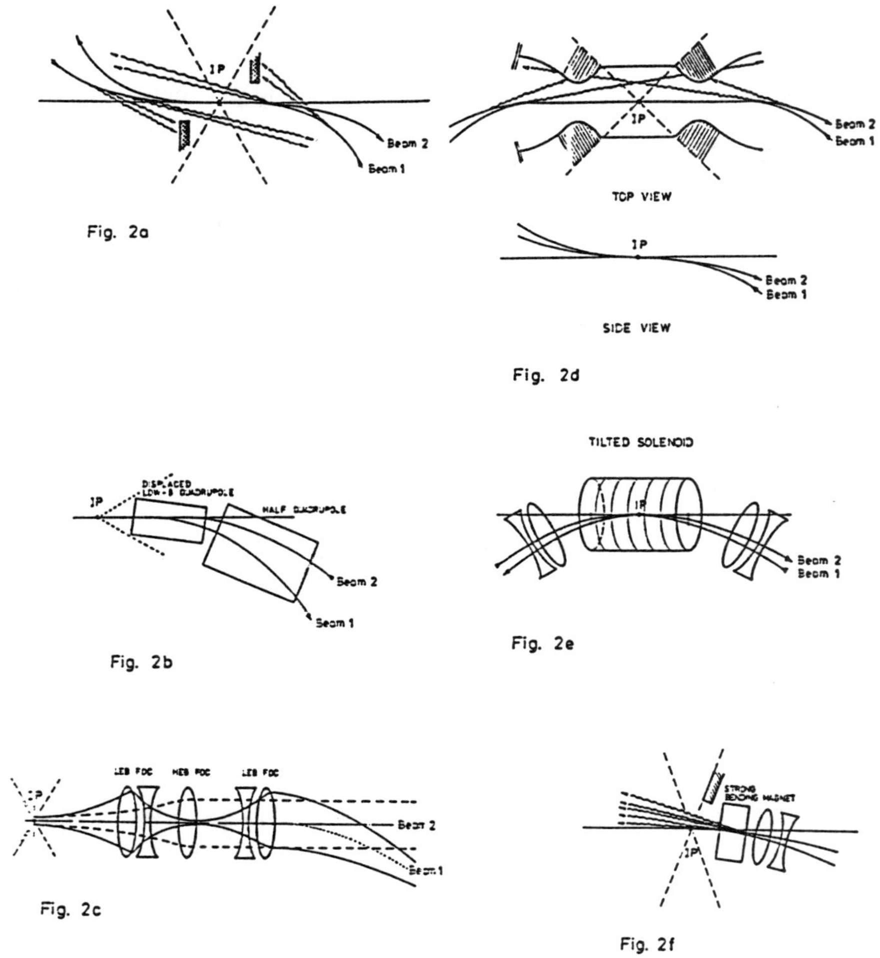

Figure 2: Beam Separation Schemes

situation where the beams are already partially separated, one has the possibility to keep the beam which causes the most important synchrotron radiation problems on or near the axis of the quadrupole magnet. Combined function magnet schemes lead to quite satisfactory solutions in case of high asymmetries [8]. However, since one cannot avoid that the central orbit of the beam passes off-centre through the quadrupoles, combined function magnets are part of almost any design.

Another idea is to put the strong focussing elements for the high-energy beam at a location where a waist occurs in the low-energy beam envelopes (see fig. 2c). This allows to focus the high-energy beam prior to separation without spoiling the focussing of the low-energy beam. The price one has to pay is an increased chromaticity (a factor of 2 in ξ^* at least).

The main beam separation can be performed in a long weak separator which is placed after the focussing lenses (seen from the IP). The scheme works best at very low asymmetries. This is used in the CERN/PSI design.

The "propeller blade scheme" is a very sophisticated arrangement which only works for very small beam emittances associated with the monochromatization scheme (see below) proposed in the Novosibirsk study. This scheme contains both horizontal and vertical deflections (see fig. 2d). The horizontal beam separation deflection by dipole or combined function magnets is symmetric. In addition to this, there are antisymmetric vertical deflections which cause the two synchrotron radiation fans from the main horizontal deflections to be separated vertically at the IP. The synchrotron radiation fan of one beam would normally hit the back of the synchrotron radiation mask for the opposite beam. This is avoided in the propeller blade scheme since the radiation fan can pass through a slot in the mask for the opposing beam. This latter is possible due to the vertical separation of the two fans. The main benefit from this arrangement is a smooth synchrotron radiation mask system which reduces both the amount of power dissipated in the collimator system and the impedance seen by the beam. A disadvantage of the scheme is that it requires extremely well-steered and stable beams at the IP.

A tilt of the detector solenoid, fig. 2e, provides a transverse magnetic field right at the IP, i.e. as early as possible. The direct synchrotron radiation from the tilted solenoid does not cause any background problems. The only disadvantage of this scheme is that it is incompatible with an antisymmetric arrangement. The end field effects of a tilted solenoid are under study [11]. Tilted solenoids are used in the CERN/PSI design.

A straightforward solution is to place a very strong bending magnet close to the IP (see fig. 2e). The large amount of sychrotron radiation generated by this magnet produces background in the detector in the form of secondary or higher-order photons. This provides early beam separation without occupying too much of the space needed for the focussing elements. Strong bend schemes have been studied in the SLAC design.

The author would like to emphasize at this point that all of these schemes are rather complementary to each other and are not alternatives. Most of the designs discussed contain more than one of the features presented.

3 Crossing Angle Geometry

In a crossing angle geometry with sufficiently large crossing angle, the beams can be separated very quickly after collision. From the point of view of parasitic beam-beam crossings, ten standard deviations of the beams (a gaussian distribution is assumed) are considered necessary and sufficient [12]. Quick beam separation allows to put many bunches in the machine. Then for a given maximum total beam current the beam emittance has to be made relatively small in order to preserve the beam-beam tune shift. A small emittance helps to keep the betatron amplitudes in the low-β quadrupoles small and this reduces the synchrotron radiation. Combining these points the required crossing angle α is therefore given by

$$\alpha = 10 \times \sqrt{\frac{\Delta l}{\beta^*}\left(1 + \frac{\beta^{*2}}{\Delta l^2}\right)} \sqrt{\frac{r_0 I}{\pi e c \gamma \Delta \nu (1+k)}}.$$

From this formula we can see that minimum emittance and minimum crossing angle are obtained for $\Delta l = \beta^*$, if one chooses the beams to be separated at a distance from the IP equal to β^*. This would allow to fill every bucket of the beam with particles. Typical values for the crossing angle would be in the order of 3.5 mrad for a B-factory with $L \approx 1 \cdot 10^{33}\ cm^{-2}s^{-1}$ ($I = 1$ Amp. $\Delta \nu = 0.03$).

Another advantage of a crossing angle geometry is that one can align the axis of the low-β quadrupoles on the trajectory of the incoming beam (the beam which moves towards the IP). The low-β quadrupoles do not have a dipole component for the incoming beam. This considerably reduces the synchrotron radiation which is directed towards the IP since the radiated power P in a quadrupole is proportional to κ ($\sigma_x^2 + \sigma_y^2 + x_0^2 + y_0^2$). σ_x, σ_y are the horizontal and vertical standard deviation of the beam size. We can assume that σ_x, σ_y are usually smaller than the offsets x_0, y_0. For the incoming low energy beam this alignment does not cause any problem. It may not always be possible to do so for the high-energy beam since the low-β quadrupoles will focus the low-energy beam back onto the high-energy beam trajectory. The pre-separation may then be insufficient at the septum necessary to achieve final separation. For low energy asymmetry, this asymmetry in the beam separation is naturally smaller and the problem is easier to solve. Crossing angle geometries are therefore more likely to be a good solution for low energy asymmetry designs.

Beam-beam interactions with a crossing angle, however, excite synchro-betatron resonances [13] which led in the case of DORIS I to a strong reduction of the maximum tolerable tune shift and thus a reduction of the luminosity [13]. More recently, the crab-crossing scheme [9] has been proposed for storage rings as a means of suppressing these synchro-betatron resonances [10]. In the original version, transverse rf resonators were to be used to provide a transverse kick in the crossing angle plane. These kicks transform after $90°$ betatron phase advance into a rotation of the bunch around the transverse axis which is perpendicular to the crossing plane. Recent studies on crab-crossing schemes [14,13] did not reveal any fundamental problems. In the working group (a joint session with the "Technology Group") an alternative scheme has been proposed [16] which uses longitudinal rf resonators which may be part of the normal rf structure. The coupling between longitudinal kicks and transverse motion is accomplished by having a large dispersion in the cavities. These need to be separated by a $180°$ β-tron phase advance from the interaction point.

The interesting question of the tolerable crossing angle has been discussed in the working group [17]. A figure of merit for the effect of a crossing angle has been given by Piwinski [13] as $\frac{1}{2}\alpha\sigma_L/\sigma^*$ (α crossing angle, σ_L bunch length, σ^* beam size at the interaction point in the plane of the crossing angle).

Since the value for DORIS I which suffered from synchro-betatron oscillations amounted to

$$\frac{\alpha\sigma_L}{2\sigma^*} \approx 0.5, \text{ we require that } \frac{\alpha\sigma_L}{\sigma^*} \ll 1.$$

Inserting the expression for the necessary crossing angle which has already been discussed, one obtains

$$10\sqrt{2}\,\frac{\sigma_L}{\beta^*} \ll 1 \text{ or } \frac{\beta^*}{\sigma_L} \gg 14.$$

This is a criterion which can apparently be satisfied, provided that one considers a small aspect ratio $\kappa = \sigma_z^*/\sigma_x^* = \frac{\beta_z^*}{\beta_x^*}$ of the beam and that the crossing angle is in the horizontal plane.

These considerations should encourage lattice designers to investigate and study crossing angle geometry designs which promise to have a large potential in solving the most crucial problems in B-factory lattice design.

4 The Round Beam Issue

The aspect ratio κ of one, $(\beta_x^* = \beta_y^*)$ and $\varepsilon_x = \varepsilon_y$ is considered provide a possibility to overcome the tune shift limit which is typically in the range of $0.02 < \Delta\nu < 0.06$ for most $e^+ - e^-$ storage rings [18]. It has also been suggested that there is an additional factor of ≈ 2 to gain because of the factor $(1 + \kappa)$ in the luminosity formula.

However, at least in case that the chromaticity, $\xi_{x,y}^*$, generated in the IR limits the minimum β^* values, the aspect ratio κ cannot be raised without raising the vertical β^* values. A thin lens analysis (a low-β doublet starting at a distance L from the IP, the two lenses spaced by L as well) shows that the vertical β decreases with decreasing κ for constant chromaticity $\zeta_x^* + \xi_y^*$.

$$\beta_y^*(\kappa) = \beta_y^*(\kappa = 1) \frac{\sqrt{2}(\kappa+1) + \kappa - 1}{2\sqrt{2}} \quad \left(\text{for } \kappa < \frac{\beta^{*2}}{4L^2}\right)$$

In going from $\kappa = 0.1$ to $\kappa = 1$, this formula tells that one has to increase β_y^* by a factor of 4.3 which is larger than the gain in the factor $1 + \kappa$. This would reduce the luminosity for round beams by a factor of ≈ 2 which would have to be overcome by an increase of the beam-beam tune shift by more than 2.

Another lattice design aspect of round beams versus flat beams is the beam divergence. In the round beam case

the beam divergence Θ which determines the betatron oscillation amplitudes in the low-β quadrupole magnets is the same in the horizontal and in the vertical plane $\Theta = \sqrt{\varepsilon/\beta^*}$. According to the formulae of the preceding paragraph, in the case of flat beams, the vertical β-function β_y^* is reduced by a factor of $(2\sqrt{2}(\kappa + 1) + \kappa - 1)/(2\sqrt{2})$, and the vertical beam emittance is reduced to $2 \kappa \varepsilon$. The ratio of beam divergence is then

$$\frac{\Theta_{\text{flat}}}{\Theta_{\text{round}}} = \sqrt{\frac{4\sqrt{2}\kappa}{2\sqrt{2}(\kappa+1) + \kappa - 1}}$$

The same ratio is obtained for the divergence in the horizontal plane. This means that the beam divergence for a flat beam with $\kappa = 0.1$ is only about half the value of the round beam. From this geometrical argument one would conclude that round beams are more disadvantageous with regard to synchrotron radiation background. The critical energy of the radiation from the quadrupoles is expected to be larger and so is the total radiated power. However, the discussions in the working group on beam-beam interaction revealed that, according to tracking calculations [19], a flat beam develops non-gaussian tails with noticeable particle densities up to amplitudes which correspond to the 10 σ value of a Gaussian distribution. Such tails have not been observed for round beams. Therefore it is not quite clear whether flat beams are really advantageous with regard to synchrotron radiation. Flat beams, however, allow quicker beam separation, especially in case of a crossing angle in the horizontal plane.

In order to produce round beams, one also has to produce a large vertical emittance $\varepsilon_y = \varepsilon_x$. This can be accomplished by having a vertical dispersion. A direct generation of vertical dispersion with vertical bending magnets involves non-planar geometries which would result in a considerable complication of the lattice design. This does not appear to be favourable. Another approach is to use skew quadrupoles to couple the horizontal dispersion into the vertical plane. A layout for CESR has been worked out [20].

Two groups of skew quadrupoles which are effectively spaced by an integer multiple of $180°$ vertical β-tron phase advance are necessary. The required skew quadrupole strength corresponds to the normal quadrupole strength in CESR.

A very elegant scheme has been proposed at SLAC [21]. It uses a so-called rotator consisting of 7 quadrupoles which form 3 FODO cells. The strengths are adjusted to produce a horizontal phase advance of $3 \times 120°$ and a vertical phase advance of $3 \times 60°$. The whole insertion is rotated by $45°$. This results in a transfer matrix

$$T = R^{-1}(45°) \begin{pmatrix} 1 & 0 & 0 & 0 \\ 0 & 1 & 0 & 0 \\ 0 & 0 & -1 & 0 \\ 0 & 0 & 0 & -1 \end{pmatrix} R(45°) = \begin{pmatrix} 0 & 0 & 1 & 0 \\ 0 & 0 & 0 & 1 \\ 1 & 0 & 0 & 0 \\ 0 & 1 & 0 & 0 \end{pmatrix}$$

which exchanges the horizontal and vertical oscillation mode (R $(45°)$ is a $45°$ rotation matrix). Two such rotators placed at opposite sides of the ring create the situation where the first oscillation mode is excited by quantum fluctuation of the synchrotron radiation in one half of the ring, and the second oscillation mode is excited in the other half.

Finally, there is also the possibility to operate on the coupling resonance $Q_x = Q_y$ with a carefully minimized width of the coupling resonance $\Delta = Q_I - Q_{II}$ for $Q_x = Q_y$. This would not require any special hardware to be installed into the lattice. Although some progress has been made recently in understanding the nonlinear beam dynamics under such operating conditions [22], most of the participants did not feel comfortable with the idea of mixing coupling and nonlinear motion.

One may conclude this section by the statement that round beams are not favourable from the point of lattice design. No fundamental problems are expected, however, in just producing the required large vertical beam emittance.

5 Monochromatization Scheme

The monochromatization scheme is the basis of the design work done at the INP Novosibirsk [6]. The idea is to have large values of the dispersion D at the interaction points with different signs in both rings so that the product of the beam energies of colliding particles

$$\sqrt{(E_1 + \Delta E_1)(E_2 + \Delta E_2)}$$

differs from the nominal $E_1 E_2$ only by second-order terms in $\frac{\Delta E}{E}$ center of mass. The center of mass energy spread depends on the transverse beam.

$$\sigma_\varepsilon = \sqrt{\frac{2\varepsilon\beta^*}{D^2} E_1 E_2}$$

and is then independent in first order of the energy spread in each beam. In this case, the beam cross-section at the IP σ^* can be made large enough to obey the tune shift limit law by making the energy spread large in the beam and making the transverse emittance small at the same time. Small emittance is obtained by having a small momentum compaction factor which also leads to a small bunch length. This, in turn, will allow a small β^* to be used. The beam divergence will be moderate nevertheless, because of the small emittance value. Both quick beam separation and tolerable synchrotron radiation background conditions may be achieved. The disadvantage of the scheme is that synchrobetatron resonances will be excited owing to the large dispersion at the IP. This question has not been addressed further in the working group.

6 Miscellaneous Subjects

The participants of the working group agreed as a very general point, that the lattice design must be sufficiently flexible that it can match the main parameters to whatever will be found to be the maximum beam current. The uncertainty of the maximum current is rather large because in the machines under consideration one is aiming for an increase of beam current by a factor of up to 10^2 with respect to what can be achieved today. To obtain maximum luminosity, one has to obtain maximum tune shifts for any given beam current. This should not be done by reducing the number of bunches since the synchrotron radiation background problem is reduced by distributing the current available always on as many bunches as possible which allows a small emittance. Therefore the beam emittance has to be variable over a large range. A change in the focussing of a FODO structure from 60°/cell to 90°/cell reduces the emittance by a factor of 3 [23], which is not sufficient. Additional devices such as a wiggler in a region with large dispersion are needed. A combined-function wiggler which changes the damping distribution might be more effective. Participants recommended that such devices should be foreseen in a B-factory lattice.

The analysis of sources of magnetic field imperfection and the sensitivity of the beam optics to these distortions was discussed. The participants are confident that orbit deviation and adjustment and stabilization of beam positions are not a big problem. There was some concern about endfield effects. In particular it was mentioned that the endfield of a tilted solenoid contains considerable octupole content which might have some influence on the beam dynamics. This problem is investigated in ref. [11]. The use of Lambertson septa for beam separation is not expected to introduce serious field imperfection problems.

It has been pointed out that coupling compensation in a double ring collider needs to be somewhat more sophisticated than just the compensation of the coupling resonance. Even with compensated resonances, the beam ellipses can be tilted at the interaction point, the tilt angles can be quite large and they may have opposite signs. This can lead to considerable reduction of luminosity, especially for flat beams.

A new scheme for compressing the bunch length has been proposed. It combines two sets of transverse rf resonators ("crab cavities") with a curved section in between. The proposed section needs to be isochronous in order to avoid the influence of synchrotron motion. A detailed discussion of this idea will be given in a separate paper [24].

7 Conclusions

The participants arrived at the conclusion that the most crucial aspect of the lattice design is the layout of the interaction region. A solution to this problem, satisfying to all of the participants has not been presented. The weakest point in all the designs considered is related to synchrotron radiation background. Since a crossing angle geometry promises advantage with regard to this problem, the participants agree that designs with finite crossing angle deserve more study and attention. Lattice designers are encouraged to work out a consistent reference design with crossing angle. Apart from IR design all the other aspects do not require more sophistication than has already been achieved in existing e^+e^- storage rings. In particular, no new technology needs to be developed.

Flat beams are more favourable from a lattice design point of view than round beams. Furthermore a large increase (> 2) in the tune shift limit would be required to balance the reduction of luminosity which is inherent in the use of round beams.

8 Acknowledgement

The author is obliged to Dr. D. Barber for discussions and for careful reading of the manuscript.

References

[1] L. Rivkin, these proceedings

[2] C. Johnson, these proc.

[3] S. Heifets, these proc.

[4] M. Tigner, these proc.

[5] Y. Funakoshi, these proc.

[6] A. Zholents, these proc.

[7] A. Hutton, these proc.

[8] K. Balewski et al., these proc.

[9] R. Palmer, SLAC PUB 4707, (1988)

[10] K. Oide, K. Yokoya, SLAC PUB 4832, (1989)

[11] E. Wilson, these proc.

[12] E. Keil, these proc.

[13] A. Piwinski, IEEE Trans. Vol. NS-32 No. 5, p. 2240 (1985)

[14] K. Oide, G.-A. Voss, SLAC PUB 5011, (1989)

[15] A. Piwinski, DESY HERA 90-04, (1990)

[16] G. Jackson, these proc.

[17] R. Schmidt, these proc.

[18] see e.g. S. Tazzari, IEEE Trans. Vol. NS-28, No. 3 (1981), pp. 2420-2424

[19] H. Chin, these proc.

[20] P. Baigley, these proc.

[21] M. Donald, these proc.

[22] G. Ripken, F. Willeke, DESY 90-01 (1990)

[23] H. Wiedemann, Nucl. Instr. Methods 172 33 (1980)

[24] Yu. Orlov, L. Schachinger, these proc.

SUMMARY OF THE WORKING GROUP ON INSTABILITIES

P. Morton
Stanford Linear Accelerator Center, Stanford, CA 94309

The formula for the luminosity can be written in the following form to indicate the desire for a large average current in the storage rings.

$$\mathcal{L} = \frac{\gamma I \xi}{4\pi K \beta_y^*}, \qquad (1)$$

where γ is the electron energy in terms of the rest mass, ξ the tune shift parameter of the colliding beams, I the average current in the ring, β_y^* the vertical beta function at the interaction point, and K a constant.

The average current is I is given by

$$I = eNkf_{rev}, \qquad (2)$$

with e the electron charge, N the number of electrons per bunch, k the number of bunches in the ring, and f_{rev} the revolution frequency. Because ξ has a maximum value, of order of 0.03, that can not be exceeded before the colliding beams become unstable and the because minimum value of β_y^* is restricted to be larger than the length of the bunch or by lattice restrictions, it is necessary to increase the current in order to increase the luminosity. This can be accomplished by increasing the number of particles per bunch or increasing the number of bunches in the ring or both.

The transverse emittance of the colliding bunches is chosen to give the maximum value of ξ by the following expression,

$$\epsilon_x = N \left(\frac{K}{\gamma \xi}\right). \tag{3}$$

The maximum value of N is generally determined by either the maximum value of the transverse emittance that can be contained inside of the vacuum chamber aperture or can be allowed for background effects in the detector, or it is determined by the maximum number of particles that can be circulated stably in a single bunch.

In this section we will consider the maximum value of N that is stable. The values of N that are required for B factories have already been achieved in many storage rings, and we have theoretical formulae that are reasonably accurately in predicting the stability properties for these rings. There are many processes that have the potential to limit the number of particles per bunch in a storage ring. One of these is intrabeam scattering; both small angle multiple scattering which increases the emittance and large angle scattering (Touschek scattering) which limits the lifetime are important in low energy storage rings. Fortunately, the minimum energy of the B factory rings is $\sim$ 3 GeV which is sufficiently high to eliminate both of these problems. Similarly, the rf voltage necessary to replace the energy lost to synchrotron radiation produces a large enough stable bucket that the quantum lifetime is not an important limitation.

The two most severe limits to the single bunch current are the turbulent bunch lengthening (often referred to as the microwave instability) and the fast head tail instability (often referred to as the mode coupling instability). The first instability is a longitudinal instability which has the effect that, above a threshold current, the length of the bunch increases. The second instability is a transverse instability which has the effect that, above a threshold current, particle loss from the bunch occurs. The values of the threshold currents for these instabilities depend upon the short-range longitudinal and transverse wake fields respectively or, equivalently, the broad-band longitudinal and transverse impedances. Thus to calculate the threshold for the single bunch current, the impedance of all the devices in the ring must be known. The general consensus of the Instability Group of the Workshop is that we now have the tools in hand to either calculate, measure or accurately estimate the impedances necessary to determine the single-bunch-current threshold, and the present theory has proven reasonably good at predicting experimental results.

The theory is incorporated into programs like ZAP and BCI. The procedure is to reduce the broad band impedances present in the ring by careful design of vacuum chamber discontinuities, rf cavities, bellow, kickers, pickup electrode and etc. Next is to find the maximum stable value of, N, the number of particles per bunch, then to pick ϵ_x to achieve the desired maximum value of ξ from equation 3. The single-bunch threshold current for a reasonable ring impedance is of the order of a few mA. With N determined, k, the number of bunches is chosen to obtain the average current necessary to achieve the desired luminosity from equations 1 and 2. This procedure will result in the number of bunches in the ring being between several hundred and a few thousand. There is very little experience with colliding-beam storage rings with large numbers of bunches and high average currents. Most of the experience with large numbers of bunches and high average currents has been with synchrotron light sources. The next-generation synchrotron light sources will expand our experience in rings with large average currents.

When multiple bunches are present in the ring the motion of a bunch is affected by the wake fields from the preceding bunches. Thus wake fields which last for times much longer than the bunch spacing (i.e. wake fields from high-Q resonant structures) are the chief source of multiple bunch instabilities. Most of the structures present in the ring are designed to have very low Q-values to avoid exciting the coupled bunch instabilities; however, one exception is the rf cavities which are explicitly designed to have a large Q-value at the main rf accelerating frequency. The Q-values at other resonant frequencies of the cavities are generally of the same order as the Q-value of the accelerating frequency unless special attention is given to lower these values. For this reason it is a reasonable assumption to consider the rf cavities as the main source of the wake fields which drive multiple bunch instabilities. Because the particle bunches are short compared to the wave length of the coupling fields, the higher order bunch shape instabilities are much less dangerous than the main dipole mode where each bunch moves as a complete unit.

The theory of predicting multiple bunch instability growth rates is well understood, and, if the wake fields are known, the theory is reasonably successful in explaining experimental observations. The present theory is adequate to predict the stability and growth rate of all of the coupled bunch modes, if the exact resonant frequency, impedance, and Q of all of the wake fields are known The main problem is that exact values and frequencies of the high-Q resonances which drive these instabilities are difficult to determine. Small changes in the resonant frequencies of these fields have a large effect on the stability and growth rates of the different coupled bunch modes. Statistical methods are often used to estimate how these fields affect stability. While it is very difficult to predict the exact resonant frequencies of all the high-Q fields, it is practical to predict the magnitude and the width or Q- value of many of the dominant wake fields The main control is to reduce the magnitude of the driving impedance by lowering the Q of the resonances. The growth rates from the resonances are then determined from the present theory and the results used to specify the strength and bandwidth of the feedback systems. It should be emphasized that finding all of the dangerous resonances and reducing their impedances is not a trivial task and will require considerable effort. Similarly, the feedback system necessary to damp the multiple bunch instabilities needs to be powerful and will require a major design effort.

There is one additional instability that has proved troublesome in electron storage rings, and it is caused by the trapping of ions produced by the collisions between the electrons in the beam and the background gas. The present theory is quite simple, and while it can qualitatively explain some of the experimental observations, it is in rather poor shape to explain anything more than the some of the most gross effects that are observed. In fact, is completely incorrect in predicting the background pressure dependence that is observed. For example, at a pressure of 2×10^{-9} torr, the total number of ions produced in one second will equal the total number of electrons in the ring. Production rates at other pressures scale linearly with pressure. Now the lifetimes of electrons in colliding beam storage rings are much longer than the ion production time so the number of trapped ions should depend only upon whether the ion is trapped or not. The effects of ions should be relatively independent of the background pressure; however, most observations find a strong dependence upon background pressure. Nevertheless, the electric field of the electrodes necessary to clear the ions can be accurately determined. It is clear that trapped ions must be removed or they will seriously degrade the luminosity. The difficulty is how to include clearing electrodes (voltages ~ 1 kV are required) without adding additional impedance into the ring which could lower the threshold currents of the other instabilities.

Several new ideas were presented at the workshop and there are papers on these ideas published in this proceedings. One idea was to use a lattice with a low momentum compaction. This has the advantage of lowering the growth rate of the longitudinal instabilities since it would require a large wake field to produce the energy spread which makes the particles move longitudinally relative to each other. The disadvantage is that the spread in the longitudinal motion due to energy spread in the beam is what produces the stabilizing effect on the beam and the threshold current also would be decreased. There was a discussion on whether one could shake the beam of electrons transversely to remove the ions, and experiments are planned to study how effective this would be for the B factory.

Report of the Working Group on Beam-Beam Effects

Eberhard Keil
CERN, CH-1211 Geneva 23, Switzerland

Abstract

This report summarizes the discussions and conclusions of the Working Group on Beam-Beam Effects during the Workshop on Beam Dynamics Issues of High-Luminosity Asymmetric Collider Rings, held at the Lawrence Berkeley Laboratory, Berkeley CA, USA, from 12 to 16 February 1990. The discussions were organized around four topics: (i) The relative advantages of round beams compared to flat beams, (ii) Energy transparency which deals with the question of which beam parameters have to be made equal although the beam energies are different in the two rings, (iii) The simulation of the beam-beam effect, (iv) Experiments in existing e^+e^- colliders to check the design concepts for future asymmetric colliders. The overall conclusions arrived at by the working group can be found at the end.

1 Introduction

The discussions of the Working Group on Beam-Beam Effects were organized around four topics in discussion groups. The topics and the participants are shown below. The discussion leaders are printed in *italics*.

- Round beams versus flat beams: *Furman*, Krishnagopal, Woods
- Energy transparency: *Chao*, Chattopadhyay, Chin, Kim, Sessler, Tennyson
- Simulation: *Courant*, Channell, Dell, Gerasimov, Hirata, Johnson, Ruggiero
- Experiments: *Seeman*, Cornelis, Kikutani, Rice, Rivkin, Schmidt

The first two topics are concerned with physics issues, the second two topics deal with tools which are used to resolve them. This summary is organized by discussion groups. The overall conclusions can be found at the end.

2 Round Beams versus Flat Beams

The physical argument in favour of round beams goes as follows: Assuming for simplicity equal horizontal and vertical rms radii of the two beams, i.e. $\sigma_x^+ = \sigma_x^- = \sigma_x$ and $\sigma_y^+ = \sigma_y^- = \sigma_y$, the luminosity L is given by:

$$L = \frac{N^+ N^- f k}{4\pi \sigma_x \sigma_y} \qquad (1)$$

© 1990 American Institute of Physics

The vertical beam-beam tune shift parameters $\xi_y^\pm$ are given by:

$$\xi_y^\pm = \frac{N^\mp r_e \beta_y^\pm}{2\pi\gamma^\pm(\sigma_x + \sigma_y)\sigma_y} \qquad (2)$$

Here, N^+ and N^- are the intensities of the positron and electron bunches, respectively, fk is the bunch collision frequency, r_e is the classical electron radius, $\beta_y^\pm$ are the vertical amplitude functions at the collision point(s) and $\gamma^\pm$ are the beam energies in units of the electron rest energy.

Eliminating one bunch intensity $N^\mp$ from Eq. (1) and introducing the beam aspect ratio $r = \sigma_y/\sigma_x$ yields the following expression which clearly displays the parameters entering into the performance of a B factory:

$$L = \frac{N^\pm fk \xi_y^\pm \gamma^\pm}{2r_e \beta_y^\pm}(1+r) \qquad (3)$$

We see that, all things being equal, the luminosity for round beams with $r = 1$ is nearly a factor of two higher than for flat beams with $r \ll 1$. This geometrical factor is one of the main arguments on favour of round beams.

Whether it is permitted to assume that other things, and in particular the beam-beam strength parameter ξ_y and the amplitude functions at the collision point(s) $\beta_y^\pm$ remain the same when the B factory design is changed from flat beams to round beams is debatable. The evidence on the maximum value of ξ_y will be discussed below. The minimum value of $\beta_y^\pm$ and many other topics related to the choice between flat and round beams were discussed by the Lattice Working Group.

2.1 Maximum value of ξ_y

For flat beams, the maximum achievable beam-beam strength parameter ξ_y was observed in all e^+e^- storage rings, and computed by several simulation programs. The proceedings of workshops in Santa Margherita di Pula, Sardinia, Italy [1] and Novosibirsk, USSR [2] contain discussions of these subjects. At this workshop, J. Tennyson reviewed incoherent beam-beam effects [3], K. Hirata reviewed coherent beam-beam effects [4], and D. Rice reviewed the present experimental data [5]. The highest experimentally observed beam-beam strength parameters fall into the range $0.02 < \xi_y < 0.04$. The results of simulation programs fall into the same range. It is generally agreed that simulation programs yield results in agreement with observations in well-tuned e^+e^- storage rings.

The only experimental data for the beam-beam strength parameter ξ_y for machines with round beams come from the DCI storage ring in Orsay, France. The observed maximum values for ξ_y fall into the range $0.035 < \xi_y < 0.041$ [6]. Further experiments on round beams will soon be done in the CESR e^+e^- storage ring at Cornell University, Ithaca NY, USA. Recent two-dimensional simulations, i.e. without synchrotron oscillations, for the case of round beams were made by Krishnagopal and Siemann [7] who find a maximum value of $\xi_y \simeq 0.1$ for a carefully optimized operating point in a machine with round beams. Chin [9] has made a comparison between a machine similar to the LBL/SLAC B factory proposal and finds $\xi_y \simeq 0.04$ with flat beams, and $\xi_y \simeq 0.035$ with round beams at the same working point. In both cases, a blow-up of the beams is observed, indicating that the maximum value of ξ_y is reached. The factor of three discrepancy between the simulation results needs clarification.

2.2 Distribution functions

It is generally accepted that the observed beam-beam limit in existing e^+e^- storage rings is related to the fact that the tails of the distribution functions, and in particular the vertical distribution function, drop less rapidly than a Gaussian distribution. The increased population of the tails in turn causes a reduction of the beam lifetime and/or an increase of the particle background in the experiments surrounding the collision point(s). Such tails have also been observed in simulations [8]. Chin [9] has compared the tails for round and flat beams, and found a striking difference: The vertical distribution for flat beams has a pronounced non-Gaussian tail while the vertical distribution for a round beam continues to have a Gaussian shape up to the beam-beam limit.

Since, for round beams, the horizontal and vertical emittance are about the same, the two transverse energies are also nearly identical. Hence, coupling between the two planes, caused e.g. by nonlinear beam-beam driven resonances, cannot transfer much energy from one plane to the other. For flat beams, the horizontal emittance is one to two orders of magnitude higher than the vertical one, and this transfer of horizontal energy into the vertical plane is indeed possible.

2.3 Miscellaneous Points

The shorter tails for round beams do not necessarily imply that the vertical aperture in a machine with round beams can be smaller than in a machine with flat beams. The consequences of the choice of the aspect ratio r on the interaction region design, the masking of synchrotron radiation, the separation of the two counter-rotating beams, etc. was studied by the Lattice Working Group.

It is unlikely that the horizontal and vertical amplitude functions β_x and β_y in a machine with round beams can both be made as small as the vertical amplitude function β_y in a machine with flat beams. This may offset any gain due to the geometrical factor of two and of a higher maximum beam-beam strength parameter ξ_y.

A particle in one beam executing synchrotron oscillations meets the oncoming bunch of the opposite beam at varying positions s along the beam direction. Even with head-on collisions, this causes a modulation of the beam-beam parameter for flat beams since $\sigma_y(s) \propto \sqrt{\beta_y(s)}$, neglecting the variation of $\beta_x(s)$ and $\sigma_x(s)$, as can be seen in Eq. (2). This tune modulation is absent for round beams because $\beta_x(s)$ and $\beta_y(s)$, and $\sigma_x(s)$ and $\sigma_y(s)$ vary such that Eq. (2) becomes independent of s.

2.4 Conclusion for round versus flat beams

The following conclusions may be drawn concerning the choice between round and flat beams:

- The beam-beam effect which seems to favour round beams over flat beams is only *one* of *many* aspects entering into the choice.

- A *fair* comparison of the results of simulation programs is needed.

- Results of more experiments in existing machines are eagerly awaited.

3 Energy Transparency

3.1 Issue

Energy transparency addresses the question what beam and machine parameters must be made equal in the two rings of an asymmetric B factory in order to achieve optimum performance although the two beams have different energies. This problem was first raised by Chin [10] who put forward the following conditions:

- Same beam radii at the interaction point:

$$\sigma_x^- = \sigma_x^+ \qquad \sigma_y^- = \sigma_y^+ \qquad (4)$$

- Same nominal beam-beam strength parameters:

$$\xi_x^- = \xi_x^+ \qquad \xi_y^- = \xi_y^+ \qquad (5)$$

- Same damping decrements $\delta^\pm$, i.e. same ratios of time between collisions and synchrotron radiation damping time:

$$\delta^- = \delta^+ \qquad (6)$$

This condition implies that the damping decrement in the low energy ring is increased beyond its natural value by adding wiggler magnets.

- Same betatron phase modulation due to synchrotron oscillations with bunchlength σ_s and tune Q_s:

$$\left(\frac{\sigma_s Q_s}{\beta_x}\right)^- = \left(\frac{\sigma_s Q_s}{\beta_x}\right)^+ \qquad \left(\frac{\sigma_s Q_s}{\beta_y}\right)^- = \left(\frac{\sigma_s Q_s}{\beta_y}\right)^+ \qquad (7)$$

Hirata and Keil [11] [12] studied the centre-of-mass motion of colliding bunches in an asymmetric collider with K^+ bunches in one and K^- bunches in the other ring, in linear approximation by computing the stability of the eigenvalues of the coupled bunch motion due to the beam-beam collisions in one interaction point, and by multi-particle tracking. They find sum resonances which are driven by the total current and occur when the tunes Q^+ and Q^- satisfy the relation:

$$K^- Q^+ + K^+ Q^- \leq n \qquad (8)$$

Here K^+ and K^- are assumed to be relatively prime, and the $\leq$ symbol indicates that the resonance occurs just below the arbitrary integer n. If there are variations in the bunch intensities within the beams, also integral and half-integral resonances are driven when the tunes satisfy the relations:

$$K^\pm Q^\pm \leq n \qquad (9)$$

$$K^\pm Q^\pm \leq n + \frac{1}{2} \qquad (10)$$

- Avoiding these closely spaced resonances when $K^{\pm} \gg 1$ requires that the two circumferences are the same, leading to the condition:

$$C^+ = C^- \qquad (11)$$

If one requires that the driving terms for synchro-betatron resonances due to the beam-beam collisions are also equal, one gets the following conditions which are considered important when the expressions involved are of order unity, and unimportant when they are small compared to unity:

- Same driving term due to beam-beam collisions at the crossing angle α, where σ_c is the rms beam radius in the plane of the crossing:

$$\left(\frac{\alpha\sigma_s}{2\sigma_c}\right)^- = \left(\frac{\alpha\sigma_s}{2\sigma_c}\right)^+ \qquad (12)$$

- Same tune modulation due to chromaticities $Q'_x = dQ_x/dp/p$ and $Q'_y = dQ_y/dp/p$:

$$\left(\frac{Q'_x \sigma_e}{Q_s}\right)^- = \left(\frac{Q'_x \sigma_e}{Q_s}\right)^+ \qquad \left(\frac{Q'_y \sigma_e}{Q_s}\right)^- = \left(\frac{Q'_y \sigma_e}{Q_s}\right)^+ \qquad (13)$$

Chin[10] finds that the high energy beam blows up the low energy beam when only Eqs. (4) and (5) are satisfied. When the damping decrements are made equal, i.e. when Eqs. (6) are satisfied, the blow up of the low energy beam is significantly reduced. When the tune modulations are also made equal, i.e. when Eqs. (7) are satisfied, the behaviour of the two beams is almost equalized, and the blow up both beams is small for nominal beam-beam strength parameters up to $\xi_0 = 0.05$. Tennyson[3] has made a comparison between machines with equal and unequal damping rates in the two rings, and finds a small difference in the luminosity. Clearly, the discrepancy between these simulation results needs to be understood.

3.2 OPINION

The following questions related to energy transparency have been examined, with the following results:

1. Is there more physics in the energy transparency conditions than reducing the problem of simulating two rings with different energies and different parameters to the problem of simulating two rings with the same energy and identical parameters? We believe that this is not the case.

2. Are the energy transparency conditions *sufficient*? We believe that they are.

3. Are the energy transparency conditions *necessary*? We don't know for sure, but our consensus is that not all conditions are necessary.

The problem of energy transparency needs further work, by simulation since no asymmetric machine exists. An approach where one starts with small deviations from energy transparency and sees how one has to change parameters such as to recuperate the good performance of transparent machines appears most promising.

4 Simulation

4.1 Weak-Strong versus Strong-Strong Simulation

The relative merits of simulating the beam-beam effect in the weak-strong and the strong-strong approximation were discussed. In the former simulation, a weak beam of test particles collides with a strong beam whose parameters, i.e. average position and beam radius, do not change with time. The strong beam only provides the forces acting on the weak beam. In the latter case, both beams exert forces on each other and their parameters are allowed to change due to the forces. Tennyson made the argument that calculating the beam parameters from the modest number of particles in the simulation, typically a few hundred, causes numerical noise which might disturb the simulation. This argument can be overcome by simply exponentially averaging the beam parameters over about a damping time.

4.2 Bunch Slicing

In order to achieve the expected performance, the vertical amplitude function β_y in modern B factory designs is no longer large compared to the bunch length σ_s. Therefore, the amplitude functions $\beta_x(s)$ and $\beta_y(s)$ and the betatron phases $\mu_x(s)$ and $\mu_y(s)$ vary significantly over the length s of the collision region. Siemann and Krishnagopal [7] have taken this into account by dividing the bunches into S slices along the beam axis. They find that the emittance of the weak beam is increased much less when the strong beam is divided into $S \geq 3$ slices. By now, bunch slicing has become general practice in beam-beam simulation programs.

4.3 Lifetime, Background, Tails

Multi-particle simulation is a good tool for obtaining data for the dynamic evolution of the beam radii, and therefore of the luminosity in e^+e^- colliders [1] [2]. However, obtaining good statistical data on lifetime, background and tails of distribution functions is very time-consuming because the particles spend only a small fraction of their time at large amplitudes.

Lifetime, background and tails are not only influenced by the beam-beam effect proper, but also by the aperture and the non-linearities of the storage ring lattice which therefore should also be taken into account. A simulation by Jackson and Siemann [13] has shown that the effect of the non-linearities on the core of the beam and the luminosity of CESR was small. Gerasimov [14,15] has used a different technique to obtain equilibrium distribution functions for oscillatory systems under the influence of quantum excitation, damping, and isolated non-linear resonances.

We were reminded rather forcefully by an experimental particle physicist that the background for experiments in a real machine is not reproducible and very sensitive to its fine tuning. Therefore, no attempts should be made to define a standard distribution function with non-Gaussian tails, e.g. for the calculation of synchrotron radiation hitting the vacuum chamber next to the interaction points.

4.4 New Material

New material was presented about the following aspects of beam-beam simulation:

- Hirata, Moshammer, Ruggiero and Bassetti presented an irreducible set of parameters and symplectic equations for the beam-beam effect in six dimensions [16].

- Channell treats the beam-beam effect as a discrete Lie-Poisson system, sets up a system of equations of motion for the moments of the distribution function, and follows them by tracking [17]. The computer code which contains this procedure runs faster than multi-particle tracking. It does not include quantum excitation and damping.

- Kikutani has a promising technique to calculate the beam-beam force from the particle distribution, based on Fourier transforms and Green functions [18]. It could replace the traditional approach of obtaining the beam-beam force from a Gaussian density distribution with variable mean and standard deviation.

- Zholents and others presented simulation results for a machine in which the horizontal beam radius is dominated by the contribution of synchrotron oscillations, i.e. $D\sigma_e \gg \sqrt{E\beta}$ [19]. They find that the beam-beam limit is about the same as for a machine where the dispersion vanishes at the interaction point.

4.5 Unanswered Questions

Standard test data describing an existing machine with a large collection of experimental data should be developed and used for all existing and future simulation programs. Programs which do not reproduce the experimental data should be rejected.

A comparison between existing simulation programs should be made, including the input data, phenomena included, algorithms, etc.

A standard simulation code should be developed which can be used by anybody who wants to simulate the beam-beam effect. It should have the reliability of standard beam-optics codes.

5 Experiments

The discussion group on experiments considered the following experiments:

5.1 Beam-Beam Strength Parameter ξ_{max} versus Energy

A very puzzling experiment in the VEPP-2M machine at Novosibirsk [20] yielded the maximum value of the beam-beam strength parameter ξ_{max} as a function of the energy for two different settings of the wiggler magnets. The observed values of ξ_{max} were nearly the same although the difference in the damping times was large. To clarify the dependance of ξ_{max} on the damping times, systematic experiments should be done in LEP between 20 and 46 GeV and in PEP between 7 and 14 GeV. Whenever possible, the damping rates should be varied with wiggler magnets located both in places where the horizontal dispersion vanishes and it places where it does not. The following quantities should be carefully recorded: Luminosity L versus current I – horizontal and vertical emittances E_x and E_y – bunchlength σ_s and relative energy spread σ_e – tunes Q_x, Q_y, and Q_s – amplitude functions β_x and β_y at the n_x interaction points – dispersions D_x and D_y and their derivatives at the interaction points – beam radii σ_x and σ_y – density distributions – nature of current limitation (beam-beam, background, lifetime, etc.) – closed orbit – frequencies of coherent modes – lifetime and shape of single beam – coherent and incoherent damping times.

5.2 Round Beam Tests

The following questions should be included in experiments on round beams in CESR and possibly other machines: Maximum beam-beam strength parameter ξ_{max} – shape of tails in distribution function – stability of saturated regime – limit on inequality of bunch currents. These experiments should be done both on the coupling resonance where the equality of beam radii is achieved by coupling, and off the coupling resonance where this equality is achieved by some other means.

If we are serious about round beams we should get an interaction region in some machine rebuilt to handle round beams in the region of parameter space considered for asymmetric B factories.

5.3 Crossing Angle and Crab Crossing

Tests in CESR, Sp$\bar{\text{p}}$S, PEP, etc. should answer the following questions:

- At what crossing angle do the beams behave differently from head-on collisions, and how does this observation compare to theoretical predictions?

- Do crab cavities work?

5.4 Beam Separation at Interaction Points

Data exist around the world, most recently at Novosibirsk, on the minimum beam separation needed at the interaction points. These data are relevant to the design of asymmetric B factories because of the close encounters between bunches near to the interaction point due to the small bunch spacing. These data should be checked and outstanding tests should be completed.

5.5 Tail Distribution

The density in transverse phase space $\rho(x, x', y, y')$ should be measured as a function of energy E, current I and beam-beam strength parameter ξ, and the following questions should be answered::

- Are there differences between round and flat beams?

- What are the growth time, overshoot and relaxation time after bringing the beams into collision?

5.6 Disruption Parameter

The disruption parameter D is the ratio of the bunch length σ_s to the focal length F of the lens representing the focusing due to the oncoming bunch. It is used in the design of linear colliders and related to the beam-beam strength parameter ξ_y by the following equation:

$$D = \frac{4\pi \sigma_s \xi_y}{\beta_y} \qquad (14)$$

The question whether D is a better indicator of the strength of the beam-beam forces than ξ_y should be answered. It enters into the synchro-betatron coupling by the beam-beam effect. The relevance of the Bassetti parameter, i.e. the ratio of the energy loss or gain due to the beam-beam collisions compared to the energy fluctuation due to the emission of synchrotron radiation between the collisions should be checked. The bunchlength σ_s and the amplitude function β_y should be varied and any changes in the beam-beam parameter ξ_{max} should be observed.

5.7 Importance of Tests

Future B factories, be they symmetric or asymmetric, aim at luminosities between one and two orders of magnitude higher than existing e^+e^- colliders. Therefore, it is important for their design, that new and untried design concepts are verified by experiments on existing machines as much as possible. Beam time for such experiments will have to be found at machines which already operate with a full schedule. Modifications of these machines needed for specific experiments will have to be funded. Therefore, beam time and funds for such experiments should be requested at a high level of laboratory management and be used at the highest possible efficiency by good preparation, coordination and collaboration.

5.8 Simulations

All of the independent simulation codes should predict the results of all these tests before they are done, for two reasons: Beam time on existing machines is expensive, and applying simulation codes to experimental situations will sharpen our understanding of the codes.

6 Conclusions

The findings of the beam-beam working group concerning the design and the performance of asymmetric B factories may be summarized as follows:

- The choice between round beams and flat beams is mainly an engineering issue. No miraculous improvements in the maximum permissible beam-beam strength parameter ξ are to be expected.

- Imposing the many conditions required by energy transparency is one way of solving the associated beam-beam problems, but most likely not the only one.

- Simulation of the beam-beam effect is again under active development in several places, after an interruption of nearly ten years.

- Experiments must be performed in existing machines to answer many of the open questions. They need the active support of the community to obtain the necessary funding and beam time.

- Beam-beam effects are but one aspect entering into the design and performance of asymmetric B factories.

This report summarizes the work of all participants in the Working Group on Beam-Beam Effects. Their names are listed in the Introduction. References to specific contributions to the Workshop are given below. It was a pleasure to chair this Working Group.

References

[1] J.M. Jowett, M. Month and S. Turner, eds., Nonlinear Dynamics Aspects of Particle Accelerators, Lecture Notes in Physics No. 247, Springer-Verlag Berlin, Heidelberg, New York, Tokyo 1986.
[2] N.S. Dikansky and D.V. Pestrikov, eds., Third Advanced ICFA Beam Dynamics Workshop, Novosibirsk 1989, to be published.
[3] J. Tennyson, these proceedings.
[4] K. Hirata, these proceedings.
[5] D. Rice, these proceedings.
[6] R. Chehab, A. Jejcic, J. Le Duff, M.P. Level, P. Marin, D. Potaux, M. Sommer, H. Zyngier, Proc. 11-th International Conf. on High-Energy Accelerators (Geneva 1980) 702.
[7] S. Krishnagopal and R. Siemann, Proc. 1989 IEEE Particle Accelerator Conf. (1989) 836.
[8] S. Myers, in ref. [1], 176.
[9] Y.H. Chin, these proceedings.
[10] Y.H. Chin, LBL Report 27665 (1989).
[11] K. Hirata and E. Keil, Physics Letters **B232** (1989) 413.
[12] K. Hirata and E. Keil, CERN/LEP-TH/89-76, to be published in Nucl. Instr. Meth.
[13] G. Jackson and R. Siemann, Proc. 1987 IEEE Particle Accelerator Conf. (1987) 1011.
[14] A.L. Gerasimov, Inst. of Nuclear Physics, Novosibirsk, Preprint 87-100 (1987).
[15] A.L. Gerasimov, Physics Letters **A135** (1989) 92.
[16] K. Hirata and F. Ruggiero, these proceedings.
[17] P.J. Channell, these proceedings.
[18] E. Kikutani, these proceedings.
[19] A. Zholents, these proceedings.
[20] P.M. Ivanov, to be published in ref. [2].

Report of the Working Group on Technology

Ferd Voelker and Glen Lambertson
LBL, Berkeley, CA

1 Introduction

The Working Group on Technology addressed the following topics:

- A conceptual design to use for the workshop.
- RF cavity design and damping of HOM.
- A practical power limit for both cavities and rf windows.
- Beam-feedback for controlling beam growth.
- A new crab-cavity design by Gerry Jackson.

2 Conceptual Design

An rf system similar to the LBL/SLAC Redbook design was chosen for discussions during the workshop. Some of the rf parameters are shown below.

RF frequency	500	MHz
Bunch frequency	250	MHz
Harmonic number	3669	
Current	3	A
Total voltage	17.5	MV
Particles/bunch	$7.5*10^{10}$	
Power to beam	10.5	MW
Revolution frequency	136.4	KHz

3 RF Cavities and HOM

The challenge of designing rf cavities for a B-Factory is how to control the higher order modes, and still have acceptable shunt impedance at the fundamental frequency. The cavity shape, that is similar to the LEP cavity shape, moves the HOM's up in frequency. There are still two longitudinal modes below cut-off of the beam pipe, one of which has a large shunt impedance. There is concern that the beam can excite power in the cavities at these frequencies. Except for a gap of perhaps 15%, every other buckets in the ring is filled with particles, and the rf spectrum contains spectral lines spaced 136 kHz apart. If lines with large amplitude fall on a cavity mode, considerable power will be generated. Estimates of HOM power did not take into account all the complexities, and we need a better estimate of HOM power, especially during the filling of the ring.

For modes above the cut-off frequency, energy can propagate down the beam pipe, reducing the shunt impedance. The conventional wisdom is that energy that leaks out will be dissipated before going all the way around the ring. However, it may not be dissipated sufficiently before getting to another cavity, and additional damping may be needed.

Dave Rubin from Cornell presented a super-conducting cavity design with a section of large-bore stainless beam pipe. Energy that is trapped in the large bore has to be absorbed. His design had additional resistive material on the wall to dissipate this trapped energy. The large bore is a desirable feature for either normal or super-conducting cavities, but the transition to the remaining beam pipe must be made carefully. If the transition from the large bore to the regular beam pipe is a gradual, the integral

$$\int E_z \exp(-ikz) \, dz$$

can be made small, resulting in a small transit time factor T. Thus $R_{sh} T^2$ can be made small even if R_{sh} isn't. This is the same as reducing the coupling of the beam to the cavity.

4   Power Levels for Cavities and RF Windows

To keep the beam impedance small in the ring, the number of cavities must be kept to a minimum. Both normal conducting and super-conducting cavities were considered. The following table lists two possible scenarios.

	normal conducting	super conducting
Number of cells	20	15
Worst HOM R/Q	14.5 ohm	2 ohm
Cavity voltage	875 kV	1.17 MV
Loaded Q of HOM	300	2900
Impedance of HOM	87 kohm	87 kohm
HOM power/cell *	10 kW	10 kW
Power into beam	700 kW	525 kW
Power into wall	137 kW	-
Power through window	672 kW	700 kW
Total rf power (all cells)	13.4 MW	10.7 MW

* From the usual calculation ignoring resonances.

After considerable discussion there was general agreement on a number of areas. It seems desirable to use single cavities, and to keep them separated from each other to reduce the number of HOM's. It seems necessary to damp HOM's, both to reduce losses at these frequencies, and to reduce the amount of feedback power needed. It was assumed that the problem of window power-level can somehow be solved, and base the number of cavities on other considerations. Power through windows in other accelerators was discussed. A table of window power is shown below.

CESR	500 MHz	350 kW	Cylindrical
TRISTAN	500 MHz	225 kW	
PETRA		120 kW	Disc in a coax
PEP	350 MHz	150 kW	
SLAC	350 Mhz	250 kW	Test cavity

The unstable modes need to be damped by at least a factor of 180 times, to hence to reach a beam impedance level of 100 kilohms for longitudinal HOM's. (The growth threshold is at about 10 kilohms.) At this impedance the feedback power is acceptable. Transverse modes were not considered in detail, but thought easier to handle than longitudinal.

Since most of the power goes to the beam, power in the walls was not thought a sufficient reason to choose superconducting cavities. A large bore tube can also be used with copper cavities to push HOM modes higher in frequency, but it would be at the expense of 50% more rf losses, or 6% more total power. Damping ports in the wall are considerably more complicated in SC cavities than in copper cavities. Either copper or SC cavities appear feasible.

Kageyama from KEK described two types of cavity damping that are being considered. One method uses multiple cavities, with four slots on each cavity that couple to waveguides. Method of construction and calculated Q's were shown, along with some measured values. External Q's as low as 30 were measured for the first transverse mode.

5. Beam Feedback

A cavity damped to a Q of 200 at 500 MHz would overlap 18 spectral lines, making it difficult to avoid exciting HOM's by tuning. The threshold impedance for growth of the beam is smaller with high current. These facts point to the need for a broad-band, all-mode feedback systems for control of both longitudinal and transverse coupled-bunch oscillations.

Most of the effort was focused on longitudinal feedback because transverse impedance is expected to need less power, and be less difficult to implement.

The parameters of a conceptual longitudinal feedback system were presented. These are listed below.

9 GEV storage ring and injection:

> Beam current = 3 A.
> Number of bunches, M = 1800.
> Inject 18000 turns, 1/10 bunch each time.
> Injection time and energy error   0.3 radian at 476 MHz.

Feedback system:

> Bandwidth $Mf_o/2$ = 125 MHz at 812 MHz.
> Max voltage kick = 18 kV/turn.
> 10 kickers, each of 7 series-connected loops.
> Power   10 x 220 watt solid-state amplifiers or TWT's.
>
>> (Alternate high-power system could be 16 stagger-tuned cavities, and sixteen 50 kW UHF TV-Klystrons)

6 A New Crab Crossing Design

The conventional design requires four transverse deflecting cavities that are placed 90 degrees in Betatron phase from the interaction region. Because the beams are close together, there isn't room for two full size cavities side

by side, requiring half cavities with the beam passing off
center, or perhaps an even more extreme design. From an rf
point of view, the new design is much easier to implement.
It requires four longitudinal accelerating cavities that
are 180 degrees from the interaction point in a region
having high dispersion.

Part II

INVITED TALKS

THE PHYSICS OF BB FACTORIES

W. Schmidt-Parzefall, DESY

ABSTRACT

Presently at many High Energy Physics laboratories including CORNELL, SLAC, CERN, DESY, KEK and Novosibirsk design studies for a high luminosity 10 GeV e^+e^- machine are being carried out. This large effort documents a wide consensus in the importance and interest in continued B-physics. Measurements of the Z^0 - pole have shown, that there are probably no more than 3 generations of quarks. The t - quark, once found, will immediately decay into a b - quark and will not provide any new information. Thus the b - quark will remain the only tool for precise studies of the weak interaction for ever.

B - physics points into two directions. Firstly a precise measurement of the Kobayashi Maskawa matrix in order to complete our knowledge of the physics within the Standard Model.
Secondly a study of rare loop induced reactions provides a window for physics beyond the Standard Model. For the KM - matrix a B - factory can make 5 basic measurements to determine and constrain its 4 parameters. In particular, the direct observation of CP violation in B - meson decays is shown to be possible. This investigation would greatly improve our understanding of this fundamental effect of nature. There also are about 5 rare loop induced reactions to be investigated. In addition to this rich field of research there are some 20 other important measurements from other fields which can be performed with a B - factory.
For details we refer to one of the numerous excellent reports on the subject.

Some of the most important measurements require the primary b - quarks to be produced in a moving centre-of-mass frame. The motion allows the time evolution of the states to be studied and also helps for background rejection. The optimum boost from the detector point of view lies between $\beta\gamma$ = 0.6 and $\beta\gamma$ 0.9.
The actual boost to be chosen will depend on the requirements of the machine to maximize the luminosity.
To make it worthwhile, a B - factory must reach a luminosity of 3.10^{33} cm^{-2} s^{-1} .

B-FACTORY LATTICE CONSIDERATIONS

A. Garren[*]

Lawrence Berkeley Laboratory

February 12, 1990

A talk given at the Workshop on Beam Dynamics Issues of High-Luminosity Asymmetric Collider Rings in Berkeley.

CHOICES

o Head-on Collisions vs. Crab Crossings

o Round Beams vs. Flat Beams

o Separation Direction: Both Horizontal and Vertical, or Horizontal, or Vertical

o Separation by Dipoles or by Shifted Quadrupoles

In this talk the emphasis will be on the first alternative for each of these choices, which represents the approach being currently investigated most actively for the APIARY design.

---

[*] Prepared for the U.S. Department of Energy under contract numbers DE-AC03-76SF00098, DE-AC03-76SF00515, and DE-AC03-81-ER40050

A FEW CONSIDERATIONS ON THE CHOICES

Head-on Collisions vs. Crab Crossings

Crabbing greatly eases masking problems

Head-on collisions makes for better LER optics, but worse HER optics

Crabbing might be incompatable with round beams

Crabbing might give geometric problems, if both rings are in the same tunnel

Round Beams vs. Flat Beams

Round beams increase luminosity by a factor of two, for the same β_y^*

It is easier to focus flat beams

There may be an increase in the beam-beam limit with round beams

Round beams could be difficult to produce, at least in the high energy ring.

Separation Direction

If round beams are desired, then vertical bends in the IR for separation can provide the vertical emittance in the low-energy ring.

Vertical separation introduces a vertical dispersion matching problem.

Horizontal+vertical separation keeps the high energy ring in the horizontal plane.

Separation by Dipoles or by Shifted Quadrupoles

It may be desirable to at least begin the separation with a dipole.

D-quadrupoles centered on the high-energy beam enhance separation.

For quickest beam separation, center F-quadrupoles on low-energy beam.

However, to minimize synchrotron radiation, center all IR quadrupoles on high-energy beam.

PROBLEMS FOR LOW-ENERGY LATTICE

o Transverse Separation of Bunches Passing Beyond IP (long range beam-beam effect)

o Separation of Bunches at Septum Magnet

o Arrangement of Magnets to enable solution of Masking Problem

o Achievement of Desired Parameters - Emittance, β^* etc.

o Matching Beta Functions and Dispersion between IR and Arc Cells

o Meeting Geometrical Constraints

PROBLEMS FOR HIGH-ENERGY LATTICE

o Controlling β_{max} and Chromaticity

o Making Beams Round

o Arrangement of Magnets to enable solution of Masking Problem

o Achievement of Desired Parameters - Emittance, β^*, etc.

o Matching Beta Functions and Dispersion between IR and Arc Cells

PRESENT APIARY DESIGN

The remainder of the talk was an exposition of the APIARY design as of the time of the workshop. It involves head-on collisions of 3.1 × 9 GeV round beams, mixed horizontal and vertical separation, with both rings vertically separated in the PEP tunnel.

CONSEQUENCES OF EQUAL BEAM SIZES AND TUNE SHIFTS

$$\beta_y/\beta_x = \varepsilon_y\varepsilon_x = \sigma_y/\sigma_x = r,$$

$$\beta_i^1/\beta_i^2 = \varepsilon_i^2/\varepsilon_i^1 = b,$$

$$b = \beta_i^1/\beta_i^2 = (E^1/E^2)(N^1/N^2).$$

$$\varepsilon_x^j = \frac{r_e N^k}{2\pi\xi\gamma^j(1+r)}, \quad \varepsilon_y^j = r\varepsilon_x^j$$

$$L = \frac{\varepsilon(1+r)}{2er_e}\left(\frac{I_\gamma}{\beta_y}\right)^{1,2} = 2.167\times10^{34}\xi(1+r)(IE/\beta_y)^{1,2} \text{ cm}^{-2}\text{sec}^{-1}$$

where *I* is in Amperes, *E* in GeV, and β_y in cm.

RECENT APIARY ROUND-BEAM PARAMETERS FOR HEAD-ON COLLISIONS

	APIARY III		AP IV a		AP IV b	
	LER	HER	LER	HER	LER	HER
C - Circumference (m)	733.3	2200	2200	2200	2200	2200
E (GeV)	3.1	9	3.1	9	3.1	9
ξ - Tune shift	0.05	0.05	0.03	0.03	0.03	0.03
β^* *cm(	2	6	3	6	3	6
ϵ (nm)	118	41	50	25	67	33.5
S_B (m)	2.5463	2.5463	1.2732	1.2732	1.6975	1.6975
I (A)	3.0	3.0	2.233	1.538	2.233	1.538
L (cm^{-2} S^{-1})	1×10^{34}		3×10^{33}		3×10^{33}	

APIARY IV LER PARAMETERS

Circumference	2200	m
Energy	3.1	GeV
$\beta^* = \beta_x^* = \beta_y^*$	3	cm
D^*	0	
$\hat{\beta}_x, \hat{\beta}_y$	91.8m 90.5	m
ν_x, ν_y	37.76, 35.78	
Chromaticity $\xi_{x,y}$	-68.5, -80.0	
Mom. Comp. α	0.00115	
Synch. Rad., U_o	1,346	MeV/turn
Damping time τ_E	17.30	ms
Bunch length σ_l	1.416	cm
Momentum width σ_p	0.00955	
Emittances $\varepsilon_x, \varepsilon_y$	67.12, 67.09	nm

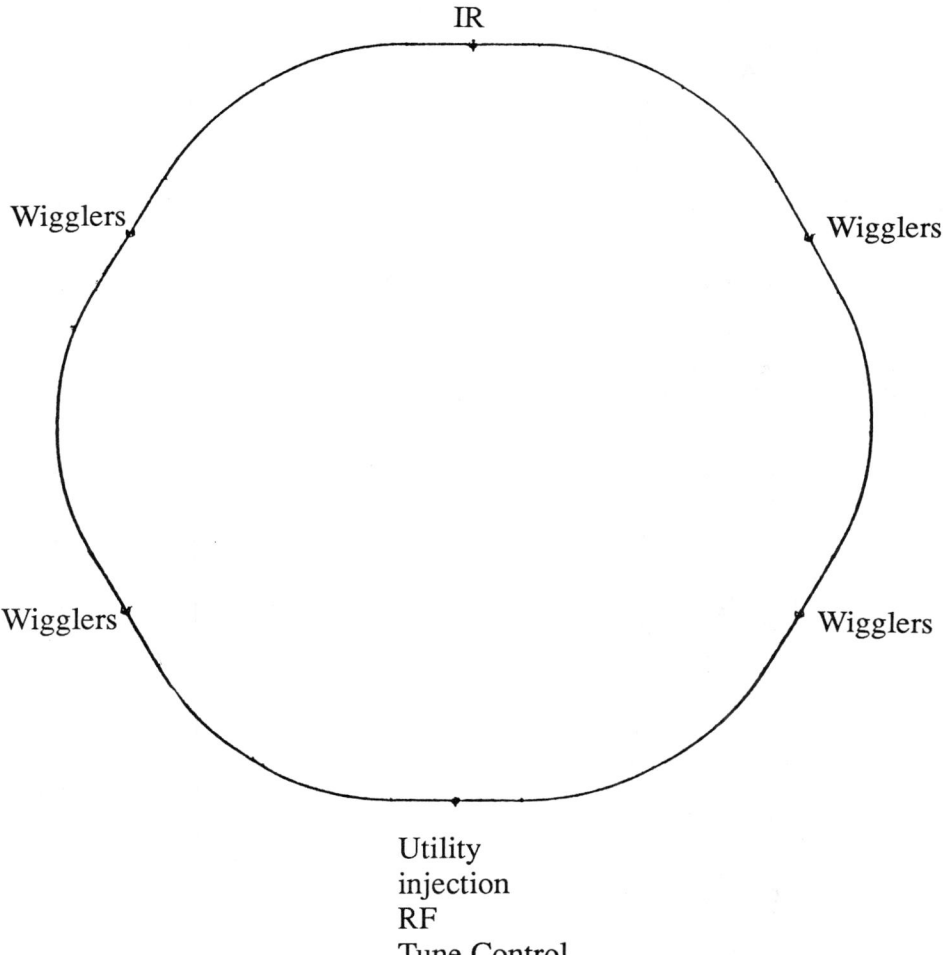

Low energy ring (LER)

50 B-Factory Lattice Considerations

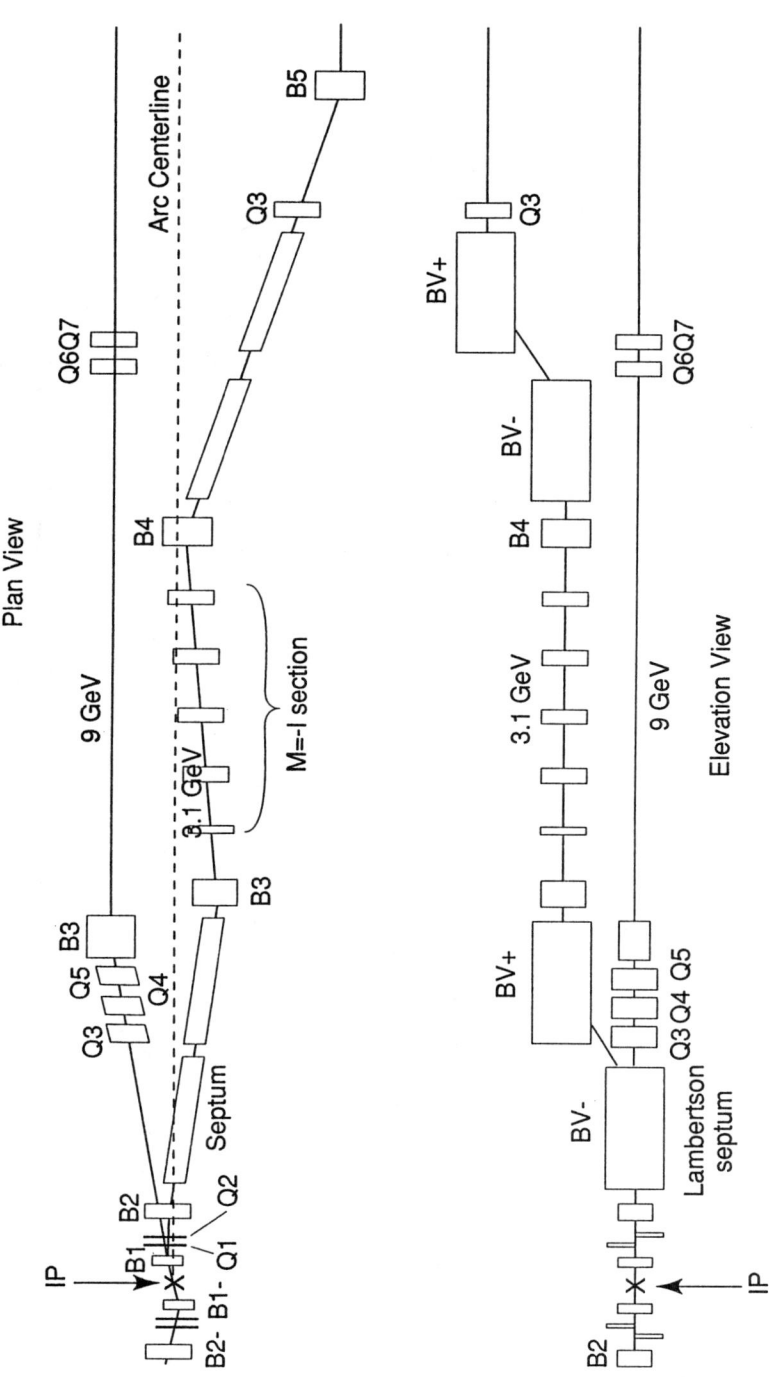

Schematic Diagram of APIARY IV Interaction Region (not to scale)

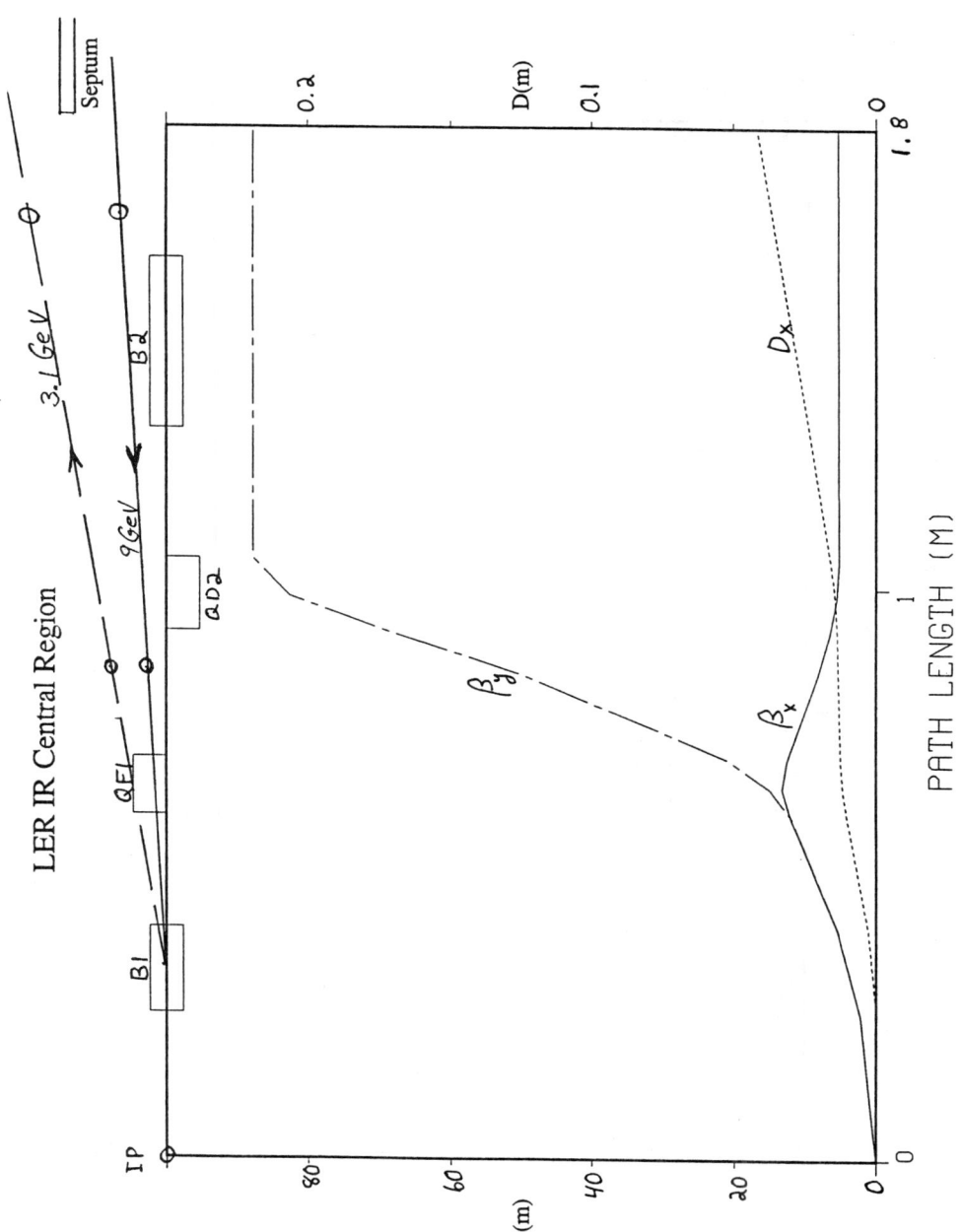

LER Separation and Dispersion Matching

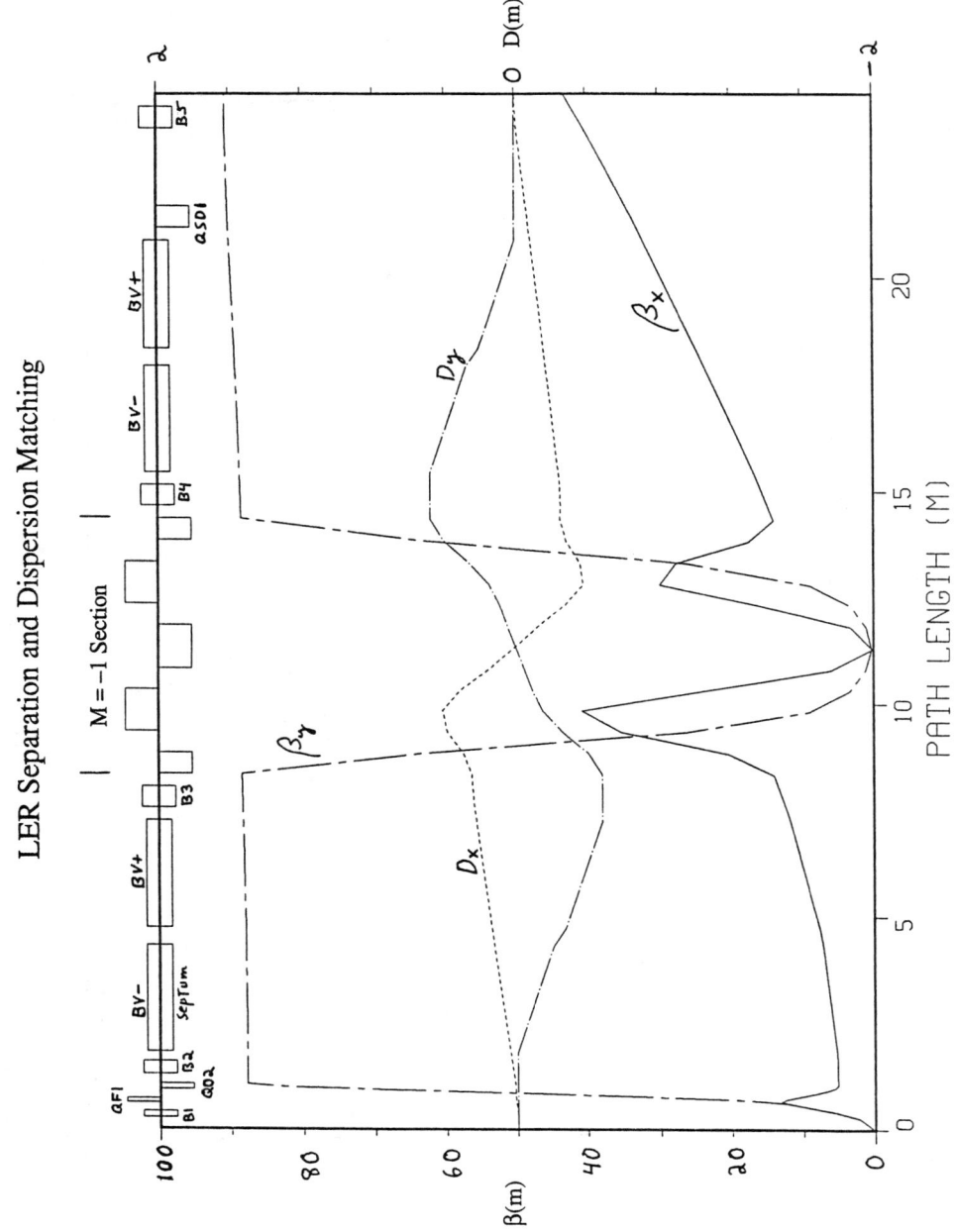

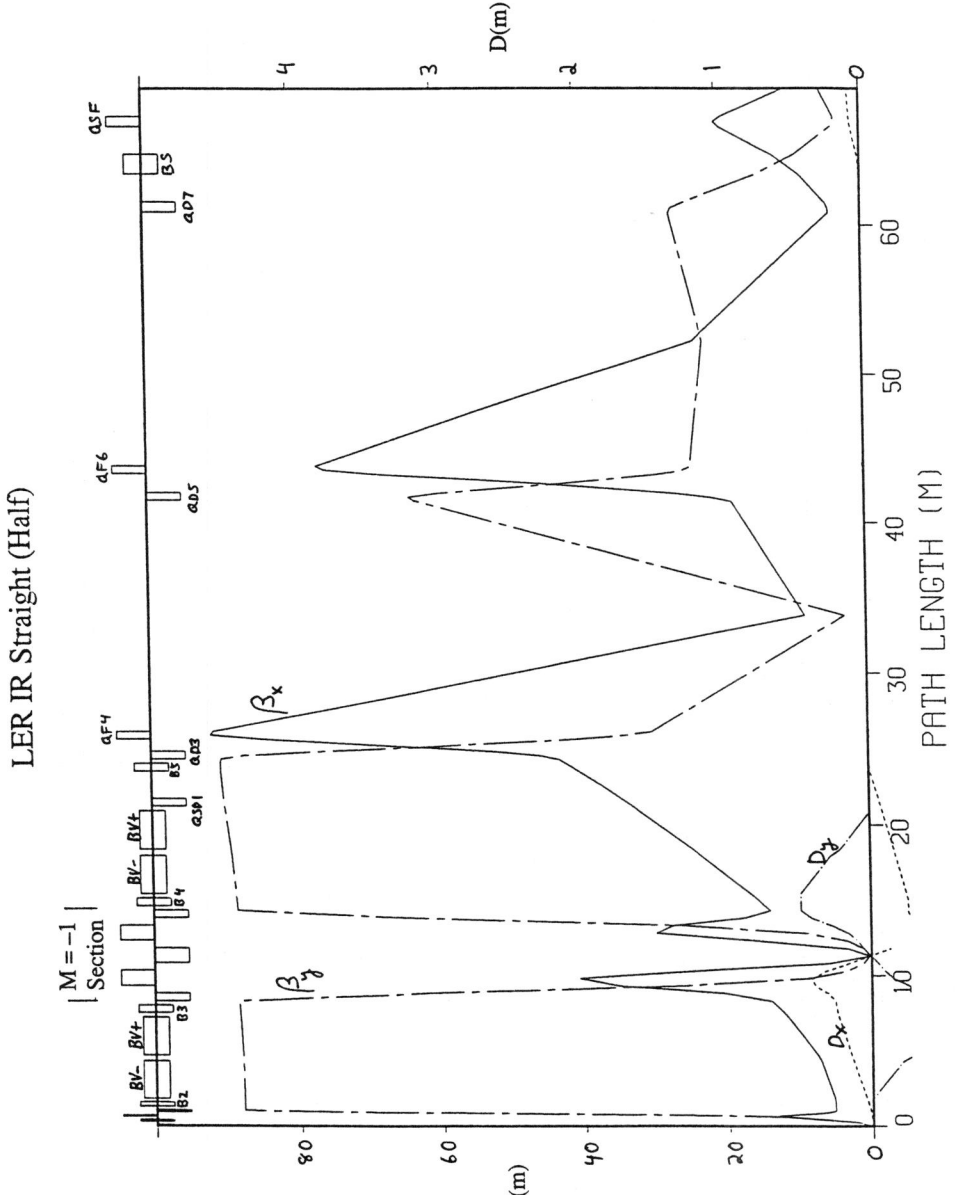

54 B-Factory Lattice Considerations

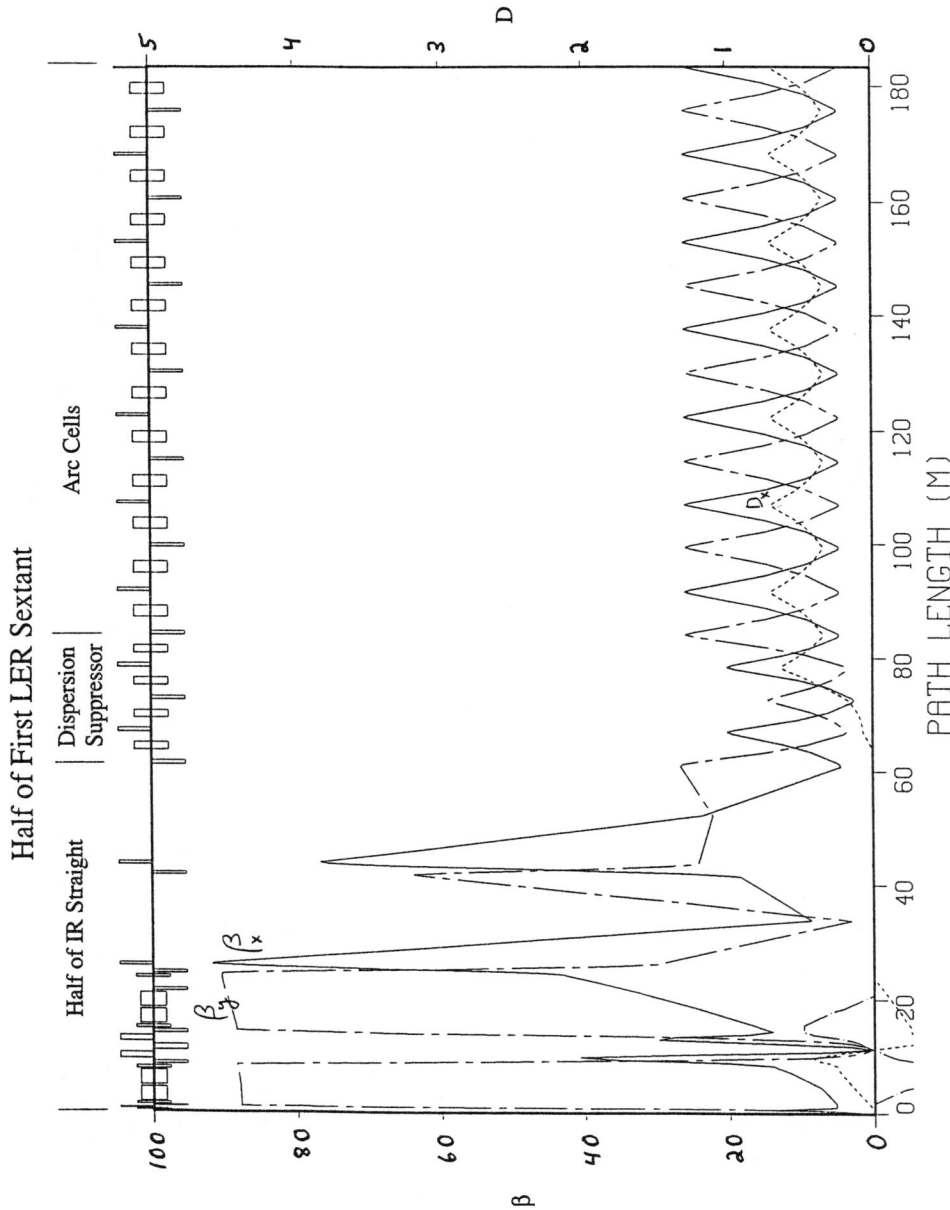

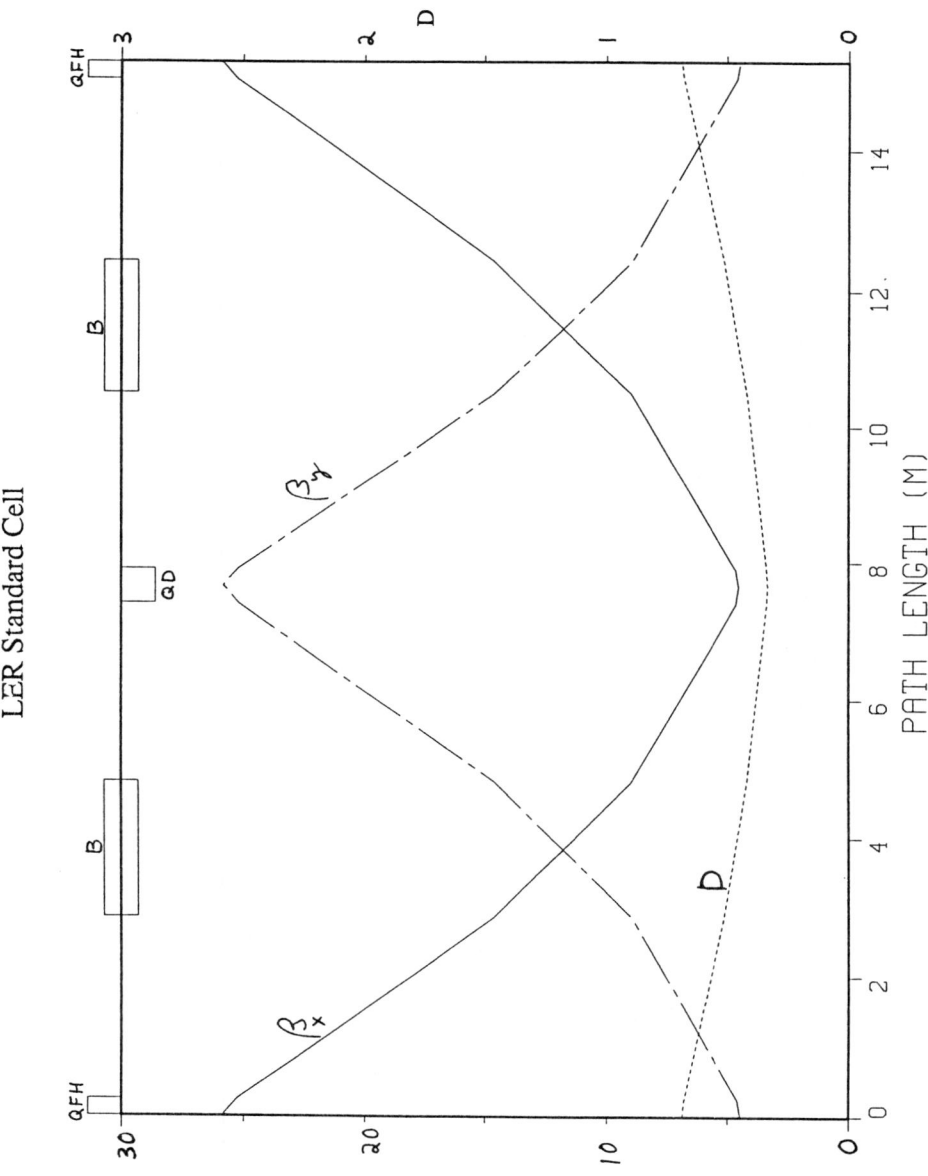

56 B-Factory Lattice Considerations

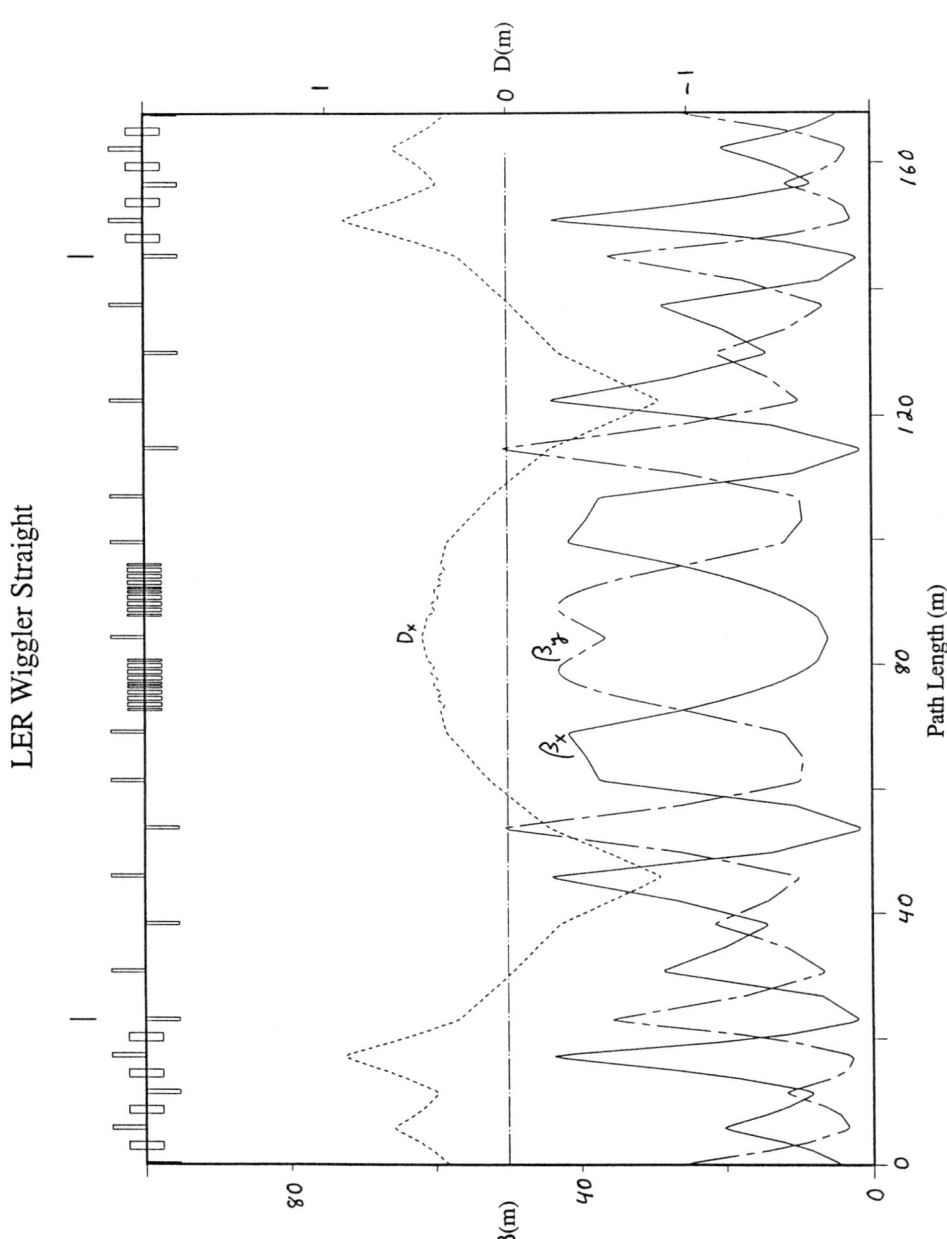

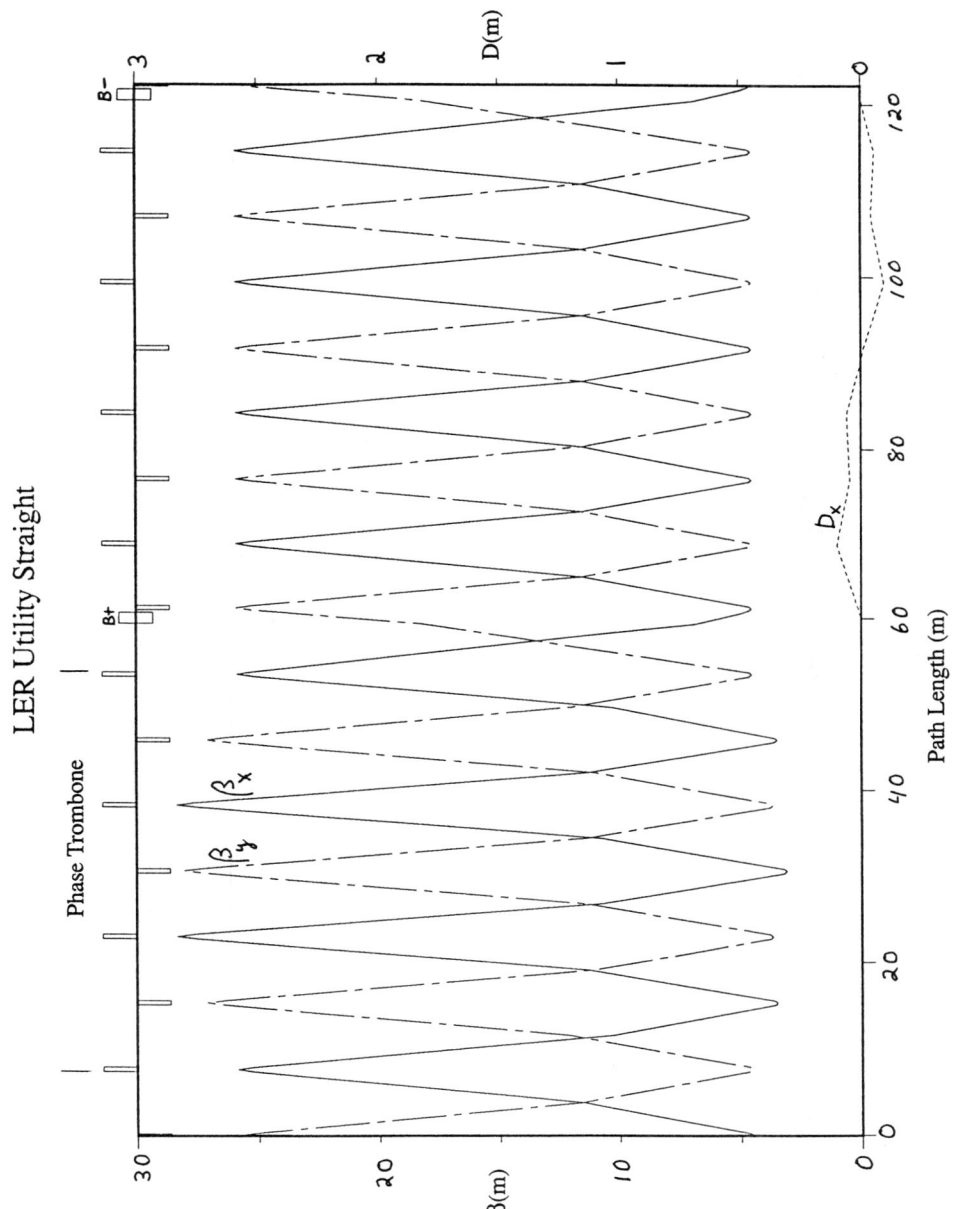

B-Factory Lattice Considerations

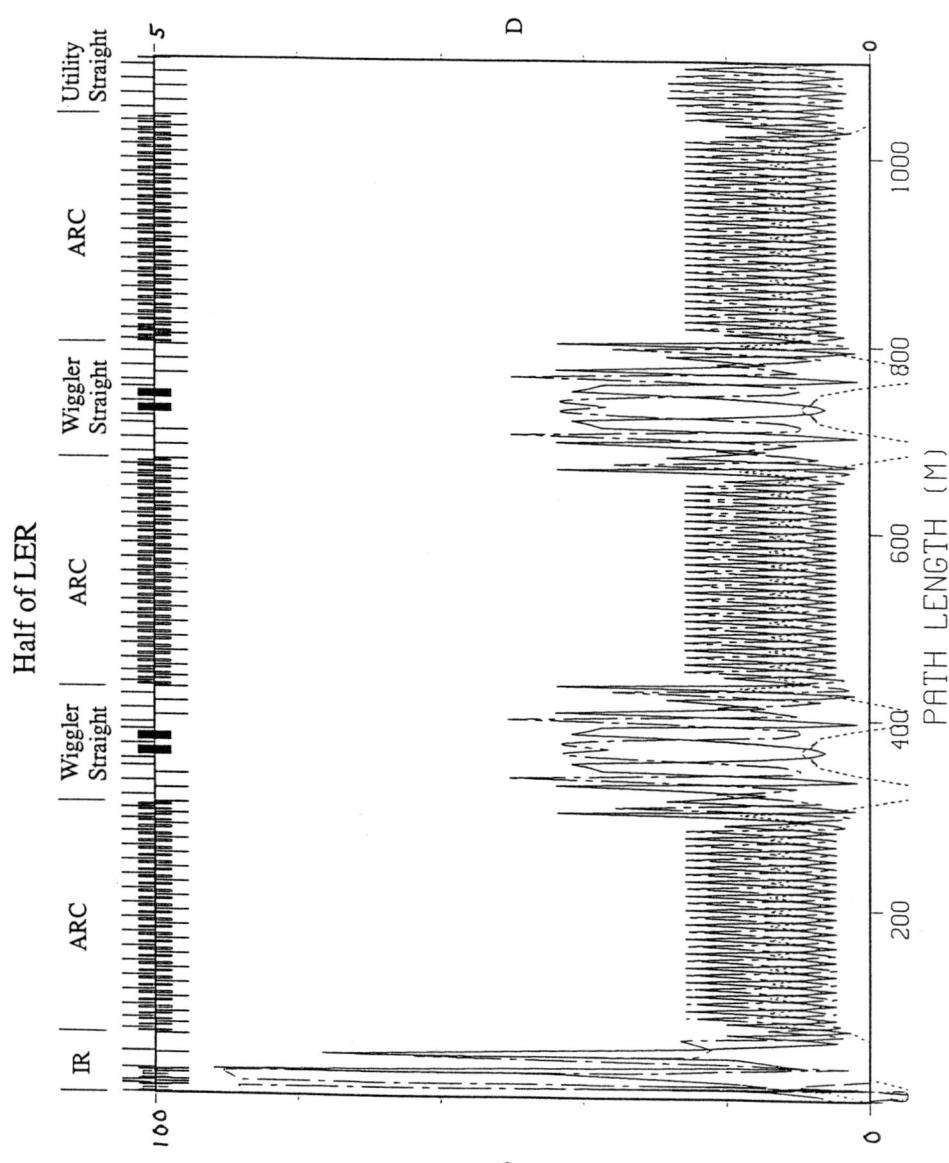

INTERACTION REGION CONSIDERATIONS*

H. DeStaebler

Stanford Linear Accelerator Center, Stanford University, Stanford, CA 94309

ABSTRACT

A number of machine-detector interface issues are mentioned, with an emphasis on detector backgrounds.

1. INTRODUCTION

The goal of the project is to observe CP violation in the $\bar{B}B$ system. This machine is supposed to be a factory for high energy physics, not an R&D project for accelerator physics. (The necessary R&D is supposed to be done before the machine is built.)

There are a number of interrelated design issues arising from the different desires of the detector and the machine, some of which are listed below.

A number of background and beampipe issues are mentioned. The emphasis is on calculations. Any satisfactory design will combine measurements on existing machines with calculations pertaining to the measurement conditions as well as to the proposed machine.

2. DETECTOR REQUIREMENTS

- Many events are required, which implies high luminosity (this means high L_{ave}, not only high L_{peak}), which, in turn, implies high current and small spots. High current is achieved with customary bunch population but much closer bunch spacing. Small spots imply small β^* which, in turn, implies short bunches and fairly large IP angles.
- Good vertex resolution is required, which implies a small, thin beampipe. The IP beampipe will be the smallest physical aperture in the machine.
- Luminosity requirements are reduced by having a moving center of mass, which implies unequal beam energies, which requires two rings. Even with equal energies, two rings are necessary to eliminate the effects of parasitic bunch crossings.
- There is competition between the detector and machine for the scarce real estate near the IP. (Don't forget space for cables and services.)
- The detector will have a solenoidal field of 1–1.5 T extending over ± 2 m.
- The detector must experience acceptable backgrounds during luminosity running. For the detector, this means a design relatively insensitive to backgrounds. For the machine, this means a number of masks (both near and far from the IP), an appropriate optics design that minimizes background problems, and a presure profile that reduces backgrounds.

*Work supported by Department of Energy contract DE–AC03–76SF00515.

- Frequent and rapid injection is required to keep L_{ave} high. This obviously constrains the machine. The detector must be insensitive to radiation damage during injection, and it must go quickly from data taking to injection and back again. Some kind of rapidly insertable and removable shielding to protect the detector against injection and poor machine performance might be useful.
- Radiation damage to the detector from commissioning and machine physics work should be small compared with the inevitable amounts during luminosity running. The detector will undoubtedly be absent during initial commissioning, but it will be present during the final stages of commissioning, since it is the best instrument to measure backgrounds, and it will be present during recommissioning following shutdowns.
- The design of the detector/machine should be flexible, for example to accomodate different headon or crossing angle geometries at the IP, or changes in the energy asymmetry.
- Special IR instrumentation is needed. For tuning, whether by operator or by computer, prompt signals proportional to background and to luminosity are needed. Radiation detectors near masks and limiting apertures would be useful in identifying sources of background, as would detectors that were sensitive to only one beam. Since some bunches might contribute more background than others, it might be very educational to be able to identify individual bunches, or at least sync the background detectors to the revolution frequency. Possibly special BPMs should be added.

3. SYNCHROTRON RADIATION BACKGROUND

Bends and quads near the IP are the main sources of synchrotron radiation that cause background problems. For headon collisions with unequal energy beams, bends are needed near the IP to separate the beams to avoid parasitic bunch collisions. Bend magnets are required to get the beams into the arcs; the final bending should be done at low field to reduce the characteristic energy of the SR.

Masks shield the IP beampipe from direct SR as well as from scattered SR. Only the higher energy photons that eventually can cause problems in the detector are of interest. The SR background can always be reduced by increasing the inner radius of the IP beampipe, but at the expense of degrading detector resolution. Almost all the SR photons go through the IP beampipe without hitting any masks, but they all eventually interact somewhere.

Appendix A contains some synchrotron radiation formulas.

3.1 SOURCES OF SR

A bend magnet produces a fan of radiation with the extreme rays being the incoming and outgoing beam trajectories. Usually the bend angle is large enough so it determines the width of the SR swath. Perpendicular to the bend plane the height of the swath is determined by the size and angular divergence of the radiating beam plus the intrinsic angular distribution of the SR photons relative to the radiating trajectory.

A quadrupole produces a much more complicated beam of SR than a bend magnet. A program commonly in use at SLAC that calculates radiation from quadrupoles, and traces the photons through a series of masks, QSRAD,[5] makes various approximations: (a) each ray of the beam emits a flat fan, (b) a single value of B characterizes the SR, derived from an average offset, and (c) the intrinsic angular spread in the beam is neglected. The four-dimensional integral over the beam distribution (x, x', y, y') is reduced to an integral over only x, y. Evenly spaced rays are traced through the system and the photons from each ray are weighted by the chosen beam density at that x, y. Or an external ray distribution file from an optics calculation may be used as input to QSRAD. A partial check on the approximations can be made by dividing physical quadrupoles into a number of shorter quads for computational purposes.

The distribution of photons from a quad depends on the transverse distribution of the beam, which often becomes more poorly known the farther one gets from the beam axis (see Sec. 6). This is especially true for the outer part of the photon beam which is most likely to hit masks and cause detector background.

3.2 MASKING AND MASK RERADIATION

Masks shadow the detector beampipe from photons coming directly from the magnets. Unfortunately, there is no such thing as a perfectly black mask; that is, every photon hitting a mask has some probability of reradiation, depending on energy, angle, material, and geometry. So frequently secondary masks shadow primary masks, and so on. Table 3.1 gives some representative reradiation probabilities calculated with EGS4[7] for forward scattering from a mask. A photon beam with $k_c = 15$ keV and width 1 cm is incident on a rectangular mask, starting from the edge. All photons are scored that scatter out with $\theta < 11.5°$. K-shell fluorescent radiation is included. All emerging photons had $k > 30$ keV.

Table 3.1

Material	$\frac{1}{n_{\text{in}}} \frac{dn_{\text{out}}}{d\Omega}$ (sr^{-1})
Ta	3×10^{-6}
Cu	4×10^{-5}
Al	7×10^{-4}

High Z masks are better because all the cross sections per atom increase with Z, and in addition the major absorption cross section (photoelectric effect) grows more rapidly with Z than the scattering cross sections. The probability of fluorescence radiation following photoelectric absorption increases with Z. Fortunately, it is not as severe for the softer spectra characteristic of B-factories as it is at linear colliders, for example. Often, masks are coated with thin layers of lower Z materials on top of higher Z to optimize the competition between absorption and reradiation. This applies as well to the inside of the IP beampipe.

4. BACKGROUND FROM LOST BEAM PARTICLES

Beam particles hitting the masks and beampipe near the detector will send degraded shower debris into the detector. As is well known, there are no black masks for high-energy beam particles. IR masks honor a beam-stay-clear that is supposed to keep beam tail particles from hitting them. This means that a distant mask system shadows the masks close to the IP. However, beam-gas interactions relatively close to the IR may cause beam particles to hit the inner masks depending on details of optics, masking, and residual pressure.

4.1 BEAM-GAS BREMSSTRAHLUNG

The cross section for fractional energy loss u by radiation is approximately[8]

$$\frac{d\sigma}{du} = 4\alpha\, r_e^2\, Z(Z+1)\,\frac{4}{3u}\,(1-u+.75u^2)\,\ell n\left(\frac{183}{Z^{1/3}}\right) \tag{4.1}$$

$$u = \frac{k}{E}. \tag{4.1a}$$

The $Z+1$ takes approximate account of radiation from the atomic electrons.[11] Note that the radiated photons themselves may be a noticeable source of background, even though their average energy is only a fraction of the energy of the beam. The angular distribution of the radiation process is usually neglected in this application. The angular distribution has characteristic angle $1/\gamma$ (that is, the transverse momentum is about mc).[13]

4.2 BEAM-GAS NUCLEAR COULOMB SCATTERING

The cross section for Rutherford scattering at polar angle θ (taken much less than 1) is[8]

$$\frac{d\sigma}{d\Omega} = 4r_e^2\, Z^2\,\frac{\left(\frac{m}{p}\right)^2}{(\theta^2+\theta_1^2)^2}, \tag{4.2}$$

$$\theta_1 = \alpha Z^{1/3}\left(\frac{m}{p}\right). \tag{4.2a}$$

The screening of the atomic electrons is accounted for by the angle θ_1. Any nuclear form factor effects are neglected, which requires $q \approx E\theta < q_{max} = 137\, m/A^{1/3}$. The energy lost by the beam particle is $q^2/2A$ which can safely be neglected.

4.3 COULOMB SCATTERING FROM ATOMIC ELECTRONS

This is Rutherford scattering of the beam particles from free electrons. Changing Z^2 to $Z(Z+1)$ in Eq. (4.2) will roughly take account of this. One might worry that the fractional energy loss on a light target, which is approximately

$$u = \frac{q^2}{2m}\frac{1}{E} = \frac{\theta^2\gamma}{2}, \tag{4.3}$$

might be a concern. However, calculations show that energy losses greater than the natural energy spread correspond to a small scattering cross section.

4.4 NUMBER OF BEAM PARTICLES HITTING MASKS

The products of beam-gas interactions are transported through the optical system to well beyond the IP. It is convenient to use the program DECAY TURTLE[9] to track the beam-gas interaction products through a system of optical elements and masks. Note that in SLC calculations it was found that including sextupoles affected the tracking results, presumably because large amplitude particles are important.[12] The source probability is weighted according to the pressure profile and the composition of the residual gas.

The rate of particles hitting masks can be estimated as follows: Take 6.25×10^{12} beam particles (corresponding to 1 A for 1 μsec) traversing 10 m of CO (37.42 g/cm^2 radiation length)[14] at a pressure of 10^{-8} Torr, which comes to 4.1×10^{-13} radiation lengths. Consider all bremsstrahlung collisions that radiate more energy than 10 times the natural energy spread in the beam, taken as 10^{-3}. The rate is $(6.25 \times 10^{12}) \times (4.1 \times 10^{-13}) \times 4/3 \times \ell n(100) = 16$.

4.5 RERADIATION OF SHOWER DEBRIS INTO THE DETECTOR

The reradiation probability into the detector is greater for particles that hit near the IP, but distant sources must be evaluated numerically. Reradiation is also greater for particles that hit the face or near the edge of a mask. The shower debris also has to be transported through the lattice, the masks, the detector magnetic field and into the detector. EGS4[7] can be used to follow the shower debris through the beampipe into the detector.

5. OTHER SOURCES OF BACKGROUND

5.1 INJECTION SHIELD

Radiation damage during luminosity running is likely to be significant, so it is important to reduce damage during injection as much as possible. An attractive idea is a massive shield that can be quickly inserted inside the main drift chamber to provide protection to all the detector elements except the silicon vertex detector. This would supplement any other possible protective measures.

5.2 MULTIPLE REFLECTIONS OF SYNCHROTRON RADIATION

The calculations in Sec. 3 typically take account of 2 or 3 photon reflections. One always worries that there is some efficient mechanism involving multiple reflections for transporting SR over long distances to the IP. I do not know of such a process. Total external reflection[15] requires exceedingly smooth surfaces and only occurs for photon energies that are quite low by our standards. Multiple forward Rayleigh scattering is diminished by the competition with photoelectric absorption; any fluorescence reradiation is isotropic. X-ray diffraction scattering from the polycrystalline wall of the beampipe is also in competition with photoelectric absorption.

5.3 GAS INTERACTIONS NEAR THE IP

Section 4 dealt with beam-gas interactions fairly far from the IP which caused beam particles to hit masks near the IP. There are also interactions near the IP (within the z acceptance of the detector) that send background particles directly into the detector. Consideration of these processes may set a restrictive limit on the IP pressure.

(a) Some convenient FORTRAN programs calculate a number of e and γ interactions on nuclei, including quasi-elastic and inelastic electron scattering and various photopion reactions.[16] These are a useful supplement to rates measured with random triggers on existing machines. The recoil proton cross sections agree with measurements.[26,27]

(b) SR photons can scatter into the detector by Compton or Rayleigh scattering; photoelectric fluorescence is not a problem from C and O because the K edges are below 1 keV. For $x \equiv k/m < 1$ Compton scattering per free electron is approximately[57]

Table 5.1

Scattered Angle (degrees)	$d\sigma/d\Omega$ (r_e^2/sr)	k'/k
0	1	1
90	$0.5/(1+2x)$	$1/(1+x)$
180	$1/(1+4x)$	$1/(1+2x)$

Rayleigh scattering in our energy range is a more complicated function of k and Z. Useful fits to the cross sections based on a Fermi-Thomas atom[17] are

$$\sigma_{tot} = \frac{8\pi}{3} Z^2 r_e^2 \left[1 + \left(\frac{B}{1.394}\right)^{1.162} \right]^{-1.628} \tag{5.1}$$

$$B = \frac{2k}{\alpha m Z^{1/3}} = \frac{k\,(\text{keV})}{1.865 Z^{1/3}}. \tag{5.1a}$$

and

$$\frac{d\sigma}{d\Omega} = Z^2 r_e^2 \frac{1+\cos^2\theta}{2} \left[1 + \left(\frac{U}{1.186}\right)^{1.199} \right]^{-2.436}, \tag{5.2}$$

$$U = B \sin\frac{\theta}{2}. \tag{5.2a}$$

The fit ranges are $0 < B < 10$ and $0 < U < 40$. The Fermi-Thomas model is pretty good for high Z. But, for example, Eq. (5.2) overestimates the cross section for C and O by about a factor of 2 at $U = 12$, and a factor of 5 at $U = 40$.

As an order-of-magnitude estimate of this background, consider a beam of 1 A for 1 μs with 1 photon per electron, a pressure of 10^{-8} Torr N_2, an acceptance of 1 m and 2π sr, and evaluate the cross sections at 90°. The results are shown in Table 5.2. A proper calculation would integrate the SR fluxes from both beams over the actual cross sections and acceptances, and include absorption in the beampipe.

Table 5.2

k (keV)	Number of Scattered Photons		
	Rayleigh	Compton	Total
5	0.9	0.7	1.6
10	0.3	0.7	1.0
20	0.06	0.7	0.8

5.4 SR-Beam Interactions

The bunches of synchrotron radiation photons and the charged beam bunches collide at the IP and at $s_b/2$. Most of the interactions are Compton scattering, like a back-scattered laser beam,[18] although some of the highest energy photons will make pairs.[19] The interaction rates are not high, and the reaction products have low p_t and make small angles with the beam axis.

5.5 Photon Radiation from a Transverse Crab Cavity

Both synchrotron radiation from the transverse kick, and Compton scattering from the RF photons in the cavity[20] are very weak.

5.6 Background from RF Cavities

At PEP, the DELCO detector experienced background from the RF cavities in a nearby straight section, apparently from field emitted electrons that were accelerated in the cavities. These were eliminated by putting a weak magnetic bump on the IP side of the closest cavity.[21] DELCO was an open detector with little self-shielding; the other PEP detectors with adjacent RF, MkII and MAC, had no such problem.

5.7 Photo Hadrons and Photo Muons

Is it possible that the effects of hadronic/muonic debris from lost particles hitting near the detector could be more severe than the electromagnetic debris, possibly in causing triggers? It doesn't seem likely, but I don't see how to rule it out. So one should check.

6. BEAM SHAPE

The beam distribution near the IP affects both the SR and lost particle backgrounds. The beam size in the final quads affects the distribution of SR photons in number, energy, and spatial extent that the masks are designed to cope with. The lost particle rate depends on the beam distribution through over-focusing of low-energy particles in the final quads. It is useful, although somewhat artificial, to divide the beam into a central Gaussian core plus a halo or tail extending to many sigma.

6.1 Gaussian Core

For a single beam, the core size is set by the emittance (SR fluctuations) and the optics. However, the beam-beam interaction increases the core size, especially in the vertical direction (for flat beams), and this is seen in the luminosity.[50] A simulation of a PEP-like machine gave an increase in sigma of 5% in the horizontal and 10% in the vertical.[47]

6.2 Halo or Tail

The halo is generated by beam-gas interactions, the beam-beam interaction, nonlinear aspects of the optics encountered at large excursions, and the resonant and tune structure. (It seems to be the conventional wisdom that when beams are first brought into collision after a fill, the backgounds get worse, implying that gas scattering alone does not set the halo.) The halo distribution is in dynamic equilibrium between the processes tending to kick particles out, and radiation damping tending to bring them back. Beam lifetime is related to the distribution near the limiting aperture.[55]

There are several approaches to arriving at a model to use for the halo distribution, but, to my mind, none is completely satisfactory. This is unfortunate because a bold SR masking scheme would depend critically on the halo distribution.

(a) Computer Tracking Simulation

One might think that the halo could be calculated since all the processes are known, with the possible exception of nonlinearities at large radius. Simulations for the core seem relatively satisfactory, but the present beam-beam codes are not designed to accurately predict the small population ($\sim 10^{-5}$) in the halo.

(b) Fit a Model to Measurements

Measurements at CESR of the vertical beam distribution clearly showed a tail.[52] This was fit to the following forms

$$\frac{1}{n}\frac{dn}{dy} = \frac{1}{\sigma\sqrt{2\pi}}\left[\exp\left\{\frac{-y^2}{2\sigma^2}\right\}, \text{ or } 3.7\times 10^{-6}\exp\left\{-1.2\left(\frac{y}{\sigma}-5\right)\right\}\right], \quad (6.1)$$

depending whether y/σ is less than or greater than 5, and used for subsequent SR calculations.

Background measurements at PEP, assumed to come entirely from SR, were used to adjust the parameters of an assumed Gaussian tail,[54] which was used for subsequent SR calculations.[56]

The problem with the first approach is the basic assumption that the vertical distribution will be the same in the new machine of interest. But the beam-beam simulations indicate significant sensitivity to various machine parameters. To use this approach with confidence, one should demonstrate scaling.

The second approach suffers the same shortcomings as the first, and in addition, it is only an integral measurement.

(c) Semiquantitative (Qualitative?) Approach

Ritson argues that the relative population in the halo should be roughly the ratio of damping time to beam lifetime, and it should fall off relatively slowly, say, as a power ~ 4 or so, rather than as an exponential or Gaussian.[49]

(d) Conservative Approach

Assume a flat background out to the limiting aperture with a population larger than is implied by beam lifetime. Since presumably this is a worst case, it is useful to at least check a mask design against it. What to do if the worst-case backgrounds are too high is another question.

6.4 LIMITING APERTURES

As masks get closer to the IP, their size in sigma units should increase. The limiting apertures should be far from the IP, designed to shadow the IP region.

7. HEATING AND COOLING

The heat loads on various beampipes, masks, and surfaces need to be specified so that adequate cooling can be provided. Possible problem areas are cooling the IP beampipe, which will decrease the IP resolution, and high SR power densities. Allowable temperature rises need to be established and the consequences of thermal expansion investigated. The final temperature of a object depends on the relative rates of heating and cooling.

7.1 SR HEATING

SR heating of masks near the IP is usually small, since a significant heat load would be an intolerable background source ($1W = 6.25\ 10^{15}$ keV/s).

Machines with headon collsions and small bunch separations (1–2 m) produce dozens of kW of SR from the bend magnets initiating the orbit separation and in beams off-axis in common quadrupoles. (The irreducible SR from the quad focusing is roughly 10 times less.) This power must be conducted to a water-cooled dump, possibly first going through a very thin window in the vacuum pipe. The transverse power density can be high, and the initial rate of energy absorption is also high.

7.2 IMAGE CURRENT HEATING

All beampipes are heated on the inside by image currents flowing in the skin depth. This is basically a bunched beam pulse heating. The appropriate formula for a Gaussian beam is[22]

$$\frac{dP}{dz} = \frac{\Gamma(3/4)}{4\pi^2 a} \frac{s_b \langle I \rangle^2}{\sigma_z^{3/2}} \sqrt{\frac{\mu Z_0}{2\,\sigma(Z)}}, \qquad (7.1)$$

$$= 2.75 \left(\frac{W}{m}\right)\left(\frac{2cm}{a}\right)\left(\frac{1cm}{\sigma_z}\right)^{3/2}\left(\frac{s_b}{1m}\right)\left(\frac{\langle I \rangle}{1A}\right)^2 \left[\frac{\mu\,\sigma(Cu)}{\sigma(Z)}\right]^{1/2}. \quad (7.1a)$$

7.3 HOM Heating

A more serious source of heating comes from the RF power radiated by the beams as they traverse changes in size and shape of the beampipe. This is frequently referred as higher order mode power (HOM) because the RF cavities are a major discontinuity in the beampipe and the radiated energy typically has higher frequencies than the cavity fundamental. There are two fairly separate parts to the problem: how much HOM power is radiated, and where is it absorbed. All the HOM power is absorbed somewhere. It's just a question of providing enough cooling at the right places.

The energy radiated when a bunch passes a discontinuity is

$$U = k\, q^2, \tag{7.2}$$

where q is the bunch charge, and k is a loss parameter depending on the geometry of the discontinuity and on the bunch length (frequency spectrum). k is usually given in pV/C. The power radiated is

$$P = f_b\, U = k\, \frac{s_b}{c}\, \langle I \rangle^2, \tag{7.3}$$

$$f_b = \frac{c}{s_b}, \tag{7.3a}$$

$$\langle I \rangle = f_b\, q. \tag{7.3b}$$

Analytic expressions for k are available for simple geometries. For more complicated (realistic) geometries, computer codes are available (for example, MAFIA, TBCI, URMEL).[34] These computer calculations are tending to replace the experimental determination of k values.

Consider a cylindrical pipe of radius a that abruptly increases to radius b for a distance g before returning to a. Approximate expressions for k are[28,29]

$$k = \Gamma\left(\frac{1}{4}\right) \frac{Z_0 c}{4\pi^{5/2}\, a} \sqrt{\frac{g}{\sigma_z}} = 1.023 \frac{Z_0 c}{2\pi^2\, a} \sqrt{\frac{g}{\sigma_z}}, \quad g < g_c \tag{7.4}$$

$$k = \frac{Z_0 c}{2\pi^{3/2}\, \sigma_z} \ln\frac{b}{a}, \quad g > g_c \tag{7.5}$$

$$g_c \approx \frac{(b-a)^2}{2\sigma_z}. \tag{7.6}$$

Equation (7.4) corresponds to a pillbox, and Eq. (7.5) to a step down in radius (there is very little loss for a step up in radius). Reference 28 finds quite good agreement between Eqs. (7.4)–(7.6) and the code TBCI. Tapering the transitions in radius reduces k, but not in a simple way.[28]

The above results are for a single bunch traversing a single cavity. For high enough Q and short enough bunch spacing, interference effect may become important.[33,28]

7.4 ABSORPTION OF HOM

All the HOM energy is absorbed somewhere, in a few skin depths on the inside surface of the beampipe. The absorption is more complicated than the generation, since it depends on the mode and frequency distribution of the energy. High frequencies can propagate down the beampipe. Low frequencies are trapped in the cavities or are rapidly attenuated in the beampipe. The critical wavelength is comparable with the diameter of the beampipe.[31,2] The propagating energy is absorbed with a characteristic $1/e$ length of roughly

$$\lambda_e \sim (100 - 300 \text{ m}) \left[\frac{\sigma(z)}{\mu\sigma(Cu)}\right]^{1/2} \left(\frac{\text{radius}}{3 \text{ cm}}\right), \qquad (7.7)$$

which depends on mode and also on frequency relative to cutoff.

The small IP beampipe could have the highest cutoff frequency, so it might absorb HOM generated far away. One might contemplate isolating the IP with a lossy section of ferromagnetic stainless steel.[32]

Billing has discussed HOM generation and absorption in the context of CESR.[35] He reports $k = 0.09$ V/pC for a 2 inch ID SR mask in a 4 inch ID pipe with 27° tapers, and $k = 0.014$ V/pC for about 3 m of IP beampipe with gently tapered (2°–5°) transitions. He makes the interesting point that configurations with large k scale roughly as σ_z^{-1}, whereas those with small k scale roughly as $\sigma_z^{-(2-4)}$. Presumably as the scale of the geometrical irregularities approaches σ_z, the dependency on σ_z becomes stronger; however, for irregularities much smaller than σ_z, one would expect k to approach zero.

7.5 ACCEPTABLE TEMPERATURE RISE

Preliminary calculations indicate the necessity of active cooling of the beampipe at and near the IP.[36] Once this big headache is accepted, it's just a question of deciding how much cooling is needed and how to supply it. It is possible that very little of the beampipe anywhere can be adequately cooled simply by convection to ambient air.

Thermal expansion and stresses set limits to temperature rise. Also, thermal desorption increases with temperature (see Sec. 9 on vacuum).

8. ACCEPTABLE DETECTOR BACKGROUND

The effects of background on the detector elements are usually divided into three categories: radiation damage, extra hits (occupancy) which confuse tracking and pattern recognition, and false triggers.

The first is cumulative; the second two accumulate over resolving times of the order of a microsecond, and depend on the details of the detector. For radiation damage, one might design for a useful life of five years (of luminosity running plus injection and machine physics) with some safety factor added.

8.1 SILICON VERTEX DETECTOR

The detector elements themselves are relatively insensitive to radiation damage with acceptable levels of the order of 1 Mrad.[41] However, the associated electronics, which is mounted on or near the detector elements, is more sensitive by a factor of 10 or more.[42] This is presently a field of active research for SSC applications and one can hope for increased radiation hardness on a time scale of interest to a B-factory.[43]

There are so many channels in a pixel detector that occupancy is not the limit, and even in a strip detector occupancy is less of a limit than radiation damage. To see this, consider a strip 25 μm wide by 20 cm long, uniformly irradiated by charged tracks. The flux that produces 0.1 Mrad/10^7s is 3.1 10^5 tracks/cm^2-s corresponding to an occuapncy of $0.016/\mu$s.

8.2 MAIN DRIFT CHAMBER

Avalanches at the sense wires cause the accumulation of deposits that degrade performance. This only occurs when the HV is on, so this is mainly a concern during luminosity running, assuming that a fast HV ramp is provided for injection. The degradation is proportional to the integrated charge density on the wire. Present limits are around 1 C/cm,[44a] and encouraging progress is being made in identifying the role of trace impurities,[44,45] so I believe one may reasonably take this as a design value for a B-factory. Note that 1 C/cm spread over a 1 m wire for five years of 10^7s each corresponds to 2 μA average current.

First, compare radiation damage and occupancy for charged tracks. Assume that a track at normal incidence gives 0.8 pC at the wire (for example, 100 ion pairs with gain 5 × 10^4; with 30 eV/ip, this corresponds to 3 keV deposited per track). Then 2 μA corresponds to 2.5 × 10^6 tracks/s or an occupancy of 250% per μs. So occupancy sets a much more severe limit than radiation damage. This result is independent of gas and cell size through the assumption of constant charge per hit. Inclined tracks give more radiation damage for the same occupancy.

For SR photons, the relationship between radiation damage and occupancy is similar to that for charged particles, but there tends to be more radiation damage per unit occupancy. This arises from the energy spectrum of the photons, which interact mainly by photoelectric effect and, at higher k, Compton scattering. Low-energy interactions always produce damage, but may not trip a discriminator for a hit. High-energy interactions can produce much more pulse height (hence damage) than necessary for a hit. Note that for a given SR flux, a He-based gas will have many fewer interactions than an Ar-based gas.

Compton scattering at low energies is not very effective at transferring energy to the recoil electron. The average kinetic energy is approximately[57]

$$\langle T \rangle = k \frac{x}{1 + \frac{11}{5} x}, \qquad x = \frac{k}{m}. \qquad (8.1)$$

8.3 CALORIMETER AND CRID

One must consider background effects in other detector elements. For example, CsI, a frequently considered material for an electromagnetic calorimeter, seems to be especially sensitive to radiation damage.

8.4 RARE EARTH PERMANENT MAGNETS

For relatively radiation hard material (Sm_2Co_{17}), the tolerable exposure is roughly 10^{10} rad.[46]

8.5 TRIGGERS

Triggers require one or two fairly high energy tracks and/or a significant energy deposition. These probably come only from lost particles. One needs a fairly detailed model of the detector to estimate the rate.

9. BEAMPIPE PRESSURE/VACUUM

Beam-gas interactions in the IR set limits on acceptable pressure. Residual gases are mostly hydrogen with a quarter to a half of CO_2 and CO. Vacuum design for a storage ring seems to be at least as much art as science. *Caveat lector.*

9.1 SOME FORMULAS

In a system, the rate of gas flow per unit pressure drop defines the conductance.[37,38] (Note: the quantity of gas is measured in Torr-liter with $3.3 \; 10^{19}$ molecules/Torr-l at 20 C.)

$$\text{Throughput} = \text{Pressure drop} \times \text{Conductance} \quad (9.1)$$

$$\dot{Q} \text{ (Torr-l/s)} = \Delta P \text{ (Torr)} \times F \text{ (l/s)} \quad (9.1a)$$

Two simple but useful examples of conductance follow. For a small hole with area A

$$F = \frac{A\bar{v}}{4} \quad \text{or} \quad F_0 = \frac{F}{A} = \frac{\bar{v}}{4}, \quad (9.2)$$

where $\bar{v}$ is the average speed of the molecules. For a Maxwellian distribution

$$F_0 = \frac{1}{4}\sqrt{\frac{8}{\pi}\frac{kT}{M}} = 11.4 \left(\frac{\text{l/s}}{\text{cm}^2}\right) \sqrt{\frac{T}{293\text{K}}\frac{28}{M}}. \quad (9.3)$$

For a round pipe with diameter d and length L

$$F = F_0 \frac{\pi}{4} d^2 \frac{4}{3}\frac{d}{L}. \quad (9.4)$$

The effective conductance of a pump (which conducts the gas to a land of no return) is called its pumping speed, S (liter/s), so a pump has associated with it a pressure drop

$$S = \frac{\dot{Q}}{\Delta P}. \quad (9.5)$$

To get the pressure at a particular point, add up the values of ΔP from the pump to the point.

9.2 Sources of Gas

(a) Thermal Desorption

The desorption rate depends on the material, how it has been cleaned, how clean it has been kept, and the temperature. A reasonable, ballpark value (at room temperature) is

$$q = 10^{-11} \text{ Torr-l/s-cm}^2 . \tag{9.6}$$

This follows an Arrhenius temperature dependence. The effective binding energy can range from 0.1 to several eV,[38] which corresponds to an increase in desorption per 10 C from 15% to a factor of 50. In a mixed Al/SS system coming off bake, I observed a factor of 1.6 corresponding to 0.35 eV.

(b) Photo Desorption by Synchrotron Radiation

SR photons desorb gas from masks and beampipes. The yield of molecules/photon depends on photon energy[39a] but usual practice seems to simply use an average value, for example,[39]

$$\eta = 5 \times 10^{-6} \text{ molecule/photon}, \tag{9.7}$$

where only photons with $k > 5$–10 eV are included.

The value of η depends on the angle of incidence of the photons (90° is perpendicular incidence). The variation measured on Al with $k_c = 3$ keV is roughly[40,39]

$$\eta(\phi) = \frac{\eta(90°)}{\sqrt{\sin \phi}} . \tag{9.8}$$

Continued irradiation by SR photons ("scrubbing") reduces η approximately as the 2/3 power of the accumulated exposure.[39]

10. OTHER ISSUES

10.1 Safety Factor

What degree of conservatism is appropriate in designing the IP region? How much insurance should be provided against tolerances, misalignment, our imperfect understanding, possible machine upgrades, and the vagaries of the real world? Should one take seriously the goal of a turnkey B-factory, which implies a brief detector commissioning period, as apparently has happened at LEP? Or should one anticipate a period of development following first operation, with the possibility of significant modifications, as has often happened in the past?

10.2 Approximations

Approximations enter in two ways, and we need to be sure they are adequate. Approximations are made in calculating a particular background process, although all the basic cross sections are well known. We also make approximations in deciding which are the dominant processes, and which can be neglected.

10.3 Comparison with Actual Experience

It is valuable, and possibly essential for a successful design, to compare our calculational techniques and procedures with data from a real detector at a real storage ring, to check whether our understanding is in tune with nature. Acceptable agreement does not assure success at a B-factory, of course, because scaling from one machine to another is imperfectly understood. But disagreement should surely cause hard thinking and lost sleep.

APPENDIX A: FORMULAS FOR SYNCHROTRON RADIATION

A.1 Bend Magnets

These formulas pertain to electrons in circular motion.[1,2,3] The average number of photons radiated in path length ds is

$$\frac{dn}{ds} = \frac{5}{2\sqrt{3}} \frac{\alpha\gamma}{\rho}, \qquad (A.1)$$

$$n = \langle n \rangle = 20.6 \, E(\text{GeV})\phi \,(\text{rad}) = 0.618 \, B \, L(\text{kG}-\text{m}) . \qquad (A.1a)$$

The spectrum is a universal function of the characteristic energy

$$k_c = \frac{3}{2}\gamma^3 \frac{\hbar c}{\rho} = \frac{3}{2}\frac{\gamma^3}{\alpha}\frac{r_e}{\rho} mc^2 , \qquad (A.2)$$

$$= 2.22(\text{keV}) \left(\frac{E}{10\text{ GeV}}\right)^3 \left(\frac{1\text{km}}{\rho}\right) , \qquad (A.2a)$$

$$= 6.66(\text{keV}) \left(\frac{E}{10\text{ GeV}}\right)^2 \left(\frac{B}{1\text{kG}}\right) . \qquad (A.2b)$$

The normalized photon energy is

$$v = \frac{k}{k_c} . \qquad (A.3)$$

The normalized number distribution of the photons is

$$\frac{1}{n}\frac{dn}{dv} = \frac{3}{5\pi} \int_v^\infty K_{5/3}(y)\,dy , \qquad (A.4)$$

$$\approx 0.4105\, v^{-2/3}(1 - 0.8438\, v^{2/3} + 0\, v^{4/3} + \ldots) \qquad , \quad v \ll 1 \quad (A.4a)$$

$$\approx \frac{3}{5\sqrt{2\pi}} \frac{e^{-v}}{\sqrt{v}} \left(1 + \frac{55}{72}\frac{1}{v} - \frac{0.9791}{v^2} + \ldots\right) \qquad , \quad v \gg 1 \quad (A.4b)$$

Half of the energy is carried above $v = 1$, by only 8.7% of the photons. Half of the photons are above $v = 0.078$.

For convenience, Table A.1 lists some values for the spectrum (A.4) and its integrals.[4] f_N is the fraction of the number of photons above v, and f_E is the fraction of the energy carried by photons above v.

Table A.1

v	$dn/n\ dv$	f_N	f_E
0.01	8.50	0.7381	0.9979
0.03	3.91	0.6277	0.9912
0.1	1.562	0.4628	0.9502
0.2	0.863	0.3483	0.9052
0.3	0.584	0.2775	0.8485
0.5	0.333	0.1896	0.7369
1.	0.1244	0.08677	0.5000
2.	0.0288	0.02326	0.2150
3.	0.00818	0.00703	0.0886
5.	0.00081	0.000737	0.0142

The energy loss per electron has average value

$$\langle U \rangle = \langle n \rangle \langle k \rangle , \quad (A.5)$$

$$= 1.267\ (\text{keV}) \left(\frac{E}{10\ \text{GeV}}\right)^2 \left(\frac{B}{10\ \text{GeV}}\right)^2 L(m) , \quad (A.5a)$$

$$= 140.8\ (\text{keV}) \left(\frac{E}{10\ \text{GeV}}\right)^4 \left(\frac{km}{\rho}\right) \phi(\text{rad}) , \quad (A.5b)$$

and variance

$$\text{var}(U) = \langle (U - \langle U \rangle)^2 \rangle = \langle n \rangle \langle u^2 \rangle , \quad (A.6)$$

$$= 11.17 (\text{keV}^2) \left(\frac{E}{10\ \text{GeV}}\right)^4 \left(\frac{B}{kG}\right)^3 L(m) , \quad (A.6a)$$

$$= 414\ (\text{keV}^2) \left(\frac{E}{10\ \text{GeV}}\right)^7 \left(\frac{km}{\rho}\right)^2 \phi(\text{rad}) , \quad (A.6b)$$

where $\langle v \rangle = 0.3079$ and $\langle v^2 \rangle = 0.4074$ have been used.[1] Equation (A.6) arises because var(U) depends on fluctuations in n as well as in k; a Poisson distribution for n has been used. The angular distribution of the radiated energy integrated over all k is

$$\frac{1}{U}\frac{dU}{dw} = \frac{21}{32}\frac{1+(12/17)w^2}{(1+w^2)^{7/2}}, \qquad -\infty \leq w \leq \infty, \qquad (A.7)$$

$$w = \gamma\psi, \qquad (A.7a)$$

where ψ is the angle perpendicular to the bend plane. For some cases of heating by bend SR, it is useful to have an approximate expression for the full double differential distribution. For fixed v, approximate the w dependence with a Gaussian. Then

$$\frac{1}{U}\frac{dU}{dv\,dw} \approx \frac{1}{U}\frac{dU}{dv} \cdot \frac{1}{\sqrt{2\pi}\sigma_w} \exp\left\{\frac{-w^2}{2\sigma_w^2}\right\}, \qquad (A.8)$$

$$\sigma_w \approx \left(\frac{0.363}{v}\right)^{0.44}. \qquad (A.9)$$

This approximation is reasonable at the 10–20% level for $0.1 < v < 3$. Outside this range, Eq. ($A.9$) overestimates the effective σ.

A.2 Quadrupoles

A photon spectrum integrated over a quadrupole field may be derived from ($A.4$).[6] This spectrum is not very useful for background calculations because only the spectrum hitting the mask is interesting, not the spectrum going down the beampipe. The moments of a quad spectrum may be interesting for power reasons. For a Gaussian beam, the scale field is B_σ, the quad field at 1 σ of the beam. The number of radiated photons and the radiated energy in units of the values for a bend magnet with B_σ are, with b the beam centroid offset in σ units:

Table A.2

Gaussian Beam	Photon Number $\langle n\rangle/n_\sigma$	Radiated Energy U/U_σ
1-D	$\sqrt{(2/\pi)} + b$	$1 + b^2$
2-D (round)	$\sqrt{(\pi/2)}$ ($b = 0$ only)	$2 + b^2$

A.3 Random Sampling from the Synchrotron Radiation Distribution

The nicest routine I know for random sampling from the spectrum Eq. ($A.4$) is due to Yokoya.[10] RN is a random number uniform between 0 and 1 representing the integral number distribution between v and infinity. The returned values of v are within 0.05% of Mack's values,[5] at least for $0.001 < v < 12$.

76 Interaction Region Considerations

```
      DATA YA1/0.5352 /, YA2/0.3053 /, YA3/0.1418 /, YA4/0.4184 /,
     % YB0/0.01192 /, YB1/0.2065 /, YB2/-0.3281 /,
     % YC0/0.003314 /, YC1/0.1927 /, YC2/0.8877 /,
     % YD0/148.3 /, YD1/675.0 /, YE0/-692.2 /, YE1/-225.5 /


      IF(RN.GT..342)THEN
      P1=1.0-RN
      P2=P1*P1
      V=(((YA4*P2+YA3)*P2+YA2)*P2+YA1) *P2*P1
      ELSEIF(RN.GT.0.0297)THEN
      V=((YB2*RN+YB1)*RN+YB0)/ (((RN+YC2)*RN+YC1)*RN+YC0)
      ELSE
      T1=-LOG(RN)
      V=T1+(YD1*T1+YD0)/((T1+YE1) *T1+YE0)
```

A.4 Short Bend Radiation

A magnet in which the bend angle is less than $1/\gamma$ is called a short magnet; the spectrum does not follow Eq. $(A.4)$ and depends on the z variation of B. The amount of energy is roughly the same as given in Sec. A.2, above but the scale or characteristic energy is greater,[3] where with $k_{c-\text{long}}$ given by Eq. $(A.2)$,

$$k_{c-\text{short}} \approx k_{c-\text{long}} \frac{\frac{2}{\gamma}}{\phi}. \qquad (A.10)$$

A.5 General SR Spectrum

The spectrum in Sec. A.1 is for $k_c \ll E$. The general case is[23,24,25]

$$\frac{1}{n}\frac{dn}{dy} = \frac{3}{5\pi}\frac{1}{(1+\xi y)^2}\left[\int_y^\infty K_{5/3}(x)\,dx + \frac{\xi^2 y^2}{1+\xi y}K_{2/3}(y)\right], \qquad (A.11)$$

$$y = \frac{\frac{k}{k_c}}{1-\frac{k}{E}} = \frac{v}{1-u} = v(1+\xi y), \qquad (A.12)$$

$$\xi = \frac{k_c}{E} = \frac{3}{2}\Upsilon, \qquad (A.13)$$

where n is given by $(A.1)$, and v only enters in the combination y. For $\xi = 0$, $(A.11)$ reduces to $(A.4)$. Υ is frequently used as a measure of k_c/E rather than ξ.[24,25] For Monte Carlo sampling of $(A.11)$ see Ref. 10.

APPENDIX B: NOTATION

a	Radius
b	Radius
c	Velocity of light
f_b	Bunch crossing frequency, $f_b = c/s_b$
k	HOM loss parameter, usually pV/C
k	Photon energy, usually keV
k'	Scattered photon energy
k_c	Characteristic energy of synchrotron radiation
m	Electron mass (energy or momentum, i.e., missing factors of c)
n	Number of radiated SR photons
p	Momentum
q	Bunch charge
q^2	(Momentum transfer)^2
r_e	Classical radius of electron, 2.82×10^{-13} cm
s	Path length along orbit
s_b	Bunch spacing
u	Energy loss normalized to E, e.g., k/E
v	Normalized photon energy, k/k_c
w	Angle nomalized to $1/\gamma$, $w = \gamma x$ angle
z	Distance along beam axis
A	Atomic weight
B	Magnetic field, usually kG
E	Beam energy
I	Beam current
L	Magnet length, usually m
SR	Synchrotron radiation
T	Kinetic energy
U	Energy radiated
Z	Atomic number
Z_o	377 ohms
α	1/137
γ	E/m
δ	Skin depth, usually μm
μ	RF magnetic permeability relative to vacuum
ϕ	Bend angle
σ	Cross section
σ_z	Rms bunch length, Gaussian parameter
$\sigma(Z)$	Dc electrical conductivity of material denoted by Z
θ	Scattering angle
ρ	Radius of curvature

References

1. J. Schwinger, Phys. Rev. **75**, 1912 (1949). Only a few of the numerous references dealing with SR are listed.
2. J. D. Jackson, Classical Electrodynamics, 2nd ed. (Wiley, 1975). Note: Jackson's k_c is two times bigger than Schwinger's.
3. R. Chrien, A. Hofmann, and A. Molinari, Phys. Report. **64**, 249 (1980).
4. R. A. Mack, Cambridge Electron Accelerator Report CEAL–1027, February 1966. Although it is not especially difficult to evaluate these SR formulas using modern computation facilities, Mack's report was quite an accomplishment in 1966.
5. A. R. Clark (LBL) wrote QSRAD in the early 1970s for use with PEP. Many versions exist now. There is no write-up, as far as I know.
6. E. Keil, Synchrotron Radiation from a Large Electron-Positron Storage Ring, CERN/ISR–LTD/76–23, June 1976.
7. W. R. Nelson, H. Hirayama, and D. W. O. Rogers, The EGS4 Code System, SLAC Report–265, December 1985. The only approximations in EGS4 that I know of that might affect B-factory calculations are no L-shell fluorescence radiation, no elastic nuclear form factors, and no quasi-elastic or inelastic nucleon/nuclear scattering of any kind.
8. B. Rossi, High-Energy Particles, Prentice-Hall, 1952. These formulas may be found in many other places.
9. D. C. Carey, K. L. Brown, and Ch. Iselin, DECAY TURTLE, SLAC–246/UC–28/Fermilab PM–31, March 1982. For beam-gas interactions, we use a version of DECAY TURTLE as modified by W. Kozanecki at SLAC.
10. K. Yokoya, A Computer Simulation Code for the Beam-Beam Interaction in Linear Colliders, KEK–Report 85–9, October 1985.
11. H. Bethe and J. Ashkin in Experimental Nuclear Physics, Vol. 1, ed. E. Segre (Wiley, NY, 1953).
12. W. Kozanecki (SLAC), private communication, March 1990.
13. H. W. Koch and J. W. Motz, Bremsstrahlung Cross-Section Formulas and Related Data, Rev. Mod. Phys. **31**, 920 (1959).
14. Y.-S. Tsai, Pair Production and Bremsstrahlung of Charged Leptons, Rev. Mod. Phys. **46**, 825 (1974); Errata, **49**, 421 (1977).
15. L. W. Jones and T. O. Dershem, p. 183, 12th Int. Conf. on High-Energy Accelerators, Fermi National Accelerator Laboratory, 1983.
16. J. W. Lightbody, Jr. and J. S. O'Connel, Modeling single-arm electron scattering and nucleon production from nuclei by GeV electrons, Computers in Physics, May/June 1988, p. 57.
17. A. T. Nelms and I. Oppenheim, Data on the Atomic Form factor: Computation and Survey, J. Res. Nat. Bureau of Standards **55**, No. 1, July 1955, pp. 53–62.
18. I. F. Ginzburg, G. L. Kotkin, V. G. Serbo, and V. I. Telnov, Colliding ge and gg Beams Based on the Single-Pass ee Colliders (VLEPP Type), Nucl. Inst. Meth. **205**, 47 (1983).
19. J. W. Motz, H. A. Olsen, and H. W. Koch, Pair Production by Photons, Rev. Mod. Phys. **41**, 581 (1969). This gives the γe cross section.
20. S. A. Heifets, Photon Beam at CEBAF, CEBAF TN/90/202, January 1990.
21. J. Kirkby (CERN), private communication, April 1990.

22. P. Morton and P. Wilson, Energy Loss and Wall Heating for a Gaussian Bunch in a Cylindrical Pipe, SLAC–AATF/79/15, November 1979.
23. K. Yokoya and P. Chen, Electron Energy Spectrum and Maximum Disruption Angle under Multi-Photon Beamstrahlung, SLAC–PUB–4935, March 1989, IEEE Particle Accelerator Conf., Chicago, March 1989.
24. A. A. Sokolov and I. M. Ternov, Synchrotron Radiation (Pergamon Press, NY, 1968); Radiation from Relativistic Electrons (American Inst. Phys., NY, 1986).
25. T. Erber, High-Energy Conversion Processes in Intense Magnetic Fields, Rev. Mod. Phys. **38**, 626 (1966). The spectrum here uses a different, but equivalent, combination of Bessel functions in Eq. ($A.11$).
26. K. V. Alanakyan et al., On the Angular Dependence of Photoprotons from Nuclei Irradiated with γ-Quanta with Maximum Energy 4.5 GeV, Nucl. Phys. **A367**, 429 (1981).
27. K. W. Chen et al., Electroproduction of Protons at 1 and 4 BeV, Phys. Rev. **135**, B1030 (1964). These authors deduced an equivalent radiator of about 0.025 radiation lengths.
28. J. J. Bisognano, S. A. Heifets, and B. C. Yunn, Loss Parameters for Very Short Pulses, CEBAF–PR–88–005, July 1988.
29. P. B. Wilson, Introduction to Wakefields and Wake Potentials, Physics of Particle Accelerators (Fermilab, 1987; Cornell, 1988), AIP Conf. Proc. 184, Vol. 1, NY, 1989.
30. Part. Accel. **25** (2–4) (1990), ed. S. Chattopadhyay, LBL Workshop on Impedance Beyond Cutoff, August 1987.
31. A. F. Harvey, Microwave Engineering (Academic Press, NY, 1963).
32. At 10 GHz, type 430 SS has μ about 10. J. Haimson (Haimson Research Corp.), private communication, April 1990.
33. P. B. Wilson, High Energy Linacs: Applications to Storage Ring RF Systems and Linear Colliders, Physics of Particle Accelerators (Fermilab 1981), AIP Conference Proceedings 87, NY, 1982.
34. A compendium of computer codes useful in accelerator physics is given in AIP Conference Proceedings 184, Vol. 2, NY, 1989.
35. M. Billing, Power Dissipation in the CLEO II Beam Pipe from Beam Heating, unpublished CESR internal note CON 87-6, May 1987; Issues Affecting Single Bunch Current Limits, B-Factory Workshop Proc., Syracuse University, September 1989.
36. J. Kent Proc. of Workshop on High Luminosity Asymmetric Storage Rings for B Physics, Report CALT–68–1552, Caltech, April 1989.
37. S. Dushman, Scientific Foundations of Vacuum Technique, 2nd edition, eds. S. Dushman and J. M. Lafferty (Wiley, NY, 1962).
38. N. B. Mistry, Ultrahigh Vacuum Systems for Storage Rings and Accelerators, Physics of Particle Accelerators (SLAC 1985), AIP Conf. Proc. 153, Vol. 2, NY, 1987.
39. B. A. Trickett, The ESRF Vacuum System, Vacuum **39**, 607 (1988).
39a. O. Groebner et al., Neutral Gas Desorption and Photoelectric Emission from Aluminum Alloy Vacuum Chambers Exposed to Synchrotron Radiation, J. Vac. Sci. Technol. **A7** (2), 223 (1989).

40. O. Groebner et al., Studies of Photon Induced Gas Desorption Using Synchrotron Radiation, Vacuum **33**, 397 (1987).
41. See articles in Proc. of 5th European Sym. on Semiconductor Detectors, Nucl. Inst. Meth. **A288** (1), 1–292 (1990).
42. J. G. Jernigan et al., Performance Measurement of Hybrid PIN Diode Arrays, SLAC–PUB–5211, May 1990; Intl. Industrial Sym. on SSC, Miami Beach, March 1990.
43. S. L. Shapiro et al., Progress Report on Use of Hybrid Silicon PIN Diode Arrays in High Energy Physics, SLAC–PUB–5212, May 1990; Vth Intl. Conf. on Instrumentation for Colliding Beam Physic , Novosibirsk, March 1990.
44. J. Kadyk et al., Anode Wire Aging Tests with Selected Gases, IEEE Trans. Nuc. Sci. **37** (2), 478 (1990).
44a. J. V. Allaby et al., The MAC Detector, Nucl. Inst. Meth. **A281**, 291 (1989).
45. J. Va'vra, Aging in Gaseous Detectors, SLAC–PUB–5207, March 1990; Vth Intl. Conf. on Instrumentation for Colliding Beam Physics, Novosibirsk, March 1990.
46. H. B. Luna et al., Bremsstrahlung Radiation Effects on Rare Earth Permanent Magnets, Nucl. Inst. Meth. **A285**, 349 (1989). (Note typo in Table 3: for 10^8 cm^{-2}, read 10^{18}.)
47. YongHo Chin (LBL), private communication, April 1990. The simulation used a code due to K. Yokoya (KEK) and applies to a PEP-like machine as described in Feasibility Study for an Asymmetric B Factory Based on PEP, LBL PUB–5244/SLAC–352/CALT–68–1589, October 1989.
48. G. Jackson and R. H. Sieman, Luminosity Performance of CESR, Nucl. Inst. Meth. **A286**, 17 (1990).
49. G. von Holtey and D. Ritson, Mini Beam Pipe at the DELPHI Interaction Region, CERN LEP NOTE 614, October 1988.
50. J. T. Seeman, Observation of the Beam-Beam Interaction, Nonlinear Dynamics Aspects of Particle Accelerators, Lecture Notes in Physics, Vol. 247 (Springer Verlag, Berlin, 1986).
51. S. Myers, Review of Beam-Beam Simulations, Nonlinear Dynamics Aspects of Particle Accelerators, Lecture Notes in Physics, Vol. 247 (Springer Verlag, Berlin, 1986).
52. G. Decker and R. Talman, Measurement of the Transverse Particle Distribution in the Presence of the Beam-Beam Interaction, IEEE Trans. Nuc. Sci., **NS–30** (4), 2188 (1983).
53. A. Jawahery, M. D. Mestayer, and R. Talman, Synchrotron Radiation and the CLEO–II Micro Vertex Detector (I), Cornell CBX–87–22, April 1987.
54. M. K. Sullivan, Synchrotron Radiation Background in the MAC and TPC Vertex Chambers, Univ. Calif. Intercampus Inst. for Research at Particle Accelerators, c/o SLAC, UC–IIRPA–88–01, May 1988.
55. M. Sands, The Physics of Electron Storage Rings: An Introduction, SLAC–121, November 1970.
56. See Feasibility Study for an Asymmetric B-factory Based on PEP, LBL PUB–5244/SLAC–352/CALT–68–1589, October 1989.
57. R. D. Evans, Encyclopedia of Physics, Vol. XXXIV (Springer Verlag, Berlin, 1958).

Influence of Collective Effects on the Performance of High-Luminosity Colliders*

Michael S. Zisman
Exploratory Studies Group
Accelerator & Fusion Research Division
Lawrence Berkeley Laboratory
Berkeley, California 94720
U.S.A.

April, 1990

Abstract

The design of a high-luminosity electron-positron collider to study B physics is a challenging task from many points of view. In this paper we consider the influence of collective effects on the machine performance; most of our findings are "generic," in the sense that they depend rather weakly on the details of the machine design. Both single-bunch and coupled-bunch instabilities are described and their effects are estimated based upon an example machine design (APIARY-IV). In addition, we examine the possibility of emittance growth from intrabeam scattering and calculate the beam lifetime from both Touschek and gas scattering. We find that the single-bunch instabilities should not lead to difficulty, and that the emittance growth is essentially negligible. At a background gas pressure of 10 nTorr, beam lifetimes of only a few hours are expected. Multibunch growth rates are very severe, even when using an optimized RF system consisting of single-cell, room-temperature RF cavities with geometrical shapes typical of superconducting cavities. Thus, a powerful feedback system will be required. In terms of collective effects, it does not appear that there are any fundamental problems standing in the way of successfully designing and building a high-luminosity B factory.

---

* This work was supported by the Director, Office of Energy Research, Office of High Energy and Nuclear Physics, High Energy Physics Division, U.S. Department of Energy, under Contract No. DE-AC03-76SF00098.

INTRODUCTION

There is presently a great deal of interest by the high-energy physics community in designing a facility for the production of copious quantities of B-mesons,[1] referred to as a "B factory." The ultimate purpose of such a facility is to study CP violations, as a means of investigating detailed predictions of the Standard Model. This will require a very high luminosity for the collider, in the neighborhood of $L = 1 \times 10^{34}$ cm^{-2}s^{-1}. Because such a luminosity is essentially two orders of magnitude beyond currently attained values, the design of a suitable collider presents many challenges to the accelerator physics community.

For a high-luminosity collider designed as a B factory, typical beam parameters are:

- total current, $I_{tot} \approx 1\text{–}3$ A
- number of bunches, $k_B \approx 1000$
- natural emittance, $\varepsilon_0 \approx 10\text{–}100$ nm·rad
- bunch length, $\sigma_l \approx 1$ cm

The high currents are necessary to achieve a high collision rate. But, to avoid difficulties with the beam-beam interaction, it is necessary to adjust the number of bunches and the beam size to keep the beam-beam tune shift below a certain maximum value that will be dictated by the storage ring itself. Because of the requirements for high beam currents and many bunches, it is necessary to store the electrons and positrons in two separate rings, irrespective of whether the two beams have the same energy (symmetric collider) or different energies (asymmetric collider). This arrangement avoids the difficulty associated with many parasitic bunch crossings at locations other than the primary interaction point, and keeps the large amount of synchrotron radiation power that must be absorbed by the vacuum chamber down to manageable levels.

In this paper, we will look at those issues related to the large beam currents required to provide a high-luminosity asymmetric collider, that is, at the *collective effects* of relevance to a B factory design. The focus here is on single-ring issues, before the beams are brought into collision. Qualitatively, the results obtained here are independent of the choice of symmetric versus asymmetric design.

A beam circulating in a storage ring interacts with its surroundings electromagnetically by inducing image currents in the walls of the vacuum chamber and other "visible" structures, such as beam position monitor electrodes, kickers, RF cavities, bellows, valves, etc. This interaction leads, in turn, to time varying electromagnetic fields that act on the beam and can give rise to instabilities. In most electron-positron colliders, single-bunch effects are the primary concern. However, different beam bunches can communicate through the narrow-band impedances in the ring, producing coupled-bunch instabilities.

Beam particles can also interact with each other or with gas molecules in the vacuum chamber, giving rise to various scattering phenomena. These include:

- intrabeam scattering (IBS), which causes emittance growth;
- Touschek scattering, which causes beam lifetime degradation; and
- gas scattering (either elastic or Bremsstrahlung), which also causes beam lifetime degradation.

We will first look at single-bunch instability thresholds and consider the growth rates of coupled-bunch instabilities. Then we will examine the possibility of emittance growth from intrabeam scattering (IBS). Finally, we will estimate beam lifetimes from Touschek scattering and gas scattering. As we will see below, the effect of the coupled-bunch instabilities is quite severe, and will likely be one of the limitations to the performance of the B factory. The results reported here were obtained with the LBL accelerator physics code ZAP.[2]

Where specific parameters are required, we use the APIARY-IV design[3] as an example. This SLAC-LBL design, which has evolved from earlier attempts[4] to produce a self-consistent B factory design, involves two equal-circumference rings in the PEP tunnel. (It is worth noting that, at this stage, we try where possible to remain faithful to known properties of the PEP ring. In particular, we take RF parameters to correspond to the presently used 353-MHz system.)

Main features of APIARY-IV include:
- initial luminosity of 3×10^{33} cm^{-2} s^{-1};
- moderate energy asymmetry (9 GeV in PEP, 3.1 GeV in the low-energy ring);
- round beams, which give a twofold (geometrical) improvement in luminosity.

Major parameters for the APIARY-IV collider are summarized in Table I.

To calculate the design luminosity, we make use of the simplified expression in Eq. (1), taken from Ref. 4:

$$L = 2.2 \times 10^{34} \, \xi \, (1+r) \left(\frac{I \cdot E}{\beta_y^*}\right)_{1,2} \quad (\text{cm}^{-2} \text{ s}^{-1}) \qquad (1)$$

where ξ is the maximum beam-beam tune shift for both beams (and in both transverse planes), r is the beam aspect ratio (r = 0 for a flat beam, r = 1 for a round beam), I is the beam current in amperes, E is the beam energy in GeV, and β_y^* is the beta function at the interaction point (IP) in cm. The subscript in Eq. (1) refers to the fact that the ratio (I·E/β^*) can be evaluated with parameters from either beam 1 or beam 2.

Parameter choices for the low-energy ring were driven to some extent by an attempt to achieve equal damping decrements in the two rings. This feature has been shown in beam-beam simulation studies[5] to be helpful in obtaining high luminosity, and it will also aid in the injection process.

INSTABILITIES

In this section, we describe some of the instabilities that are relevant to the design of an asymmetric B factory. Numerical evaluations will be presented to indicate the seriousness of the particular effect for a typical B factory design. Before doing so, we digress briefly to define the beam impedances that drive the various instabilities.

Impedances

Beam instabilities can occur in either the longitudinal or transverse phase planes. Longitudinal instabilities are driven by voltages induced via interactions of the beam with its environment. The strength of the interaction can be characterized by the ring impedance, $Z_{\|}(\omega)$, in ohms, which is defined in Eq. (2):

$$V_{\|}(\omega) = -Z_{\|}(\omega) \cdot I_b(\omega) \qquad (2)$$

where $V_{\|}(\omega)$ is the longitudinal voltage induced in the beam per turn arising from a modulation of the beam current at some particular angular frequency, $I_b(\omega)$.

Transverse instabilities arise from the transverse dipole wake field, which gives a force that increases linearly with transverse distance from the electromagnetic center of the vacuum chamber and is antisymmetric in sign about that center. The transverse impedance (in Ω/m) is defined by

$$Z_{\perp}(\omega) = \frac{-i \int_0^{2\pi R} F_{\perp}(\omega, s)\, ds}{e \Delta I_b(\omega)} \qquad (3)$$

where $F_{\perp}$ is the transverse force, integrated over one turn, experienced by a charge e having transverse displacement Δ. Explicitly, $F_{\perp}$ is given by

$$F_{\perp} = e\hat{\theta}\,(E_{\theta} + B_r) + e\hat{r}\,(E_r - B_{\theta}) \quad . \qquad (4)$$

In a typical storage ring, the impedance seen by the beam can be loosely characterized as being either broadband or narrow-band. As illustrated schematically in Fig. 1, sharp discontinuities in the vacuum chamber act as local sources of wake fields. These fields have a short time duration, which means that they include many frequency components. Thus, we refer to this impedance as a broadband impedance.

For instability calculations performed in the frequency domain (e.g., with ZAP), such impedances are typically represented with a so-called Q=1 resonator, whose analytical form is given in Eq. (5) for both the longitudinal and transverse cases.

$$Z_\parallel^{BB}(\omega) = \frac{R_s}{\left[1 + i \left(\frac{\omega_c}{\omega} - \frac{\omega}{\omega_c}\right)\right]} \tag{5a}$$

$$Z_\perp^{BB}(\omega) = \left(\frac{\omega_c}{\omega}\right) \frac{R_T}{\left[1 + i \left(\frac{\omega_c}{\omega} - \frac{\omega}{\omega_c}\right)\right]} \tag{5b}$$

This representation has convenient analytical properties and exhibits qualitatively the correct behavior for the actual impedance of a storage ring. In particular, the modulus of the longitudinal impedance, $|Z_\parallel|$, is proportional to frequency up to a cutoff frequency, ω_c, after which it falls off as $1/\omega$ with increasing frequency. In the calculations of longitudinal instabilities described below, we make use not of $|Z_\parallel|$ but of the related quantity $|Z_\parallel/n|$, where $n \equiv \omega/\omega_0$ is the harmonic of the revolution frequency ω_0. This quantity remains essentially constant up to the cutoff frequency, beyond which it decreases as $1/\omega^2$. As can be seen from inspection of Eq. (5b), the frequency dependence of the transverse impedance follows that of $|Z_\parallel/n|$.

The other category of impedance-producing objects in a typical storage ring consists of cavity-like objects, represented schematically in Fig. 2. Such objects can trap electromagnetic energy and exchange it with the beam. The wake field from a cavity oscillates for a long time, and thus gives a narrow spectrum in the frequency domain. These impedances are represented in calculations as narrow-band (i.e., high-Q) resonators, as given in Eq. (6).

$$Z_\parallel(\omega) = \frac{R_s}{\left[1 + i\, Q\left(\frac{\omega_r}{\omega} - \frac{\omega}{\omega_r}\right)\right]} \tag{6a}$$

$$Z_\perp(\omega) = \left(\frac{\omega_r}{\omega}\right) \frac{R_T}{\left[1 + i\ Q\left(\frac{\omega_r}{\omega} - \frac{\omega}{\omega_r}\right)\right]} \qquad (6b)$$

Typical values for Q lie in the range of 10^2–10^5, with parasitic modes of the RF cavities being closer to the upper end of the range (unless special procedures are used to de-Q them). As a result of the relatively long duration of these wake fields, trailing beam bunches feel the effects of the bunches that preceded them. The motion of the many bunches in the ring thus becomes coupled, and can become unstable for certain patterns of relative phase between bunches. This topic will be investigated later in this paper.

Longitudinal Microwave Instability

The first instability we consider is the longitudinal microwave instability, sometimes referred to as turbulent bunch lengthening. This instability, which has been seen in numerous storage rings (both proton and electron rings), is not a "fatal" instability, in the sense that it does not lead to beam loss. The instability causes an increase in both the bunch length and the momentum spread of a bunched beam, as illustrated in Fig. 3. The threshold (peak) current for the instability is given by

$$I_p = \frac{2\pi\ |\eta|\ (E/e)\ (\beta\sigma_p)^2}{\left|\frac{Z_\parallel}{n}\right|_{eff}^{BB}} \qquad (7)$$

where $|Z_\parallel/n|_{eff}$ is the effective broadband impedance of the ring and $\eta = \alpha - 1/\gamma^2$ is the phase-slip factor.

We refer to an "effective" impedance here to account for the fact that the bunch samples the storage ring impedance weighted by its power spectrum, $h(\omega)$, which is the square of the Fourier spectrum of the bunch. A short bunch—one having a frequency spectrum that extends well beyond the cutoff frequency of the broadband impedance—does not sample the broadband impedance fully, as can be seen in Fig. 4. To evaluate the effective impedance, we calculate the summation given in Eq. (8)

$$\left|\frac{Z_\parallel^{BB}}{n}\right|_{eff} = \frac{\left|\sum_{p=-\infty}^{\infty} h(\omega_p) \frac{Z_\parallel(\omega_p)}{(\omega_p/\omega)}\right|}{\sum_{p=-\infty}^{\infty} h(\omega_p)} \qquad (8)$$

where

$$h(\omega_p) = \exp\left[-\left(\frac{\omega_p \sigma_\ell}{\beta c}\right)^2\right] \qquad (9)$$

and $\omega_p = p\omega_0$. The result of such an calculation is shown in Fig. 5.

This reduction in effective impedance can be modeled in calculations by making use of the "SPEAR Scaling" ansatz[6] for $\sigma_\ell < b$:

$$\left|\frac{Z_\parallel^{BB}}{n}\right|_{eff} = \left|\frac{Z}{n}\right|_0 \left(\frac{\sigma_\ell}{b}\right)^{1.68} \qquad (10)$$

where b is the chamber radius. (In terms of the discussion above, the dependence on b in Eq. (10) results from our estimate of the cutoff frequency of the broadband impedance to be $\omega_c = c/b$.) The result of the impedance roll-off for short bunches is that the bunch lengthening threshold will be increased, as shown schematically in Fig. 6. The fact that experimental data from PEP[7] are in good agreement with the SPEAR Scaling estimates, as can be seen in Fig. 7, provides verification that the phenomenological model has some validity.

It is worth noting that the expression given in Eq. (10), which was determined phenomenologically, is in reasonable agreement with the behavior expected from a simple Q = 1 resonator. In Fig. 5 we can see that, in the short bunch length regime, the effective impedance does follow a power-law dependence. If we fit this region to determine the power law, as shown in Fig. 8, we obtain a value of about 2 (as expected for a Q = 1 resonator). However, the *measured* bunch length data correspond to a more restricted range of σ_ℓ/b, between 0.1 and 1.0. Confining the power law fit to this range, we obtain (Fig. 9) a value of 1.58, in good agreement with the SPEAR Scaling estimate.

Given that the actual broadband impedance in a storage ring is not likely to be exactly a Q = 1 resonator shape, the above argument should not be taken as a "proof" of the SPEAR Scaling law, but rather as a justification that the general trend of SPEAR Scaling—the decrease in effective impedance for short bunches—is reasonable. Obviously, the actual roll-off of the broadband impedance in any storage ring will depend on the details of the particular vacuum chamber hardware. Indeed, in modern storage rings that are specifically designed to minimize the broadband impedance it may well be that the impedance is dominated by a few discrete items, making the concept of an amorphous broadband impedance somewhat suspect.

To evaluate what happens for a typical B factory scenario, we use parameters from Table I. The bunch lengths for the high- and low-energy APIARY rings are shown in Fig. 10 as a function of RF voltage. To achieve a natural bunch length of 1 cm requires V_{RF} = 25 MV in the high-energy ring, and V_{RF} = 10 MV in the low-energy ring. Thresholds for bunch lengthening have been estimated for both rings, based on a low-frequency broadband impedance of $|Z_\parallel/n|_0$ = 1.5 Ω (i.e., half that of the present PEP ring); the results are summarized in Fig. 11. We see that, for our chosen parameters, the required current is well below threshold for both rings. In our calculations we have ignored the effect of potential-well distortion, which—for short bunches—is predicted to reduce the bunch length; this effect is estimated to be minor.

From these estimates, we conclude that there are no problems associated with the longitudinal microwave instability provided the impedance of the ring can be kept as low as 1.5 Ω. It is clear, however, that the low-energy ring could become a problem if we were envisioning considerably fewer bunches or much higher currents than proposed for the APIARY collider.

Transverse Mode Coupling

In contrast to the longitudinal single-bunch instability discussed above, the transverse mode-coupling instability is a "fatal" instability, in the sense of leading to beam loss. The typical manifestation of this instability is a limitation on the current that can be injected into a single bunch. The instability arises because the imaginary part of the transverse broadband impedance causes frequency shifts of the synchrotron sidebands of the betatron motion. When the frequency shifts are sufficient to cause two sidebands to cross, an instability develops. For the electron case with which we are concerned, the typical situation is that the m = 0 and m = -1 synchrotron sidebands cross.

For long bunches, the dependence of the threshold (average) current scales with the storage ring parameters according to

$$I_b = \frac{4\,(E/e)\,\nu_s}{\langle \text{Im}(Z_\perp)\,\beta_\perp\rangle R} \frac{4\sqrt{\pi}}{3}\sigma_\ell \qquad (11)$$

with

$$\nu_s = \frac{1}{\beta}\left[\frac{-h\eta V_{RF}\cos\phi_s}{2\pi\,(E/e)}\right]^{1/2}. \qquad (12)$$

From this scaling we see that, for the same ν_s value, the threshold current will be lower for a larger ring. It is also generally true that large rings have more impedance producing hardware, such as RF cavities, than do small rings. Note that it is the *beta-weighted* impedance that determines the threshold, so a significant gain can be made by "hiding" the devices contributing the transverse impedance in low-beta regions of the ring.

In the short-bunch regime of relevance for a B factory, the mode-coupling threshold is expected to increase again, as illustrated in Fig. 12. The reason for this behavior is related to the roll-off of the impedance for short bunches discussed earlier for the longitudinal case: short bunches do not fully sample the broadband impedance of the ring. In Table II we show how typical B factory parameters compare with those for a PEP configuration in which the transverse mode-coupling threshold was determined[8] to be 8.4 mA. In PEP, the transverse impedance is dominated by the RF system, which consists of 120 cells. For a B factory, we envision a much-reduced RF system (having about 20 single-cell cavities), with a proportionate decrease in broadband impedance. Moreover, the smaller number of RF cells can be located in a lower beta region of the ring, so the reduction in beta-weighted transverse impedance will be even greater. Taking these factors into account, we expect an increase in threshold current of about a factor of three for the high-energy ring of a B factory. Even at its comparatively low beam energy, the low-energy ring is expected to have a higher threshold current than PEP. For the parameters of Table I, the required single-bunch currents for the high- and low-energy rings are 1.2 mA and 1.7 mA, respectively, so we have a comfortable margin.

Although we appear to be safe in terms of the RF contribution, we must take note of the other transverse impedance—especially for the low-energy ring—to make sure that it does not grow too large. In a B factory, for example, we will require complicated masking to shield the detector from both synchrotron radiation and scattered beam particles. This can contribute significantly to the transverse impedance. By way of warning, we show in Fig. 13 a prediction[9] of the mode-coupling threshold for the PEP low-emittance optics (developed for the synchrotron radiation users of the PEP ring). Although a threshold of 2.7 mA was predicted based on the supposedly well-known transverse impedance of the PEP ring, the experimental results, shown in Fig. 14, gave a lower threshold. It is likely that at least some of this discrepancy is related to additional transverse impedance associated with synchrotron radiation masks.

Coupled-bunch Instabilities

As discussed earlier, wake fields trapped in high-Q resonant objects can affect the motion of trailing bunches. If the decay time of the wake fields is long compared with the interbunch spacing (as is usually the case), the overall strength of the effect scales with the *total* current in the ring, and the instability growth rates are not very sensitive to the bunch pattern itself. The coupled-bunch motion in synchrotron phase space can be described as dipole (a = 1), quadrupole (a = 2), etc., as illustrated in Fig. 15. Longitudinal instability obviously requires some synchrotron motion, so the lowest-order longitudinal mode is the a = 1 mode. Transverse instability, in contrast, can occur even in the absence of synchrotron motion (denoted the a = 0 or "rigid dipole" mode).

In the case of a bunched beam in a storage ring, the bunch frequency line spectrum is given by

$$\omega_\parallel^{coh} = (pk_B + s + a v_s) \omega_0 + \Delta\omega_\parallel^{coh}$$
$$= v_p \omega_0 + \Delta\omega_\parallel^{coh} \tag{13a}$$

and

$$\omega_\perp^{coh} = (pk_B + s + v_\beta + a v_s) \omega_0 + \Delta\omega_\perp^{coh}$$
$$= v_p^\perp \omega_0 + \Delta\omega_\perp^{coh} \tag{13b}$$

where the index s = 0, 1, 2,..., k_B-1 labels the normal modes of the k_B bunches in terms of their bunch-to-bunch phase shift, i.e.,

$$\Delta\phi = 2\pi \frac{s}{k_B} . \tag{14}$$

The physical meaning of the index s is illustrated for a simple case in Fig. 16, taken from Ref. 10.

The evaluation of coupled-bunch instability growth rates involves the calculation of the complex frequency shift $\Delta\omega_\parallel$ or $\Delta\omega_\perp$. In the Wang formalism,[11] the longitudinal frequency shift is expressed as

$$\Delta\omega_{s,a}^{\parallel} = i \frac{I_b \omega_0^2 \eta\, k_B}{2\pi\, \beta^2\, (E/e)\, \omega_s} \frac{(\sigma_L/R)^{2(a-1)}}{2^a\,(a-1)!} (Z_{\parallel})_{\text{eff}}^{s,a} \tag{15a}$$

where

$$(Z_{\parallel})_{\text{eff}}^{s,a} = \sum_{p=-\infty}^{\infty} (pk_B + s)^{2a} \exp\left\{-(pk_B + s)^2 (\sigma_L/R)^2\right\} \left[\frac{Z_{\parallel}(v_p \omega_0)}{v_p}\right]. \tag{15b}$$

Transverse shifts are calculated in a similar manner with the expressions:

$$\Delta\omega_{s,a}^{\perp} = -i \frac{I_b\, c\, k_B}{4\pi\, (E/e)\, v_\beta} \frac{(\sigma_L/R)^{2a}}{2^a\, a!} (Z_{\perp})_{\text{eff}}^{s,a} \tag{16a}$$

and

$$(Z_{\perp})_{\text{eff}}^{s,a} = \sum_{p=-\infty}^{\infty} \left(pk_B + s + v_\beta - \frac{\xi}{\eta}\right)^{2a} \exp\left\{-\left(pk_B + s + v_\beta - \frac{\xi}{\eta}\right)^2 (\sigma_L/R)^2\right\} \left[Z_{\perp}(v_p^{\perp}\omega_0)\right]. \tag{16b}$$

The time dependence of either instability is $e^{-i\Delta\omega t}$, so the real part of $\Delta\omega$ gives the coherent frequency shift and the imaginary part gives the growth rate. It can be seen from the above expressions that the growth rate is related to the real part of the impedance itself. In the typical situation, about half of the bunch modes (index s) grow and half are damped. Even for the modes that have a positive growth rate, many will grow more slowly than the radiation damping rate, and thus will not present a problem.

To evaluate frequency shifts and instability growth rates quantitatively with Eq. (15) or Eq. (16), it is necessary to know the frequency dependent impedance $Z_{\parallel}(v_p\omega_0)$ or $Z_{\perp}(v_p\omega_0)$. In particular, we must have information on the higher-order parasitic modes of the RF cavity. Using the representation in Eq. (6), we need to know R_s or R_T, ω_r, and Q for each parasitic mode. These values can be estimated reasonably well with electromagnetic codes such as URMEL,[12] or they can be measured in the actual cavity if it already exists.

For the case of a B factory, most designs are based on the use of many single-cell cavities to produce the required voltage. Although the cavities are all nominally identical, dimensional tolerances and temperature effects will generally conspire to move the parasitic modes to somewhat different frequencies in the different cells. The evaluation of coupled-bunch instabilities, then, requires that these different modes be considered in some way. The straightforward approach would be simply to take all the parasitic modes from the many RF cells and use them in the calculation. This probably requires some sort of statistical approach to assigning the parasitic mode frequencies, unless all cells are separately measured. The practical difficulty here is that the calculations become quite time consuming when many cells and many modes are involved.

To minimize the calculational effort—especially at the early design stage when no cavity exists—an alternative approach is to take the nominal modes from a single cell and de-Q them somewhat to represent the fact that the mode frequencies will vary from cell to cell, as illustrated in Fig. 17. The total strength of each mode, proportional to $n_{cell} \cdot R/Q$ (where n_{cell} is the number of RF cells), should be kept fixed in this approach. This de-Qing is only a calculational technique, and does not imply any actual changes in the modes themselves.

To reduce or eliminate problems with coupled-bunch instabilities, another type of de-Qing procedure, which involves physically reducing the Q of a high-order mode by means of a coupler or damping antenna, is sometimes undertaken. The helpfulness of this technique depends to some extent on where the modes land with respect to the rotation harmonics. In Fig. 18 we illustrate several cases. The lowest-frequency mode is sitting essentially at a beam rotation harmonic; de-Qing it then reduces the impedance sampled by the beam at that frequency and would reduce the instability growth rate. In contrast, the middle mode is not initially sampled by the rotation harmonic, but after de-Qing the impedance seen by the beam is worse than it was originally, which would increase the instability growth rate. The highest frequency mode in Fig. 18 is beyond the frequency band sampled by the beam bunch, and so is essentially invisible as far as instabilities are concerned. Because the power spectrum of the beam, $h(\omega)$, extends to ever higher frequencies as the bunch get shorter, a short bunch can sample higher frequencies. This tends to produce higher growth rates for the regime of interest for a B factory. Furthermore, we expect minimal Landau damping (from synchrotron tune spread) for short bunches.

To minimize the instability growth rates ($1/\tau$), and thus the demands on the feedback system (for which the power requirement scales as $1/\tau^2$), it is best to try to eliminate the impedance at its source. One goal of a B factory design should be to try to use the minimum number of RF cells to provide the needed voltage (about 25 MV in the high-energy ring and about 10 MV in the low-energy ring for the case treated here). There are several implications of this choice:

- a high voltage per cell is needed, which means that a lot of RF power ($\approx$280 kW with the parameters considered here) must be put through the RF input window;
- the beam power lost to synchrotron radiation power must be replenished with few cells, again requiring high power through the RF input window.

As a result, an RF window power requirement of more than 500 kW arises—this has not yet been reached in an operating accelerator and will require R&D.

It is also desirable to choose RF cells with low impedance and the fewest possible number of parasitic modes. To accomplish this, it will probably be necessary to adopt a cell with a smooth shape and a large beam aperture, i.e., a geometry typical of a superconducting cavity. There are more or less suitable designs presently available at frequencies of about 350 MHz and 500 MHz; the design used for the estimates given here has only two longitudinal modes and one transverse mode trapped in the cell.

The choice of room-temperature versus superconducting cells is not completely clear. The superconducting option minimizes the overall power requirement, since power is no longer needed to generate the voltage, but it probably complicates the power input and removal, which must make a transition from a room-temperature to a cryogenic environment. It is also worth noting that there is presently no operating experience with superconducting RF in the high-current regime of interest to a B factory.

To get a feeling for the seriousness of the coupled-bunch instabilities, calculations have been performed for typical B factory parameters. For these calculations, it was assumed that 20 single-cell cavities for the high-energy ring and 10 such cavities for the low-energy ring were used. Other relevant parameters can be found in Table I. Two sets of calculations were performed for each ring, the first taking the nominal parasitic mode Q values generated by URMEL, and the second assuming a reduction in Q by a factor of 200. The results for the fastest growing unstable modes are summarized in Tables III and IV for the high- and low-energy ring, respectively.

The calculations predict rapid growth for the lowest synchrotron mode both longitudinally (a = 1) and transversely (a = 0). Because the rates are well beyond the radiation damping rate, a powerful feedback system is clearly needed. Moderately rapid growth is also predicted for the next higher synchrotron mode (a = 2 longitudinally; a = 1 transversely) in the case where the cavity modes are not de-Qed. However, de-Qing by a factor of 200 reduces the predicted growth rates to values comparable to or below the radiation damping rate. From the results in Tables III and IV, it appears that the benefits of de-Qing are stronger in the longitudinal than in the transverse plane. It is not clear, however, that this result can be generalized to other cases.

It is worth noting one other feature of the calculations that depends on the ring size. The 2200-m circumference assumed here corresponds to a revolution frequency of only 136 kHz. This means that the individual bunch harmonic lines are quite close together (equivalent to $Q \approx 5000$ at a typical frequency of 750 MHz), making it difficult to avoid *any* parasitic modes.

SCATTERING PROCESSES

In this section, we will describe the various scattering processes that can occur, including intrabeam scattering (IBS), Touschek scattering, and gas scattering. The first of these can cause a blowup of the beam emittance, whereas the latter two effects cause a loss of particles and thus degrade the beam lifetime.

<u>Intrabeam Scattering</u>

The emittance growth due to IBS is a result of multiple, small-angle Coulomb scattering within a beam bunch. The collision probability is inversely proportional to the phase-space volume density of the bunch:[13]

$$g_{IBS} \propto \frac{1}{\Gamma} = \frac{1}{(2\pi \beta\gamma m_e)^3 \varepsilon_x \varepsilon_y \sigma_\ell \sigma_p} \tag{17}$$

which means that the B factory requirement for short bunches is a disadvantage. As a result of the collision, there is an exchange of energy among the three phase planes. In the bunch rest frame, the particle motion is treated nonrelativistically. Because of the distribution of particle momenta, collisions can occur. The rms momentum spreads in the three phase planes are given by

$$H: \sigma_{x'} p_0 \tag{18a}$$

$$V: \sigma_{y'} p_0 \tag{18b}$$

$$L: \frac{\sigma_p p_0}{\gamma} \,. \tag{18c}$$

In general, the scattering event will transfer momentum from the transverse to the longitudinal plane. However, in dispersive regions of the lattice, the change in longitudinal momentum excites a horizontal betatron oscillation because of the change in the closed orbit for the off-momentum particles, that is

$$x \rightarrow x - D_x \left(\frac{\delta p}{p}\right) \qquad (19a)$$

$$x' \rightarrow x' - D'_x \left(\frac{\delta p}{p}\right) . \qquad (19b)$$

The measure of the emittance growth is

$$\mathcal{H} = \left[\gamma_x D^2 + 2\alpha_x DD' + \beta_x D'^2\right] . \qquad (20)$$

This is the same parameter as for quantum excitation, which determines the natural emittance of the lattice. One distinction between the IBS process and quantum excitation, however, is that the former can occur anywhere in the lattice, that is, it is not restricted only to the dipoles. To evaluate the equilibrium emittance, we add an additional growth term, g_{IBS}, to the standard expression that includes only radiation damping and quantum excitation:

$$\left[g_{IBS}(\varepsilon) - g_{SR}\right]\varepsilon + g_{SR}\,\varepsilon_0 = 0 \qquad (21)$$

where

$$g \equiv \frac{1}{\varepsilon}\frac{d\varepsilon}{dt} . \qquad (22)$$

This equation is transcendental, since g_{IBS} is a function of the emittance. Nonetheless, it is possible with ZAP to calculate the IBS growth rate at many lattice points around the ring and obtain a weighted average value of g_{IBS} that can be used to solve Eq. (21). In Fig. 19 we present typical results for the B factory parameter regime, taken from Ref. 14. We see that there is no significant emittance growth for either ring. This has, up to now, been true in all cases examined, so IBS emittance growth is not expected to be a significant problem for a B factory.

Touschek Scattering

Touschek scattering is also an intrabeam scattering process but, in contrast to the IBS multiple scattering process described above, it involves large-angle single Coulomb scattering events. In these (relatively rare) events, the momentum deviation of the scattered particle can exceed the acceptance of the storage ring. The momentum acceptance limit can come from either the longitudinal or the transverse plane.

In the longitudinal plane, there is a limit on the momentum deviation at which a particle can still undergo stable synchrotron oscillations. This is referred to as the RF acceptance (or "bucket height"), given by:

$$\left(\frac{\Delta p}{p}\right)_{bucket} = \frac{1}{\beta} \left\{ \frac{-V_{RF}\left[\frac{2}{\pi}\cos\phi_s - \left(1 - \frac{2}{\pi}\phi_s\right)\sin\phi_s\right]}{h\,\eta\,(E/e)} \right\}^{1/2} \qquad (23)$$

where V_{RF} is the RF voltage, ϕ_s is the synchronous phase, and h is the harmonic number.

In the transverse plane, the limitation arises because, in dispersive regions of the lattice, the change in momentum excites a large betatron oscillation at the scattering location. At any other lattice location (denoted i), the maximum particle amplitude resulting from the original Touschek event is

$$\Delta x_i = \left[|D_i| + \sqrt{\beta_i\,\mathcal{H}_s}\right]\left(\frac{\delta p}{p}\right) \qquad (24)$$

where $\mathcal{H}_s$ represents the function defined in Eq. (20), evaluated at the scattering location. Touschek scattered particles can thus hit the vacuum chamber wall or exceed the dynamic aperture of the ring. For a high-luminosity collider with low-beta optics, the latter possibility is a real concern. It is important to note that, in addition to a large betatron amplitude, the scattered particles are also far off momentum—in general a bad combination. To properly assess the dynamic aperture limitation, it is necessary to track the off-momentum particles (including synchrotron oscillations and the full lattice nonlinearities).

The Touschek lifetime is a very strong function of the momentum acceptance, scaling roughly as

$$\tau_T \propto \left(\frac{\Delta p}{p}\right)^3 \qquad (25)$$

so the battle for Touschek lifetime is won or lost here. The lifetime also scales inversely with the bunch volume

$$V_B \propto 8\pi^{3/2} \sigma_x \sigma_y \sigma_l \qquad (26)$$

so a large bunch is helpful in this regard.

Estimates of the momentum acceptance for typical B factory parameters are shown in Fig. 20. For both the high- and low-energy rings, the acceptance is expected to be limited transversely. (The RF acceptances are unnecessarily large because the voltages are chosen to be high to maintain short bunches. In principle, the "extra" voltage is not good for the lifetime, as it serves only to decrease the bunch volume, but this is the price we must pay for short bunches.) Fortunately, the Touschek lifetime is predicted not to be a problem in this parameter regime, as demonstrated in Fig. 21. At the nominal operating energies of 9 GeV and 3.1 GeV, the Touschek lifetimes for the parameters of Ref. 14 are 400 hours and 100 hours, respectively.

Gas Scattering

Another limitation on the beam lifetime arises from interactions between electrons and residual gas atoms in the vacuum chamber. The collisions can be either elastic or inelastic (referred to as Bremsstrahlung), and both processes are important. As the electron beam energy increases, the elastic process becomes progressively less important compared with the Bremsstrahlung, so the latter typically dominates the lifetime in high-energy storage rings.

In the elastic scattering case, the electron undergoes a single Coulomb scattering that changes its angle sufficiently that it either hits the chamber wall or becomes dynamically unstable. In the ring, the motion of a particle is defined by an invariant phase-space area ("emittance") given by

$$\varepsilon_z = \frac{1 + \alpha(s)^2}{\beta(s)} z^2 + 2\alpha(s)\, z \cdot z' + \beta(s)\, z'^2 \tag{27}$$

where z represents either the x or y coordinate. If the z' from the scattering is too large, the emittance exceeds the acceptance of the lattice, and the scattered particle is lost somewhere in the ring. The physical acceptance is defined as

$$A_\perp = \min\left(\frac{b^2}{\beta_\perp}\right) \tag{28}$$

i.e., the smallest value in each transverse plane of the invariant emittance corresponding to a chamber half-aperture b. For a uniform chamber aperture, the acceptance limit will occur at the maximum beta function, but this is usually not the case in an actual ring. To derive the loss probability, we integrate over the Rutherford cross section and obtain[15]

$$\sigma_{el} = \frac{2\pi r_e^2 Z^2}{\gamma^2} \frac{\langle \beta_\perp \rangle}{A_\perp} \tag{29}$$

where the brackets denote an average over the ring circumference.

In the case of Bremsstrahlung interactions, the inelastic scattering leads to an energy loss for the scattered particle. If the resultant momentum deviation, δp/p, exceeds the acceptance of the ring, the particle is lost. As discussed for Touschek scattering, the momentum acceptance can be limited either longitudinally or transversely. For the B factory parameters considered here, the limit is transverse. The loss probability, obtained by integrating over energy loss, is given by[15]

$$\sigma_{Br} = \frac{4}{3} \frac{4 r_e^2 Z^2}{137} \ln\left(\frac{183}{Z^{1/3}}\right) \left[\ln\left(\frac{1}{(\Delta p/p)_{lim}}\right) - \frac{5}{8} \right] . \quad (30)$$

The combined loss rate for the two gas scattering processes is given by[15]

$$\frac{1}{\tau_g} \equiv \frac{1}{N_b} \frac{dN_b}{dt} = 3.22 \times 10^{22} \, n_Z \, P \, \beta c \left(\sigma_{el} + \sigma_{Br}\right) \quad (31)$$

where n_Z is the number of atoms of species Z per molecule, and P is the pressure in Torr.

To mitigate the effects of gas scattering, there are several options:
- make the apertures large;
- keep the beta functions low;
- increase the momentum acceptance;
- maintain a low residual gas pressure in the ring, especially for high-Z species.

The first option is straightforward, but is always costly because magnet aperture is expensive. Low beta functions serve to minimize the emittance increase of a scattered particle (see Eq. (27)) and to increase the acceptance of the ring for a given aperture (see Eq. (28)). From Eq. (30), we can see that the momentum acceptance has a relatively weak (logarithmic) influence on the lifetime, so improving it has little beneficial effect. The requirement for a low gas pressure is obvious, but is not so easily achieved in a ring that must accommodate up to 3 A of beam current. Present rings operate with average pressures of about 10 nTorr (N_2-equivalent), which we take here to be an achievable, though difficult, goal.

At a pressure of 10 nTorr, the beam lifetimes from gas scattering are about 3 hours in either the high- or the low-energy ring, as summarized in Table V.

SUMMARY

In this paper we have examined the important collective effects influencing the performance of high-intensity storage rings of the type required to serve as a B factory. To achieve a high luminosity, it appears inevitable that many bunches will be used in each ring. This permits the parameters to be chosen in such a way as to minimize any possible difficulties with single-bunch effects. On the other hand, the coupled-bunch instabilities are severe and are likely to be the main performance-limiting feature of a B factory. It is likely that the success of such a collider will depend rather strongly on the skills of the feedback system designer.

The limitation on the total current that can be stored in the collider rings will come from one or more of the following:

- total synchrotron radiation power into the vacuum chamber walls;
- tolerable background gas pressure in the ring;
- coupled-bunch instability growth rates.

All of these issues are, in a sense, technology rather than physics constraints. The synchrotron radiation power can, in principle, be handled by proper design of the vacuum chamber and its cooling system. Designs to handle a linear power density of 20 kW/m appear to be practical, and this is sufficient for the typical B factory parameter regime. The second issue becomes ultimately one of pumping speed per meter of ring circumference. To maintain a pressure of 10 nTorr in the face of 3 A of circulating beam requires a pumping speed on the order of 1000 ℓ/s per meter. Achieving this is possible, but will require careful attention to the design. In terms of the coupled-bunch instabilities, the main questions concern how much the rates can be reduced by proper design of the RF cavities, and how much growth rate can be handled by a feedback system. It seems possible, based on present technology, to reduce the instability growth rates to below 1000 s^{-1}, a rate that can be handled with a feedback system of reasonable power (a few kW). Because the feedback power requirement scales as the square of the growth rate that must be damped, however, a growth rate of 10^4 s^{-1} may be unmanageable.

For the parameter regime presently being contemplated in most B factory designs, single-bunch limitations will come mainly from the beam-beam interaction. This situation results from a conscious choice to push the instability problems into the multibunch arena, rather than picking an in-between regime in which both single- and coupled-bunch instabilities are severe. Based on the experience at existing rings, it should be possible to maintain the longitudinal impedance to a level of $|Z/n| \approx 1 \text{ }\Omega$, which will not lead to problems for the short bunches required for a B factory. Even if the impedance were somewhat higher, the bunch lengthening would probably not be a major problem. The one possible concern is that the low-energy ring is vulnerable to the transverse mode-coupling instability if the transverse impedance gets too large. It is nontrivial to minimize the impedance contribution from the many masks that will be needed to protect the detector region from synchrotron radiation and scattered particle backgrounds. In our favor, of course, is the fact that we are concerned with the beta-weighted transverse impedance, and the beta functions in this region are reasonably low.

Because the coupled-bunch instability growth rates scale with the total current and are essentially independent of the bunch pattern, the limit on the number of bunches is really dictated by the ability to separate the closely spaced bunches near the IP. A secondary issue is the bandwidth of the feedback system; this, however, is not believed to be a strong constraint on the design.

Beam lifetimes will be typical of modern colliders—on the order of a few hours—provided an adequate vacuum system can be designed. There is no reason to believe that solutions cannot be found, although they may be rather costly.

Thus, it seems fair to conclude that, in terms of collective effects, nothing yet seems to preclude the possibility of building a successful high-luminosity B factory.

ACKNOWLEDGMENTS

It is a pleasure to thank A. Garren and M. Donald for providing the lattices used in this study, and S. Chattopadhyay, Y. Chin, W. Davies-White, A. Hutton, G. Lambertson and P. Oddone for many beneficial discussions on B factory design.

REFERENCES

1. See, e.g., "Proposal for an Electron Positron Collider for Heavy Flavour Particle Physics and Synchrotron Radiation," Paul Scherrer Institute Report PR-88-09, July, 1988; "The Physics Program of a High-Luminosity Asymmetric B Factory at SLAC," Stanford Linear Accelerator Center Report SLAC-353, Lawrence Berkeley Laboratory Report LBL-27856, California Institute of Technology Report CALT-68-1588, October, 1989; A.N. Dubrovin, A.N. Skrinsky, G.N. Tumaiki, and A.A. Zholents, "Conceptual Design of a Ring Beauty Factory," European Particle Accelerator Conference, Rome, June, 1988, p. 467; H. Nesemann, W. Schmidt-Parzefall, and F. Willeke, "The Use of PETRA as a B Factory," European Particle Accelerator Conference, Rome, June, 1988, p. 439.

2. M.S. Zisman, S. Chattopadhyay, and J.J. Bisognano, "ZAP User's Manual," Lawrence Berkeley Laboratory Report LBL-21270, December, 1986.

3. "Investigation of an Asymmetric B Factory in the PEP Tunnel," Lawrence Berkeley Laboratory Report LBL PUB-5263, Stanford Linear Accelerator Center Report SLAC-359, California Institute of Technology Report CALT-68-1622, March, 1990.

4. A.A. Garren, et al., "An Asymmetric B-meson Factory at PEP," in Proc. of the Particle Accelerator Conf., Chicago, March, 1989, p. 1847; "Feasibility Study for an Asymmetric B Factory Based on PEP," Lawrence Berkeley Laboratory Report LBL PUB-5244, Stanford Linear Accelerator Center Report SLAC-352, California Institute of Technology Report CALT-68-1589, October, 1989.

5. Y.H. Chin, "Beam-Beam Interaction in an Asymmetric Collider for B Physics," Lawrence Berkeley Laboratory Report LBL-27665, August, 1989; Proc. of XIV Intl. Conf. on High Energy Accelerators, to be published.

6. A.W. Chao and J. Gareyte, "Scaling Law for Bunch Lengthening in SPEAR-II," Stanford Linear Accelerator Center Note SPEAR-197, PEP-224, December, 1976.

7. M. Donald, et al., "Feedback Experiment at PEP," CERN Report LEP-553, January, 1986.

8. L. Rivkin, "Collective Effects in PEP," in Proc. of Workshop on PEP as a Synchrotron Radiation Source, October 20–21, 1987, p. 139.

9. M.S. Zisman, "Collective Effects in the PEP Low-Emittance Configuration," in Proc. of Workshop on Accelerator Physics Issues Relating to the Use of PEP as a Synchrotron Radiation Source, Stanford Synchrotron Radiation Laboratory Report 88/06, November, 1988.

10. A.W. Chao, "Coherent Instabilities of a Relativistic Bunched Beam," in Physics of High Energy Particle Accelerators, American Institute of Physics Conf. Proc. 105, 1983, p. 353.

11. J.M. Wang, "Longitudinal Symmetric Coupled Bunch Modes," Brookhaven National Laboratory Report BNL-51302, December, 1980; S. Chattopadhyay and J.M. Wang, unpublished note (1986).

12. T. Weiland, "On the Computation of Resonant Modes in Cylindrically Symmetric Cavities," Nucl. Instr. and Meth. 216, 329 (1983).

13. J. Bjorken and S.K. Mtingwa, Part. Accel. <u>13</u>, 115 (1983).

14. M.S. Zisman, "Study of Collective Effects for the APIARY Collider," Lawrence Berkeley Laboratory Report LBL-27676, October, 1989; Proc. of Workshop on B-Factories and Related Physics Issues, Blois, June 26 – July 1, 1989, to be published.

15. J. LeDuff, "Current and Density Limitations in Existing Electron Storage Rings," Nucl. Instr. and Meth. <u>A239</u>, 83 (1985).

Table I
APIARY-IV Major Parameters

	High-energy	Low-energy
Energy, E [GeV]	9	3.1
Circumference, C [m]	2200	2200
Number of bunches, k_B	1296	1296
Particles per bunch, N_B [10^{10}]	5.44	7.88
Total current, I [A]	1.54	2.23
Emittance,[a] ε_x [nm-rad]	33	66
Bunch length, σ_l [mm]	10	10
Relative momentum spread, σ_p [10^{-4}]	6.1	9.5
Damping time		
$\tau_{x,y}$ [ms]	37.0	32.3
τ_E [ms]	18.5	17.3
Beta functions at IP		
β_x^* [cm]	6	3
β_y^* [cm]	6	3
Betatron tune		
horizontal, ν_x	21.28	37.76
vertical, ν_y	18.20	35.79
Synchrotron tune, Q_s	0.053	0.039
Momentum compaction, α	0.00245	0.00115
RF parameters		
frequency, f_{RF} [MHz]	353.2	353.2
voltage, V_{RF} [MV]	25	10
Nominal beam-beam tune shift		
ξ_{0x}	0.03	0.03
ξ_{0y}	0.03	0.03
Luminosity, $\mathcal{L}$ [cm^{-2} s^{-1}]	3×10^{33}	

[a] Equal horizontal and vertical emittances.

Table II

Transverse Mode-Coupling Threshold Scaling

	Low-energy	PEP	High-energy
E [GeV]	3.1	14.5	9.0
$\beta_\perp$ [m]	20	87	40
R [m]	350	350	350
Q_s [10^{-2}]	3.9	4.6	5.3
$Z_\perp$ [MΩ/m]	0.4	0.8	0.4
Relative factor[a]	1.6	1	3.1
Observed threshold [mA]	—	8.4	—

[a]Scaling factor is $F = (E\, Q_s/\beta_\perp Z_\perp R)$.

Table III

Coupled-Bunch Growth Rates for APIARY-IV

High-Energy Ring

(9 GeV; τ_E = 18.5 ms; τ_x = 37 ms)

Longitudinal

no de-Q		Q/200	
$\tau_{a=1}$ (ms)	$\tau_{a=2}$ (ms)	$\tau_{a=1}$ (ms)	$\tau_{a=2}$ (ms)
0.4	22	4	390

Transverse

no de-Q		Q/200	
$\tau_{a=0}$ (ms)	$\tau_{a=1}$ (ms)	$\tau_{a=0}$ (ms)	$\tau_{a=1}$ (ms)
1.2	38	2	66

Table IV
Coupled-Bunch Growth Rates for APIARY-IV Low-Energy Ring
(3.1 GeV; τ_E = 17.3 ms; τ_x = 32.3 ms)

Longitudinal

no de-Q		Q/200	
$\tau_{a=1}$ (ms)	$\tau_{a=2}$ (ms)	$\tau_{a=1}$ (ms)	$\tau_{a=2}$ (ms)
0.3	20	3	290

Transverse

no de-Q		Q/200	
$\tau_{a=0}$ (ms)	$\tau_{a=1}$ (ms)	$\tau_{a=0}$ (ms)	$\tau_{a=1}$ (ms)
1	57	0.5	110

Table V

Gas Scattering Lifetimes

(P = 10 nTorr, N_2-equivalent)

	Low-energy	High-energy
Elastic [h]	10	8
Bremsstrahlung [h]	5	5
Total [h]	3	3

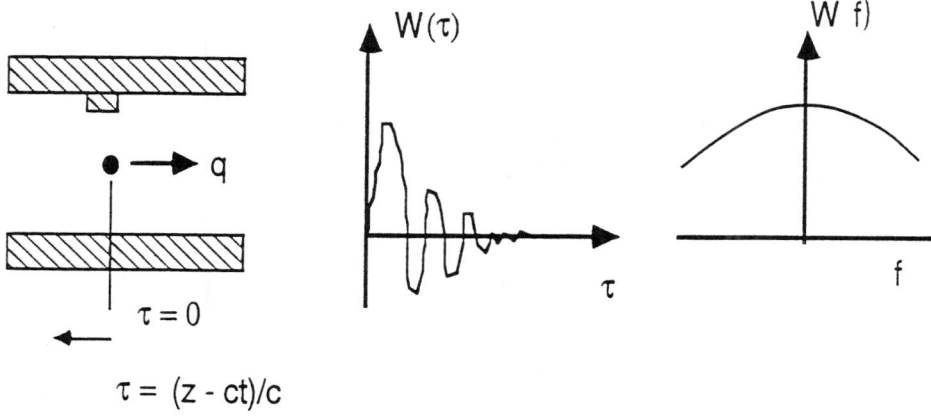

Fig. 1. Impedance properties of a sharp discontinuity in a storage ring vacuum chamber.

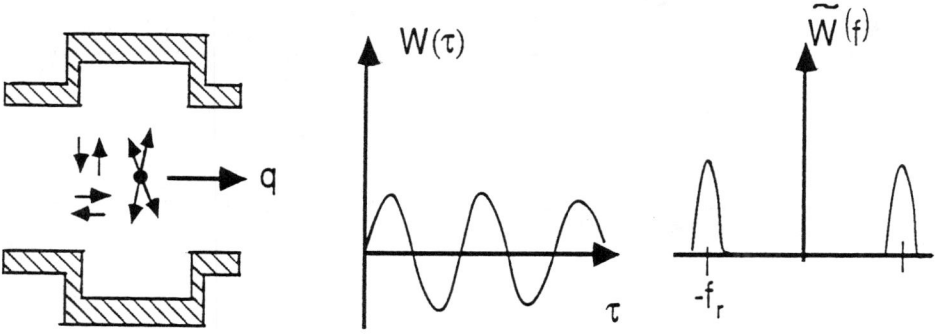

Fig. 2. Impedance properties of a cavity-like object in a storage ring vacuum chamber.

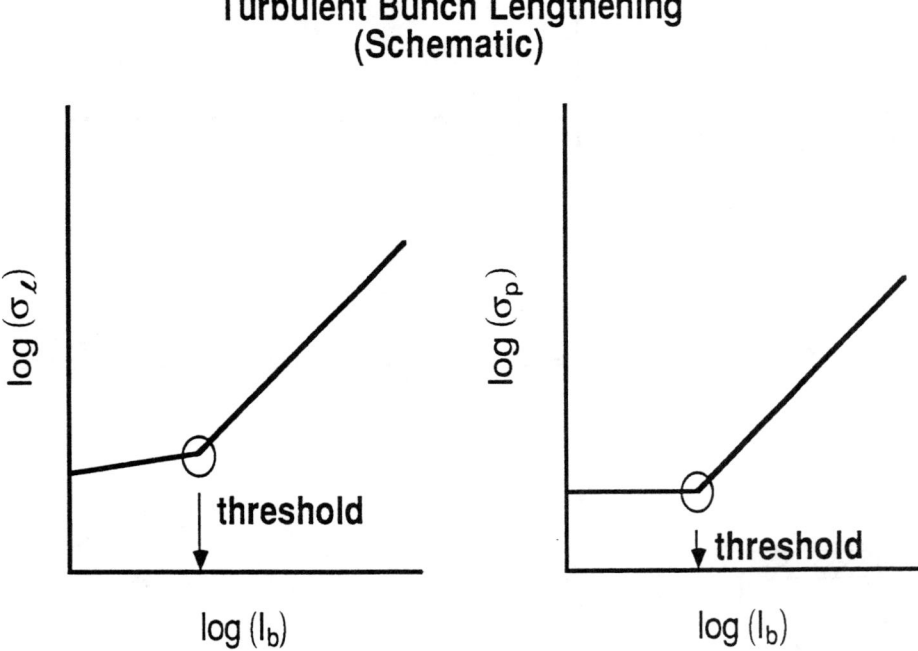

Fig. 3. Schematic of bunch lengthening and widening due to the longitudinal microwave instability.

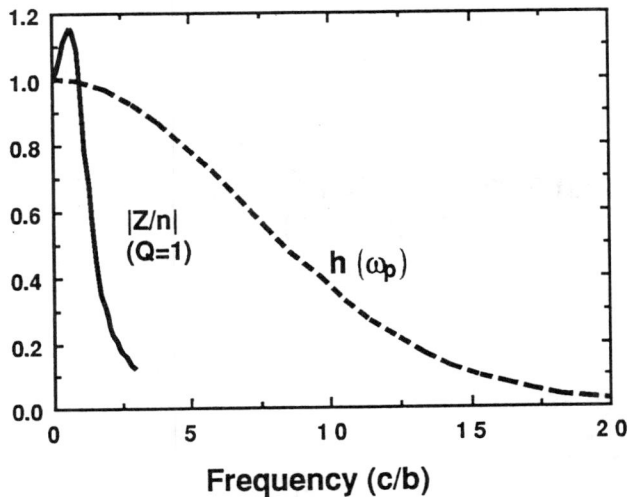

Fig. 4. Sampling of broadband impedance with short bunches.

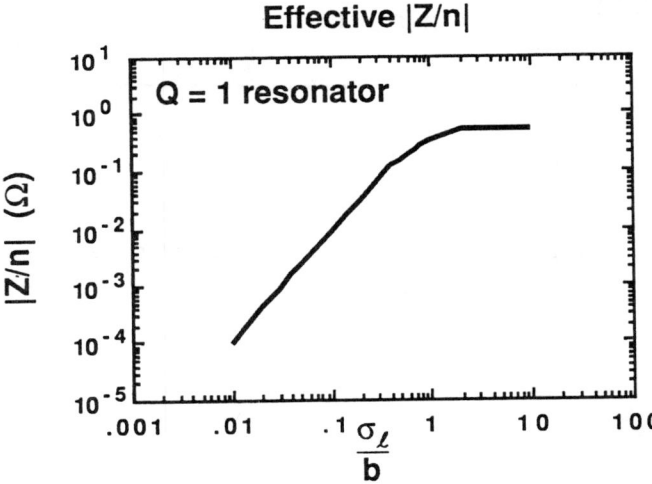

Fig. 5. Effective broadband impedance, as a function of bunch length, for a $Q = 1$ broadband resonator model.

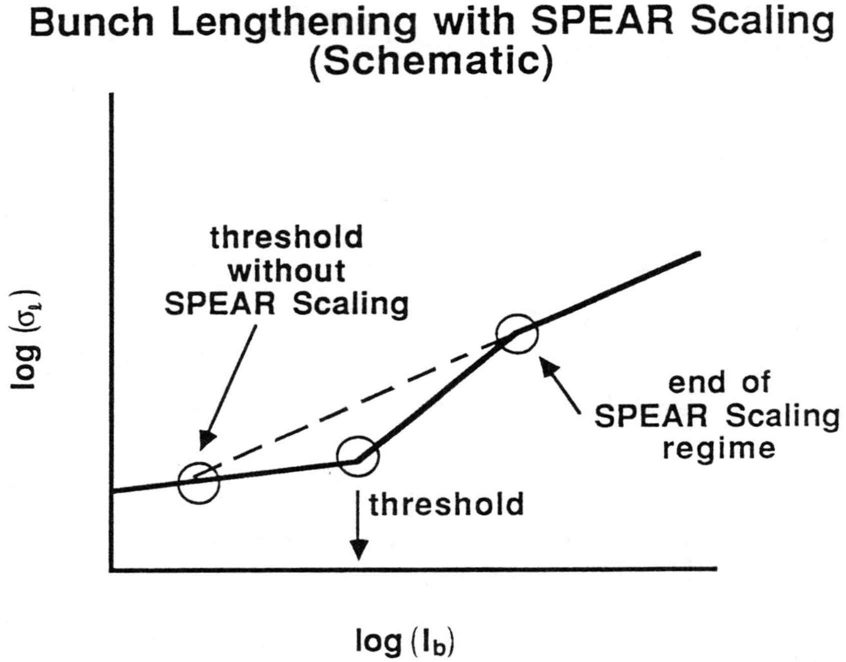

Fig. 6. Expected change in bunch lengthening behavior in the short-bunch regime.

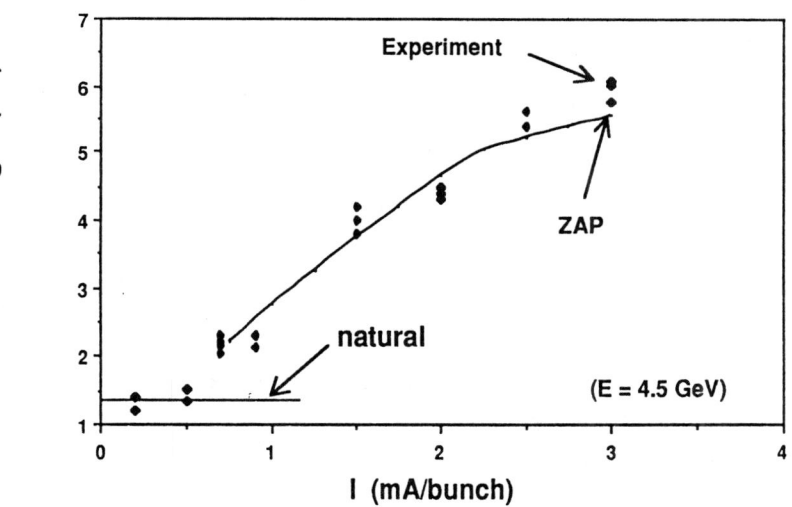

Fig. 7. Comparison of PEP bunch length measurements at 4.5 GeV with predictions based on SPEAR Scaling.

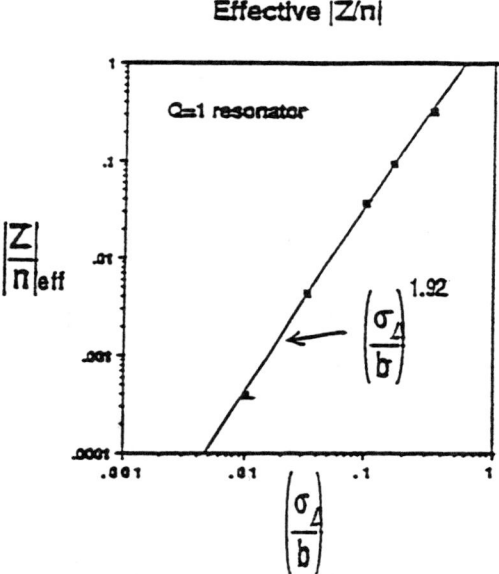

Fig. 8. Power law fit to short bunch effective impedance calculation from Fig. 5.

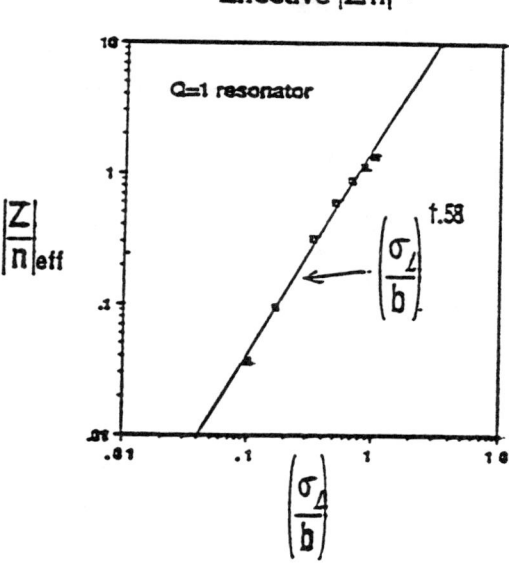

Fig. 9. Power law fit to short bunch effective impedance calculation from Fig. 5 for the restricted bunch length range $0.1 \leq \sigma_l/b \leq 1$ from which the SPEAR Scaling phenomenology was originally derived. The resultant power law is in reasonably close agreement with the value of 1.68 determined experimentally.

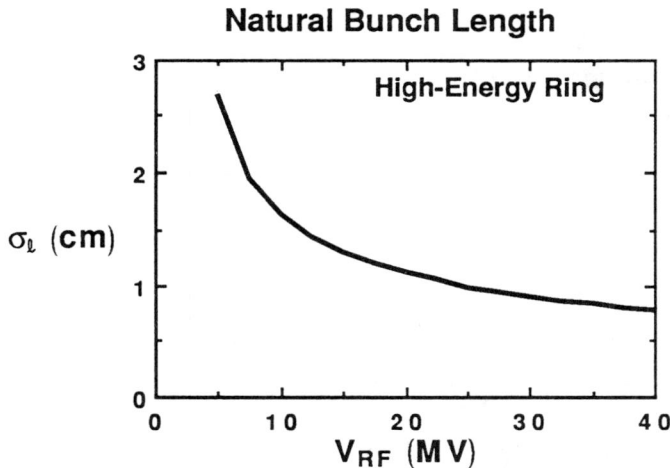

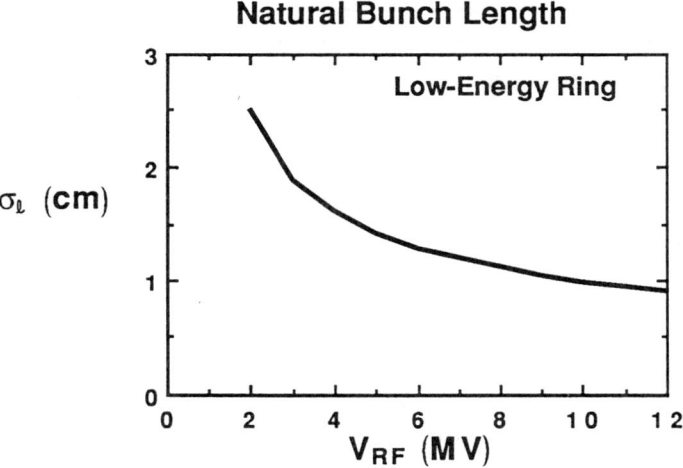

Fig. 10. Natural bunch lengths for the high-energy and low-energy APIARY-IV storage rings as a function of RF voltage.

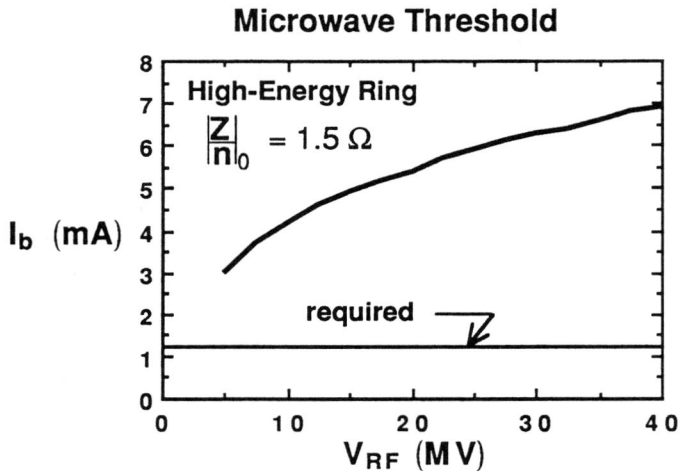

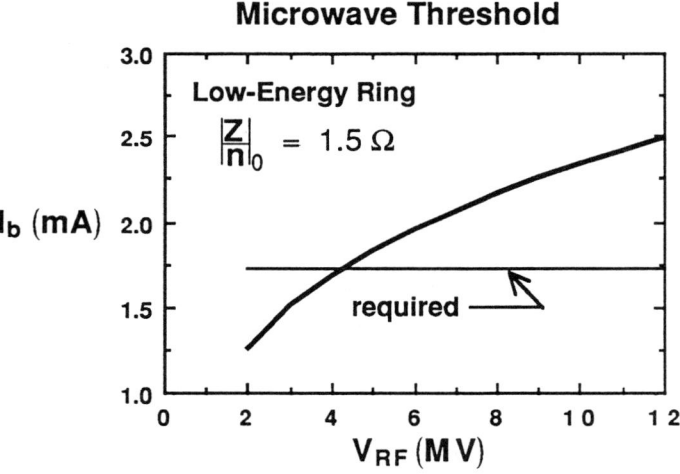

Fig. 11. Bunch lengthening thresholds for the high-energy and low-energy APIARY-IV storage rings as a function of RF voltage.

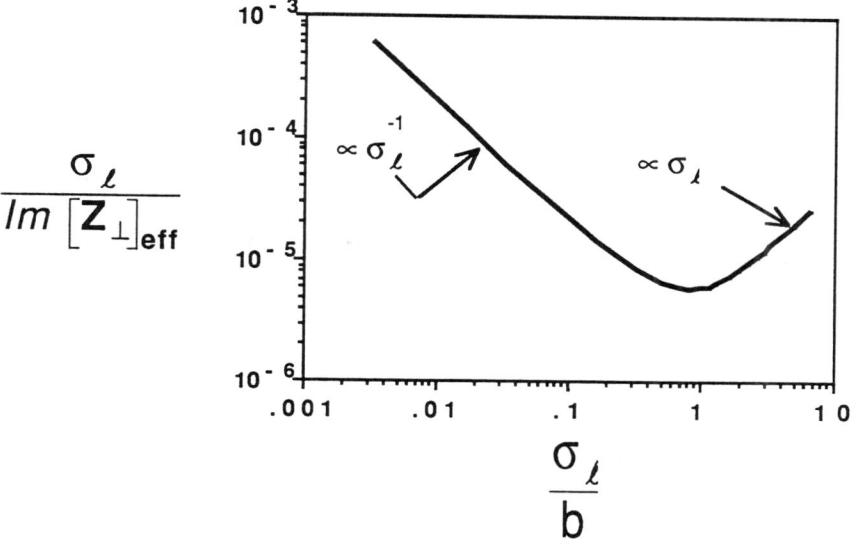

Fig. 12. Expected behavior of transverse mode-coupling threshold as a function of bunch length. In the short bunch length regime, the threshold is expected to increase because the broadband impedance is not sampled fully.

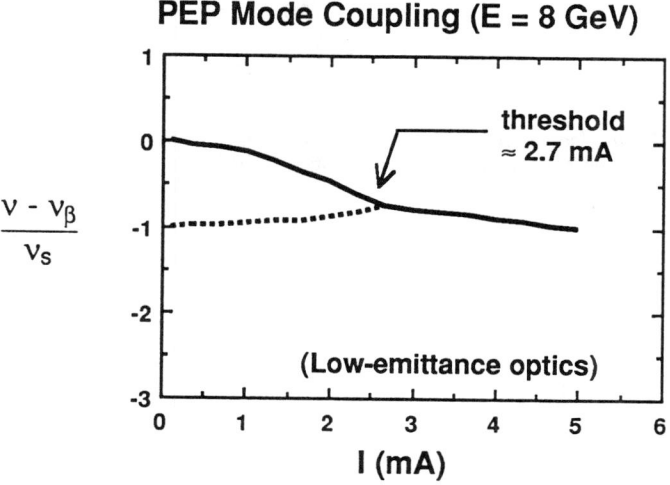

Fig. 13. Predicted mode-coupling threshold for the PEP low-emittance optics. The $m = 0$ and $m = -1$ modes cross at 2.7 mA.

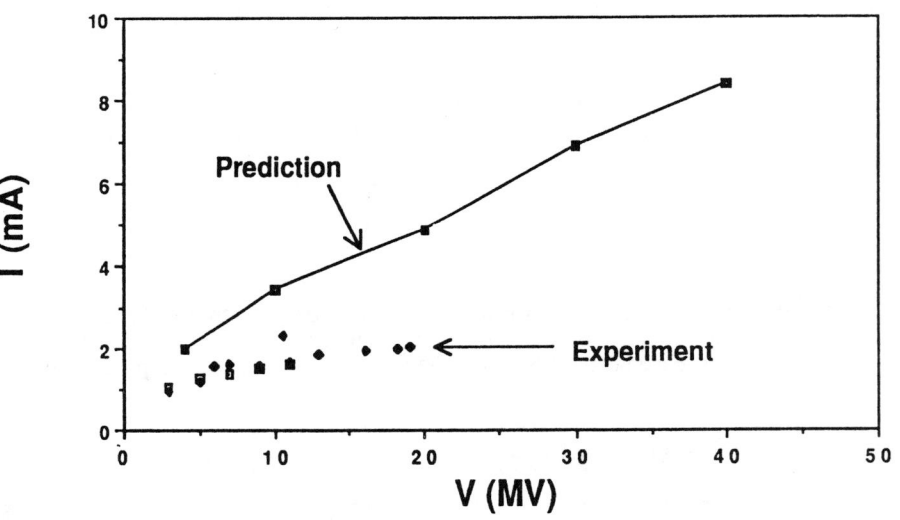

Fig. 14. Comparison between predicted mode-coupling threshold for the PEP low-emittance optics and experimental results.

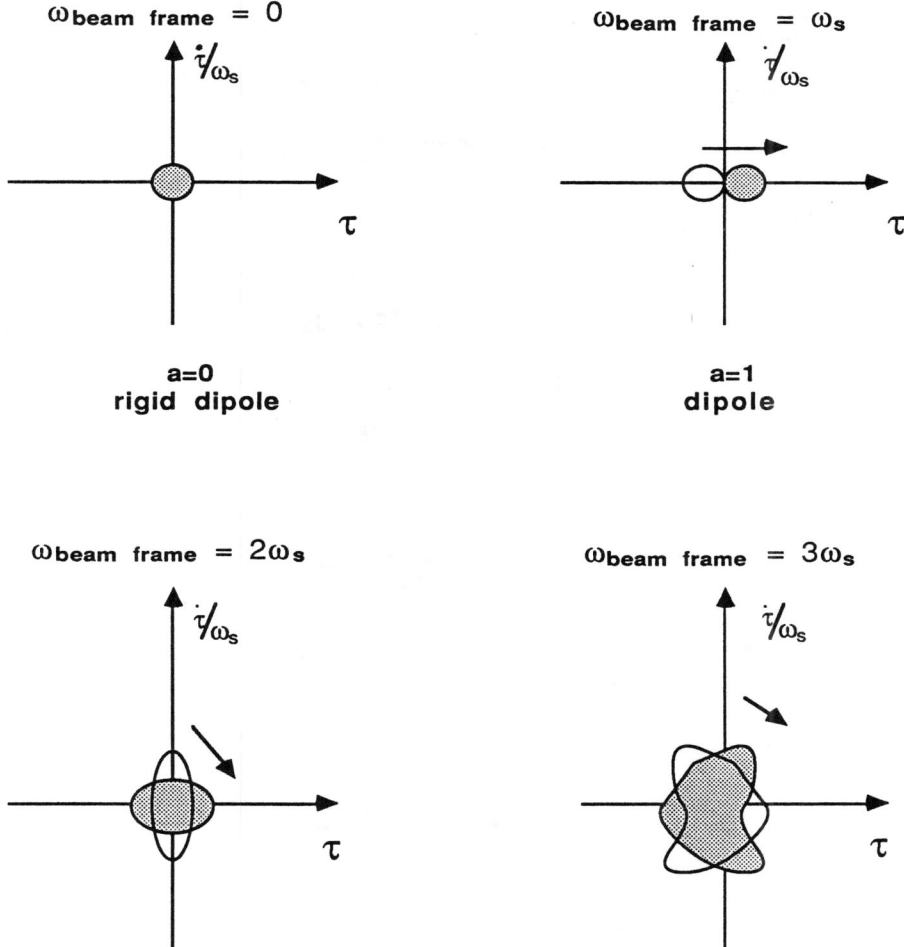

Fig. 15. Schematic diagram of coupled-bunch synchrotron modes. For the longitudinal case, the lowest mode that can give rise to an instability is the a = 1 mode, whereas for the transverse case the a = 0 mode can also be unstable.

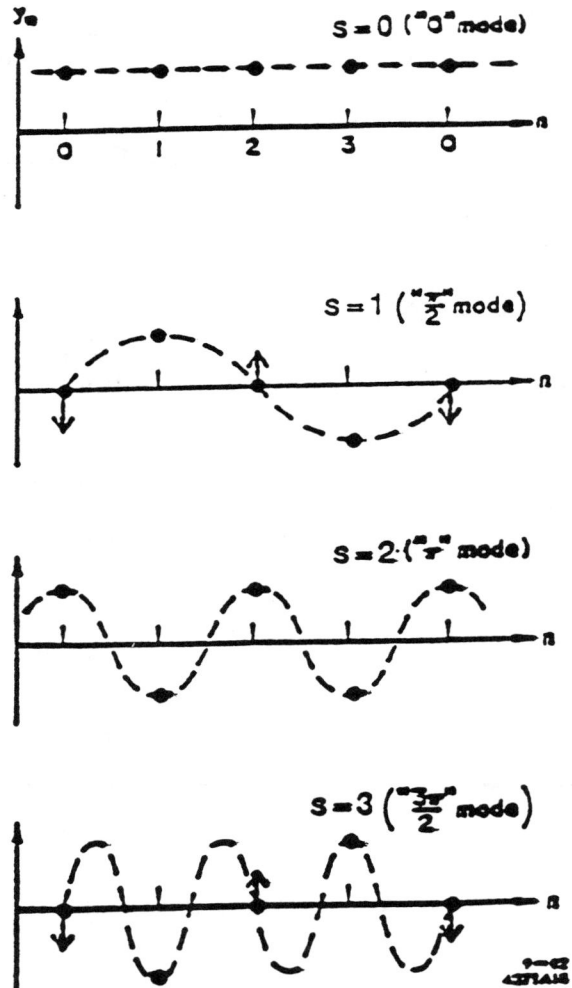

Fig. 16. Normal modes of coupled-bunch motion for the four-bunch case (taken from Ref. 10).

De-Qing Procedure
(schematic)

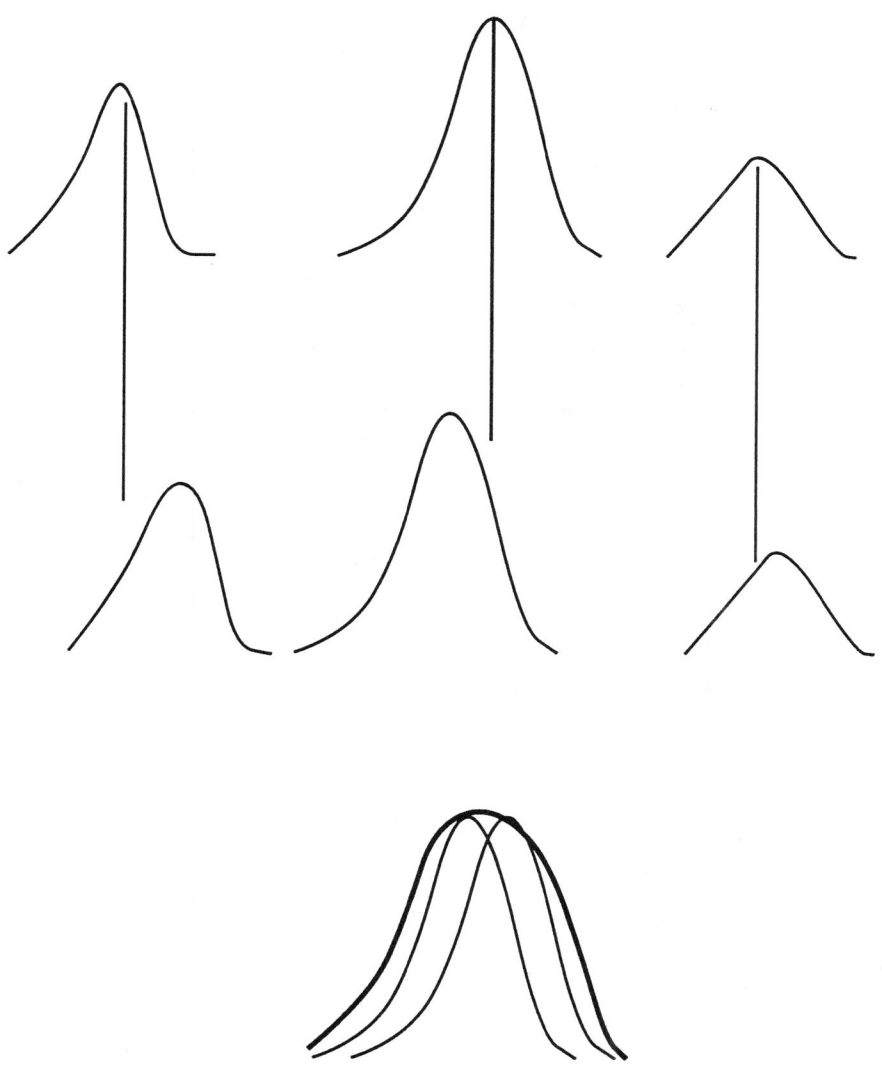

Fig. 17. Schematic picture of the de-Qing procedure to minimize calculational time.

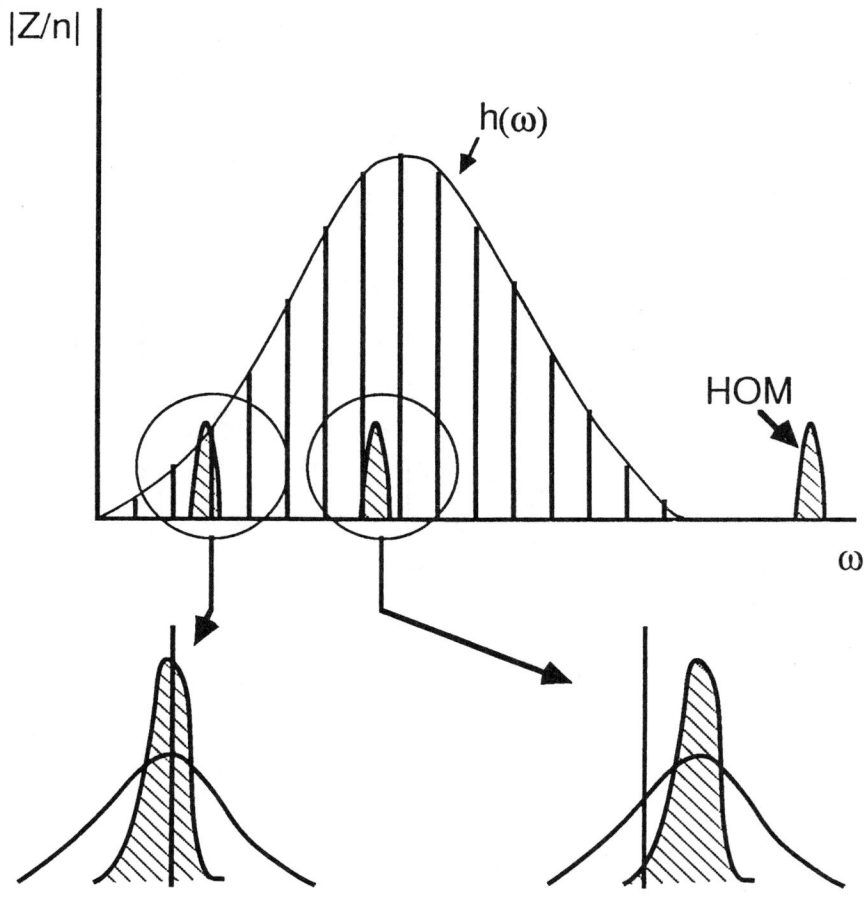

Fig. 18. Possible results of physical de-Qing of higher-order parasitic modes. If the mode is originally close to a rotation harmonic, reducing the Q will reduce the growth rate (leftmost mode). If the mode was initially between harmonics, de-Qing can give increased impedance at the rotation harmonic, and thus increased growth rate (central mode). If the higher-order mode is beyond the bunch frequency cutoff (rightmost mode), it is not sampled by the beam.

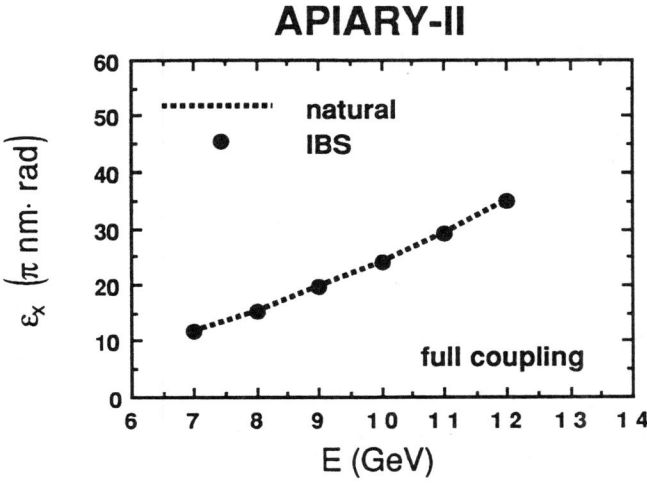

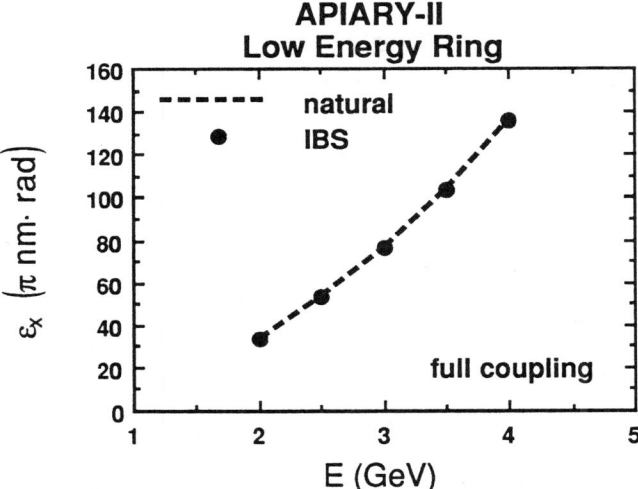

Fig. 19. Predicted emittance growth from intrabeam scattering for the B factory parameters of Ref. 14. The dots are calculated equilibrium emittance values, obtained from solving Eq. (21).

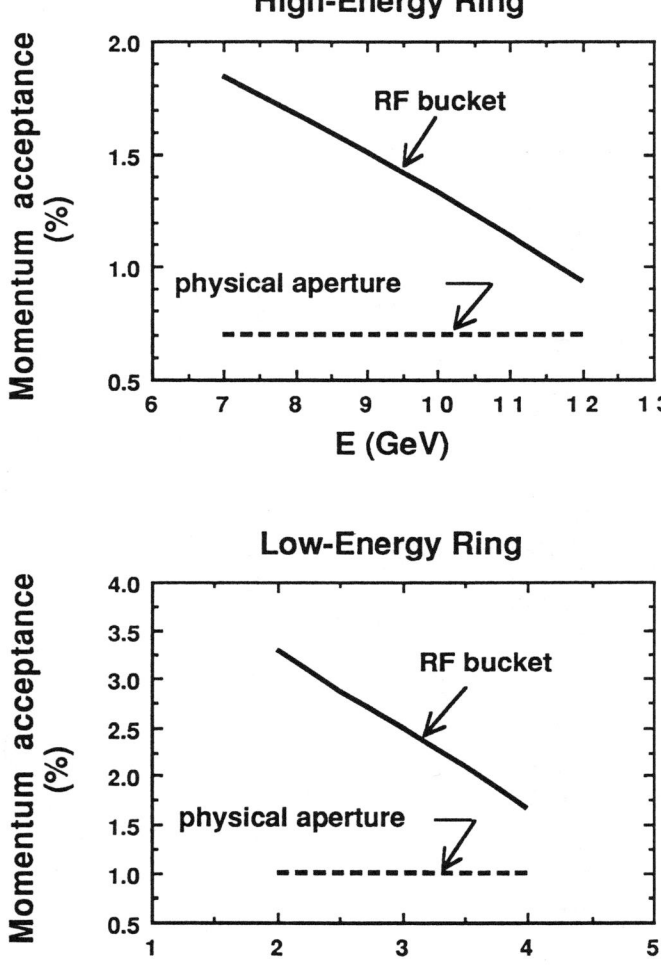

Fig. 20. Calculated momentum acceptance for the APIARY-IV rings described in Ref. 3. The transverse (physical aperture) limit is the more restrictive.

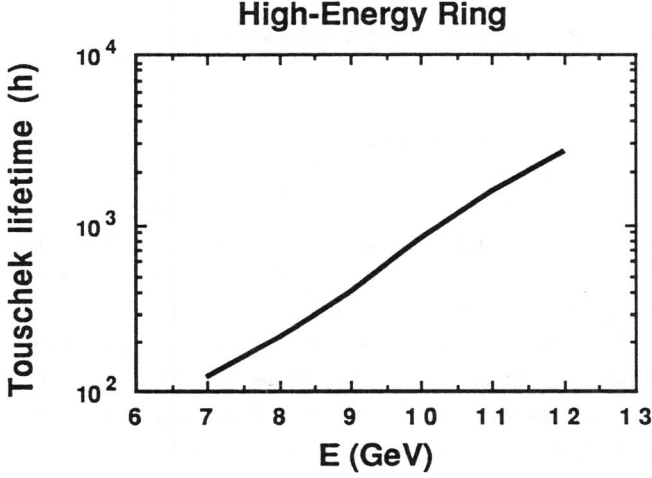
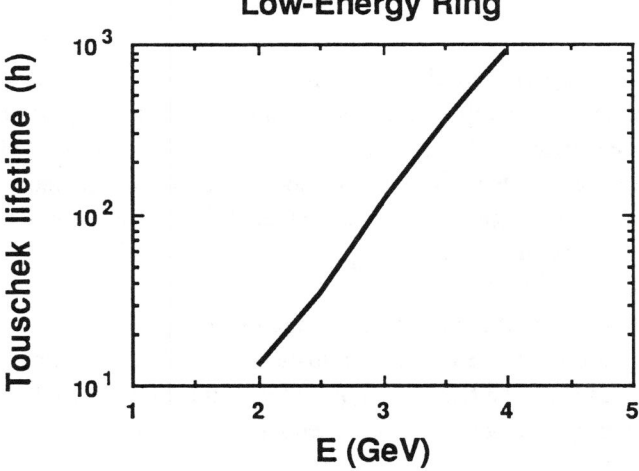

Fig. 21. Calculated Touschek lifetimes for the B factory parameters of Ref. 14.

The Beam-Beam Limit in Asymmetric Colliders: Optimization of the B-Factory Parameter Base

J.L. Tennyson [†]
California Institute of Technology, Pasadena CA 91125

ABSTRACT

This paper presents a general theory of the beam-beam limit in symmetric and asymmetric lepton ring colliders. It shows how the beam-beam limit in these accelerators affects the maximum attainable luminosity and presents a specific algorithm for parameter base optimization. It is shown that the special problems inherent in asymmetric colliders derive not from the asymmetry, but from the fact that the two beams must be in different rings. Computer simulation experiments are used to demonstrated the various phenomena discussed in the theory.

1. Introduction

The B-factory parameter base must provide as much luminosity as possible, subject to cost and detector constraints. The complexity of luminosity optimization requires that a well defined algorithm be adopted. This algorithm should be practical; it should utilize analysis techniques that are available and reliable. It should also have a solid theoretical foundation; one must have confidence that it will in fact result in the highest luminosities possible with present technology.

This report describes the theory of the beam-beam interaction in asymmetric colliders and suggests a specific algorithm for selecting a parameter base yielding maximum luminosity. Section two describes the beam-beam parameter set and the general optimization problem. A simple model of the asymmetric beam-beam system is presented in section three.

Section four examines a more realistic model of the beam system, the numerical model used in most simulation codes. This model is far more "robust" than the simple model. The realistic model exhibits three distinct beam-beam limits: one due to beam blow up,

---

[†] Present address, BIN 26, SLAC, P.O. Box 4349, Stanford CA 94305

a second due to the flip-flop effect, and a third resulting from particle loss to the aperture. The limiting tune shift is parameter-dependent in the realistic model and is in general different for the two beams.

Theoretical arguments in section four show that the realistic model suffers from a condition called "balance-point sensitivity": if both beams are operating at their respective beam-beam limits, any independent variation of the natural emittance of either beam will force one beam over its beam-beam limit. This appears to be the primary detrimental effect of having the two beams in separate rings. It is shown that this condition is not serious and can always be mitigated.

The dependence of the luminosity on the damping times of the two rings is explored with simulation in section five. It is shown that the luminosity depends approximately on the product of the two damping times, at least for the round beam case explored.

One B-Factory design has adopted constraints known collectively as "energy transparency"[1] which reduce the size of the available parameter space. These constraints tend to symmetrize the beam-beam interaction (to make the microscopic dynamics nearly identical for the two beams). The purpose of energy transparency is to safeguard against possible unknown effects which might result from beam-beam asymmetry.

It is shown here that there is no evidence, either theoretical, experimental, or numerical, that beam-beam asymmetry is harmful or that the energy transparency constraints will improve the performance of the machine. These constraints do not help solve the problem of balance point sensitivity since the latter results not from the asymmetry, but from the more fundamental fact that the two beams are in separate rings.

The analysis presented here assumes that the two beams are round and operate on the main coupling resonance. This assumption implies that $\beta_x^* = \beta_y^*$ for both beams and that the beams remain round, even when blown up. A similar though more complex analysis can be performed for flat beams.

2. Beam-Beam Parameter Space

The beam-beam parameter space **P** is quite large. The instantaneous luminosity is determined by the following fixed parameters:

a) The natural beam radii at the IP r_{o1}, r_{o2}
b) The numbers of particles per bunch N_1, N_2
c) The beta functions at the IP β_1, β_2
d) The nominal energies γ_1, γ_2
e) The damping decrements δ_1, δ_2
f) The transverse machine tunes ν_1, ν_2
g) The longitudinal machine tunes ν_{s1}, ν_{s2}
h) The bunch lengths σ_{s1}, σ_{s2}
i) The chromaticities C_1, C_2
j) The dispersions at the IP D_1, D_2
k) The orbit displacements at the IP Δ_1, Δ_2

The beam-beam lifetime is assumed to depend on all of the above parameters plus

l) The size of the hard aperture

and possibly others, including the passing separation of bunches in the wings of the IR and lattice nonlinearities (which determine the dynamic aperture). The parameters a) - l) will be referred to collectively as the full parameter set {**p**}. The manifold on which these parameters are coordinates will be called the full parameter space **P**. A point in **P**, which corresponds to a particular realization of the set {**p**}, is denoted **p**.

The problem faced by accelerator designers is to determine the optimal value for each of the parameters in this 20+ dimensional space. The resulting optimal parameter set provides the minimum overall cost for the required luminosity of $L = 10^{34}$. If a luminosity of 10^{34} is unreachable in a single accelerator with today's technology, this minimum cost will be infinite.

The optimization algorithm is straightforward: calculate the luminosity $L(\mathbf{p})$ and cost $C(\mathbf{p})$ as functions in the parameter space **P**. Then either minimize the cost subject to a fixed luminosity, or maximize the luminosity subject to a fixed cost.

3. A Simple Model for Optimization

There are certain features of the beam-beam system that help to reduce the complexity of the optimization procedure. These features can most easily be understood by considering a simplified model[†] of the beam-beam system.

The parameter space **P** of the simple model is only eight-dimensional with the following parameters:

1. The number of particles per bunch for each beam N_1, N_2
2. The radius of each beam at the IP r_1, r_2
3. The beta functions at the IP for each beam β_1, β_2
4. The energies of the two beams γ_1, γ_2

The cost function $C(\mathbf{p})$ in the simple model is treated as an infinite barrier: it is assumed that one cannot afford currents N_1, N_2 above a certain maximum for each ring. Likewise, there are minimum affordable βs, minimum and maximum affordable r_1, r_2, and the energies are fixed. The function $C(\mathbf{p})$ is then defined to be:

$C(\mathbf{p}) = 0$ if

$$N_1 < N_{M1}$$
$$N_2 < N_{M2}$$

$$r_{M1} > r_1 > r_{m1}$$
$$r_{M2} > r_2 > r_{m2}$$

$$\beta_1 > \beta_{m1}$$
$$\beta_2 > \beta_{m2}$$

$$\gamma_1 = \gamma_{o1}$$
$$\gamma_2 = \gamma_{o2}$$

(1)

$C(\mathbf{p}) = \infty$ otherwise.

The luminosity function in the simple model is given by the standard analytic expression for round beams when **p** is below the beam-beam limit, and by zero when it

---
[†] This model suggested by A. M. Sessler

is operating above the beam-beam limit:

$$L = \frac{N_1 N_2}{2\pi (r_1^2 + r_2^2)} \quad \text{both } \xi_1 < \xi_{lim} \text{ and } \xi_2 < \xi_{lim}$$

$$L = 0 \quad \text{either } \xi_1 > \xi_{lim} \text{ or } \xi_2 > \xi_{lim} \tag{2}$$

This expression implies that there is no beam blow-up in the simple model. The cause of the beam-beam limit is not specified. The tune shift ξ_{lim} at the beam-beam limit is considered a strict constant in the simple model. The tune shifts are given by

$$\xi_1 = \frac{r_e N_2 \beta_1}{4\pi \gamma_1 r_2^2} \qquad \xi_2 = \frac{r_e N_1 \beta_2}{4\pi \gamma_2 r_1^2} \tag{3}$$

Where r_e is the classical electron radius. Equations (3) can be used to eliminate r_1 and r_2 from (2)

$$L = \frac{2\xi_{lim}}{r_e \left(\frac{\beta_1}{\gamma_1 N_1} + \frac{\beta_2}{\gamma_2 N_2}\right)} \tag{4}$$

or to eliminate N_1 and N_2 from (2)

$$L = \left(\frac{\xi_{lim}}{r_e}\right)^2 \frac{8\pi \gamma_1 \gamma_2}{\beta_1 \beta_2 \left(\frac{1}{r_1^2} + \frac{1}{r_2^2}\right)} \tag{5}$$

It should be noted that the partial derivatives of the luminosity depend on the choice of dependent variables and will be different for the three equations (2), (4) and (5). Since these derivatives play a central role in optimization, it's important to keep them straight.

In what follows, we adopt the notation $\bar{i} = 2,1$ when $i = 1,2$. The simple model leads to a straightforward recipe for optimization. Let r_{Li} be the radius of beam #i at the beam-beam limit $\xi_{\bar{i}} = \xi_{lim}$ when N_i and $\beta_{\bar{i}}$ are set equal to the extreme values allowed by cost (1). From (3)

$$r_{L1} = \sqrt{\frac{r_e \beta_{m2} N_{M1}}{4\pi \gamma_{o2} \xi_{lim}}} \qquad r_{L2} = \sqrt{\frac{r_e \beta_{m1} N_{M2}}{4\pi \gamma_{o1} \xi_{lim}}} \tag{6}$$

Let N_{Li} be the number of particles per bunch at the beam-beam limit when $r_i = r_{Mi}$ and $\beta_i = \beta_{mi}$

$$N_{L1} = \frac{\xi_{lim}\gamma_{o2} 4\pi r_{M1}^2}{r_e \beta_{m2}} \qquad N_{L2} = \frac{\xi_{lim}\gamma_{o1} 4\pi r_{M2}^2}{r_e \beta_{m1}} \qquad (7)$$

The optimal parameters are then expressed in an "optimization theorem".

Optimization Theorem: Maximum luminosity (at finite cost) is attained in the simple model when $\beta_i = \beta_{mi}$ and if

Case 1	$r_{Li} < r_{mi}$	when $N_i = N_{Mi}$ and $r_i = r_{mi}$
Case 2	$r_{mi} < r_{Li} < r_{Mi}$	when $N_i = N_{Mi}$ and $r_i = r_{Li}$
Case 3	$r_M < r_{Li}$	when $N_i = N_{Li}$ and $r_i = r_{Mi}$

The proof of the optimization theorem for the simple model is given in Appendix A.

The optimization theorem says that maximum luminosity is always realized in a configuration where the beta functions are at their minimum values allowed by cost, and the two beams are each operating at the maximum tune shifts attainable (subject to cost constraints and their respective beam-beam limits). Furthermore, if either of the beam-beam limits gets in the way, the beam radii should always be increased from their minimum values (all the way to their maximum values, if necessary) before the currents are decreased from their maximum values.

The three optimization cases are represented graphically in figure 1 where the cost constraints, beam-beam limit, and contours of constant luminosity are shown in the (N, r) sub-plane. The three limiting tune shift contours may be interpreted as corresponding to three different values of β_i. These three contours give optimal points p_{opt} for the three different cases of the optimization theorem. In case **1**, the cost constraints prevent the system from reaching the beam-beam limit. In case **2**, the limiting tune shift contour intersects both the beam radius barrier and the current barrier. Maximum luminosity is always found at the current barrier intersection. In case **3**, the limiting tune shift contour intersects both the upper and lower beam radius barriers. Maximum luminosity is found at the intersection with the upper barrier. In both case **1** and **2**, the beam radius should be maximized subject to the constraint that the beams stay at the beam-beam limit.

Equations (4) and (5) imply that any parameter variation that leaves the factor

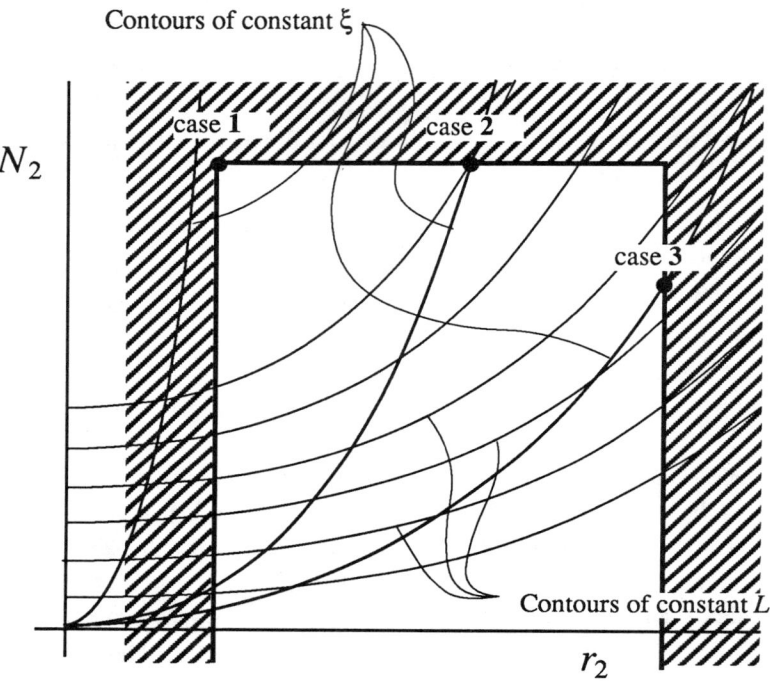

Figure 1.
The three optimization cases for the simple model. Contours of constant luminosity, and three contours of constant tuneshift are shown in the current-radius plane. The infinite cost barriers are represented by the shaded region. In case **1**, the tuneshift contour does not penetrate the accessible region and the luminosity is limited only by the cost constraints. In case **2**, the maximum luminosity is at the current cost barrier, but between the two radius cost barriers. In case **3**, maximum luminosity is at the upper radius cost barrier but below the current cost barrier. The optimal points in each case are marked with a spot ● .

$$\left(\frac{\beta_1}{\gamma_1 N_1} + \frac{\beta_2}{\gamma_2 N_2}\right) \tag{8}$$

or, equivalently, the factor

$$\beta_1 \beta_2 \left(\frac{1}{r_1^2} + \frac{1}{r_2^2}\right) \tag{9}$$

unchanged also leaves the luminosity at the beam-beam limit unchanged. An asymmetry in one parameter pair, say γ_1, γ_2, can thus be "compensated" by an asymmetry in another parameter pair, e.g. N_1, N_2.

In actual practice, one usually starts by assuming case 2 and treating the beam radii (r_1, r_2) as dependent variables determined by the constraints $r_i = r_{Li}$. There are three reasons for this: the first is that it results directly in the calculation of r_{Li} and thus allows one to determine which of the three cases actually holds. The second is that the beam radii in the realistic model (described in the next section) cannot be directly controlled. The beams can be blown up by the beam-beam interaction and the beam radii therefore make poor control parameters. The third is that case 2 is the predominant case in e^+ - e^- colliders.

It's useful to define a reduced parameter set $\{\mathbf{p'}\}$ which represents the full parameter set $\{\mathbf{p}\}$ less the two energies γ_1, γ_2 and the two radii r_1, r_2. Fixing the energies and beam radii

$$\gamma_1 = \gamma_{o1}$$
$$\gamma_2 = \gamma_{o2}$$

$$r_1 = r_{L1}$$
$$r_2 = r_{L2}$$

one defines a codimension-four subsurface of the full parameter space **P**. This subsurface will be called the reduced parameter space **P'**. Note that a realization **p'** of the reduced parameter set does not correspond directly to a point on **P'**; a point on **P'** implicitly assigns values to γ_1, γ_2, r_1, and r_2 while **p'** does not. The reduced parameter set $\{\mathbf{p'}\}$ can be used as a coordinate system on **P'**, and is particularly useful in case 2. In case 3, it is more useful to include the beam radii and exclude the currents in the coordinate system on **P'**.

For every realization **p'** of the reduced parameter set there is a characteristic portrait of the beam-beam limit in the (r_1, r_2) plane. An example for the simple model is shown in figure 2 where the beam-beam limits are horizontal and vertical lines which depend on the energies γ_1, γ_2, the currents N_1, N_2, and the beta functions β_1, β_2. The luminosity contours, from (2), are quarter circles. It's clear from this picture that the maximum luminosity is always found at the coincident beam-beam limit.

Figure 3 shows a similar picture for the current plane (N_1, N_2). Here the limits are again horizontal and vertical lines, but now they depend on the energies γ_1, γ_2, radii r_1, r_2, and beta functions β_1, β_2. The luminosity contours in figure 3 are curves

$$N_1 = \frac{K}{N_2} \tag{10}$$

where K parameterizes the contours. In both figures 2 and 3, the luminosity is clearly maximized when both beams are operating at the beam-beam limit, regardless of where the limits are.

Note that because no assumption has been made about beam-beam symmetry, the energies, currents, beam sizes, and beta functions may be different for the two beams. This means that, in the general case, optimal luminosity is attained when the two beams have different sizes. More specifically, the two beams will be the same size at the optimal operating point only if the limit lines in figures 2 and 3 intersect on the diagonal, i.e. only if, from (6)

$$\frac{\beta_{m1}}{\gamma_{o1} N_{M1}} = \frac{\beta_{m2}}{\gamma_{o2} N_{M2}} \tag{11}$$

There is no *a priori* reason that the cost constraints in an asymmetric collider should satisfy this condition.

4. A Realistic Model for Optimization

The above analysis is based on the simple model. In reality, the beam-beam limit ξ_{lim} is not a constant and is complicated by blow-up at the limit, self-consistency conditions, fixed and dynamic apertures, the flip-flop effect, and other sundry details. How do these various real-life complications affect the above analysis and the validity of the optimization theorem?

The realistic model discussed here is essentially identical to the model used in modern

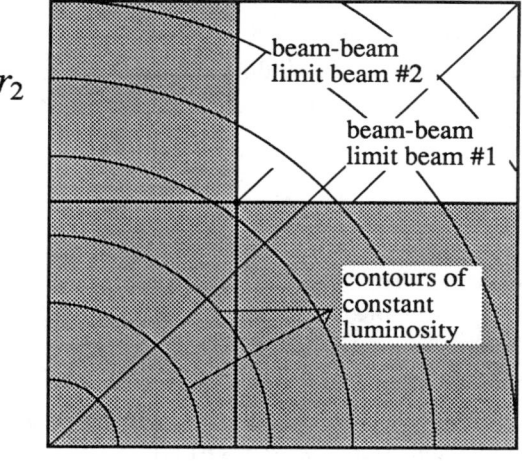

Figure 2. The beam-beam limit for the simple model. The plot characterizes a point in the reduced parameter set. The shaded area is above the beam-beam limit and therefore inaccessible. Maximum luminosity is found at the coincident beam-beam limit, which is rarely on the diagonal if the parameters are not symmetric.

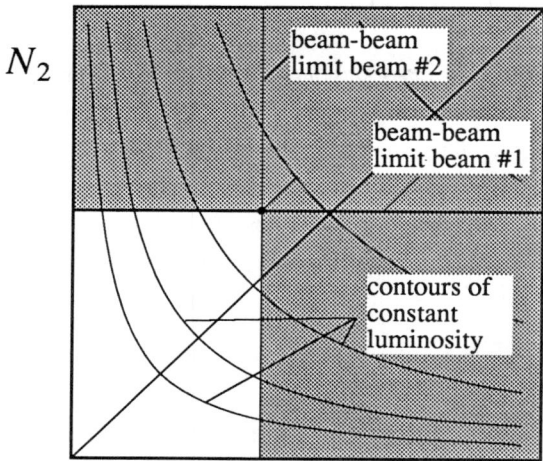

Figure 3. Another representation of the beam-beam limit for the simple model. This plot characterizes a particular choice of energies, beta functions, and beam radii. As in figure 2, the shaded area is above the beam-beam limit and therefore inaccessible. Maximum luminosity is again found at the coincident beam-beam limit.

beam-beam tracking codes.[2] These have three degrees of freedom with radiation excitation and damping, a transverse beam-beam force, chromaticity, dispersion, finite bunch lengths, finite apertures, machine nonlinearities, and other errors.

There are five important differences between the simple and realistic models. The first has to do with the size of the parameter space, the second third and fourth are related to the luminosity function $L(\mathbf{p})$, and the fifth concerns the actual nature of the cost function $C(\mathbf{p})$.

The Realistic Parameter Space

The realistic parameter space is much larger than that of the simple model. The full set of parameters affecting the luminosity and lifetime were listed in section two and number more than twenty (excluding the energies). The additional parameters do not affect the luminosity expression (2), but do affect the values of the tuneshifts at the coincident beam-beam limit. These values must now be expressed as functions of the reduced parameter set $\{\mathbf{p'}\}$. The limiting tune shifts for beams #1 and #2 may now be different. The reduced parameter space $\mathbf{P'}$ is still a codimension-four subsurface of $\mathbf{P}$ and still represents the coincident beam-beam limit; but it no longer corresponds to fixed values of the tune shifts and cannot, in the general case, be described by an analytic function.

Parameter Dependence of the Beam-Beam Limit

In the realistic model, the limiting tune shifts ξ_{lim_1} and ξ_{lim_2} are functions of the reduced parameter set $\{\mathbf{p'}\}$ and vary in the reduced parameter base $\mathbf{P'}$. For example, the damping decrement δ_i increases ξ_{lim_i} while bunch length, chromaticity, dispersion, and other errors reduce it. Changes in tune can cause large non-monotonic variations of ξ_{lim_i}. The fact that ξ_{lim_i} changes with $\mathbf{p'}$ does not alter the validities of equations (2), (3), (4) and (5). However, it could alter the signs of the derivatives (A4), and thereby require a modification of the optimization theorem. This possibility is discussed below.

Cause of the Beam-Beam Limit

In the simple model, the cause of the beam-beam limit was not specified. In the realistic model, there are three distinct limiting phenomena. Each of these three phenomena has its own beam-beam limit. The actual beam-beam limit is the most restrictive of the three candidate limits. The three limiting phenomena are blow-up, flip-flop instability, and particle loss to the aperture. None of these limits look like the vertical and horizontal lines of figures 2 and 3.

Blow-up Limit

The blow-up limit results from transverse enlargement of the beam due to the accumulation of beam-beam kicks at high tuneshifts. In figure 4a, a representative family of curves shows the dependence of the radius of beam #1 on the radius of beam #2 for a number of different "natural radii" of beam #1 (assuming the current in beam #2 remains constant). There's a limit to how small beam #2 can be, even when the natural size of beam #1 approaches zero. This limit to the smallness of beam #2 limits the luminosity. A similar limit exists for beam #1, as shown in figure 4b. The two limits together, shown in figure 4c, determine the blow-up limitation on the luminosity of the system.

Flop-flop limit

The flop-flop limit is more complicated.[3] Briefly, there are some regions of the (r_1, r_2) plane that are unstable to flip-flop. These regions can be found using "weak beam size function" curves similar to those shown in figure 4. In figure 5a, a family of curves are shown which, again, represent the size of beam #1 as a function $S(r_2; \mathbf{p}_1)$ of the size of beam #2.

$$r_1 = S(r_2; \mathbf{p}_1) \tag{12}$$
$$\mathbf{p}_1 = \{\nu_1, \nu_{s1}, \beta_1, \eta_1, \tau_1 = 20000, \ r_{o1}, N_2, ..., \text{etc.}\}$$

These curves are smoothed simulation results[4] obtained from the model and parameters of section five when the damping time of beam #1 is 20000 turns. A similar set of curves is shown in figure 5b for beam #2 with a damping time of 1200 turns.

$$r_2 = S(r_1; \mathbf{p}_2) \tag{13}$$
$$\mathbf{p}_2 = \{\nu_2, \nu_{s2}, \beta_2, \eta_2, \tau_2 = 1200, \ r_{o2}, N_1, ..., \text{etc.}\}$$

The self-consistent equilibria for a pair of natural beam sizes (r_{o1}, r_{o2}) are given by the intersections of the corresponding curves. There are typically either one or three equilibria associated with each pair of natural beam sizes. A self-consistent equilibrium is stable[3] if the product of the slopes of the two functions (12) and (13) at the equilibrium is less than one.

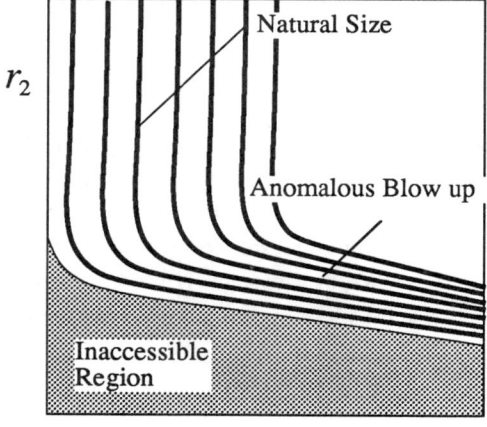

Figure 4a.
Weak beam size functions for beam #1. The size of beam #1 is plotted against the size of beam #2 for several natural radii of beam #1. These graphs are representative of actual functions. When the size of beam #2 falls below a certain value, beam #1 begins to blow up. The blow up makes certain areas of the plane inaccessible

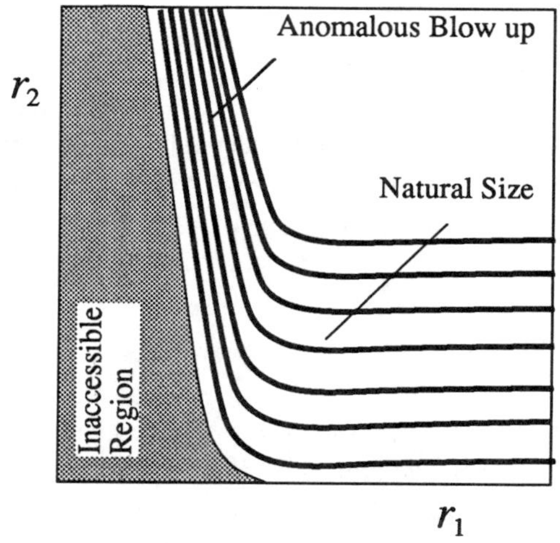

Figure 4b.
Weak beam size functions for beam #2. This figure is identical to 4a except that the roles of beam #1 and #2 have been interchanged.

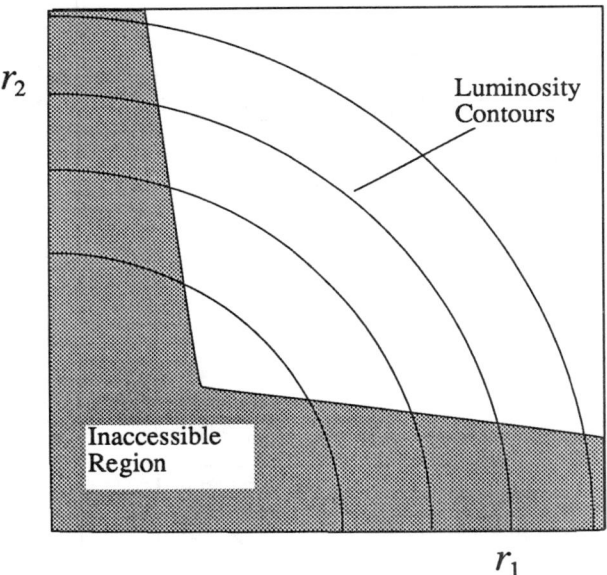

Figure 4c.
Combining figures 4a and 4b shows the total inaccessible area due to beam blow-up. Beam blow-up alone places a limit on the maximum attainable luminosity.

144 Optimization of the B-Factory Parameter Base

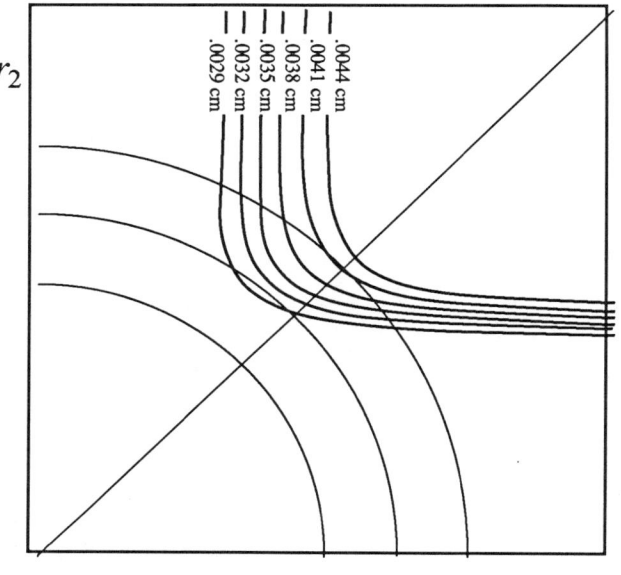

Figure 5a.
Weak beam size functions for beam #1. These curves are derived from numerical simulation using the parameters listed in section 5. The damping time of beam #1 is 20000 turns.

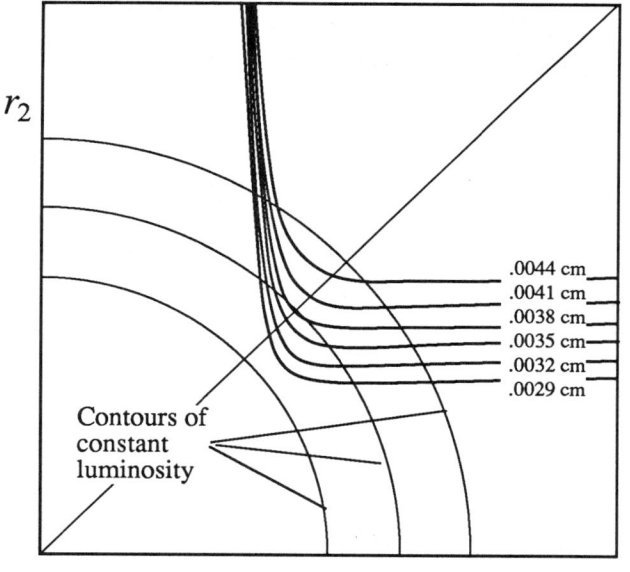

Figure 5b.
Weak beam size functions for beam #2. The damping time of beam #2 is 1200 turns. The natural radii for each curve is shown. This beam clearly blows up at a lower tune shift than beam #1 (shown in the preceding figure).

$$\frac{dS(r_2;\mathbf{p}_1)}{dr_2} \frac{dS(r_1;\mathbf{p}_2)}{dr_1} < 1 \qquad (14)$$

If there is only one equilibrium, it is always stable. If there are three equilibria, the middle equilibrium is unstable while the two satellites are stable. The superposition of figures 5a and 5b are shown in figure 5c. Each intersection is labeled with either a white or black dot. The white dots mark equilibria that satisfy (14) and are therefore stable. The black dots mark unstable equilibria. A thick curve separates the white dots from the black.

The thick curve in figure 5c is called the "flip-flop" bump. The area of the (r_1,r_2) plane below the flip-flop bump is unstable and therefore inaccessible. Note that the inaccessible region is the region of highest luminosity. Thus, in the absence of lifetime limitations, the flip-flop phenomena almost always reduces the maximum attainable luminosity.

The flip-flop limit is represented by a single curve in the (r_1,r_2) plane while the blow-up limit is represented by two separate curves. This is because the blow-up limits are independent for the two beams; the parameters $\mathbf{p}_1$ that determine the radius of beam #1 do not affect the radius of beam #2 and vice versa. The flip-flop limit, on the other hand, is a self-consistent effect and is therefore dependent on all the parameters affecting the blow-ups of both beams.

Note that the luminosity limits affected by blow-up and flip-flop are independent of the size of the aperture. One cannot improve these limits by increasing the size of the aperture. These limits are unique to lepton colliders since they are related to the equilibrium sizes of the two beams; the beams in a hadron collider are never in equilibrium and therefore do not exhibit these limits. It should also be noted that each point in the (r_1,r_2) plane defines not only a pair of actual radii (r_1,r_2) but also a pair of natural radii (r_{o1},r_{o2}) which are almost always smaller.[†]

Particle Loss Limit

The third luminosity limit is imposed by particle loss to the aperture. As with blow-up, there is a separate particle loss limit for each beam. Figure 6 shows simulation results for the same system configuration shown in figure 5. In Figure 6a the black triangles mark points in the (r_1,r_2) plane where the lifetime of beam #1 (again with a damping

---

[†] The only exception to this is when the focusing effect of the beam-beam force lowers the effective beta function at the IP. This occurs when the fractional part of the tune (per interaction) is between 0 and .25 or between .5 and .75.

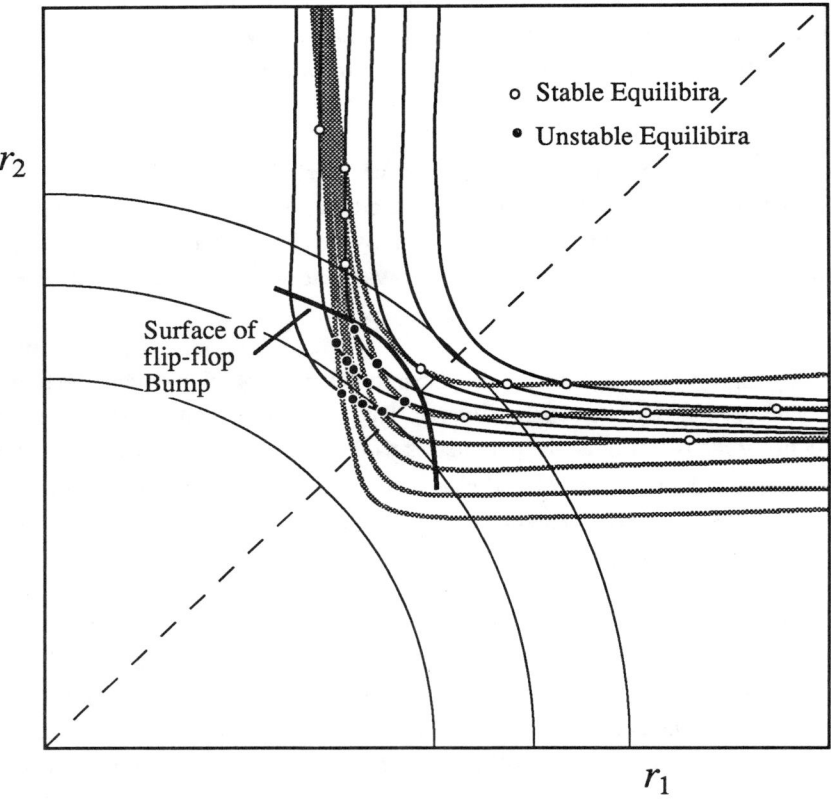

Figure 5c.
Superposition of figures 5a and 5b shows the regions of flip-flop stability and instability. Each crossing of a grey and black line has one of two orientations. Crossings with one orientation correspond to stable equilibria and crossings with the other orientation correspond to unstable equilibria. The regions of stable and unstable equilibria are well separated and their interface is represented here with a thick curve. Because the boundary is characteristically convex, the unstable region is referred to as the "flip-flop bump".

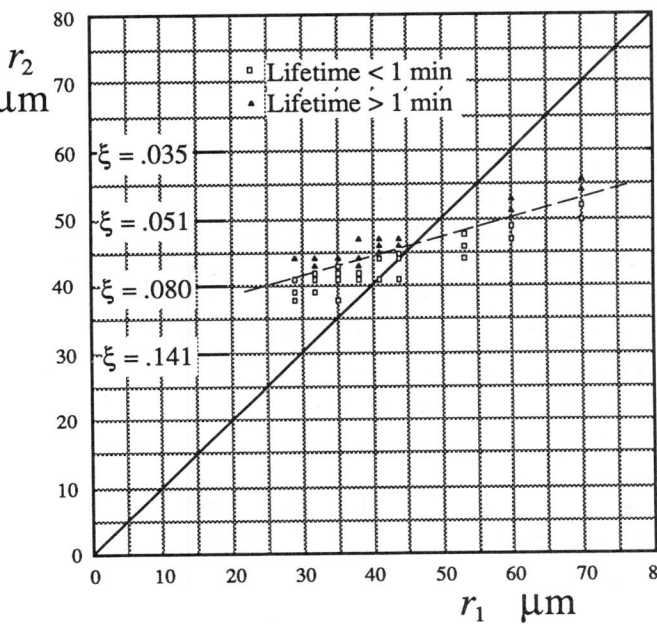

Figure 6a. Results of simulation runs which measured the lifetimes of beam #1 for several different values of the natural beam sizes. Beam size pairs with lifetimes greater than one minute are marked with triangles; those with lifetimes less than one minute are marked with squares. The aperture is at 500 microns, so the beam-beam limit is fairly straight in this plane. The damping time of this beam (#1) is 20000 turns.

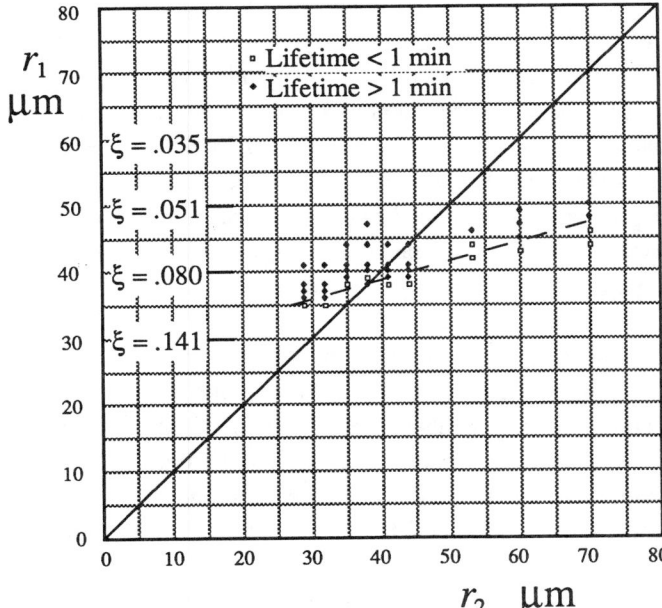

Figure 6b. Results of simulation runs which measured the lifetimes of beam #2 for several different values of the natural beam sizes. The aperture is at 500 microns. The damping time of this beam (#2) is 1200 turns. Note that the limiting tune shift on the diagonal is about $\xi = .087$ compared to $\xi = .060$ for the highly damped beam in the previous figure.

time of 20000 turns) is longer than one minute. Hollow squares mark points where the lifetime is less than one minute. A dashed line is drawn separating the two families of points. The corresponding points and separating line are shown in figure 6b for beam #2 (again with a damping time of 1200 turns). The combined picture is shown in figure 6c.

The lifetime limit curves are fairly straight, but unlike those in figure 2, they're sloped. The area of highest luminosity therefore sits at a "vertex" characterized by an acute angle. The limit lines are sloped for two reasons. One is that the aperture is fixed; as a beam gets larger, the relative size of its aperture shrinks and the average particle density at the aperture increases. The other reason is that the limiting tune shift naturally decreases as the opposing beam gets smaller with constant tune shift. This is due to microscopic effects that cause particle stability to decrease with amplitude relative to the RMS width of the opposing beam.

When the apertures of the two beam systems are finite, they bend the beam-beam limit lines up and close the accessible area of the (r_1, r_2) plane. This is shown in figure 6d. The finite apertures cause lifetimes to decrease when the beam sizes get too large.

It may seem curious to some readers that there can be significant particle loss with little or no blowup. This phenomena is characteristic of round beams and is due to the fact that particles escape to the aperture through a leak in the phase space. The leak is through a specific region of phase space where both the synchrotron and betatron amplitudes are large (two or three sigma), and where the horizontal and vertical oscillations are almost exactly in phase. The motion in this narrow region of phase space is dynamically unstable. The leak provides a channel for particles to move rapidly from a relatively compact (not blown up) core to the aperture. It is difficult to see in simulations because the phase relations between the horizontal and vertical motions are generally hidden. A similar leak exists in flat-beam machines, though in a more apparent region of phase space.

The lifetime vertex is not on the diagonal for most "asymmetric" configurations. In the case of figure 6c, the only asymmetry is in the damping times of the two rings. The beam sizes at the beam-beam limit differ by about 10%. Note that if the apertures of the two rings are different, this by itself causes an asymmetry in the lifetime limits of the two beams and shifts the vertex off the diagonal. Thus, the blow-up and flip-flop limits can be perfectly symmetric, even when the lifetime limits are quite asymmetric.

The luminosity and cost functions are symmetric in the full parameter space **P**. These symmetries are broken by fixing the energies of the two beams at different values

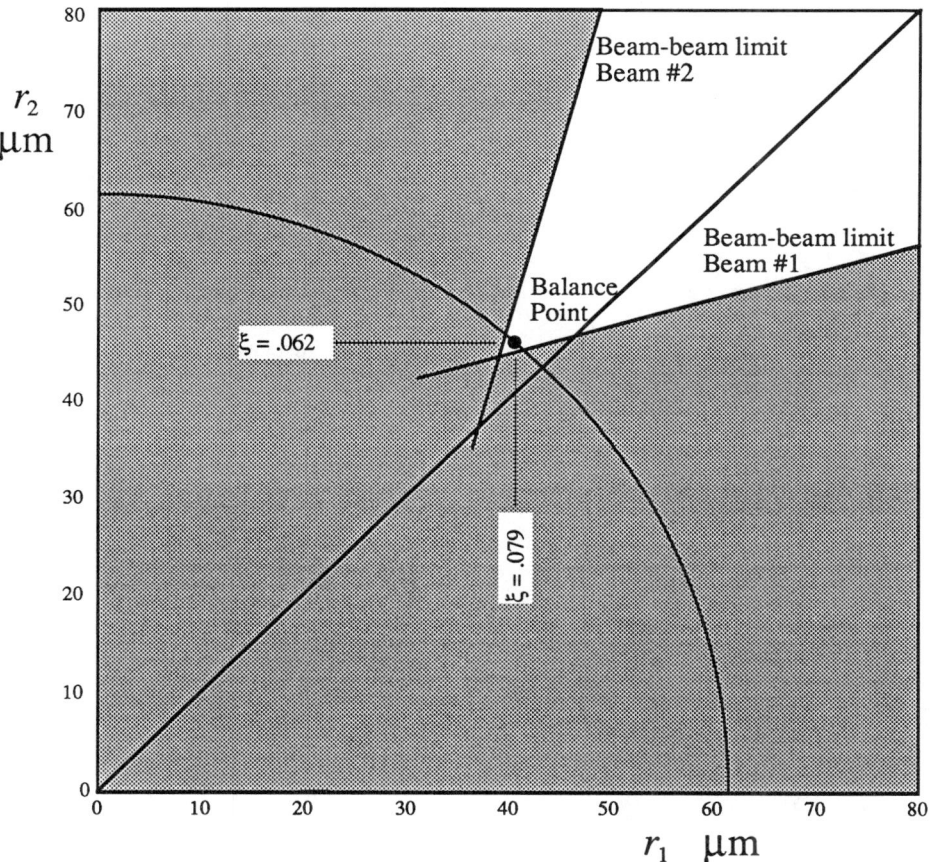

Figure 6c.
Superposition of figures 6a and 6b. The shaded area is the region in which the lifetime of one or both beams is less than one minute. Maximum luminosity is found close to the vertex. Note that the limiting tuneshift for the weakly damped beam is some 30% below that of the strongly damped beam.

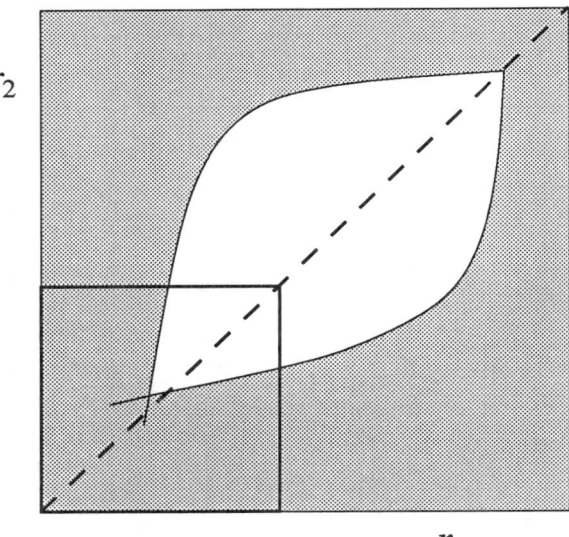

Figure 6d.
The lifetime limits actually bend back and enclose the accessible region when the apertures are finite. The closing point is 1/6-1/5 of the aperture for each beam. This is a representation figure (not derived from simulation). In case **3**, the size of the accessible region goes to zero. The box is roughly the area shown in figure 6c

▓ Inaccessible

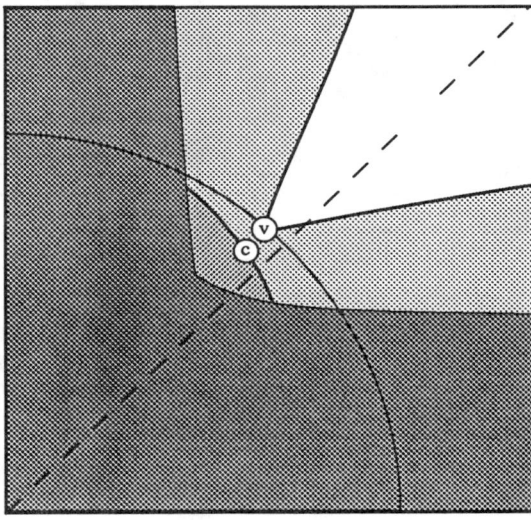

Figure 7.
Superposition of the three limits. In this case, the lifetime vertex is outside the beam-beam bump and therefore determines the actual beam-beam limit.

ⓥ Lifetime Vertex
ⓒ Critical Point
▓ Excluded by Blow-up
▓ Excluded by Flip-flop
▓ Excluded by Particle Loss

(e.g. at 3 GeV and 9 GeV). The resulting asymmetry virtually guarantees that the optimal parameter configuration will also be asymmetric. For example, the two beam radii at the optimal parameter base will almost always be different. It should be pointed out, however, that although the beam sizes are generally different at maximum luminosity, they are rarely radically different. This is because some particles in the larger beam do not participate in the collision and are therefore "wasted". This "waste" is only economical if the cost of current in one beam (usually the high energy beam) is large compared to the cost of current in the other beam, or if one beam (usually the low energy beam) is more sensitive to the beam-beam interaction. Even in these situations, one would not expect a size difference of more than 50%. In the B-Factory case, the higher sensitivity of the low energy beam (illustrated in figure 6c) partly compensates the higher cost of current in the high energy beam; one therefore expects beam sizes to be nearly equal in the optimal configuration.

The three luminosity limits are shown together in figure 7. For this system configuration **p'**, the lifetime vertex is outside the beam-beam bump. Particle loss is therefore responsible for the actual luminosity limit. Note that by enlarging the aperture (it was 10 sigma for both beams in these runs), it should be possible to move the lifetime vertex to higher luminosities, possibly high enough to cross the flip-flop bump. Then the luminosity would be limited by the flip-flop effect.

In general, the beam-beam limit is always determined either by the flip-flop instability or by particle loss. The lifetime limit may be larger than either the blow-up limit or the flip-flop limit. The flip-flop limit, on the other hand, is always lower than the blow-up limit.

Balance Point Sensitivity

If the lifetime vertex is outside the flip-flop bump, as is usually the case with round beams and often the case with flat beams, then lifetime determines the beam-beam limit. In this case, the vertex of the realistic model, figure 7, is loosely referred to as the "balance point"; it is the operating point with the highest attainable luminosity. The acute opening angle at the vertex creates an illusion of "instability", especially if one is blindly probing the space with a strong-strong simulation. This pseudo-instability will be called "balance point sensitivity". If one starts from a point just inside the vertex and varies the size of either beam in either direction, one of the beams immediately hits the beam-beam limit and its lifetime drops below the acceptable level. Clearly, the closer the operating point is to the vertex, the more sensitive the balance. If balance point sensitivity is a problem, one must back away from the vertex (and thus reduce the luminosity) until the sensitivity falls to a tolerable level.

If the lifetime vertex is inside the flip-flop bump, then luminosity is limited by the flip-flop instability. The points on the surface of the flip-flop bump correspond to bifurcation points. All but one of them are tangent bifurcations points (first order phase transitions); the single exception is a critical point. The critical point is a pitchfork bifurcation and is analogous to a second order phase transition. When the flip-flop instability is the limiting phenomena, the balance point is associated with the critical point rather than with the lifetime vertex. The critical point does not always have the highest accessible luminosity in the (r_1, r_2) plane; because the flip-flop bump is generally convex, there may be other (tangent bifurcation) points on its surface that have higher luminosities. However these other points are dangerous. If one operates too close to a tangent bifurcation, fluctuations can momentarily push the system out of the local minima and a "catastrophe" may occur. The catastrophe appears as a sudden irreversible change in the sizes of the two beams. The occurrence of a catastrophe is usually accompanied by a precipitous drop in the luminosity and the lifetime of one beam.

A catastrophe is less dangerous close to the critical point since the region surrounding the critical point corresponds either to a single (marginally stable) equilibrium or to three separate but very tightly bunched equilibria (for practical purposes they're a single point).

Close to the critical point, the beam sizes are extremely sensitive to small changes in any of the beam-beam parameters $\{p\}$; for example $\partial r_1/\partial r_{o1}$ becomes very large (the sensitivity is infinite at the critical point). This means that one can expect high balance-point sensitivity at the coincident beam-beam limit even when the lifetime vertex is well inside the flip-flop bump.

Other phenomena characteristic of operation close to a critical point include long relaxation times and large fluctuation response functions. The former can be tens or hundreds of times longer than a damping time. Generally, the damping time determines the weak-strong relaxation time: the strong-strong or self-consistent relaxation times are determined by the stability of the equilibrium. Relaxation times that are much longer than the damping time have been observed in experiment[5]. The second effect, large sensitivity to fluctuations, are typically more important in simulations than in actual machines. In simulations, super-particle discreteness drives fluctuations that become very large close to a critical point. For this reason, it is almost impossible to investigate flip-flop criticality with a standard strong-strong beam-beam simulation code.

Since the beam-beam balance at both the lifetime vertex and the critical point are extremely sensitive to any change in the value of any beam-beam parameter (any change pushes the system over the beam-beam limit), it is always necessary to back away from the true balance point (ideally, by slightly increasing the natural beam sizes). This reduces both the sensitivity and the luminosity. The loss of luminosity associated with this "backing away" is likely to be small, probably less than 10%.

Balance point sensitivity exists in symmetric and asymmetric configurations, and in both one-ring and two-ring accelerators. In one-ring machines, the problem is ameliorated by the fact that virtually all beam-beam parameters, other than the currents, are constrained to be the same for the two beams. Any imbalance can then be attributed to current mismatch and can be removed by adjusting the currents. In addition, because the fluctuations in other parameters are constrained to be the same in the two rings, one-ring machines are expected to be less sensitive to fluctuations than two-ring machines.

In a two-ring machine, there is no practical advantage to running the beams in a symmetric configuration; the symmetric and asymmetric configurations have the same balance point sensitivities. Even if a two-ring machine is designed to be symmetric, it will not be balanced when it is first turned on (or following any changes in either lattice). Although imbalance in a two-ring accelerator can come from any of several parameter asymmetries, imbalance in one pair of parameters can be corrected by introducing a compensating imbalance in another pair of parameters. Being able to compensate in this way means that the beams in a two-ring accelerator can be both asymmetric and balanced.

In the conventional one-ring machine where imbalance can only be corrected by adjusting the currents, there is no need (even if the possibility existed) to control the emittances of the two rings independently. Even in two-ring machines, one can always balance the beams at the beam-beam limit by appropriately adjusting the two currents. However, a two-ring collider that could only be balanced by changing the two currents would be a poor performer. Balancing the beams would be difficult because current can only be removed from, and not added to, an accelerated beam. Furthermore, luminosity could not be optimum because at least one beam would not be operating at its maximum current capacity. For this reason, a two-ring collider should have knobs capable of varying the beam emittances, at least within some limited range of values. These knobs would be the principle instruments used to balance the two beams. As pointed out above, they are no less desirable in a symmetrically designed machine than in an asymmetrically designed machine. For a round beam collider, they would control the emittances in both the vertical and horizontal planes. For a flat beam collider, they would control the vertical emittances only.

Smooth Cost Functions

The fourth important difference between the simple and realistic models has to do with the cost function $C(\mathbf{p})$. Real cost functions are smooth, not infinite barriers. If a point $\mathbf{p}_{opt}$ in $\{\mathbf{p}\}$ gives maximum luminosity for a given cost, or minimum cost for a given luminosity, then it must hold at $\mathbf{p}_{opt}$ that

$$dL + \lambda dC = 0 \tag{15}$$

for any variation $d\mathbf{p}$ and some fixed constant λ (Lagrange multiplier). It follows directly that

$$\frac{dC}{dL} = \frac{1}{\lambda} \tag{16}$$

for any variation $d\mathbf{p}$. A variation in which only the parameter p_i is changed is denoted

$$\left.\frac{dC}{dL}\right|_{p_i}$$

Equation (16) then gives the optimization theorem for the realistic model:

Optimization Theorem (Realistic Model) : At the optimal parameter base $\mathbf{p}_{opt}$, i.e. the configuration that gives the maximum luminosity at a given fixed cost C_f and for a given cost function $C(\{\mathbf{p}\})$,

$$\left.\frac{dC}{dL}\right|_{p_1} = \left.\frac{dC}{dL}\right|_{p_2} = \ldots = \left.\frac{dC}{dL}\right|_{p_n} \tag{17}$$

for all parameters p_i in the full parameter space $\{\mathbf{p}\}$.

Equation (17) says that, in the realistic model, the increase in luminosity that would be gained by spending $1 more to raise the current in the high energy ring should be equal to the luminosity that would be gained if $1 more were spent to raise the current in the low energy ring. These should also be equal to the luminosities that would be gained by spending $1 to increase the damping decrement in either ring, decrease the bunch length/ beta function ratio in either ring, etc. This is the only scientifically defendable criteria for setting parameters in the realistic model.

Note that this (more general) optimization theorem takes into account the fact that the luminosity drops abruptly to zero at the beam-beam limit. The derivatives at the limit are degenerate and correspond to values greater than minus infinity and less than the derivative just below the limit.

Because the cost function in the realistic model is not defined as explicitly as in the simple model, one cannot be quite as explicit in describing the optimal solutions. However, due to the nature of the realistic cost function $C(\mathbf{p})$, the solutions of (17) fall approximately into catagories corresponding to the cases **1, 2**, and **3** of the simple model (this is why the simple model is useful). More specifically, case **1** is still clearly defined in the realistic model (due to the fact that the beam-beam limit is very sharp). Cases 2 and 3 still exist, but only approximately since there is no longer a fine line separating them. The correspondence between the realistic and simple models arises because the derivatives (17) either increase or decrease monotonically with almost every parameter p_i in $\{\mathbf{p}\}$.

$$\frac{\partial}{\partial p_i} \frac{dC}{dL}\bigg|_{p_i} > 0 \qquad p_i = N_i, \gamma_i, \delta_i$$

$$\frac{\partial}{\partial p_i} \frac{dC}{dL}\bigg|_{p_i} < 0 \qquad p_i = \beta_i, C_i, D_i, \Delta_i$$

(18)

There are two exceptions: one is the dependence on the tunes and the other is the dependence on the beam radii. But since tune is essentially free, it does not enter into cost optimization at all. A beam radius typically has a minimum cost at some finite value; the cost increases monotonically with r above this value and decreases montonically with r below it. This, together with (18), means that there is a correspondence between the cost barrier (1) of the simple model and the point $\mathbf{p}_{opt}$ at which all the derivatives (17) are equal to $1/\lambda$. The value λ depends on the total cost C_f that one can afford.

The situation where the optimal parameter base $\mathbf{p}_{opt}$ is not at the beam-beam limit corresponds to the previous case 1). Here, the beam-beam effect can be ignored, and optimization is performed in the standard way. Note that without the beam-beam effect, many parameters (such as the damping times) no longer enter into consideration and one can generally retreat to the parameter space of the simple model. Unfortunately, it is not possible to confirm that a particular cost function corresponds to case 1) without first performing the optimization with the beam-beam effect included.

If the optimal point $\mathbf{p}_{opt}$ turns out to be at the beam-beam limit, then one is typically pushing hard against either current costs, case 2), or beam aperture costs, case 3), but rarely both.

In the simple model, the luminosity increases with beam radius when the beams are constrained to their respective beam-beam limits. This is because the derivative (A4) of the luminosity at constant tune shift and beta function is positive

$$\frac{\partial L}{\partial r_i}\bigg|_{\xi_i,\beta_i} > 0 \qquad (19)$$

The fact that the tuneshift limit ξ_{lim_1} now depends on the size of beam #1 threatens the validity of this assertion for the realistic model. It effectively introduces an (r_1, r_2) dependence into the numerator of equation (5). In particular

$$\frac{\partial \xi_{lim_1}}{\partial r_1}, \frac{\partial \xi_{lim_2}}{\partial r_2} < 0 \qquad (20)$$

If the absolute values of these derivatives are large enough, the condition (19) can be violated, and the theorem becomes invalid. The derivatives (20) can be estimated from the slopes of the particle loss limits, figure 6. A critical slope can be calculated; slopes above this critical value violate the optimization theorem, slopes below this level conserve the theorem. It is shown in Appendix **B** that the critical slope is .5. Since slopes greater than .5 correspond to complete elimination of the accessible region (at least in symmetric machines), they can be discounted. The bottom line is that (19) almost always holds and that, as with the simple model, one should always increase the beam sizes before reducing the beam currents when the luminosity is limited by the beam-beam effect.

The optimization theorem for the realistic model provides a specific algorithm for optimizing the B-Factory parameter base. The algorithm is not easy to execute: the luminosity derivatives are generally obtained from simulations of questionable reliability and the cost derivatives can be even harder to estimate since they're not always expressible in currency units. Part of the cost, for example, may be associated with risk and an associated probability that can only be guessed at. Nevertheless, its worthwhile trying to calculate the values of these derivatives and their errors, if only to rationalize the more intuitive design decisions that typically characterize machine development.

The cost derivative can be separated into two parts

$$\left.\frac{dC}{dL}\right|_{P_i} = \frac{\partial C}{\partial p_i}\left(\frac{\partial L}{\partial p_i}\right)^{-1} \tag{21}$$

The first derivative on the right hand side is estimated from considerations other than the beam-beam interaction. For case **1** situations, the second factor on the RHS is just the derivative of the simple formula (2). In cases **2** and **3**, the two beams are constrained to their respective beam-beam limits, so the derivatives need only be calculated in the reduced parameter space **P'**. This is the codimension-4 surface in **P** corresponding to both beams being at the beam-beam limit with fixed energies. When derivatives are taken in this space, two parameters must be treated as dependent variables. In case **2** situations, it is convenient to use the beam radii as the dependent variables since the costs of the radii in this case are usually negligible. In case **3**, it is convenient to give this role to the beam currents since, in this case, current is the parameter with negligible cost. Note that, except at the optimal point $\mathbf{p}_{opt}$, the value of the derivative will depend on which parameters are treated as dependent variables. A derivative in **P'** will be denoted

$$\left.\frac{dC}{dL}\right|_{p_i,r_1,r_2} \quad \text{or} \quad \left.\frac{dC}{dL}\right|_{p_i,N_1,N_2}$$

depending on the whether the radii or currents are taken as the dependent variables. This notation means that the three qualifiers are allowed to change during the variation (the opposite of convention). These derivatives break into three terms;

$$\left.\frac{dC}{dL}\right|_{p_i,r_1,r_2} = \frac{\partial C}{\partial p_i}\left.\frac{dp_i}{dL}\right|_{p_i,r_1,r_2} + \frac{\partial C}{\partial r_1}\left.\frac{dr_1}{dL}\right|_{p_i,r_1,r_2} + \frac{\partial C}{\partial r_2}\left.\frac{dr_2}{dL}\right|_{p_i,r_1,r_2} \tag{22}$$

or

$$\left.\frac{dC}{dL}\right|_{p_i,N_1,N_2} = \frac{\partial C}{\partial p_i}\left.\frac{dp_i}{dL}\right|_{p_i,N_1,N_2} + \frac{\partial C}{\partial N_1}\left.\frac{dN_1}{dL}\right|_{p_i,N_1,N_2} + \frac{\partial C}{\partial N_2}\left.\frac{dN_2}{dL}\right|_{p_i,N_1,N_2} \tag{23}$$

In case **2**, the last two terms on the RHS of (22) are small and can usually be neglected. The same is true of (23) in case **3**. So optimization corresponds approximately to the conditions

$$\frac{\partial C}{\partial p_1}\left.\frac{dp_1}{dL}\right|_{p_1,r_1,r_2} = \frac{\partial C}{\partial p_2}\left.\frac{dp_2}{dL}\right|_{p_2,r_1,r_2} = \ldots = \frac{\partial C}{\partial p_n}\left.\frac{dp_n}{dL}\right|_{p_n,r_1,r_2} \quad \text{case } \mathbf{2} \tag{24}$$

$$\left.\frac{\partial C}{\partial p_1}\frac{dp_1}{dL}\right|_{p_1,N_1,N_2} = \left.\frac{\partial C}{\partial p_2}\frac{dp_2}{dL}\right|_{p_2,N_1,N_2} = \ldots = \left.\frac{\partial C}{\partial p_n}\frac{dp_n}{dL}\right|_{p_n,N_1,N_2} \quad \text{case 3}$$

The partial derivatives (the cost derivatives) are the same for all three cases and must be estimated. Simulation is use to determine the derivatives

$$\left.\frac{dp_1}{dL}\right|_{p_1,r_1,r_2}, \left.\frac{dp_2}{dL}\right|_{p_2,r_1,r_2}, \ldots, \left.\frac{dp_n}{dL}\right|_{p_n,r_1,r_2}$$

or equivalently,

$$\left.\frac{dL}{dp_1}\right|_{p_1,r_1,r_2}, \left.\frac{dL}{dp_2}\right|_{p_2,r_1,r_2}, \ldots, \left.\frac{dL}{dp_n}\right|_{p_n,r_1,r_2} \quad \text{case 2} \quad (25)$$

$$\left.\frac{dL}{dp_1}\right|_{p_1,N_1,N_2}, \left.\frac{dL}{dp_2}\right|_{p_2,N_1,N_2}, \ldots, \left.\frac{dL}{dp_n}\right|_{p_n,N_1,N_2} \quad \text{case 3}$$

The method currently used to determine these derivatives is fairly primitive: a reference point **p'** in the reduced parameter set is selected using either coordinate system. The values of the dependent parameter pair (r_1,r_2) or (N_1,N_2) at the beam-beam interaction is determined using simulation. The value of the luminosity at the coincident beam-beam limit is then recorded for that point in {**p'**}. The same procedure is used to find the luminosity at a nearby point in {**p'**} which differs from the first only by small change Δp_i in the value of the single parameter p_i. The quantity $\Delta L/\Delta p_i$ for the two points then gives the approximate value of the derivative $dL/dp_i|_{p_1,r_1,r_2}$.

Additional Constraints

The analysis presented in this paper introduced two constraints. In the most common situation, case **2**, these constraints determined the natural sizes of the two beams from the many other parameters of the system; the natural sizes are always such that both beams are at their respective beam-beam limits. These constraints were justified by the demonstration that the optimal parameter base always satisfies them.

It has been suggested that certain additional constraints be adopted that allow some types of asymmetry but not others. Examples are the "energy transparency" conditions of Chin[6] and the "equal footprint" conditions of Siemann[7]. These additional constraints include keeping the tunes shifts, beam sizes, and damping decrements equal for the two beams (Chin), and keeping the bunch lengths and beta functions equal (Siemann). It is easy to see that these constraints are not in general satisfied by the optimal parameter bases associated with the two models described in this work.

Theoretically there is no way that the particle dynamics of one beam can know the tune shift, damping decrement, tune, chromaticity, etc. of the opposing beam. The behavior of a particle in one beam depends only on the other beam via its particle distribution; its behavior does not depend on the distribution of its own beam. Thus, the two beams are entirely uncoupled (with the technical exception related to long bunch lengths). This is demonstrated in the equations (12) and (13) where it is apparent that none of the parameters affecting the size of one beam affects the size of the other beam (the sets $\{p_1\}$ and $\{p_2\}$ do not intersect). Furthermore, the dependences of beam size on lifetime, damping decrement and most other parameters (excepting the tunes) are monotonic. Therefore, if one plots maximum luminosity vs. the damping decrement or tune shift of one beam (for example), there can be no special behavior (in particular, no peak) when the tune shifts or damping decrements of the two beams are equal.

A fixed energy asymmetry (e.g. a 3 x 9 GeV split) results in certain parameter values being more expensive in one ring than in the other (damping decrement, for example). This means that the optimal working point, the point giving the most luminosity for the least money, is very unlikely to satisfy any partial symmetry constraints. The fact that the costs associated with the many beam-beam parameters are different for the two rings is enough, by itself, to preclude the validity of adding additional constraints.

Finally, the optimal parameter configurations found with simulation almost never satisfy either energy transparency or equal foot print constraints. Examples are given in the next section.

5. Simulation Examples

The example described in this section is taken from another work[4] to demonstrate the behavior of the "realistic model", its characteristic behavior, and the various principles presented above.

In this example, the dependence of round-beam luminosity on the damping decrements of the two rings is determined by simulation. The position of the critical point, as well as the shape and position of the lifetime vertex, is calculated for several pairs of damping time values (T_1, T_2). The values of all parameters other than the damping decrements and beam sizes (the tunes, energies, beta functions, energy spreads, synchrotron frequencies and chromaticities) are held fixed. The luminosities at these balance points are recorded.

160 Optimization of the B-Factory Parameter Base

The simulation includes oscillation, quantum fluctuations, and damping in all three degrees of freedom. The vertical and horizontal damping times are the same for both beams and the longitudinal damping time is half the transverse damping time. When the damping times are varied, the quantum excitations are also varied to keep the desired natural beam sizes. There is a fair amount of synchro-betatron coupling from chromaticity; no other "errors" are included. Although the energies and beta functions are different for the two beams, these asymmetries exactly cancel. Thus, the only asymmetry here is that due to different (for some of the cases) damping times.

Parameters:

Beam radii	$r_x = r_y$ (varied)	(both rings)
Tunes	$v_x = v_y = .08$	(both rings)
Chromaticities	$C_x = C_y = 10$ units	(both rings)
Dispersions	$\eta_x = \eta_y = 0$	(both rings)
Particles/Bunch	$N = 1.7\ e11$	(both rings)
Energy spread	$\Delta p/p = .001$	(both rings)
Synchrotron Period	$P_s = 25$ turns	(both rings)
Energies	$E_1 = 9$ GeV $E_2 = 3$ GeV	
Beta functions	$\beta_{x1} = \beta_{y1} = 6$ cm $\beta_{x2} = \beta_{y2} = 2$ cm	
Apertures	$A_1 = A_2 = .05$ cm at IP	(both rings)

No lattice nonlinearities or long range forces
One IP and one bunch per beam

The damping times of both beams are varied between 1200 turns and 20000 turns. For each pair of damping times (T_1, T_2) a number of runs were performed with different natural radii to determine the boundaries of the various regions shown in figure 7. An efficient method using weak-strong simulation was used to establish the rough boundaries. Strong-strong simulation was then used to verify and position more precisely the lifetime vertices. It should be stressed that either strong-strong or weak-strong simulation (or both) can be used to do this. The result does not depend on the method used as long as superparticle fluctuations are adequately suppressed in the strong-strong runs.

The critical points for three cases,

(T_1, T_2) = (1200,1200)
= (1200,20000)
= (20000,20000)

and the corresponding weak-beam size functions are shown in figure 8a. The lifetime vertices for the same cases are shown in figure 8b. It is seen in figure 8a that these round beams hit the flip-flop instability before they can blow-up by more than about 15%. However, it is seen from figure 8b that they also hit the lifetime limit before they can reach flip-flop instability. For example, when both beams have damping times of 1200 turns, the tune shift at the flip-flop limit is about $\xi=.091$ and about $\xi=.084$ at the lifetime vertex. For equal damping times of 20000 turns, these are $\xi=.071$ and $\xi=.068$ respectively. At both limits, the luminosity falls by 25% to 30% as the damping time increases from 1200 to 20000 turns. From many simulation studies with apertures at 10 sigma, it appears that this is a typical situation for round beams: the beam-beam limit is determined by the lifetime and there is very little blow-up at the beam-beam limit.

One also notices from figures 8a and 8b that the strongly asymmetric case $(T_1, T_2) =$ (1200,20000) is better than the symmetric case (20000,20000) but worse than the symmetric case (1200, 1200). The mere fact that the damping times are different does not affect the luminosity as long as the beam sizes are made slightly asymmetric (by about 15% in this case). This is shown in detail in figure 9 where the luminosity at the coincident beam-beam limit (lifetime limit) is plotted against the damping times of the two beams for fifteen pairs of damping times. This plot shows that the luminosity at (5000,5000) is nearly equal to (in fact, slightly less than) the luminosity at (20000,1200) and (10000,2500). The increase in the damping time of one beam is compensated by the decrease in the damping time of the other beam.

Because

$$\frac{\partial L}{\partial T_1} < 0, \quad \frac{\partial L}{\partial T_2} < 0 \tag{26}$$

for all T_1, T_2, and since there is no evidence that either of the derivatives (26) behave unusually when $T_1 = T_2$, there appears to be no advantage to having both damping times equal. It is clear that the luminosity obtained when $(T_1, T_2) = (5000,5000)$ can be maintained by simultaneously raising the damping time in beam #2 and lowering it

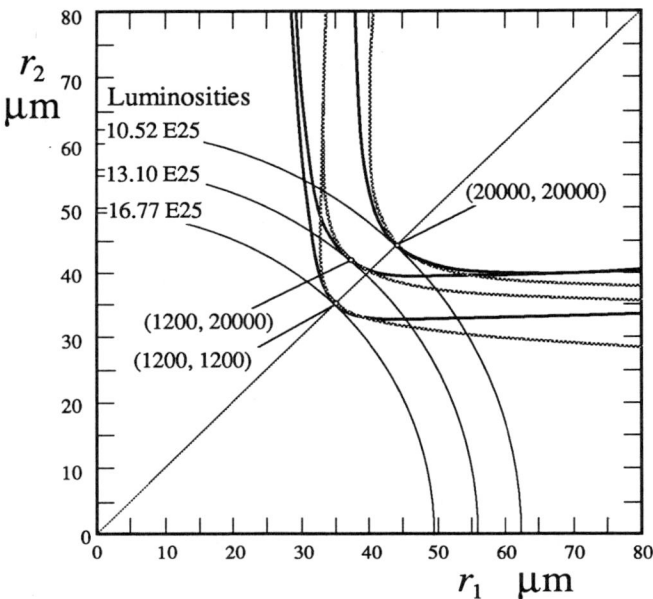

Figure 8a. Critcal points for three damping time pairs. The black and grey lines represent the weak-beam size functions for beams #2 and #1 respectively. The numbers in parentheses are the damping times of beams #2 and #1 respectively. For the two symmetric cases, the increase in damping time from 1200 to 20000 turns loses almost 40% of the luminosity. These are the flip-flop limits. Note that there is very little blow up at the critical point.

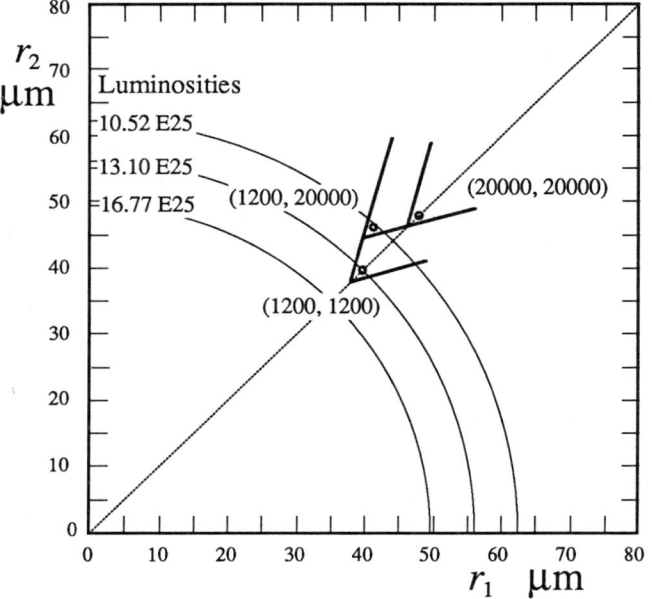

Figure 8b. Lifetime vertices for the cases shown in figure 8a. The pattern is very similar to the critical point pattern, but the lifetime limits are significantly below the flip-flop limits. Thus, in this case, the beam beam limit is determined by particle loss. Blowup at the beam-beam limit is negligible. The lifetime limits here correspond to threshold lifetimes of one minute and apertures at 500 μm.

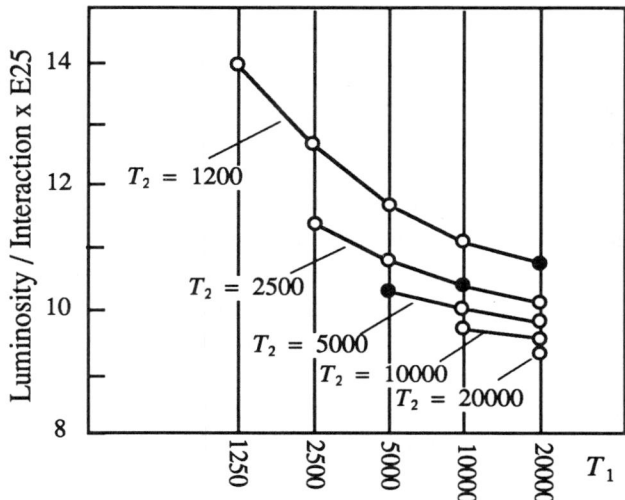

Figure 9.
The luminosity at the lifetime limit is plotted for 15 pairs of damping times. The parameters used in the simulation are those listed in section five. The three black dots represent three cases where the products of the damping times are the same, but their ratios are 1, 4, and 16. There is a slight increase in the luminosity as the asymmetry increases.

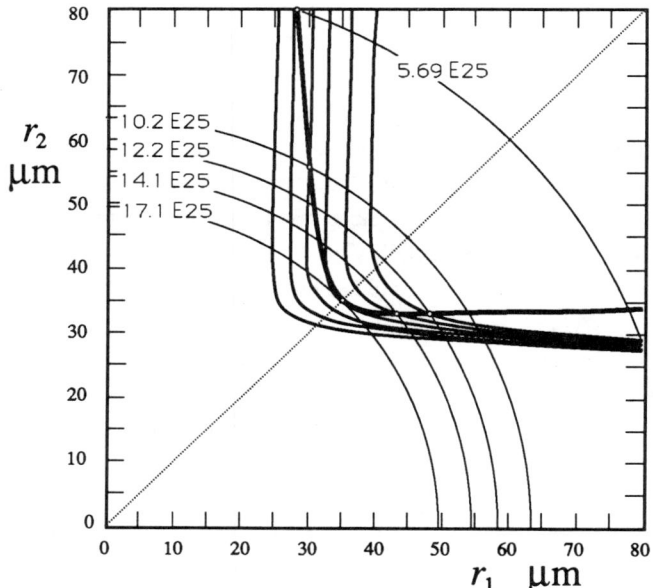

Figure 10a.
This plot is used to determine the sensitivity of beam balance close to the critical point. Both beams have damping times of 1200 turns. The natural size of beam #1 is varied between about 25 and 40 microns in six descrete jumps. The positions of the stable equilibria and their luminosities are shown. The luminosity is plotted explicitly against the size of beam #1 in figure 10b.

appropriately in beam #1. The simulation results thus agree with the theoretical treatment in the previous section; the luminosity of the system always increases if the damping in one ring is strengthened (regardless of the damping in the other ring).

It also appears from these figures that there is no significant difference between the balance point sensitivity of symmetric configurations and asymmetric configurations, or between strongly damped and weakly damped configurations.

Figures 10a and 10b illustrate balance point sensitivity at the flip-flop limit. The damping time of both beams is 1200 turns. The natural radius of beam #2 is fixed at 35 microns while the natural radius of beam #1 is varied in six descrete jumps from 25 to 40 microns. One weak-beam size function for beam #1, and six weak-beam size functions for beam #1 are plotted in figure 10a. The radius chosen for beam #1 is the optimal value, i.e. the value that it has at the coincident beam-beam limit. The locations of the self-consistent equilibria are shown. The high sensitivity of the equilibrium to the natural size of beam #2 is due to the fact that the weak-beam size functions are nearly symmetrical with respect to reflection across the diagonal. The sensitivity is shown more explicitly in figure 10b where the luminosity is plotted against the natural size of beam #1.

6. Summary and Comments

The beam-beam limit usually corresponds to tune shifts between $\xi = .02$ and $\xi = .07$. The precise value depends on several parameters including the transverse tunes, synchrotron tunes, chromaticities, dispersions at the IP, damping times, aperture sizes, and various "errors". There are three separate phenomena that limit the maximum attainable tune shift: beam blow up, flip-flop instability, and particle loss to the aperture. The actual beam-beam limit is determined either by particle loss or flip-flop.

If the two rings have different characteristics, the tune shift limits may be different in the two rings. This could be due to different damping times, different aperture sizes, etc. Maximum luminosity will typically correspond to unequal currents and unequal beam sizes.

The principle problems associated with energy asymmetry result from the fact that the two beams are in separate rings, not from the fact that the parameters in the two rings may be different. The separate rings create the necessity to explicitly balance the two beams against each other. The balance can be very sensitive to small changes in the machine parameters. This sensitivity increases as one approaches the coincident beam-beam limit. Because one can always reduce this sensitivity by backing away from the

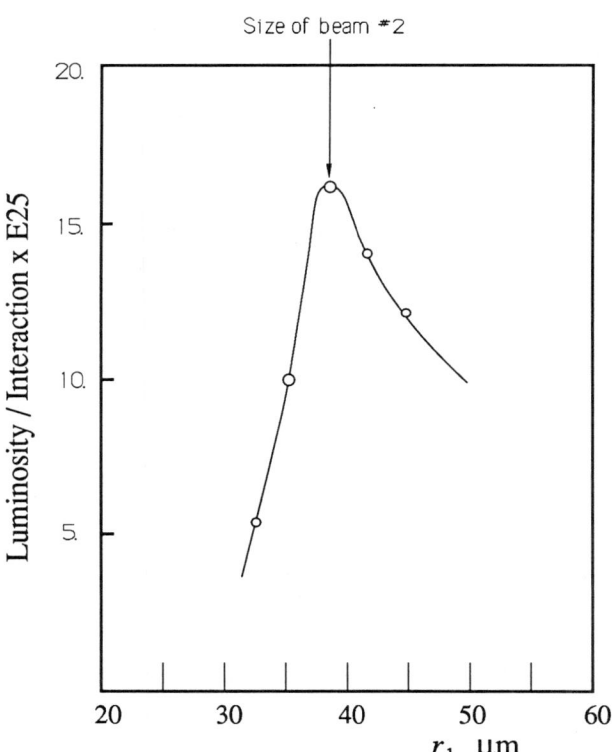

Figure 10b.
The luminosity determined in figure 10a is plotted explicitly against the size of beam #1. There is a fairly narrow peak at the point where the two beams are of equal size. This figure illustrates balance point sensitivity close to the critical point. If the damping times of the two beams had been different, the peak would have been displaced from the size of beam #2.

beam-beam limit, the sensitivity of the balance point is not a fundamental obstacle to the operation of two-ring accelerators.

An optimal parameter base will typically correspond to one of three cases: **1** the beam-beam effect cannot be reached and does not limit the luminosity, **2** the beams run at the maximum current allowed by cost and the beam sizes are adjusted to put both beams at the beam-beam limit, **3** the beam is run with the maximum radius allowed by cost and the beam currents are adjusted to put both beams at the beam-beam limit. The most common of these is case **2**.

The optimal parameter base in case **2** is found by neglecting the cost of the beam radii (the emittances) and by optimizing on the reduced parameter space **P'**. Optimization is performed by estimating the costs $\partial C / \partial p_i$ of the different parameters in $\{p\}$, numerically determining the derivatives $dL / dp_i |_{p_1, r_1, r_2}$, and then finding a point **p** at which

$$\frac{\partial C}{\partial p_1} \frac{dp_1}{dL}\bigg|_{p_1, r_1, r_2} = \frac{\partial C}{\partial p_2} \frac{dp_2}{dL}\bigg|_{p_2, r_1, r_2} = \ldots = \frac{\partial C}{\partial p_n} \frac{dp_n}{dL}\bigg|_{p_n, r_1, r_2}$$

for all p_i. This condition says that the cost of increasing the luminosity by changing any parameter is the same as the cost of increasing it by changing any other parameter; one must be pushing equally hard on all parameters. The cost of beam radii is insignificant, by definition of case **2**, so these are simply adjusted to keep the two beams at the beam-beam limit. It is always better to increase the sizes of the two beams rather than decrease the currents when the luminosity is beam-beam limited.

As long as the machine in running in the case **2** regime, i.e. as long as the size of the physical aperture is not constraining the beam size, then all asymmetries can in principle be balanced with four beam-size knobs. If the beams are round and operate on the coupling resonance, only two knobs are necessary.

The utility of these knobs is independent of whether the machine is designed to be "energy transparent" or not. Even an energy transparent machine will have to be balanced. Fine tuning the balance can be done with the currents, but at a loss of luminosity and convenience.

Asymmetric configurations per se are not detrimental. For example, in the case investigated in section 5, the luminosity of the machine is almost independent of the ratio of the damping times ratio provided the product of the damping times remains constant. For example, the luminosity of a $(T_1, T_2) = (5000, 5000)$ split is approximately the same as that of a $(T_1, T_2) = (20000, 1200)$ split.

There is no evidence, either theoretical or numerical, that "energy transparency" is necessary or even desirable for an energy asymmetric collider. Because it places unjustified constraints on the parameter base, energy transparency is generally not optimal. In almost all cases, asymmetric configurations with coincident beam-beam limits give higher luminosities at a fixed cost, or lower costs at a fixed luminosity. Furthermore, energy transparent configurations are no more "stable" than "energy opaque" configurations. The balance point sensitivities are comparable for the two cases.

Simulations can be used to estimate the derivatives of the luminosity with respect to the many different beam-beam parameters. The derivatives obtained from simulation, unlike the luminosity itself, should be quite reliable. These derivatives are necessary for cost-setting the parameters in the reduced parameter base.

This analysis has been performed for round beams operating on the coupling resonance. It is not clear how relevant it is to flat beams. A separate study must be made. The flat beam analysis will be more complex due to the additional degree of freedom.

A careful beam-beam analysis is highly desirable for the design of a B-factory, but it is not essential. Since the limiting tuneshift values are between .03 and .06 in practically every colliding beam accelerator built to date, it would probably be safe to simply set them equal to the constant value .03, use the simple model, and not worry about an in-depth analysis. The value of an in-depth analysis is two-fold: it mitigates the concern that asymmetric machines might for some reason have significantly lower tune shift limits than symmetric machines and it provides the possibility of increasing the luminosity by a factor of 2 to 4 by choosing the machine parameters wisely. Thus a superficial, incorrect, or nonexistent analysis would cost at most a factor of four in luminosity and most likely no more than a factor of two.

Experiments with symmetric colliders are strongly advised. The beam-beam interaction is not fundamentally different for asymmetric colliders, and these measurements can be very useful. In particular, they can be used to calibrate or verify the simulation results.

Acknowledgements

I would like thank A.M. Sessler and F. Porter for helpful discussions and encouragement. The bulk of this work was supported by the Caltech Physics Department, with SLAC contributing to the preparation of the final draft. The simulations used in this study were performed on a CRAY Y-MP8/864 at the San Diego Super Computer Center. Machine time was provided by the San Diego Super Computer Center with funds from the National Science Foundation. The simulation code was developed in the Exploratory Studies Group at the Lawrence Berkeley Laboratory.

References

1. **Feasibility Study for an Asymmetric B Factory Based on PEP**, LBL PUB-5244, SLAC-352, CALT-68-1589 (October 1989) p.2-8.
2. S. Myers, **Nonlinear Dynamics Aspects of Particle Accelerators**, eds. J.M. Jowett, M. Month, S. Turner, (Springer Verlag 1985), p. 176.
3. J.L. Tennyson, "Flip-Flop Modes in Symmetric and Asymmetric Colliding Beam Storage Rings", LBL preprint 28013 (1989).
4. J.L. Tennyson, "Luminosity dependence in Round Beam Lepton Colliders: a Simulation Study", in preparation.
5. Private communication; Martin Donald & Ewan Paterson.
6. Y-H. Chin, "Symmetrization of the Beam-Beam Interaction in an Asymmetric Collider", this proceedings.
7. S. Krishnagopal and R. Siemann, "The Beam-Beam Interaction in Asymmetric e^+e^- Storage Rings", Cornell preprint.

Appendix A: Proof of the Optimization Theorem for the Simple Model

We consider a six dimensional space $\{\mathbf{p}\}$ with coordinates N_i, r_i, β_i ($i=1,2$). There are four functions on this space, $L(\mathbf{p})$, $C(\mathbf{p})$, $\xi_1(\mathbf{p})$, and $\xi_2(\mathbf{p})$; these are defined for the simple model in section three of the text.

As long as $\xi_{\bar{j}} < \xi_{lim}$, the derivatives of the luminosity (2) with respect to N_i and r_i are always positive and negative, respectively.

$$\left.\frac{\partial L}{\partial N_j}\right|_{N_i, r_i, \beta_i} > 0 \qquad \left.\frac{\partial L}{\partial r_j}\right|_{N_i, r_j, \beta_i} < 0 \qquad \left.\frac{\partial L}{\partial \beta_{\bar{j}}}\right|_{N_i, r_i, \beta_j} = 0 \qquad (i=1,2) \quad (A1)$$

Since ξ_i depends only on $N_{\bar{i}}$, $r_{\bar{i}}$ and β_i, the space $\{\mathbf{p}\}$ separates into two subspaces $\{\mathbf{p}_1\} = \{N_2, r_2, \beta_1\}$ and $\{\mathbf{p}_2\} = \{N_1, r_1, \beta_2\}$. The two subspaces can be optimized independently. The parameter base $\mathbf{p}_i$ that gives the most luminosity at finite cost must then satisfy either

Case A: $N_{\bar{i}} = N_{M\bar{i}}$ and $r_{\bar{i}} = r_{m\bar{i}}$ when $\xi_i < \xi_{lim}$

or

Case B: $\xi_i = \xi_{lim}$

Case A is just case 1 of the optimization theorem. The value of β_i does not affect the luminosity as long as it is small enough that $\xi_i < \xi_{lim}$.

In case B, the condition $\xi_i = \xi_{lim}$ is not sufficient to determine the optimal point $\mathbf{p}_i$. Two additional conditions must be found. The derivatives of L with respect to β_i are always negative

$$\left.\frac{\partial L}{\partial \beta_i}\right|_{\xi_i, r_i} < 0 \qquad \left.\frac{\partial L}{\partial \beta_i}\right|_{\xi_i, N_i} < 0 \qquad (A2)$$

Thus, in case B, the condition $\beta_i = \beta_{mi}$ is always satisfied at the optimal point. The third condition in case B is contingent on the maximum values of the radius $r_{M\bar{i}}$ and current $N_{M\bar{i}}$. Fixing $\xi_i = \xi_{lim}$ and $\beta_i = \beta_{mi}$ we find that

$$\left.\frac{\partial L}{\partial N_{\bar{i}}}\right|_{\xi_i,\beta_i} > 0 \qquad (A3)$$

or similarly

$$\left.\frac{\partial L}{\partial r_{\bar{i}}}\right|_{\xi_i,\beta_i} > 0 \qquad (A4)$$

This means that $N_{\bar{i}}$ and $r_{\bar{i}}$ should be maximized subject to the additional constraints $r_{\bar{i}} < r_{M\bar{i}}$ and $N_{\bar{i}} < N_{M\bar{i}}$. One of these is more constraining than the other.

If the latter is more constraining than the former, then the maximum value of $N_{\bar{i}}$ is determined by its own cost limit. This corresponds to case 2 of the optimization theorem with the third condition $N_{\bar{i}} = N_{M\bar{i}}$.

If the former is more constraining than the latter, then case 3 is realized and the third condition is $r_{\bar{i}} = r_{M\bar{i}}$.

Appendix B: Optimization Theorem in the Realistic Model

The purpose of this appendix is to determine whether the luminosity of the realistic model increases or decreases with beam size when the beam is constrained to the beam-beam limit. The fact that the beam-beam limit depends on the aperture means that the treatment of Appendix A needs to be modified.

The luminosity is assumed to be given by

$$L = \frac{N_1 N_2}{2\pi \left(r_1^2 + r_2^2\right)} \qquad \text{both } \xi_1 < \xi_{lim} \text{ and } \xi_2 < \xi_{lim} \tag{B1}$$

$$L = 0 \qquad \text{either } \xi_1 > \xi_{lim} \text{ or } \xi_2 > \xi_{lim}$$

The tune shift of beam #1 is given by

$$\xi_1 = \frac{r_e N_2 \beta_1}{4\pi \gamma_1 r_2^2} \tag{B2}$$

The tune shift at the beam-beam limit ξ_{lim1} is now considered to depend on the beam size r_1 in such a way that, at the beam-beam limit,

$$A r_1 + B = r_2 \tag{B3}$$

where A and B are functions of the reduced parameter set $\{\mathbf{p'}\}$. This is equivalent to assuming that the particle-loss beam-beam limit curve for beam #1 in (r_1, r_2) space, figure 6, is a straight line. It is assumed that the cost of aperture is zero below a certain width and infinite above that width. Furthermore, it is assumed that for a fixed value of r_1, the limiting tune shift does not depend on N_2, γ_1, or β_1. Squaring (B3), inverting, and multiplying by a factor K gives

$$\frac{K}{(A r_1 + B)^2} = \frac{K}{r_2^2} \tag{B4}$$

If

$$K = \frac{r_e N_2 \beta_1}{4\pi \gamma_1} \tag{B5}$$

then the right side of (B4) is the limiting tune shift for beam #1. To ensure the desired lack of dependence on N_2, γ_1, and β_1

$$A = C_{a1}\sqrt{K} = C_{1a}\sqrt{\frac{r_e N_2 \beta_1}{4\pi\gamma_1}} \tag{B6}$$

$$B = C_{b1}\sqrt{\frac{r_e N_2 \beta_1}{4\pi\gamma_1}} \tag{B7}$$

where C_{a1}, C_{b1} may depend on parameters other than $N_1, N_2, \gamma_1, \gamma_2, \beta_1, \beta_2$, or r_1, r_2. In particular, one would expect them to depend on the aperture of beam #1. The limiting tune shift for beam #1 is then given by (B4), (B6) and (B7)

$$\xi_{lim\ 1} = \frac{1}{(C_{a1}r_1 + C_{b1})^2} \tag{B8}$$

with

$$C_{a1} = \frac{A}{\sqrt{\frac{r_e N_2 \beta_1}{4\pi\gamma_1}}} \tag{B9}$$

$$C_{b1} = \frac{B}{\sqrt{\frac{r_e N_2 \beta_1}{4\pi\gamma_1}}} \tag{B10}$$

where A and B are measured in simulation. The constraint that beam #1 be at the beam-beam limit is expressed by

$$\xi_1 = \xi_{lim\ 1}$$

or from (B2) and (B8)

$$\frac{r_e N_2 \beta_1}{4\pi\gamma_1 r_2^2} = \xi_{lim\ 1} = \frac{1}{(C_{a1}r_1 + C_{b1})^2} \tag{B11}$$

If (B11) is solved for N_2 and substituted into (B1), and an the equivalent substitution is made for N_1, then the luminosity becomes

$$L = \frac{\xi_{lim\,1}\,\xi_{lim\,2}}{r_e^2\,\beta_1\beta_2\left(\frac{1}{r_1^2}+\frac{1}{r_2^2}\right)} \frac{8\pi\gamma_1\gamma_2}{}$$

or, using (B11)

$$L = \frac{1}{(C_{a1}r_1+C_{b1})^2}\frac{1}{(C_{a2}r_2+C_{b2})^2}\frac{8\pi\gamma_1\gamma_2}{r_e^2\,\beta_1\beta_2\left(\frac{1}{r_1^2}+\frac{1}{r_2^2}\right)}$$

Rearranging gives

$$L = \frac{8\pi\gamma_1\gamma_2}{r_e^2\,\beta_1\beta_2}\frac{1}{(C_{a1}r_1+C_{b1})^2(C_{a2}r_2+C_{b2})^2\left(\frac{1}{r_1^2}+\frac{1}{r_2^2}\right)} \tag{B12}$$

The derivative with respect to r_1 is

$$\frac{\partial L}{\partial r_1} = \frac{8\pi\gamma_1\gamma_2}{r_e^2\,\beta_1\beta_2}\left[\frac{-2C_{1a}}{(C_{a1}r_1+C_{b1})^3(C_{a2}r_2+C_{b2})^2\left(\frac{1}{r_1^2}+\frac{1}{r_2^2}\right)} +\right.$$

$$\left. + \frac{\frac{2}{r_1^3}}{(C_{a1}r_1+C_{b1})^2(C_{a2}r_2+C_{b2})^2\left(\frac{1}{r_1^2}+\frac{1}{r_2^2}\right)^2}\right] \tag{B13}$$

Every quantity in this expression is positive. The first term in the brackets represents the correction due to the finite aperture; it is negative. The second term is the usual derivative (from the simple model) and is positive. The strengths of the two terms must be estimated to determine the sign of the sum. For purposes of estimation, it is assumed that {**p'**} is such that the particle loss limits are identical for the two beams

$$C_a \equiv C_{a1} = C_{a2}$$
$$C_b \equiv C_{b1} = C_{b2}$$

and that the vertex is on the diagonal in the (r_1, r_2) plane, i.e. that

$$r_1 = r_2 = \sigma$$

at the vertex. This implies, from (B3), that

$$A\sigma + B = \sigma \tag{B14}$$

The tune shift at the vertex is defined to be

$$\xi_o \equiv \frac{r_e N_2 \beta_1}{4\pi\gamma_1\sigma^2} = \frac{r_e N_1 \beta_2}{4\pi\gamma_2\sigma^2}$$

So that (B9), (B10) reduce to

$$C_a = \frac{A}{\sqrt{\xi_o\sigma^2}} \qquad C_b = \frac{B}{\sqrt{\xi_o\sigma^2}} \tag{B15}$$

Using (B14) and (B15), the derivative (B13) at the vertex $r_1 = r_2 = \sigma$ collapses to

$$\frac{\partial L}{\partial r_1} = \frac{8\pi\gamma_1\gamma_2\xi_o^2\sigma}{r_e^2\,\beta_1\beta_2} [.5 - A] \tag{B16}$$

Thus, as long as the slope A of the beam-beam limit line in figure 6 is less than .5, the derivative (B16) is positive and the optimization theorem holds (at least to the extent that the assumptions made here are representative of the actual situation). The slope A is never greater than .5 (since in this case, there could be no vertex). Therefore, the luminosity always increases with beam size in the realistic model when beams are constrained to the beam-beam limit and the cost of aperture can represented as an infinite barrier.

The Beam-Beam Interaction: Coherent Effects

Kohji HIRATA
KEK, National Laboratory for High Energy Physics
Oho, Tsukuba, Ibaraki 305, Japan

ABSTRACT

Coherent phenomena and recent theoretical approaches to them are reviewed for the symmetric and asymmetric e^+e^- colliding storage rings.

1. Introduction
2. The Dipole Mode: Rigid Gaussian Model
   2.1 Linear Analysis: Identical Circumference
   2.2 Nonlinear Analysis: Spontaneous Separation
   2.3 Beam-Beam Mode as a Diagnostic Tool
3. Quadrupole Modes: Soft Gaussian Model
   3.1 Gaussian Approximation
   3.2 The Model
   3.3 Cusp Catastrophe and Flip-Flop Hysteresis
   3.4 Other Issues
4. Higher Order Moments: Beyond Gaussian Approximation
   4.1 Perturbative Approach
   4.2 Nonperturbative Approach
5. Discussion
Appendix A Linear Analysis: Different Circumferences
Appendix B Comparison between Round and Flat Beams
Appendix C Different Damping Rates

1. Introduction

We summarize the "coherent approach" to the beam-beam effects[1,2,3] in the e^+e^- colliding storage rings. The simulations and experiments will also be discussed as long as the coherent approach is concerned. With the beam-beam interaction, some coherent effects appear, such as the coherent dipole mode and the change of the beam sizes. A single particle dynamics tells us very little on these effects. We should consider a distribution function of the $e^\pm$ bunch

$$\psi_\pm(\mathbf{x}),$$

(and the number of particles $N_\pm$ contained in it). Here $\mathbf{x}$ is the vector of the phase space coordinates (2, 4 or 6-dimensional) and ψ is normalized to unity:

$$\int d\mathbf{x}\psi(\mathbf{x}) = 1.$$

In the coherent approach, we try to understand the behaviour of ψ in terms of some lower order coherent modes (moments). The most important coherent modes are the dipole (first order) moment

$$\bar{x}_i = <x_i>,$$

and the quadrupole (second order) moment

$$M_{ij} = <(x_i - \bar{x}_i)(x_j - \bar{x}_j)>.$$

Here $<>$ is the average over ψ:

$$<A> = \int d\mathbf{x} A\psi(\mathbf{x}),$$

and i runs from 1 to $2N_D$, ($N_D \equiv 1, 2,$ or 3). For example, M_{11} is the transeverse beam size squared.

When the beams are Gaussian, as in the case without the beam-beam effects, $\bar{x}_i$ and M_{ij} are sufficient to characterize ψ:

$$\psi(\mathbf{x}) = G(\mathbf{x}; \bar{\mathbf{x}}, M)$$
$$\equiv \frac{1}{(2\pi)^{N_D} \det M} \exp\left\{-\frac{1}{2}\sum_{i,j=1}^{2N_D} M_{ij}^{-1}(x-\bar{x})_i(x-\bar{x})_j\right\}. \tag{1}$$

Even with the beam-beam effects, the Gaussian form G is still a good first approximation. In this paper, we follow the following line of thoughts:

1. **Rigid Gaussian Model** First, we discuss the coherent dipole oscillation, or the beam-beam mode[4-13]: bunches of e^+ and e^- beams couple to each other and form coherent dipole oscillation modes. To study it, it is almost enough to assume that the beams always remain Gaussian with fixed variances: we consider the $\bar{x}_i$ only. That is, we assume

$$\psi = G(\mathbf{x}; \bar{x}_i, M_{ij}), \qquad (2)$$

with M's being fixed[8-13].

2. **Soft Gaussian Model** The transverse sizes of the beams are enlarged by the beam-beam interaction[2]. This reduces the luminosity. In order to understand how the beam sizes change due to the beam-beam interaction, we use Eq.(2) again but regard M_{ij} as variable[14-18].

3. **Beyond Gaussian** The results of the rigid and soft Gaussian models should be corrected by the deviation of ψ from a Gaussian. To study it, we need higher order moments

$$M_{ij \cdots k} = <(x-\bar{x})_i(x-\bar{x})_j \cdots (x-\bar{x})_k>.$$

As we go higher and higher orders, we can be more and more accurate and become closer and closer to incoherent approach. In the coherent approach, however, we should stop at some order. A possible method will be discussed[19-22].

The situation is summarized as follows:

Dipole	Quadrupole	Higher
Beam-Beam Mode	Beam Size Increase	Nonlinear Resonance
$\sigma - \pi$ splitting	ξ-saturation	Non-Gaussian Tail
Spontaneous Separation	Flip-Flop (Hysteresis)	
Rigid Gaussian Model	Soft Gaussian Model	AOCD
⟵ coherent	$\cdots$	incoherent ⟶
⟵ simple	$\cdots$	complicated ⟶
⟵ better understood	$\cdots$	less understood ⟶

Here, the first block shows the characteristic phenomena and the second the typical theoretical approaches. "AOCD" stands for "All One Can Do", that is, using Vlasov[23-26] or Fokker-Planck equations, Stratonovich expansion[19-22], weak-strong approximation[22,27-30] and all that.

We will discuss these issues. Since lower order moments are easier to study, we go from left to right in the table.

2. The Dipole Mode: Rigid Gaussian Model

When two bunches collide head on, each particle in one of them feels a kick.

$$\delta y' + i\delta x' = -\frac{N_* r_e}{\gamma} f(x - \bar{x}_*, y - \bar{y}_*; \psi_*), \tag{3}$$

where x (x') and y (y') is the horizontal and vertical coordinates (slopes), N_* is the number of particles in the encountering bunch and the complex function f depends on the distribution function of the encountering bunch. The kick Eq.(3) is called the incoherent kick. If the encountering bunch is a Gaussian, the kick is expressed by the Bassetti-Erskine formula[31]:

$$\begin{aligned} f = f_G(x, y; \sigma_x, \sigma_y) = & \sqrt{\frac{2\pi}{\sigma_x^2 - \sigma_y^2}} \\ & \times \left\{ w\left(\frac{x + iy}{\sqrt{2(\sigma_x^2 - \sigma_y^2)}}\right) - \exp\left(-\frac{x^2}{2\sigma_x^2} - \frac{y^2}{2\sigma_y^2}\right) w\left(\frac{\frac{\sigma_y}{\sigma_x}x + i\frac{\sigma_x}{\sigma_y}y}{\sqrt{2(\sigma_x^2 - \sigma_y^2)}}\right) \right\}. \end{aligned} \tag{4}$$

Here w is the complex error function and ψ_* is represented by σ_x^* and σ_y^*.

The kicks to the dipole moments, that is, the coherent kicks, can be obtained by averaging the incoherent kick, Eq.(3), over the distribution function ψ:

$$\delta \bar{y}' + i\delta \bar{x}' = <\delta y' + i\delta x'> = -\int dx \frac{N_* r_e}{\gamma} f_G(x - \bar{x}_*, y - \bar{y}_*; \sigma_x^*, \sigma_y^*) \, \psi(\mathbf{x}).$$

In the rigid Gaussian model, $\psi = G$, it was shown[8] that

$$\delta \bar{y}' + i\delta \bar{x}' = -\frac{N_* r_e}{\gamma} f_G(\bar{x} - \bar{x}_*, \bar{y} - \bar{y}_*; \Sigma_x, \Sigma_y), \tag{5}$$

where Σ's are the effective beam sizes

$$\Sigma_x \equiv \sqrt{(\sigma_x^+)^2 + (\sigma_x^-)^2}, \quad \Sigma_y \equiv \sqrt{(\sigma_y^+)^2 + (\sigma_y^-)^2}. \tag{6}$$

The experiment in SLC[32] shows that Eq.(5) is quite accurate. In Fig.1, the deflection angle $\delta \bar{y}'$ is shown as a function of the off set $\bar{y}_+ - \bar{y}_-$. The solid line is the theoretical results based on Eq.(5).

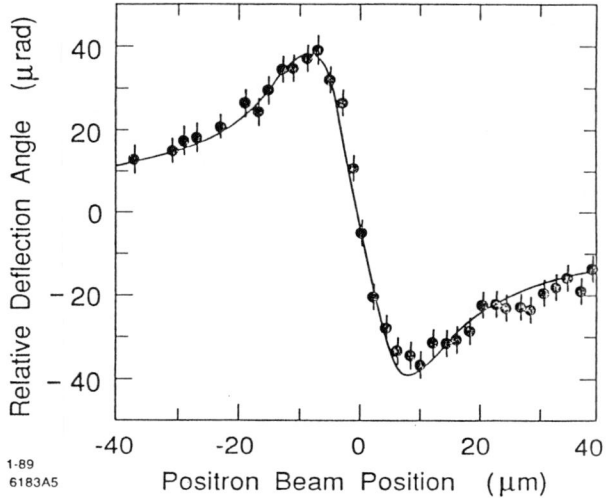

Fig. 1. A direct observation of the coherent dipole kick.

2.1. LINEAR ANALYSIS: IDENTICAL CIRCUMFERENCE

When $x - \bar{x}_*$ and $y - \bar{y}_*$ are small enough compared to Σ's, we can linearlize Eq.(3) obtaining

$$\delta z' = \frac{-2N_* r_e}{\gamma \sigma_z (\sigma_x + \sigma_y)} (z - \bar{z}_*) = -\frac{4\pi \xi_z}{\beta_z} (z - \bar{z}_*), \qquad (7)$$

Here z stands for either x or y and ξ is the (incoherent) beam-beam parameters:

$$\xi_z = \frac{N_* r_e}{\gamma} \frac{\beta_z}{2\pi \sigma_z (\sigma_x + \sigma_y)}.$$

For the coherent kick, according to Eq.(5), we have

$$\delta \bar{z}'_\pm = \frac{-2N_\mp r_e}{\gamma_\pm \Sigma_z (\Sigma_x + \Sigma_y)} (\bar{z} - \bar{z}_*) = -\frac{4\pi \Xi_z^\pm}{\beta_z^\pm} (\bar{z} - \bar{z}_*). \qquad (8)$$

Here Ξ's are the coherent (or effective) beam-beam parameter

$$\Xi_z^\pm = \frac{N_\mp r_e}{\gamma_\pm} \frac{\beta_z^\pm}{2\pi \Sigma_z (\Sigma_x + \Sigma_y)}, \qquad (9)$$

Note that the averaging $< >$ and the linearlization **do not commute**: if we averaged Eq.(7), we would obtain $\Xi_\pm = \xi$. It is wrong: when $\sigma^+ = \sigma^-$, for

example, we have
$$\Xi_\pm = \xi/2.$$

In the linear approximation, the ring complex for asymmetric collider is characterized by

1. The circumferences of both rings $C_\pm$,
2. The coherent beam-beam parameter $\Xi_\pm$. (In the linear approximation, the differences in γ's, N's, β's and σ's can be integrated into Ξ's.)
3. The damping decrement[33] $\delta_\pm$ or its inverse

$$T_\epsilon^\pm = 1/\delta_\pm = \frac{\text{transeverse damping time}}{\text{revolution time}}.$$

4. Tune $\nu_\pm$.

For simplicity, we here consider rings with the same circumferences ($C_+ = C_-$) and share only one interaction point (IP). (The case with $C_+ \neq C_-$ is treated in Appendix A). Since one e^+ bunch collides always with the identical e^- bunch, the number of the bunches in each beam is irrelevant. Thus, we can reduce the problem to the case where there are one e^+ bunch and one e^- bunch colliding at one IP, which is refered to as the case of

$$1 \oplus 1 = 1.$$

In order to use essential variables only, we employ the following set of canonical variables for the dipole modes,

$$Z_\pm = \sqrt{N_\pm \gamma_\pm}\, \frac{z_\pm}{\Sigma_z} \quad \text{and} \quad P_\pm = \sqrt{N_\pm \gamma_\pm}\, \frac{\beta_z^\pm z'_\pm}{\Sigma_z}.$$

Here $\sqrt{N\gamma}$ plays the role of the bunch mass.

The beam-beam kick and the successive map for the betatron oscillation can be represented by a 4×4 matrix M, acting on the 4−vector $(Z_+, P_+, Z_-, P_-)^t$:

$$M = DVR,$$

with

$$D = \exp - \begin{pmatrix} \delta_+ & & & \\ & \delta_+ & & \\ & & \delta_- & \\ & & & \delta_- \end{pmatrix} \qquad (10)$$

$$V = \begin{pmatrix} U(\mu_+) & 0 \\ 0 & U(\mu_-) \end{pmatrix}, \quad U(\alpha) = \begin{pmatrix} \cos\alpha & \sin\alpha \\ -\sin\alpha & \cos\alpha \end{pmatrix} \tag{11}$$

where $\mu = 2\pi\nu$, and

$$R = I - 4\pi \begin{pmatrix} 0 & 0 & 0 & 0 \\ \Xi_+ & 0 & -\sqrt{\Xi_+\Xi_-} & 0 \\ 0 & 0 & 0 & 0 \\ -\sqrt{\Xi_+\Xi_-} & 0 & \Xi_- & 0 \end{pmatrix},$$

with I being the 4×4 unit matrix.

We study the stability of M. For this purpose, we can neglect the damping term D. Its effect can be included easily. Then $M = VR$ is a 4×4 symplectic matrix so that we can calculate its eigenvalues straightforwardly. Let ρ be the eigenvalue. Then we have

$$1/\rho + \rho = \cos\mu_+ + \cos\mu_- - 2\pi\Xi_+ \sin\mu_+ - 2\pi\Xi_- \sin\mu_- \pm \sqrt{Q}, \tag{12}$$

$$Q = (\cos\mu_+ - \cos\mu_- - 2\pi\Xi_+ \sin\mu_+ + 2\pi\Xi_- \sin\mu_-)^2 + 16\pi^2\Xi_+\Xi_- \sin\mu_+ \sin\mu_-.$$

The stability condition is that the r.h.s. of Eq.(12) is real and lies between -2 and 2. When the motion is stable, the l.h.s. can be written as $2\cos\bar{\mu}_{\omega_\pm}$, and $\bar{\nu}_{\omega_\pm} \equiv \bar{\mu}_{\omega_\pm}/2\pi$, is called the eigentune of $\omega_\pm$ mode.

<u>Identical Tunes</u> In the particular case of $\nu_+ = \nu_- \equiv \nu$, Eq.(12) becomes quite simple:

$$\cos\bar{\mu}_{\omega_\pm} = \cos\mu - \pi(\Xi_+ + \Xi_-)(\sin\mu \pm |\sin\mu|). \tag{13}$$

These two modes are often referred to as σ and π modes:

$$\sigma - \text{mode}: \quad \cos\mu_\sigma = \quad\quad\quad \cos\mu$$
$$\pi - \text{mode}: \quad \cos\mu_\pi = \quad \cos\mu - 2\pi(\Xi_+ + \Xi_-)\sin\mu.$$

When we measure the tune by deflecting the beams, as we do usually for $e^\pm$ rings, we see two peaks of the frequency response corresponding to these modes. The σ mode gives the nominal tune and the π mode gives a peak roughly at

$$\nu_\pi \simeq \nu + \Xi_+ + \Xi_-, \quad (\nu_+ = \nu_-, \ 1 \oplus 1 = 1). \tag{14}$$

When $\Sigma_+ = \Sigma_-$ so that $\Xi_+ = \Xi_- = \xi/2$, the coherent tune shift is, thus, just ξ.

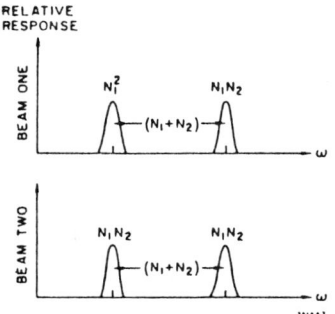

FIG. 2--Coherent response detected versus driving frequency.

The $\sigma - \pi$ splitting was first observed in SPEAR[4]: See Fig.2. Here N_1 and N_2 are N_+ and N_-, respectively. Note that $(N_+ + N_-) \propto (\Xi_+ + \Xi_-)$, so that it is consistent with Eq.(14). From Eq.(13), it can be seen that the instability occurs when and only when

1. [1/2 resonance] $\nu \lesssim 1/2 +$ integer. $(\cos \mu_\pi < -1)$, i.e.

$$1 < 2\pi(\Xi_+ + \Xi_-) \tan \frac{\mu}{2},$$

2. [2/2 resonance] $\nu \lesssim$ integer. $(\cos \mu_\pi > 1)$, i.e.

$$1 < -2\pi(\Xi_+ + \Xi_-) \cot \frac{\mu}{2},$$

In Fig.3, we show the stable region in $(\nu, \Xi_+ + \Xi_-)$ space. Only these two resonances can occur when $\nu_+ = \nu_-$.

When $\nu_+ \neq \nu_-$, the analysis is more complicated.

<u>Different Tunes</u> When $\nu_+ \neq \nu_-$, Q in Eq.(12) is no longer a complete square and can become negative. This leads a new instability, the sum resonance instability. From Eq.(12), the instability occurs when and only when

1. [1/2 resonance] $\nu \lesssim 1/2 +$ integer. The condition is $\cos \mu_\pi < -1$, i.e.

$$1 < 2\pi \left(\Xi_+ \tan \frac{\mu_+}{2} + \Xi_- \tan \frac{\mu_-}{2} \right),$$

2. [2/2 resonance] $\nu \lesssim$ integer. The condition is $\cos \mu_\pi > 1$, i.e.

$$1 < -2\pi \left(\Xi_+ \cot \frac{\mu_+}{2} + \Xi_- \cot \frac{\mu_-}{2} \right),$$

3. [Sum resonance] $\nu_+ + \nu_- \lesssim$ integer. The condition is $Q < 0$.

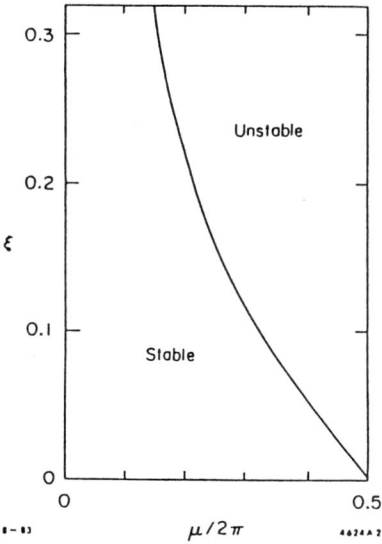

Fig. 3. The stability region in $(\nu, \Xi_+ + \Xi_-)$ space for the case of $\nu_+ = \nu_-$ and $1 \oplus 1 = 1$. The vertical axis is $\Xi_+ + \Xi_-$, rather than ξ. They, however, can be identified when two beams are symmetric. The diagram repeats with period $\mu = \pi$.

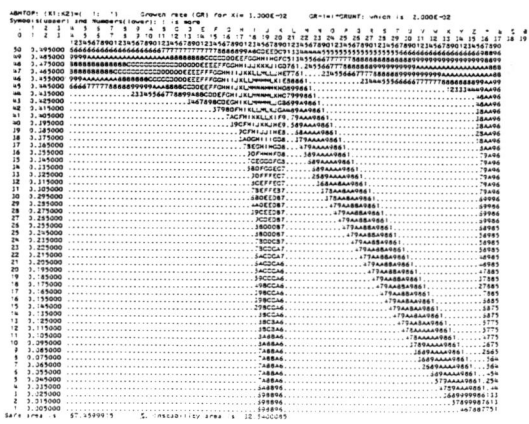

Fig. 4. The unstable region in (ν_+, ν_-) space for the case of $1 \oplus 1 = 1$. The horizontal axis is ν_+ and varies form 0 to 1 and the vertical is ν_- from 0 to 0.5: $(\nu_+, \nu_-) \to (\nu_+ + 1/2, \nu_- + 1/2)$ does not change eigenvalues of M.

The unstable region in (ν_+, ν_-) space is shown in Fig. 4.

2.2. NONLINEAR ANALYSIS: SPONTANEOUS SEPARATION

What happens when a linear instability occurs? The linear consideration in the previous section would predict infinite divergence of the barycentre variables with constant growth rate. Actually, however, once they begin to grow, the nonlinear nature of the force becomes important and the linear model is no longer reliable. Since the nonlinear part of the beam-beam force contains terms which couple horizontal and vertical motion, the full analysis is complicated. Instead, we will study it in terms of a simpler model, called nonlinear rigid Gaussian model[13].

Let us assume that the beams are flat

$$\Sigma_x \gg \Sigma_y,$$

and consider the vertical motion only. That is, we assume the horizontal beam-beam force weak enough. If we assume that $\bar{x}_\pm = 0$ and that $\bar{y} \ll \Sigma_x$, we can replace the first w contained in Eq.(4) by unity. We thus get

$$\delta \bar{y}'_\pm = -\frac{N_\mp r_e}{\gamma_\pm} \frac{\sqrt{2\pi}}{\Sigma_x} \mathrm{erf}\left(\frac{\bar{y}_\pm - \bar{y}_\mp}{\sqrt{2}\Sigma_y}\right).$$

When $\bar{y}_+ - \bar{y}_- \ll \Sigma_y$, we recover Eq.(8) for y. If the separation becomes so large that $\bar{y}_+ - \bar{y}_- \sim \Sigma_x$, the force should begin to decrease. Such a case is not considered here.

Using this model, we can show[13] that the beams tend to separate from each other at the IP and reach an equilibrium (in this case, we need the damping term), when one of the linear instability condition is satisfied. We call it the "Spontaneous Separation" of beams (SSB). There are three kinds of the SSB, corresponding to the three different ways of the linear instabilities. They are summarized in Table 2

SSB Mode	1	2	3
Cases	2/2 resonance	1/2 resonance	Sum resonance
Behaviour	period-one fixed point closed orbit	period-two fixed point double closed orbit	limit cycle no closed orbit
Bifurcation	Pitchfork	Period doubling	Hopf

Table 2.   Three kinds of the spontaneous separation.

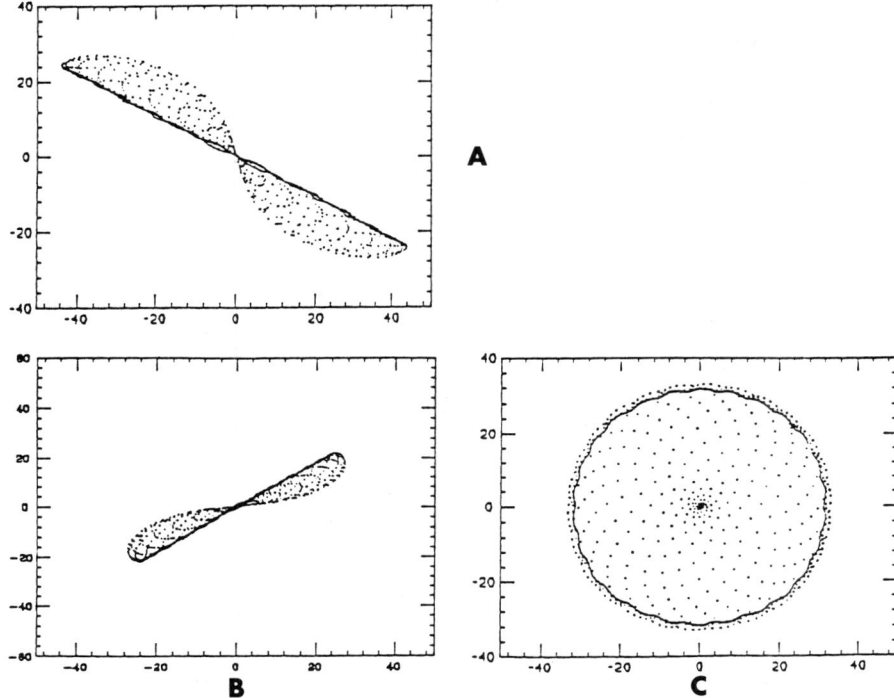

Fig. 5. The SSB of the third kind. The Poincaré surface plot at the IP of (A) $(\hat{Y}_+ - \hat{Y}_-, \hat{P}_+ - \hat{P}_-)$, (B) $(\hat{Y}_+ + \hat{Y}_-, \hat{P}_+ + \hat{P}_-)$ and (C) $(\hat{Y}_+, \hat{P}_+)$. The vector $(\hat{Y}_+, \hat{P}_+, \hat{Y}_-, \hat{P}_-)$ forms a one dimensional circle.

In the first mode of the SSB, each bunch has a different closed orbit at the IP. In the second mode, each bunch has two closed orbits at the IP and passes them alternately. The third kind is more dynamical. It appears that the system is in a one-dimensional limit cycle in the 4-dimensional space $(\hat{Y}_+, \hat{P}_+, \hat{Y}_-, \hat{P}_-)$. See Fig.5. The separation becomes larger for larger Ξ's. Since the luminosity is

$$L = \frac{f_{rev}}{2\pi} \frac{N_+ N_-}{\Sigma_x \Sigma_y} \exp\left[-\frac{(\bar{x}_+ - \bar{x}_-)^2}{2\Sigma_x} - \frac{(\bar{y}_+ - \bar{y}_-)^2}{2\Sigma_y}\right], \quad (15)$$

the SSB reduces L at these resonances.

<u>Multiparticle Tracking</u> The results presented so far depend on the linear- and/or nonlinear-rigid-Gaussian models. These results should be checked by simulations as long as it is possible. In particular, we ignore the possible change of the beam sizes and nonlinear horizontal-vertical coupling. We can see how these are serious by comparing it with a multiparticle tracking.

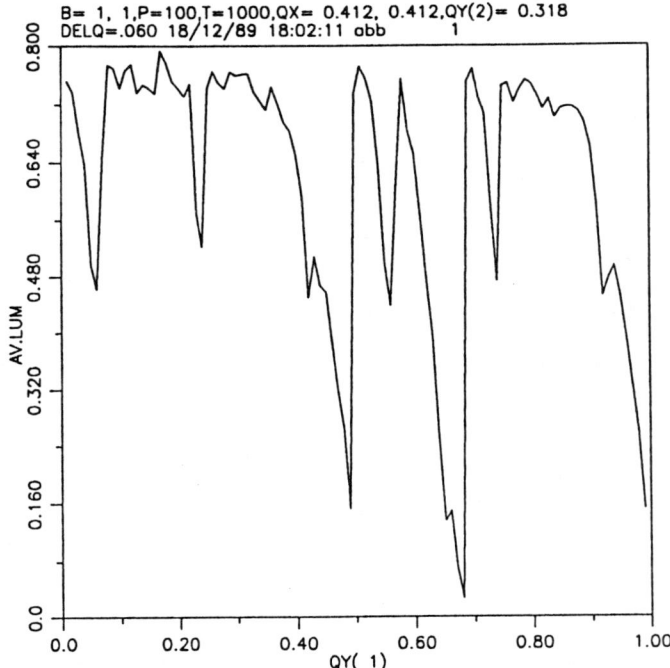

Fig. 6. Results of the multiparticle tracking program AB. The luminosity (normalized by the nominal value) as a function of ν_y^+. Parameters are $\nu_x = (0.412, 0.412)$, $\nu_y^- = 0.318$, $\Xi = (0.03, 0.03)$, $\delta_y = (0.005, 0.005)$.

The computer program AB[13] is an adaptation of the similar multiparticle tracking program BB[34] to asymmetric colliders. In Fig.6, we show its results for the case of $1 \oplus 1 = 1$. The luminosity is estimated as a function of ν_y^+. The reduction of the luminosity comes either from the SSB or the change of the beam size. All the dips in luminosity can be identified with resonances from left to right:

1. Coupling resonance $2\nu_x^+ + 2\nu_y^+ = 1$ below $\nu_y^+ \simeq 0.088$.
2. Nonlinear resonance $4\nu_y^+ = 1$ below $\nu_y^+ \simeq 0.25$.
3. Coupling resonance $2\nu_x^+ - 2\nu_y^+ \simeq 0$ above $\nu_y^+ = 0.412$.
4. Half integer resonance $2\nu_y^+ = 1$ below $\nu_y^+ \simeq 0.5$.
5. Coupling resonance $2\nu_x^+ + 2\nu_y^+ = 2$ below $\nu_y^+ \simeq 0.588$.
6. The sum resonance $\nu_y^+ + \nu_y^- = 1$ below $\nu_y \simeq 0.682$.
7. Nonlinear resonance $4\nu_y^+ = 1$ below $\nu_y^+ \simeq 0.75$.
8. Coupling resonance $2\nu_x^+ - 2\nu_y^+ = 1$ below $\nu_y^+ = 0.912$.
9. Integer resonance below $\nu_y^+ = 1$.

Among all resonances, the sum resonance causes the largest reduction of the luminosity. The qualitative results of the model are confirmed by multiparticle tracking with the simulation program AB. In particular, it is found that:

1. The behaviour listed in Table 2 is valid also in the multiparticle tracking.
2. In the cases with spontaneous separation, the change of the beam sizes are rather minor so that the assumption used in the rigid-Gaussian model is permissible.

2.3. BEAM-BEAM MODE AS A DIAGNOSTIC TOOL

We have discussed the $1 \oplus 1 = 1$ case. In the conventional machines, the case is

$$K \oplus K \leq 2K,$$

that is, we have K e^+ bunches and K e^- bunches and they collide at $2K$ or less IP's. (TRISTAN is $2 \oplus 2 = 4$ and LEP is $4 \oplus 4 = 4$). In $K \oplus K = N_{IP}$ case, some pairs of bunches do not collide with each other. Then, how many beam-beam modes does the system have? In many cases, the problem can be reduced to the case with two IP's in distance of s, (denoted as $(0, s)$, where s is an integer less than $2K$ and the whole ring is counted as $2K$). In this case, the whole e^+ (or e^-) bunches are divided into N_F families. Bunches in different family do not interact through the beam-beam force. It can be shown that[35]

$$N_F = G.C.D.(K, s),$$

where $G.C.D.$ stands for the greatest common divisor. Each family has K/N_F bunches so that the number of the beam-beam modes is $2K/N_F$ and the tune difference between σ mode (unperturbed) and the π mode (most perturbed) is roughly $2K\xi/N_F$.

In the general cases, it is difficult to find the eigenvalues of the revolution matrix M analytically. The BBMODE code[11] gives them for the reasonably general cases. (For example, the phase advance between IP's can be different for different arcs and different beams.)

In these cases, the beam-beam dipole instability can be avoided easily by choosing right tunes. These dipole modes are rather an useful diagnostic tool. (They are not so for the two-ring colliders with different circumferences. See Appendix A). From

$$L = \frac{f_r}{r_e} \frac{N_+\gamma_+ N_-\gamma_-}{N_+\gamma_+ + N_-\gamma_-} \left(\frac{\Xi_x^+ + \Xi_x^-}{\beta_x} + \frac{\Xi_y^+ + \Xi_y^-}{\beta_y} \right), \qquad (16)$$

we can maximize the luminosity by observing and maximizing Ξ's. This was used in PETRA[5] and more quantitatively in TRISTAN[9,36]. The measurement

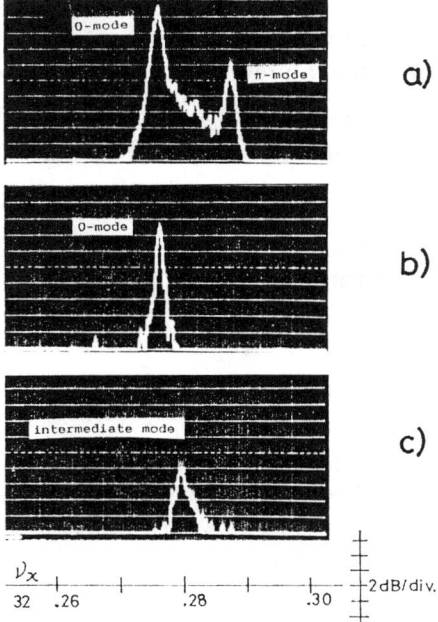

Fig. 7. Beam-Beam modes in TRSISTAN. a) to c) differ from each other in the weights for the signal from each bunch.

is quite quick. The measurement gives directly the emittances of the beams. From Eq.(9), we can extract Σ_x and Σ_y from Ξ_x and Ξ_y. (But see Sect.4.1).

In Fig.7, the frequency response of $2 \oplus 2 = 4$ case is shown[9]. We add the four signals coming from four bunches with appropriate complex weight in order to select one particular mode.

In the tune measurement, when the deflection is large, the system shows a nonlinear response: the shift of the peak, double valuedness of the response and hysteresis[8,10]. This phenomenon can be useful to study the nonlinear nature of the beam-beam force.

The rigid Gaussian model can also give a clear calculation[37] of how the four beam compensation schemes[38] leads a rapid instability.

As long as the beams remain not far from Gaussian, the measurement of the Ξ's can also indicate the possible change of Σ's due to the beam-beam interaction. In Fig.8, Ξ is shown as a function of the current. As is clear from Eq.(9), Ξ increases in proportion to the current. When the current exceeds a certain value, however, Ξ begins to decrease suddenly. This rapid decrease of Ξ_y implies the sudden increase of Σ_y according to Eq.(9). This will be explained by the soft Gaussian model in the next chapter.

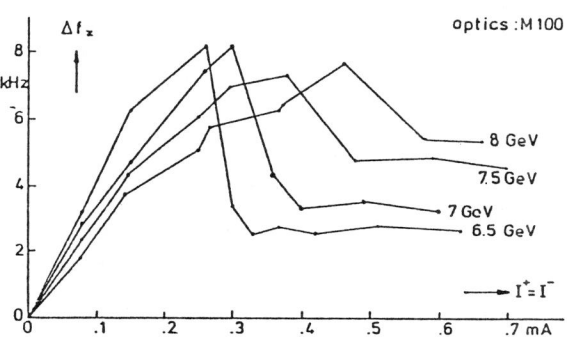

Fig. 8,   The coherent tune shift measurement at PETRA[5] showing the rapid growth of Σ.

3. Quadrupole Modes: Soft Gaussian Model

In the conventional machines, one can easily avoid the dipole instabilities by choosing right tunes. But he should encounter quadrupole moment instability: the beam sizes $\sigma_{x,y}^{\pm}$ increase due to the beam-beam interaction. This reduces the luminosity a lot through the relation Eq.(15).

The beam sizes (more precisely, Σ_x and Σ_y) can be well measured by observing the beam-beam modes as shown in the end of the previous chapter and Fig.8. The sudden decrease of Ξ can be interpreted as a flip-flop effect: one of the beams is blown up and the other is not much affected. In Fig. 9 (a), the flip-flop phenomenon is shown by means of the synchrotron light image. When there is a controllable asymmetry between beams, such as the tiny difference in energies, a hysteresis appears at the flip-flop phenomenon as shown in Fig.9 (b).

By the soft Gaussian model where the beam is assumed to be always Gaussian and $\bar{x}_i$ and M_{ij} are only changeable variables, we will show that these phenomenon are interpreted as the cusp catastrophe[41-44].

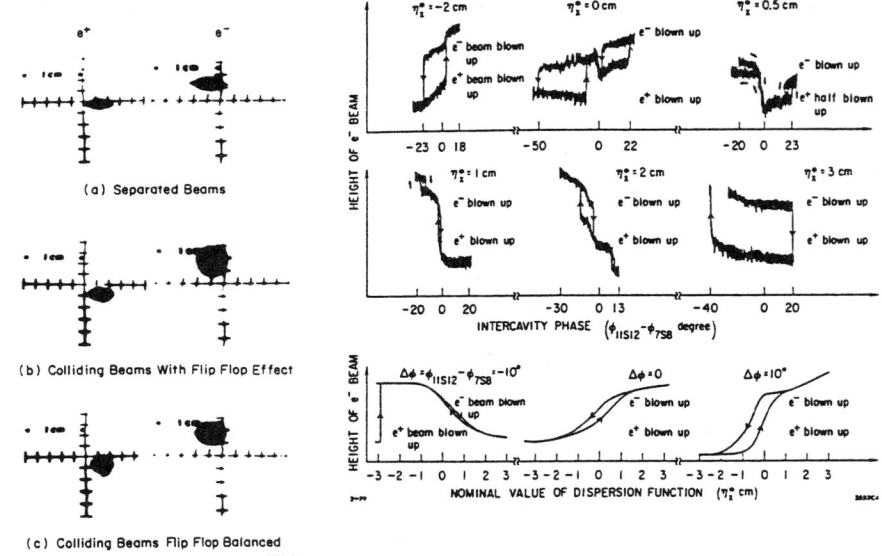

Fig. 9. The beam size observations at SPEAR. a) Flip-Flop effect[39]. b) flip-flop hysteresis[40].

3.1. GAUSSIAN APPROXIMATION

All of the information of the beams is contained in the distribution function in the phase space, $\psi_\pm(x, x')$ for $e^\pm$ bunch. The ψ_+ and ψ_- change independently through the arc. They influence each other by the beam-beam interaction.

We can divide ψ uniquely into Gaussian and non-Gaussian parts:

$$\psi_\pm = \psi_\pm^G + \psi_\pm^N,$$

where ψ^G is the Gaussian approximation of ψ. That is, ψ^G is the $G(\mathbf{x}; \bar{\mathbf{x}}, M)$ in Eq.(1) using $\bar{\mathbf{x}}$ and M_{ij} of ψ as follows:

$$\begin{array}{c} \psi(\mathbf{x}) \\ \downarrow \\ \psi_G(\mathbf{x}) \end{array} \text{Gaussian Approximation} \equiv \left\{ \begin{array}{c} \bar{\mathbf{x}} = \int d\mathbf{x}\, \mathbf{x}\psi(\mathbf{x}) \\ M_{ij} = \int d\mathbf{x}\, x_i x_j \psi(\mathbf{x}) \\ \downarrow \\ \psi_G \equiv G(\mathbf{x}; \bar{\mathbf{x}}, M). \end{array} \right. \quad (17)$$

The incoherent kick for an $e^\pm$ particle depends on $\psi_\mp$ as Eq.(3). We can divide the kick $F_\pm[\psi_\mp]$

$$x' \longrightarrow x' + F(x;\psi_*),$$

also into two parts:

$$F_\pm[\psi_\mp] = F_\pm^G[\psi_\mp^G] + F_\pm^N[\psi_\mp^G, \psi_\mp^N], \qquad (18)$$

where F^G is the kick assuming the Gaussian distribution for the encountering bunch, *i.e.*Eq.(4), whereas F^N is the correction due to the deviation from a Gaussian. The term F^N is quite complicated and difficult to study analytically. Even in the multiparticle tracking[34,45,46], F^N is usually ignored because of calculation time consideration. There, however, is no reasoning of ignoring F^N. An idea of rapid numerical calculation was developed[47] and a trial is going on[48].

Even when we ignore F^N, the effect of F^G alone can produce ψ^N, because the change of $\psi_\pm$ is as follows:

$$\psi_\pm(x,x') \longrightarrow \psi_\pm\left(x, x' - F_\pm^G(x;\psi_\mp)\right), \qquad (19)$$

the latter being a highly nonlinear function even if it was a Gaussian before. We have,

$$\begin{array}{ccccc} \psi^G & \Longrightarrow & F^G & \Longrightarrow & \psi^G \\ & \searrow & \times & & \\ \psi^N & \longrightarrow & F^N & \longrightarrow & \psi^N \end{array} \qquad (20)$$

In the Gaussian model, we follow the line indicated by $\Longrightarrow$ in Eq. (20): we ignore both ψ^N and F^N. That is, from the r.h.s. of Eq. (19), we reconstruct a new ψ^G by means of the Gaussian approximation, Eq.(17).

3.2. THE MODEL

For the sake of simplicity, we assume that

1. the beams are either completely round or extremely flat so that the dynamics is reduced to be one dimensional. (Since the discussion goes parallel, we use the round beam (RB) limit mainly and show the results for the flat beam (FB) limit briefly),
2. the dipole moment $\bar{x}_i$ is stable so that negligible,
3. the motion of a particle through the arc is the linear betatron oscillation perturbed by a constant damping and a Gaussian quantum noise,
4. the nominal (*i.e.*without beam-beam effect) emittances ε, β values at the IP are the same for both beams,

5. but, two beams can have different $\gamma_\pm$ and $N_\pm$, $\delta_\pm$ and $\nu_\pm$.

The beam-beam kick felt by an individual particle in a bunch, now, can be expressed as

$$\delta x' = -\frac{2N_* r_e}{\gamma} \frac{1}{x} \left\{ \exp(-\frac{x^2}{2\sigma_*^2}) - 1 \right\}, \quad \text{(RB)}$$

$$\delta y' = -\frac{N_* r_e}{\gamma} \frac{\sqrt{2\pi}}{\sigma_x^*} < \exp(-\frac{x^2}{2(\sigma_x^*)^2}) >_x \operatorname{erf}(\frac{y}{\sqrt{2}\sigma_y^*}), \quad \text{(FB)}$$
(21)

Here, in the FB case, $< >_x$ means the average over the horizontal distribution, which is assumed to be Gaussian with the nominal beam size.

The controllable parameters are

1. the nominal (or unperturbed) beam-beam parameters η_+ and η_- defined by

$$\eta_\pm = \begin{cases} \dfrac{N_\mp r_e}{4\pi\gamma_\pm\varepsilon}, & \text{(RB)}, \\ \dfrac{N_\mp r_e}{2\pi\gamma_\pm} \dfrac{\beta_y}{\sigma_x^0 \sigma_y^0}, & \text{(FB)}, \end{cases}$$
(22)

where ε is the nominal emittance which is regarded as same to either beams. (We use ξ for the real beam-beam parameter and η for the nominal value of ξ, i.e. without beam-beam phenomenon, in order to avoid a confusion). Correspondingly, we introduce an asymmetry parameter χ and the effective nominal beam-beam parameter H as

$$\chi = \frac{\eta_-}{\eta_+} = \frac{\gamma_+ N_+}{\gamma_- N_-}, \quad H = \sqrt{\eta_+ \eta_-}.$$
(23)

Conversely, we have

$$\eta_+ = \frac{H}{\sqrt{\chi}}, \quad \eta_- = H\sqrt{\chi}.$$
(24)

2. Tunes, $\nu_\pm$.
3. Damping decrements $\delta_\pm$.

The most important observable quantity is the luminosity, Eq.(15). A corresponding observable quantity is the coherent beam-beam parameter $\Xi_\pm$ defined by Eq. (9),

$$\Xi_\pm = \frac{N_\mp r_e}{\gamma_\pm} \begin{cases} \dfrac{\beta_x}{4\pi \Sigma^2}, & \text{(RB)} \\ \dfrac{\beta_y}{2\pi \Sigma_x \Sigma_y}, & \text{(FB)} \end{cases}$$
(25)

In order to describe the motion of an individual particle, it is convenient to use the following canonical set of variables:

$$X = \frac{x}{\sqrt{\beta\varepsilon}}, \quad P = \frac{\alpha x + \beta x'}{\sqrt{\beta\varepsilon}}. \tag{26}$$

$$X_1 \equiv X, \quad X_2 \equiv P.$$

We redefine M_{ij} as

$$M_{ij} = <X_i X_j>.$$

Without the beam-beam interaction, we have $M_{ij} = \delta_{ij}$. (the Kronecker's delta).

Let us define the R ratio

$$R = \frac{M_{11}^-}{M_{11}^+}, \tag{27}$$

Now, the changes of the canonical variables during one turn can be written as a successive operations of the following three mappings:

O(*oscillation*):
$$\begin{pmatrix} X' \\ P' \end{pmatrix}_{\pm} = U(\mu_{\pm}) \begin{pmatrix} X \\ P \end{pmatrix}_{\pm},$$

B(*beam − beam force*): $\quad X' = X, \quad P' = P + K_{\pm} F_{\pm}(X),$

where

$$K_{\pm} = \begin{cases} 8\pi\eta_{\pm} & \text{(RB)}, \\ 2\pi^{3/2}\eta_{\pm} & \text{(FB)}, \end{cases} \tag{28}$$

$$F_{\pm}(X) = \begin{cases} \dfrac{1}{X}\left[\exp(-\dfrac{X^2}{2M_{11}^{\mp}}) - 1\right], & \text{(RB)}, \\ -\mathrm{erf}(\dfrac{X}{\sqrt{2M_{11}^{\pm}}}), & \text{(FB)}, \end{cases} \tag{29}$$

R(*radiation*): $\quad X' = X, \quad P' = \lambda_{\pm} P + \sqrt{(1-\lambda_{\pm}^2)}\,\hat{r},$

where $\lambda_{\pm}$ is the damping rate defined by

$$\lambda = \exp(-2/T_e) \tag{30}$$

and $\hat{r}$ is a Gaussian noise with unit standard deviation. Without **B**, $M_{ij}^{\pm}$ approaches to δ_{ij} in roughly T_e turns.

In the above, we have treated the effect of radiation as if it works only on P, for some technical reason. It is a little unphysical but harmless in the present case[49,50].

Mapping Equations for Moments

Let us track the changes of $M_{ij}^{\pm}$ under the mappings

$$\cdots \longrightarrow \mathbf{B} \longrightarrow \mathbf{R} \longrightarrow \mathbf{O} \longrightarrow \cdots.$$

The changes under **O** and **R** are straightforwardly obtained as

O: $\quad M_{ij}^{\pm\prime} = (U_{\pm} M_{\pm} U_{\pm}^{t})_{ij}$,

R: $\quad M_{11}^{\pm\prime} = M_{11}^{\pm}, \quad M_{12}^{\pm\prime} = \lambda M_{12}^{\pm}, \quad M_{22}^{\pm\prime} = \lambda_{\pm}^{2} M_{22}^{\pm} + (1 - \lambda_{\pm}^{2}).$

Here U^t is the transpose of U. Note that the effect of **R** can now be treated as deterministic process. The change under **B** is described as

B: $\quad M_{11}^{\pm\prime} = M_{11}^{\pm}, \quad M_{12}^{\pm\prime} = M_{12}^{\pm} + K_{\pm} < XF_{\pm}(X) >_{\pm},$

$$M_{22}^{\pm\prime} = M_{22}^{\pm} + 2K_{\pm} < PF_{\pm}(X) >_{\pm} + K_{\pm}^{2} < F_{\pm}(X)^{2} >_{\pm}.$$

By elementary integrations, we can calculate these averages explicitly. Let us track the change of M_{ij} on the Poincaré surface built just before **B**. It appears that, after sufficient number of iterations, M_{ij} always falls into a fixed point on the surface, which is denoted as $\tilde{M}_{ij}$. This is the solution of

$$M = \mathbf{ORB} \circ M. \tag{31}$$

After some algebra, we obtain

$$\tilde{M}_{11}^{\pm} = \begin{cases} D_{\pm} + \sqrt{D_{\pm}^{2} + E_{\pm}}, & \text{(RB)} \\ [D_{\pm} + \sqrt{1 + D_{\pm}^{2} + E_{\pm}}]^{2}, & \text{(FB)} \end{cases}$$

where

$$D_{\pm} = \begin{cases} K_{\pm} \dfrac{\lambda}{1+\lambda} A(\tilde{R}^{\pm 1}) \cot \mu + \dfrac{1}{2}, & \text{(RB)} \\ K_{\pm} \dfrac{2\lambda}{1+\lambda} A(\tilde{R}^{\pm 1}) \cot \mu, & \text{(FB)} \end{cases} \tag{32}$$

$$E_{\pm} = \begin{cases} \dfrac{K_{\pm}^{2} \lambda^{2}}{1-\lambda^{2}} \left[B(\tilde{R}^{\pm 1}) - \dfrac{2\lambda}{1+\lambda} A(\tilde{R}^{\pm 1})^{2} \right], & \text{(RB)} \\ \dfrac{4 K_{\pm}^{2} \lambda^{2}}{1-\lambda^{2}} \left[B(\tilde{R}^{\pm 1}) - \dfrac{2\lambda}{1+\lambda} A(\tilde{R}^{\pm 1})^{2} \right], & \text{(FB)}. \end{cases} \tag{33}$$

and

$$A(R) = \begin{cases} \sqrt{\dfrac{R}{1+R}} - 1, \\ -\dfrac{1}{\sqrt{2\pi(1+R)}}, \end{cases} \quad B(R) = \begin{cases} 2\sqrt{1+\dfrac{1}{R}} - \sqrt{1+\dfrac{2}{R}} - 1, & \text{(RB)} \\ \dfrac{1}{4} - \dfrac{1}{\pi}\arcsin\sqrt{\dfrac{R}{2(1+R)}}, & \text{(FB)} \end{cases}$$

Both in RB and FB cases, everything is expressed as a function of $\tilde{R}$, which in turn is a solution of the consistency condition

$$R = h(R), \tag{34}$$

$$h(\tilde{R}) = \frac{\tilde{M}_{11}^-}{\tilde{M}_{11}^+} = \begin{cases} \dfrac{D_- + (D_-^2 + E_-)^{1/2}}{D_+ + (D_+^2 + E_+)^{1/2}}, & \text{(RB)}, \\[2mm] \dfrac{[D_- + (1 + D_-^2 + E_-)^{1/2}]^2}{[D_- + (1 + D_-^2 + E_-)^{1/2}]^2}, & \text{(FB)}, \end{cases} \tag{35}$$

Let us examine this solution.

3.3. Cusp Catastrophe and Flip-Flop Hysteresis

The roots of Eq.(34) are illustrated in Fig. 10 for the case of $\delta_+ = \delta_-$ and $\nu_+ = \nu_-$. This is the well-known diagram of the cusp catastrophe. Let us start at the point $R = 1$, $\chi = 1$ and $H = 0$. If H is increased, the root of Eq.(34) bifurcates at a certain value of H, which is called H_b. If $H > H_b$, we have threefold roots. The upper and the lower are stable but the middle is unstable. Stable states are asymmetric: $R \neq 1$. This is a kind of the spontaneous symmetry breaking. When it occurs, either σ_+ or σ_- becomes much larger than the other so that the $\Xi_+ + \Xi_-$ and the luminosity decrease a lot.

The case $\chi > 1$ means that e^+ beam is stronger than e^- beam (more energetic or more intense). It is thus quite natural to expect that e^- beam is blown up ($\tilde{R} > 1$). There, however, can be a stable state where the opposite occurs. Let us call the former case 'natural' and the latter 'unnatural'. Let the state be unnatural with some χ: then as H decreases, the state 'jumps' to the natural one.

For $H > H_b$ and $\chi \simeq 1$, there are two stable values of $\tilde{R}$, while there can be only one stable state when χ is to some extent different from 1. A oscillation of χ around 1 with enough amplitude results in a hysteresis loop in $R - \chi$ plane. Necessary amplitude depends on $H - H_b$.

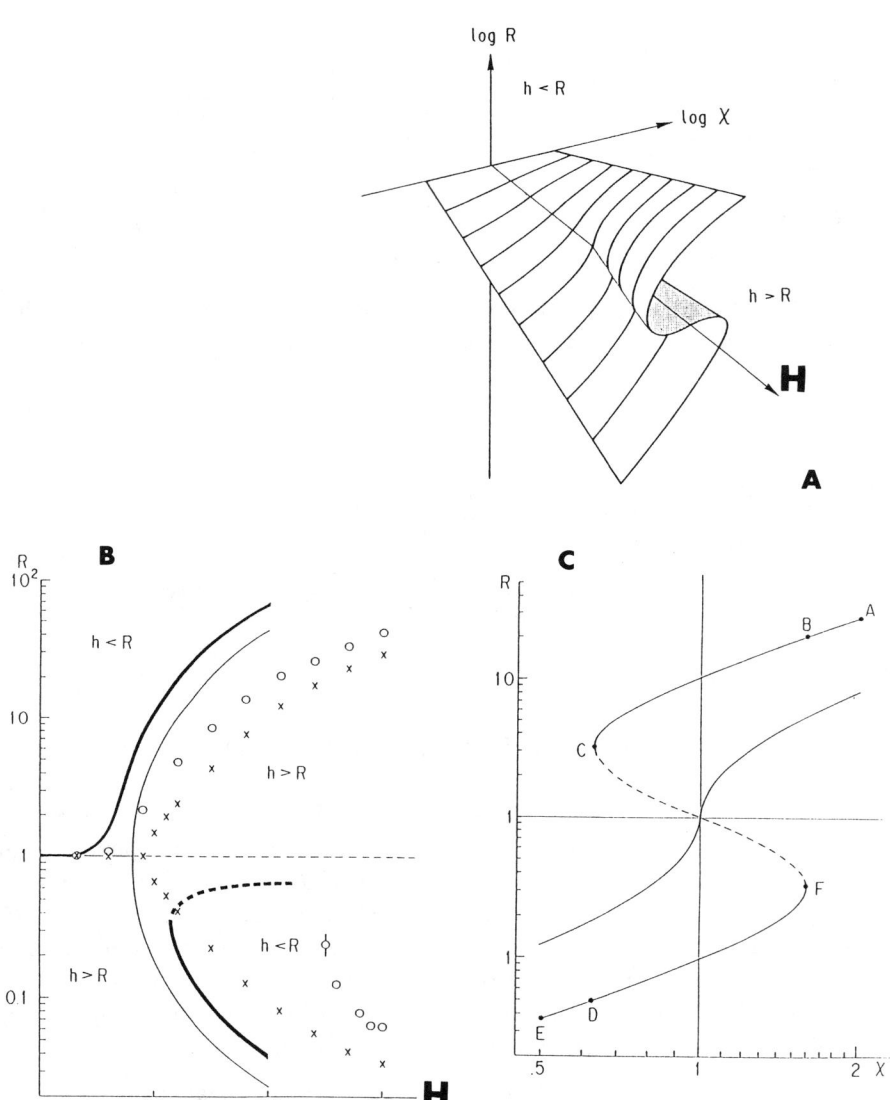

Fig. 10. (A) Conceptual illustration of the root $\tilde{R}$ as a function of H and χ. (B) Cross sections at constant χ's: $\chi = 1$ and $\chi = 1.5$. (C) Cross sections at constant H's: $H < H_b$ and $H > H_b$.

In the present case, the controllable parameters are only H and χ. This, however, does not restrict the generality of its results, since this behavior is not a special one, but rather a generic one. According to the general theory of bifurcation,[42] only this type of behavior is the structurally stable for the system with two parameters. We thus expect that the same structure will occur also from other types of asymmetry. There are many systems which have the same state diagram.[43,44]

Some characteristic results of the model with identical δ and ν are

1. the spontaneous symmetry breakdown is inevitable at some H, even if the set up is completely symmetric between beams ($\chi = 1$)

2. when χ is slightly different from 1, flip-flop hysteresis is generic in e^+e^- colliding rings provided enough H is available

3. the unnatural state provides larger luminosity than the natural state.

The typical values of H_b in this model are ($\nu = 0.2$)

$T_e =$	142.8	1428	14280	142800	
$H_b =$	0.0683	0.014	0.0043	0.0015	(RB)
	0.082	0.0257	0.008	0.0025	(FB)

The results of the model has been compared to those of the Multiparticle tracking. All the rough qualitative results are confirmed by it.

Smoothed Catastroph

For some tunes, however, the bifurcation at $H = H_b$ is much smoother according to a multiparticle tracking. This discrepancy could be understood by the lack of degrees of freedom in the soft Gaussian model: we need infinitely many degrees of freedom to have a smooth transition. When the catastrophe is smooth, it appears as if the the spontaneous symmetry breakdown does not occur within the reasonable values of H. See Fig.11. In this case, we can reach a higher luminosity. We can apply our model also to this case: we pose $R = 1$ as a restriction. Let us consider the case where $\chi = 1$. The subscript $\pm$ for η, K etc. can be omitted. In this case, we can express the beam sizes and Ξ explicitly:

$$\xi = \Xi_+ + \Xi_- = \begin{cases} \dfrac{H}{D_1 + \sqrt{D_1^2 + E_1}}, & \text{(RB)} \\ \dfrac{H}{D_1 + \sqrt{1 + D_1^2 + E_1}}, & \text{(FB)} \end{cases} \qquad (36)$$

where D_1 and E_1 are $D(1)$ and $E(1)$ in Eqs.(32) and (33).

As easily seen, (i) when H is small, $\xi = H$, (ii) when H becomes large, ξ

approaches to a limit ξ_∞, called the beam-beam limit:

$$\xi_\infty = \frac{1}{d_1 + (d_1^2 + e_1)^{1/2}} \times \begin{cases} \frac{1}{8\pi}, & \text{(RB)} \\ \frac{1}{2\pi^{3/2}}, & \text{(FB)} \end{cases} \qquad (37)$$

where

$$d_1 = \frac{\lambda}{1 + \lambda \tan \mu} \frac{A_1}{} \times \begin{cases} 1 & \text{(RB)} \\ 2 & \text{(FB)} \end{cases}, \qquad (38)$$

$$e_1 = \frac{\lambda^2}{1 - \lambda^2}(B_1 - \frac{2\lambda}{1+\lambda}A_1^2) \times \begin{cases} 1 & \text{(RB)} \\ 4 & \text{(FB)} \end{cases}. \qquad (39)$$

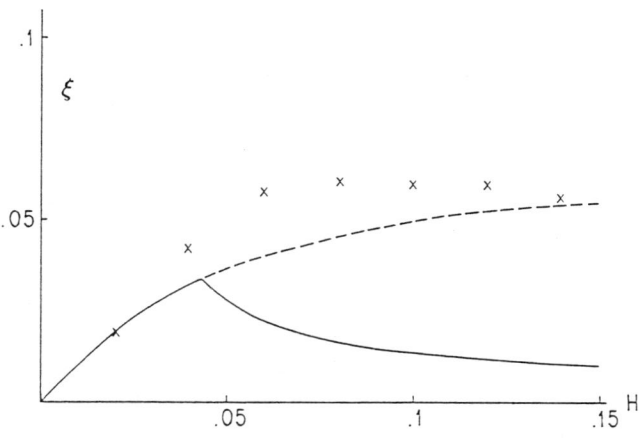

Fig. 11. Smoothed catastrophe seen by a multiparticle tracking. The × is the multiparticle tracking results, solid line is the model prediction with the bifurcation and the dashed line is the model with the restriction of $R = 1$.

Some characteristic points of this model with the restriction of $R = 1$ are

1. the ξ_∞ increases as $\nu \longrightarrow 0^+$. This is consistent with the observation.[55] It, however, does not necessarily imply that ν should be close to zero. The H_b is smaller for $\nu \longrightarrow 0^+$, for the RB case.

2. the dependance of ξ_∞ on T_ϵ is quite weak: the model implies

$$\xi_\infty \propto 1/\sqrt{T_\epsilon}.$$

Typical values of ξ_∞ are ($\nu = 0.2$)

$T_\epsilon =$	142.8	1428	14280	142800	
$\xi_\infty =$	0.068	0.021	0.0065	0.002	(RB)
	0.13	0.04	0.012	0.0039	(FB)

It is interesting to see that the same dependance was predicted by Chao[1] by quite different consideration.

3. In the $K \oplus K = N_{IP}$ case, T_ϵ should be multiplied by N_{IP} and ν is divided by N_{IP}. It follows that

$$\xi_\infty \propto \frac{1}{\sqrt{N_{IP}}}, \qquad (40)$$

which is the same prediction as that given by Wiedeman[39] and by Keil and Talman[33].

4. The linear tune shift limit Δ_ν^∞, defined as

$$\cos 2\pi(\nu + \Delta_\nu^\infty) = \cos\mu - 2\pi\xi_\infty \sin\mu, \qquad (41)$$

is more universal than Ξ_∞. This agrees with an observation by Seeman[3].

3.4. OTHER ISSUES

<u>Linear Soft Gaussian Model</u> The Gaussian approximation ignores the ψ^N. See Eq.(20). This causes a relatively large numerical disagreement between the model and the multiparticle trackings. This shortcoming does not exist if the force F was linear. In this case, the '$\Longrightarrow$' line in Eq.(20) holds exactly. A trial[17], however, shows that the beam sizes $M_{11}^\pm$ becomes chaotic after some threshold of H. That is, $M_{11}^\pm$ differs every turn in a chaotic manner at the IP. Such a behaviour has not been observed in the accelerators nor in multiparticle trackings. The linear approximation of the beam-beam force seems to be an over-simplification.

<u>Self-Consistent Equilibria</u> As long as observed so far, in one dimensional simulation, we have always an equilibrium state eventually. If we are interested in equilibrium state alone, we can replace the strong-strong simulation by a set of weak-strong ones[18]. This calculates, by simulation, the equilibrium value of the $\sigma_\pm$ as a function of $\sigma_\mp$

$$\sigma_\pm = f(\sigma_\mp), \qquad (42)$$

and find a consistent sets (σ_+, σ_-).

By this, we can include the arrow $F^G \longrightarrow \psi^N$ in Eq.(20). This method seems to produce all the results obtained by the present model with much higher accuracy. When, however, we want to include F^N, it seems quite difficult to go along this way of thought, since we should replace Eq.(42) by

$$\rho_\pm = f[\rho_\mp],$$

where ρ is the coordinate distribution (density function) and $f[\ ]$ represents a functional.

4. Higher Order Moments: Beyond Gaussian Approximation

We consider the non-Gaussian part, ψ^N.

4.1. Perturbative Approach

The perturbative approach here means an approach studying the limit of

$$\eta \longrightarrow 0.$$

In the actual machines, η is at most 0.1 so that this approach appears to be permissible. On the other hand, the flip-flop effect, for example, does not seem to be a perturbative effect so that the applicability of this approach is limited.

Almost all the weak-strong studies[27-30] fall into this category. This case has been studied much better than the strong-strong case but it seems quite difficult to extend the formalism to the strong-strong case. (The self-consistent equilibria[18] can be a candidate).

The perturbative approaches to the strong-strong case were

1. study of the resonance structure based on the water-bag model[23,24]. This is also based on the Vlasov equation, rather than the Fokker-Planck equation. It is useful for proton machines but not for the electron rings, because it tells little to the equilibrium beam size and the luminosity.
2. study of the non-Gaussian effect to the beam-beam mode[25,26]. The Vlasov equation was linearized around the nominal Gaussian distribution. The $\sigma - \pi$ splitting was obtained by an eigenvalue analysis. The results are compared to the other estimates[1,5,7,8] for the $1 \oplus 1 = 1$ case as follows:

$\sigma - \pi$	Piwinski	Chao	Talman	Hirata	MSYFKKU
horizontal	$2\xi_x$		$2/\sqrt{6}\xi_x$	ξ_x	$\Lambda(r)\xi_x$
vertical	$2\xi_y$	$\sqrt{2}\xi_y$	ξ_y	ξ_y	$\Lambda(1-r)\xi_y$

Here MSYFKKU stands for Meller- Siemann- Yokoya- Funakoshi- Kikutani-

Koiso- Urakawa. The Λ is

$$\Lambda(r) = 1.33 - 0.37r + 0.279r^2, \quad \left(r = \frac{\sigma_y}{\sigma_x + \sigma_y}\right),$$

and gives 1.33 (horizontal) and 1.24 (vertical) for the flat beam limit. This correction factor was measured at the TRISTAN accumulation ring. A result is shown in Fig.12. The MSYFKKU factor seems to be consistent with the observation. For the time being, the following questions remain unanswered: 1) whether the factor should be applied to the kick or to the tune shift, (see Fig.12). 2) how the factor depends on the separation between beams (long range force), 3) how to calculate the factor for fully asymmetric cases.

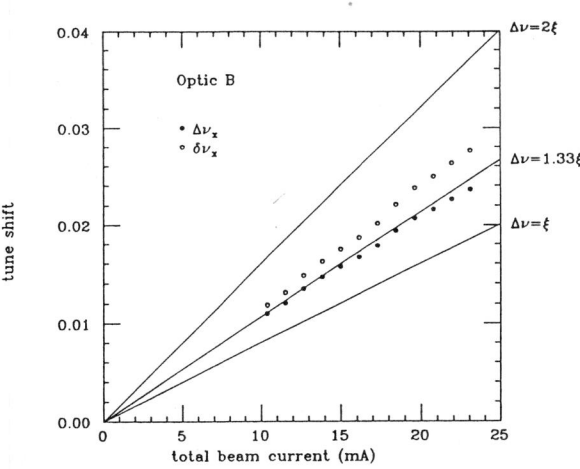

Fig. 12. Horizontal $\sigma - \pi$ split[26]. The • is the observation results, which fits well the line of $1.33\xi_x$. The ○ is expected value based on the assumption that the correction factor should be multiplied to the dipole kick.

4.2. NONPERTURBATIVE APPROACH

In order to discuss the nonperturbative effects, such as the flip-flop effect, the soft Gaussian model was the simplest. To go beyond that approximation, it is natural to start from this model. We propose a method[20-22] based on Stratonovich expansion[19] and truncating it at a finite order. In the lowest order approximation, it is identical with the soft Gauissian model. In the next-to-lowest order approximation, the model presents improved agreement with the multiparticle tracking.

For the simplicity, we will discuss the weak-strong case first and touch on the strong-strong case later. The following formalism is directly applied to this case.

General Stratonovich Expansion

As canonical variables in the 2-dimensional phase-space, we use the same variable as Eq.(26). In this chapter, however, we write as

$$X^1 = \frac{x}{\sqrt{\beta\varepsilon}}, \qquad X^2 = \frac{\alpha x + \beta x'}{\sqrt{\beta\varepsilon}}.$$

General Stratonovich expansion[19], is an expansion of two dimensional distribution function $\psi(X^1, X^2)$ around the two dimensional Gaussian distribution,

$$G(\vec{X}; g) = \frac{1}{2\pi\sqrt{\det g}} \exp -\phi, \qquad \phi = \frac{1}{2} g_{\alpha\beta} X^\alpha X^\beta,$$

where $g_{\alpha\beta}$ is the inverse of $g^{\alpha\beta}$ and $\det g = g^{11}g^{22} - (g^{12})^2$. Here $g^{\alpha\beta}$ is **any symmetric positive definite matrix.** Here and in what follows, we employ **Einstein's summation convention**; when the same symbol appears in both upper and lower indices simultaneously, a summation with respect to the symbol from 1 to 2 is implied. The $g^{\alpha\beta}$ will be called **metric** because of the similarity to Riemanian geometry. **Note that A^α is different from A_α, which is $g_{\alpha\beta}A^\beta$.**

We start from the following lemma:
Lemma 1, Any distribution function $\psi(\vec{X})$, which is symmetric in phase-space, $\psi(-\vec{X}) = \psi(\vec{X})$, which is normalized to unity, and which falls exponentially at infinity, can be expanded as

$$\psi(\vec{X}) = G(\vec{X}; g) P(\vec{X}; g, Q), \tag{43}$$

$$P(\vec{X}; g, Q) = 1 + \sum_{\substack{n=2 \\ \text{even}}} \frac{1}{n!} Q^{\alpha_1 \alpha_2 \cdots \alpha_n} H_{\alpha_1 \alpha_2 \cdots \alpha_n}(\vec{X}; g). \tag{44}$$

Here the sum extends over all even numbers from 2 to infinity.

Here, H is the generalized Hermite polynomial,

$$H_{\alpha_1\alpha_2\cdots\alpha_n}(\vec{X};g) = e^\phi \prod_{i=1}^{n}(-\partial_{\alpha_i})e^{-\phi}.$$

The Q's are called **quasi-moments**:
Lemma 2, For given ψ, the quasi-moments are obtained as

$$Q^{\alpha_1\alpha_2\cdots\alpha_n} = <H^{\alpha_1\alpha_2\cdots\alpha_n}(\vec{X};g)>,$$

where $<\ >$ is the expectation value with respect to ψ. Thus ψ is equivalent to the set of (g, Q).

The quasi moments can be related to the usual moments

$$M^{\alpha_1\alpha_2\cdots\alpha_n} = <X^{\alpha_1}X^{\alpha_2}\cdots X^{\alpha_n}>.$$

Using Eq.(43), we have
Lemma 3, A moment M of even order can be expanded in terms of the quasi-moments of even order as

$$\begin{aligned} M^{\alpha_1\alpha_2\cdots\alpha_n} &= Q^{\alpha_1\alpha_2\cdots\alpha_n} \\ &+ \sum_{apc} g^{\alpha_1\alpha_2} Q^{\alpha_3\cdots\alpha_n} \\ &+ \sum_{apc} g^{\alpha_1\alpha_2} g^{\alpha_3\alpha_4} Q^{\alpha_5\cdots\alpha_n} \\ &+ \cdots\cdots \\ &+ \sum_{apc} g^{\alpha_1\alpha_2} g^{\alpha_3\alpha_4} \cdots g^{\alpha_{n-3}\alpha_{n-2}} Q^{\alpha_{n-1}\alpha_n} \\ &+ \sum_{apc} g^{\alpha_1\alpha_2} g^{\alpha_3\alpha_4} \cdots g^{\alpha_{n-1}\alpha_n}. \end{aligned} \quad (45)$$

Here

$$\sum_{apc}$$

means a sum over **all possible combination** of indices but restricted not to reproduce the same expression. For example,

$$\sum_{apc} g_{\alpha\beta}g_{\gamma\delta} = g_{\alpha\beta}g_{\gamma\delta} + g_{\alpha\gamma}g_{\beta\delta} + g_{\alpha\delta}g_{\beta\gamma}.$$

In the same manner, we have
Lemma 4, A quasi-moment Q of even order can be expanded in terms of the

moments of even order as

$$
\begin{aligned}
Q^{\alpha_1 \alpha_2 \cdots \alpha_n} = &\, M^{\alpha_1 \alpha_2 \cdots \alpha_n} \\
&- \sum_{apc} g^{\alpha_1 \alpha_2} M^{\alpha_3 \cdots \alpha_n} \\
&+ \sum_{apc} g^{\alpha_1 \alpha_2} g^{\alpha_3 \alpha_4} M^{\alpha_5 \cdots \alpha_n} \\
&- \cdots \cdots \\
&+ \sum_{apc} (-g^{\alpha_1 \alpha_2})(-g^{\alpha_3 \alpha_4}) \cdots (-g^{\alpha_{n-3} \alpha_{n-2}}) M^{\alpha_{n-1} \alpha_n} \\
&+ \sum_{apc} (-g^{\alpha_1 \alpha_2})(-g^{\alpha_3 \alpha_4}) \cdots (-g^{\alpha_{n-1} \alpha_n}).
\end{aligned} \quad (46)
$$

In the above, Q and M of only even order are considered. Inclusion of the odd order M's and Q's are straightforward, but irrelevant for the present problem.

<u>Gauge Degree of Freedom</u>   In the above, we did not specify $g^{\alpha\beta}$. Of course, $M^{\alpha\beta}$, for example, has definite meaning, which can be expressed as

$$ M^{\alpha\beta} = g^{\alpha\beta} + Q^{\alpha\beta}. \quad (47) $$

Unless we specify $g^{\alpha\beta}$, Q's are not defined.

We can freely specify g. This freedom is called Gauge freedom. It is related to how to express ψ, but has nothing to do with physical contents of ψ, **provided we do not truncate the expansion**. Once we specify the metric g, the gauge is fixed.

There are some characteristic gauges:

proper gauge	$g^{\alpha\beta} = M^{\alpha\beta}$
nominal gauge	$g^{\alpha\beta} = \delta^{\alpha\beta}$
Gaussian gauge	$g^{\alpha\beta} = g_G^{\alpha\beta}$

In the proper gauge, we use real second moments itself for g, so that $Q^{\alpha\beta} = 0$ by definition. In the nominal gauge, we use the nominal value of the second-order moment. This gauge is the simplest of all, but the convergence is the worst. In the Gaussian gauge, we use the result of the soft-Gaussian model, g_G. The first merit of this gauge is that the changes of Q's around the ring can be expressed as linear transformation. (See below). It is easy to relate different gauges, if we use the fact that the moment M is gauge independent. Once they are related to moments, using Eqs.(45) and (46), the relation becomes clear.

Weak-Strong Model Here, we study the weak-strong case of the round beam.

In the present notations, the soft Gaussian model can be summarized as follows: we put all Q's zero so that g is the second moment.

In the arc, the beam undergoes

$$g_{new} = U[\Lambda g_{old} \Lambda^t + (1 - \Lambda^2)]U^t,$$

$$Q_{new}^{\alpha_1 \alpha_2 \cdots \alpha_n} = (U\Lambda)^{\alpha_1}{}_{\beta_1}(U\Lambda)^{\alpha_2}{}_{\beta_2} \cdots (U\Lambda)^{\alpha_n}{}_{\beta_n} Q_{old}^{\beta_1 \beta_2 \cdots \beta_n}, \qquad (48)$$

where

$$g = \begin{pmatrix} g^{11} & g^{12} \\ g^{21} & g^{22} \end{pmatrix}, \quad \Lambda = \begin{pmatrix} 1 & 0 \\ 0 & \lambda^2 \end{pmatrix}. \qquad (49)$$

The beam-beam kick, at the IP, is represented by the following:

$$g_{new}^{\alpha\beta} = <(X+F)^\alpha (X+F)^\beta>_G. \qquad (50)$$

$$Q_{new}^{\alpha_1 \alpha_2 \cdots \alpha_n} = <H^{\alpha_1 \alpha_2 \cdots \alpha_n}(\vec{X} + \vec{F}; g_{new})>. \qquad (51)$$

Here, $\vec{F} = (0, F)$ is the beam-beam kick

$$F(X) = 8\pi\eta \frac{1}{X}[\exp(-X^2/2) - 1].$$

The $<\ >_G$ stands for the average with respect to the Gaussian part $G(\vec{X}; g)$: to evaluate the r.h.s. of Eqs.(50), we use $G(\vec{X}; g_{old})$. To evaluate Q_{new}, we should truncate the expansion. Let us truncate it at $2N_{max}$-th order.

We denote

$$Q[N,n] \equiv Q^{\overbrace{111\cdots 1}^{2N-n}\overbrace{222\cdots 2}^{n}}, \quad H[N,n] = H\underbrace{{}_{111\cdots 1}}_{2N-n}\underbrace{{}_{222\cdots 2}}_{n}.$$

for the sake of brevity.

Then we can define vectors

$$(\vec{Q})^k = Q[N,n], \quad (\vec{H})_k = H[N,n],$$

where

$$k = k(N,n) = N^2 + n. \quad (1 \le N \le N_{max}, \ 0 \le n \le 2N)$$

That is,

$$\vec{Q} \equiv \begin{pmatrix} Q^{11} \\ Q^{12} \\ Q^{22} \\ Q^{1111} \\ \vdots \end{pmatrix}, \quad \vec{H} \equiv \begin{pmatrix} H_{11} \\ H_{12} \\ H_{22} \\ H_{1111} \\ \vdots \end{pmatrix}.$$

These vectors are $k_{max} \equiv N_{max}^2 + 2N_{max}$ dimensional.

Now, we can show that, under the beam-beam kick, $\vec{Q}$ changes as

$$\vec{Q}_{new} = A(g,g')\vec{Q}_{old} + \vec{B}(g,g'),$$

where g (g') is the second order moments in the Gaussian approximation just before (after) the beam-beam kick and A is a $k_{max} \times k_{max}$ matrix. Since we have an explicit expressions of g and g', A and $\vec{B}$ are also expressed explicitly.

Also under the betatron oscillation and radiation damping, Q are transformed linearly as

$$\vec{Q}_{new} = O(\mu,\lambda)\vec{Q}_{old}.$$

Now we have a complete set of mappings for Q. If we observe Q at the Poincaré section just before the beam-beam kick, it changes every turn as

$$\vec{Q}_{(n+1)\text{-th turn}} = O(A\vec{Q}_{n\text{-th turn}} + \vec{B}).$$

If the ψ is to fall into a fixed distribution after many turns, (although it is not assured), we expect also that $\vec{Q}$ will eventually be a single fixed point in the k_{max} dimensional vector space. This fixed point is expressed as

$$\vec{Q} = \frac{1}{1 - OA} O\vec{B}. \tag{52}$$

As is well known, the mapping converges into the fixed point if and only if no eigenvalue of OA has the absolute value larger than 1.

<u>Fixed Point</u> A Fortran program called SBS (Strong-weak Beam-beam interaction in Stratonovich expansion) solves Eq.(52). Due to the memory size restriction, $N_{max} \leq 10$, for the time being.

In the table below, we show the numerical results of SBS on M^{11}, which is obtained from the solution of Eq(52) and the relation Eq.(47). We compare results with different N_{max}. Since it can be shown that, in this gauge, when $N_{max} = 1$, $\vec{B}$ vanishes and consequently $\vec{Q} = 0$, we do not show this case in the Table. In some cases, the large eigenvalue occurs so that the result of SBS is not reliable (indicated by !). In some other cases, we cannot use SBS because of the overflow in the numerical process. The latter is less difficult a problem. We also show the Multiparticle tracking results in the Table. As N_{max} becomes larger, the agreement is better, except for the case of large eigenvalues.

N_{max}	$\eta = 0.03$	$\eta = 0.05$	$\eta = 0.06$	$\eta = 0.1$	$\eta = 0.3$	$\eta = 0.5$
(Gauss)	1.08	2.06	3.60	12.5	83.3	179.7
2	0.92	1.50	2.52	9.89	75.1	165.8
3	0.89	1.14	1.81	8.00	69.7	157.6
4	0.88	0.97	1.40	6.69	66.2	152.5
6	0.88	0.87	1.01	5.03	61.1	144.9
8	0.89	!	0.87	4.05	56.6	137.4
9	0.89	!	0.81	3.71	*	*
(MPT)	~0.9	~0.8	~0.8	1.95	36.0	59.5

Table 3.   Second moment M^{11} for some values of N_{max}. Parameters are $T_{\varepsilon} = 142.8$ and $\nu = 0.15$. (Gauss) and (MPT) stand for the Gaussian approximation and the Multiparticle tracking results, respectively. The * means that the calculation fails due to overflow. The ! means that the mapping has too large eigenvalue(s).

It is clearly seen that the agreement with the multiparticle tracking results become better and better for larger N_{max}. It is observed that there is some domain in the parameter space (ν, λ, η) where some of the eigenvalues become larger than unity. This region seems to be connected with the beginning of the rapid increase of M^{11}.

<u>Strong-Strong Case</u> The strong-strong case was studied in Ref.20 in the prpper gauge. The tracking method was applied to (g, Q) and only $N_{max} = 2$ case was successful. The agreement with the multiparticle tracking was improved. The

difficulty of using N_{max} more than 2 is, presumably, due to the loss of digit in numerical procedure.

We should also include the effect of F^N, see Eq.(18). This can be done within the present formalism. For the time being, the results are quite limited but they seem to be sufficient to convince us that the method is promising. Further investigation is planned.

5. Discussion

The coherent approaches to the beam-beam interaction have been helpful to get insight into the complicated phenomena. The ability of them are still quite limited. Since we need more knowledge and understanding of the beam-beam effects, much more effort should be made. The simulation, in particular, the multiparticle tracking will help us a lot but it cannot be a major source of understanding. All the theory should be checked by the multiparticle tracking and, at the same time, all the results of the multiparticle tracking should be understood by a theory.

Let us make a list of what is necessary for the future investigation.

1. The dipole mode is the best understood. It is almost the exact science now. In order to use it as a reliable diagnostic tool, however, we need more careful comparison between the observation and the theory. In particular, the factor for the $\sigma - \pi$ splitting should be understood more. See comments given for the MSYFKKU factor in the previous chapter. The nonlinear response of the transeverse kick can also be useful for more detailed study of the beam-beam interaction.

2. The quadrupole mode issues are still not settled. In order to understand the beam blow up, perhaps we need to consider two transverse directions at the same time and also the synchrotron motion[51].

3. The higher order effects are little understood. In particular, we even do not know how the effect of F^N is important.

The beam-beam interaction is an old problem but still can provide many interesting, challenging and fruitful questions.

Acknowledgements: Some parts of the work presented here were done when the author stayed at CERN. He thanks the members of LEP theory group for the hospitality. In particular, E. Keil is gracefully acknowledged for his contribution and many discussions.

APPENDIX A

Linear Analysis: Different Circumference

Let us consider rings with different circumferences[12,13]:

$$C_+ \neq C_-.$$

If the energies are different, it appears more natural to use smaller circumference for the beam with the lower energy, not only for economical reasons but also for rapid damping which may help some instabilities. However, we will show that there is a severe restriction on the choice of the circumference ratio. In some cases, the unstable region due to the beam-beam mode becomes quite dense in the tune diagram, quite contrary to the existing conventional machines where the beam-beam mode is more useful as a diagnostic tool than a cause of dangerous instability, because the unstable region in the tune diagram can be avoided easily if $C_+ = C_-$.

<u>The Dynamics</u> We will consider a pair of asymmetric rings with different circumferences. The ratio of circumferences of both rings is K_+/K_-. Without loss of generality, we can assume that $K_+ \leq K_-$ and that K_+ and K_- have no common factor except for unity. The values of $\gamma_\pm$, $\beta^\pm_{x,y}$, $\sigma^\pm_{x,y}$ and $\nu^\pm_{x,y}$ may be different for the two beams, as before. The ring for the $e^\pm$ beam can have $K_\pm K_0$ bunches, which are spaced equally and contain $N_\pm$ particles each, where K_0 is an arbitrary integer. While an e^+ bunch completes K_- turns in its ring, the e^- bunch completes K_+ turns in its ring. This period is called '**super period**'. At the IP, the e^+ and e^- bunches are coupled by the beam-beam force. A common factor $K_0 > 1$ is irrelevant for the dynamics of the beam-beam interaction, because an $e^\pm$ bunch is coupled to only $K_\mp$ $e^\mp$ bunches among a total of $K_\mp K_0$ bunches. We will assume $K_0 = 1$.

We label the bunches as $n = 1, 2, \cdots, K_+$ for e^+ beam and $m = 1, 2, \cdots, K_-$ for e^- beam. The collision of one particular set of bunches (n, m) occurs once in every super period. In this connection, it is useful to define the 'super tunes',

$$\tilde{\nu}_\pm \equiv K_\mp \nu_\pm,$$

which correspond to the number of betatron oscillations of each bunch per one super period.

<u>Resonance Structure</u> Let us first consider the extreme case that particles are only in the first bunches in both beams ($n = 1$ and $m = 1$), while the other bunches are empty. In this case, we can apply the same argument as we did for the $(K_+, K_-) = (1, 1)$ case. The only difference is that the collision occurs only

once in a super period. Thus all we have to do is to replace the tunes by the super tunes,
$$\nu_{\pm} \longrightarrow \tilde{\nu}_{\pm},$$
in expressions obtained for the $(K_+, K_-) = (1,1)$ case. It follows that resonance lines becomes $K_+ K_-$ times denser than the $(1,1)$ case. That is,

1. the sum resonance lines appear as
$$\tilde{\nu}_+ + \tilde{\nu}_- \lesssim \text{integer} \quad \text{i.e.} \quad K_-\nu_+ + K_+\nu_- \lesssim \text{integer},$$

2. the 1/2 resonances appear as
$$\tilde{\nu}_+ = K_-\nu_+ \lesssim 1/2 + \text{integer} \quad \text{or} \quad \tilde{\nu}_- = K_+\nu_- \lesssim 1/2 + \text{integer}.$$

3. the 2/2 resonances appear as
$$\tilde{\nu}_+ = K_-\nu_+ \lesssim \text{integer} \quad \text{or} \quad \tilde{\nu}_- = K_+\nu_- \lesssim \text{integer}.$$

For more general cases, where each bunch has some particles, the above structure still exists but the strength of each resonance changes.

Detailed Structure   The precise eigenvalues of the revolution matrix are calculated with a computer code ABMODE, derived from the BBMODE code[11]. In Fig. 13, we show the variation of the instability region and the growth rate due to the inequality of the number of particles between bunches. In the equally populated case between bunches in the same beam, (d), some resonances disappear. In summary, we have a complete list of the linear instabilities where the beams are separated spontaneously.

Resonance	Definition	Remark
Sum	$\tilde{\nu}_+ + \tilde{\nu}_- \lesssim$ integer	Dangerous (dense)
Integral tune	$\nu_+ \lesssim$ integer or $\nu_- \lesssim$ integer	Dangerous but avoidable
Half-integral tune	$\nu_+ \lesssim 1/2 +$ integer or $\nu_- \lesssim 1/2 +$ integer	avoidable (not dense)
Integral super tune	$\tilde{\nu}_+ \lesssim$ integer or $\tilde{\nu}_- \lesssim$ integer	Dense but weak (absent in
Half-integral super tune	$\tilde{\nu}_+ \lesssim 1/2 +$ integer or $\tilde{\nu}_- \lesssim 1/2 +$ integer	the equally populated case)

Among these resonances, the sum resonances are the most dangerous: the luminosity reduction is the worst. In addition, the sum resonances are more and more closely spaced in the tune diagram for colliders with larger and larger $\sqrt{K_+^2 + K_-^2}$.

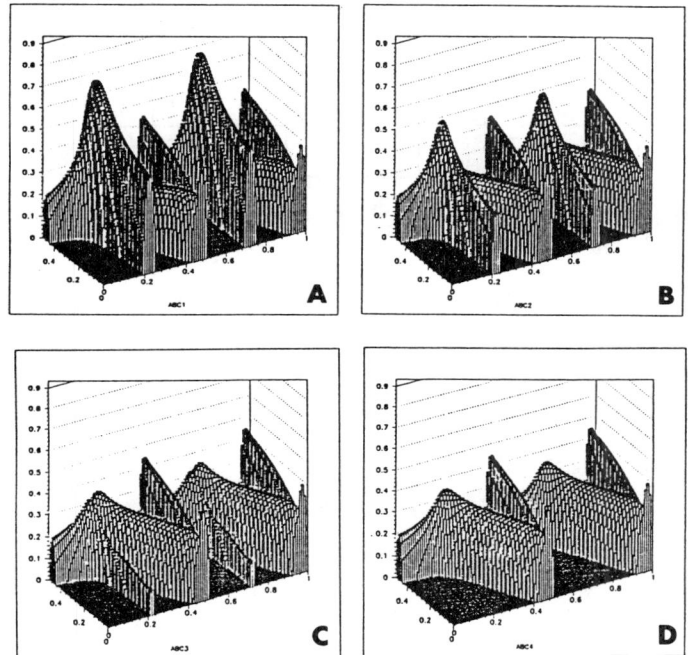

Fig. 13. The growth rate in (ν_+, ν_-) space for $(K_+, K_-) = (1, 2)$ for $\Xi_- = 0.03$ and Ξ_+'s are a) $(0.06, 0)$, b) $(0.05, 0.01)$, c) $(0.04, 0.02)$ and d) $(0.03, 0.03)$.

Restriction to the Circumference Ratio These resonances, in particular the sum resonances, pose a severe restriction on the choice of K_+ and K_-, because enough space in the tune diagram is needed for the operation of the rings. The following table shows the instability region (in%) in the tune space (ν_+, ν_-) for different set of (K_+, K_-), for the case of the equal population between bunches of the same beam and $\Xi = 0.03$.

	1	2	3	4	5	6	7
1	32.5	34.9	41.9	43.2	44.0	48.6	51.9
2	-	-	48.4	-	55.3	-	61.6
3	-	-	-	57.1	61.7	-	67.9
4	-	-	-	-	66.6	-	73.6
5	-	-	-	-	-	74.8	78.2
6	-	-	-	-	-	-	82.6

If we require that the instability region in the tune space is not larger than 0.5, we find that for $\Xi = 0.03$

$$(K_+, K_-) = (1,1),\ (1,2),\ (1,3),\ (1,4),\ (1,5),\ (1,6)\ \text{and}\ (2,3)$$

are the only acceptable choices of $K_\pm$.

In this connection, it is interesting to see that the shift of the super tunes is the maximum where $\tilde{\nu}_+ = \tilde{\nu}_-$ and it is

$$\tilde{\nu} \longrightarrow \tilde{\nu} + K_-\Xi_+ + K_+\Xi_-.$$

In this case, we have

$$\nu_+ \longrightarrow \nu_+ + \Xi_+ + \frac{K_+}{K_-}\Xi_-,$$
$$\nu_- \longrightarrow \nu_- + \Xi_- + \frac{K_-}{K_+}\Xi_+,$$

Thus, when $K_+ \ll K_-$, e^- ring suffers from an enormous tune shift. Considering that there are many nonlinear resonance lines in addition to the beam-beam resonances, such a large tune shift poses another restriction.

APPENDIX B

Comparison between Round and Flat Beams

Based on the soft Gaussian model[52], we compare the round beam (RB) and the flat beam (FB). The RB here implies the case where the emittance, beta and tune are identical between horizontal and vertical directions. As clear from Eq.(16), RB appears to have a merit of larger luminosity: terms for x and y contributes equally. This '2' is called the geometrical factor. The model, however, implies that this merit is cancelled by the stronger beam-beam effect for the RB.

In order to compare RB and FB, we assume

1. $[\sigma^2]_{\rm RB} = [\sigma_x \sigma_y]_{\rm FB}$: two cases have the same nominal luminosity.

2. $\beta_{\rm RB} = \beta_{\rm FB}^y$: the RB case can enjoy its merit if Ξ's are identical.

These assumptions are quite favourable to the RB. Actually, it does not seem easy to achieve them, if we consider chromaticity corrections, for example.

We considered two cases for the luminosity limitation. (For simplicity, we assume $\chi = 1$.)

1. In the case of the spontaneous symmetry breakdown, the luminosity is limited by H_b. From Eqs. (15) and (22) and the above assumptions, we have

$$L_{max} \propto H_b^2 \begin{cases} 4, & ({\rm RB}) \\ 1, & ({\rm FB}) \end{cases},$$

where $\Xi_+ + \Xi_- \simeq H$, was used. As we have seen, however, H_b of the RB is almost always the half of that of FB. Thus, RB and FB give similar L. More precise calculation[52] gives

$$\frac{L_{\rm RB}}{L_{\rm FB}} \simeq 1.4, \quad {\rm for} \quad T_\epsilon \simeq 5000.$$

2. In the case, where the catastrophe is smooth, L is not limited but becomes

$$L \propto N\xi_\infty \begin{cases} 2, & ({\rm RB}) \\ 1, & ({\rm FB}) \end{cases}.$$

Since ξ_∞ of the RB is almost the half of that of the FB, we have

$$\frac{L_{\rm RB}}{L_{\rm FB}} \simeq 1, \quad {\rm for\ the\ same\ current.}$$

Thus, in spite of the assumptions which is a little too much favourable to the RB, RB could not give remarkably larger luminosity. This agrees with the results of a multiparticle tracking by Chin[53].

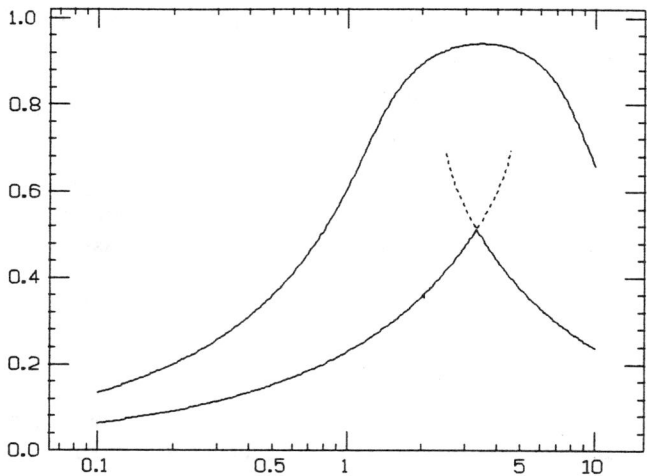

Fig. 14. the luminosity reduction factor D as a function of χ for the case of $(T_\epsilon^+, T_\epsilon^-) = (1428, 142.8)$, $(\nu_+, \nu_-) = (0.2, 0.2)$ and $H = 0.03$ and 0.06.

APPENDIX C
Different Damping Rates

When two beams have different energies, it is not easy to make their damping rates λ_+ and λ_- identical, because the two rings should not have quite different circumferences as shown in Appendix A. One possibility is that we install a lot of damping wigglers in the low energy ring. It may cause a problem of chromaticity correction due to the strong nonlinearity of the wigglers. Here, we discuss how to maximize the luminosity when $\lambda_+ \neq \lambda_-$ based on the soft Gaussian model.

We introduce an asymmetry of δ's to the model. The beam sizes at the IP are still assumed to be identical. In Fig. 14, we show a result of the soft Gaussian model for this case. We set different values for the damping rates,

$$(T_\epsilon^+, T_\epsilon^-) = (1428, 142.8),$$

and see the luminosity reduction factor.

$$L = D \times \text{nominal luminosity},$$

which is given by the present model as

$$D = \begin{cases} \dfrac{2\varepsilon}{M_{11}^+ + M_{11}^-}, & \text{(RB)}, \\ \sqrt{\dfrac{2\varepsilon}{M_{11}^+ + M_{11}^-}}, & \text{(FB)}, \end{cases}$$

Since H is identical on each line, the nominal luminosity is also identical on it. For $H = 0.06$, we have an extra line (shown by the dotted line) which represents a case of the unnatural state (see Fig. 10).

From some similar results, we deduce a semi-empirical scaling law: the luminosity is maximum when

$$\chi = \frac{\eta_-}{\eta_+} = \sqrt{\frac{T_\epsilon^+}{T_\epsilon^-}},$$

that is,

$$\frac{N_+ \gamma_+}{\sqrt{T_\epsilon^+}} = \frac{N_- \gamma_-}{\sqrt{T_\epsilon^-}}.$$

This seems to agree with the results of Tennyson[54] based on a multiparticle tracking.

With different values of T_ϵ, the catastrophe structure remains the same but the Fig. 10 becomes tilted. As long as T_ϵ's are fixed, our parameters are still two (H and χ or equivalently η_+ and η_-). It is quite interesting to decide the catastrophe type for four (H, χ, T_ϵ^+ and T_ϵ^-) or more parameters.

REFERENCES

1. For example, A.W. Chao, in *Physics of High Energy Particle Accelerators-1983*, edited by M. Month, P.F. Dahl and M. Dienes, AIP Conference Proceedings No.127 (American Institute of Physics,New York,1985),p.201, and references therein.

2. For example, *Nonlinear Dynamics and the Beam-Beam Interaction-1979*, edited by M.Month and J. C. Herrera, AIP Conference Proceedings No. 57 (American Institute of Physics,NewYork,1979).

3. J.T. Seeman, *Observations of the Beam-Beam Interaction of Particle Accelerators-1985*, edited by J.M. Jowett, M. Month and S. Turner, Lecture Notes in Physics No.247(Springer, Berlin Heiderberg New York Tokyo,1986).

4. The SPEAR group: R. H. Helm, M. J. Lee, M. Matera, P. L. Morton, J. M. Paterson, B. Richter, A. P. Saberski, P. B. Wilson, M. A. Allen, J. E. Augustin and G. E. Fischer, in *9th International Conference on High Energy Accelerators*, Stanford, 1974 p.66

5. A. Piwinski, I.E.E.E.Trans.Nucl.Sci. **NS-26**,4268(1979).

6. A. Chao and E. Keil, *Coherent Beam-Beam Effect*, CERN-ISR-TH/79-31(1979).

7. R. Talman, *Measurement, Analysis and Modification of Multi-Particle Phenomena in Accelerators*, Cornel Report CLNS-84/610(1980).
8. K. Hirata, Nucl. Instrum. Methods Phys. Res. **A269**,7 (1988).
9. T. Ieiri, T. Kawamoto and K. Hirata, Nucl. Instrum. Methods Phys. Res. **A265**,364 (1988).
10. T. Ieiri and K. Hirata, *Observation and Simulation of Nonlinear Behavior of Betatron Oscillations during the Beam-Beam Collision*, KEK Preprint 89-10 (1989), submitted to the 1989 Particle Accel. Conf. Accel. Science and technology, March 20-23 ,1989, Chicago,U.S.A.
11. K. Hirata and E. Keil, *A Program for Computing Beam-Beam Modes*, CERN/LEP-TH/89-57(1989).
12. K. Hirata and E. Keil, Phys.Lett. **B232**,413 (1989).
13. K. Hirata and E. Keil, *Barycentre Motion of beams due to Beam-Beam Interaction in Asymmetric Ring Colliders*, CERN/LEP-TH/89-76(1989). To appear in Nucl. Instrum. Methods Phys. Res.
14. K. Hirata, Phys.Rev.Lett.**58**,25 (1987);**58**,1798(E)(1987).
15. K.Hirata, Phys.Rev.,**D37**,1307 (1988).
16. K.Hirata, *Solvable Model, Flip-Flop Hysteresis and Catastrophe in e^+e^- Colliding Storage Rings*, KEK report, No.86-102 (1987).
17. M. A.Furman, K. Y. Ng and A. W. Chao, *A Symplectic Model of Coherent Beam-Beam Quadrupole Modes*, preliminary version of SSC report, SSC-174 (1988).
18. J. F. Tennyson,*Flip-Flop Modes in Symmetric and Asymmetric Colliding-Beam Storage Rings*, LBL-28013 (1989) and in this proceedings.
19. R.L.Stratonovich, *Topics in the Theory of Random Noise*.
20. K. Hirata, *Beyond Gaussian Approximation for Beam-Beam Interaction —an Attempt—*, CERN/LEP-TH/88-56(1988).
21. K. Hirata, *Stratonovich Expansion and Beam-Beam Interaction*, CERN/LEP -TH/89-14, submitted to the 1989 Particle Accel. Conf. Accel. Science and technology, March 20-23 ,1989, Chicago,U.S.A.
22. K. Hirata, *Higher Order Stratonovich Expansion in Weak-Strong Beam-Beam Interaction*, CERN/LEP-TH/89-35, presented at the Third Advanced ICFA Beam Dynamics Workshop, Novosobirsk, U.S.S.R., 29 May-3 June 1989.
23. A. W. Chao and R. Ruth, Particle Accel. **16**, 201 (1985).
24. E. Forest, Particle Accel. **21**, 133 (1987).

25. R.E. Meller and R.H. Siemann, IEEE Trans. Nucl. Sci. **NS-28** (1981) 2431.
26. K. Yokoya, Y. Funakoshi, E. Kikutani, H. Koiso and J. Urakawa, KEK Preprint 89-14 (1989).
27. J. F. Schonfeld, Ann. Phys. (N.Y.) **160**, 149(1985); **160**, 194(1985); **160**, 241(1985).
28. S.Kheifets, Part.Accel.**15**,153(1984).
29. F.Ruggiero,Ann. Phys. (N.Y.) **153**,122(1984).
30. Y.H.Chin,*Renormalized beam-beam interaction theory*, KEK report, No.87-143 (1988).
31. M. Bassetti and G. Erskine, CERN/ISR-TH/80-06 (1970).
32. P. Bambade et al., Phys. Rev. Lett. **62** 2949 (1989).
33. E. Keil and R. Talman, Part. Acc. **14** 109 (1983).
34. E.Keil, Nucl. Instrum. Methods Phys. Res. **188** 9 (1981).
35. K. Hirata and E. Keil, to be published.
36. K. Satoh, in Proc. 6-th Symp. Accel. Sci. Technology (Tokyo, 1987) 18.
37. E. Keil, *Four Beam Compensation Schemes*, CERN-LEP TH/89-37 (1989), submitted to the Third Advanced ICFA Beam Dynamics Workshop, Novosobirsk, U.S.S.R., 29 May-3 June 1989.
38. G. Arzelier et al., Proc. 8th Int. Conf. High Energ. Accel. Geneva, 1971, p150.
39. H. Wiedemann in Ref.2, p.84.
40. M.H.R. Donald and J.M. Paterson, IEEE Trans. Nucl. Sci. **26**,3580(1979).
41. R. Tohm, *Stabilité et Morphogénèse: Essai d'une théorie générale des modèles*, Benjamin,New York(1972).
42. H. Whitney, Ann.Math.**62**,374(1955) and Ref.41.
43. J.M.T. Tohompson, *Instabilities and Catastrophes in Science and Engineering*,Wiley,Chichester.
44. E. C. Zeeman, *Catastrophe Theory*, Addison-Wesley,Reading,Mass(1977).
45. A.Piwinski, in Proc. 11 th Int. Conf. on High Energy Accelerators, Genève-July 1980, 751 (Birkhäuser,1980).
46. S. Myers, Nucl. Instum. Methods Phys. Res. **211**,263(1983).
47. K. Oide and K. Yokoya, private communication.
48. E. Kikutani, this proceedings.
49. K. Hirata and F. Ruggiero, *Treatment of Radiation in Electron Storage Rings*, CERN Report, LEP-note-611(1988).

50. K. Hirata and F. Ruggiero, *Treatment of Radiation for Multiparticle Tracking in Electron Storage Rings*, CERN/LEP-TH/89-43 (1989), presented at the XIV Int. Conf. on High Energy Accelerators, Tsukuba, 1989.
51. K. Hirata, F. Ruggiero, H. Moshammer and M. Bassetti, *Synchro-Beam Interaction* CERN SL-AP/90-02 (1990) and this proceedings.
52. K. Hirata, *Round and Flat Beams in the Phi-Factory*, INFN-FRASCATI report, ARES-15 (1989).
53. Y. H. Chin, this proceedings.
54. J. F. Tennyson, in this proceedings.
55. SPEAR Group, IEEE Trans. Nucl. Sci. **20**,838(1973).

BEAM-BEAM INTERACTION: EXPERIMENTAL

David H. Rice
Newman Laboratory of Nuclear Studies[†], Cornell University, Ithaca, NY 14853

ABSTRACT

Performance limitations of electron-positron circular colliders are reviewed with emphasis on examining beam-beam limits through experimental data collected from operating storage rings. Particular emphasis has been put on reviewing scaling laws for the beam-beam parameter, or "tune-shift."

Observational data on colliders with asymmetric energy are not readily available. Several measurements of the dependence of luminosity and beam-beam parameter on operating energy are presented to stimulate interest in further experiments.

BASIC RELATIONS

The basic equations[1] which aid in understanding beam-beam interactions evolve from fundamental definitions (equations 1) to an operationally useful form (equations 2) to a conceptual equation (3) which, if used with appropriate caution, helps to break down luminosity performance into one possible set of component parts.

In equation 1a the luminosity is defined as the proportionality constant between the counting rate of events and the event cross section, σ_T. The linear tune shift, or beam-beam parameter, is equivalent to the tune shift for a "small" gradient error as shown in (1b). In this case, "small" implies that there is negligible change in the ß function as a result of this error. (All symbols are defined in the Appendix.)

$$\dot{n} = L \cdot \sigma_T \qquad (1a)$$

$$\xi \equiv -\frac{1}{4\pi}\frac{\beta}{f} \qquad (1b)$$

More often one starts from the operational form where real machine numbers may be directly inserted to calculate luminosity or beam-beam parameter as shown in equations 2.

$$L = \frac{f^* \, N^+ N^-}{4\pi\sigma_x\sigma_y} \qquad (2a)$$

[†] Work supported by the U.S. National Science Foundation.

$$\xi_i = \frac{N_b r_e \beta_i^*}{2\pi\gamma(\sigma_x+\sigma_y)\sigma_i} \qquad (2b)$$

The subscript i in (2b) refers to either x or y as appropriate. Eliminating the beam sizes in the equations above gives the familiar conceptual form (equation 3) permitting us to break down luminosity performance into optics parameters (r and β_y^*), beam-beam limit (ξ_y), and current limit (I). This form must be used with caution, since a horizontal beam-beam limit or vertical aperture limit may be hidden in the limitation of the total beam current, I.

$$L = 2.17 \times 10^{32} \, E_o(1+r) \, \frac{\xi_y}{\beta_y^*} \, I \qquad (3)$$

As mentioned above, the linear tune shift is given by (1b) and (2b) only when the effect is small enough so that β^* is not significantly perturbed. If this is not the case, the matrix formalism may be used to calculate the true tune shift, ΔQ, and change in β^*. This procedure yields equations 4a and 4b.

$$\cos(\mu+\Delta\mu) = \cos(\mu) - \frac{\beta_o^*}{2f}\sin(\mu) \qquad (4a)$$

$$\frac{\beta^*}{\beta_o^*} = \frac{\sin(\mu)}{\sin(\mu+\Delta\mu)} \qquad (4b)$$

The effects of equation 4a are shown in Figure 1. As the fractional tune between interaction points approaches a half-integer multiple from above, the beam-beam space charge focussing effect decreases ß at the interaction point, lowering the actual ΔQ. As the tune approaches a half-integer multiple from below, ß is increased by the beam-beam effect and ΔQ is larger than ξ. Eventually the single-turn lattice is unstable, shown by the diverging values of ΔQ.

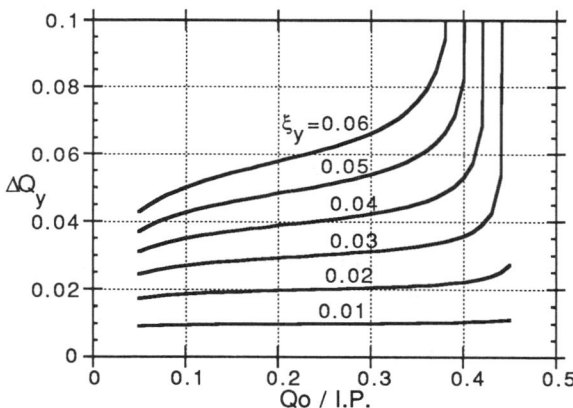

Fig. 1 - Coherent beam-beam tune shift vs. fractional betatron tune for several values of beam-beam space charge parameter.

PERFORMANCE LIMITATIONS - GENERAL

Equation 3 will be the starting point for the analysis of performance limitations of electron-positron colliders. While looking at the achieved values of the various terms in this relation, we must keep in mind that the goal of any accelerator built for high energy physics is to achieve maximum integrated luminosity in a given period of time. In many situations a strategic compromise has been made in one parameter when the result is higher average luminosity.

Although emphasis will be placed on the beam-beam limitations expressed in equation 3, a brief overview of the other limits is presented for completeness.

The first two terms are usually determined in early design considerations. E_o, the **beam energy**, is determined by the size of the machine or by desired physics. The **beam aspect ratio**, r, is fixed by machine floor plan or, in some cases, by limitations to bunch currents combined with large horizontal emittance which put a premium on minimizing vertical beam size at the interaction point.

Limitations to Beam Current - The total beam current, I, is limited by either **single beam** or **beam-beam** effects.

Particularly in larger storage rings, beam current is limited by **instabilities driven by wall impedances**. These may be divided into a 2x2 matrix of single or coupled bunch, longitudinal or transverse instabilities. A B factory storage ring will be limited at some point by single beam instabilities. This topic is covered in several other papers at this workshop.[2,3]

High beam currents cause both **heating** and degradation of **vacuum**. Both may result either from synchrotron radiation power striking the vacuum chamber (power/length $\propto I_{beam}E^4/\rho^2$) or from higher order mode heating (power $\propto R_{parasitic} \cdot \Sigma(i_{bunch})^2$). In the first case the power is given by design decisions and the task is to effectively dissipate it. In the second case, good design may lower $R_{parasitic}$, and dividing the beam current into many smaller bunches will lower higher order mode power, although at the expense of a lower threshold for coupled bunch instabilities. **RF power** may limit beam current near the maximum operating energy of a storage ring. Designing RF cavities to couple adequate power to the beams will be a critical task in planning a B factory.

There are two principal **beam-beam** limits to (bunch) current. The most common limit is the **growth of effective beam emittance** to the point where particles hitting an aperture, either physical or dynamic, cause unacceptably short beam lifetime or high detector background.

Measurements of the vertical aperture required for good beam lifetime have been performed at many storage rings.[4] A common method of making such a measurement is to use a scraper, or calibrated movable aperture. With colliding beams near the maximum current, a scraper is moved in toward the beam centerline until lifetime becomes unacceptably short. If the beam emittance and beta function (and energy spread and dispersion if appropriate) are known at the scraper, the distance of the scraper from the beam centerline may be expressed in units of rms beam size (σ_x or σ_y). The beam vertical emittance may be measured with a profile monitor[5] or calculated from the measured luminosity and lattice functions using equation 2a. In the horizontal plane, a direct calculation from the dipole fields and optics functions is generally sufficient.

A summary of measurements (from several machines) for the aperture required in the vertical plane is shown in Table 1. An average of $20 \times \sigma_y$ is required to maintain acceptable lifetime. If the transverse density distribution were gaussian, 6σ clearance would give about an 8 hour lifetime. There are many more particles in the "tails" of the vertical density distribution than would be expected in a single beam.

Table 1 - Measurements of vertical beam-beam aperture limits (from Ref. 4)

Minimum vertical half aperture of storage ring expressed in σ_y units (calculated from luminosity)	Number of measurements
14	1
15-19	3
20-24	7
25-29	2
30-34	2

The horizontal plane is not immune from non-gaussian tails either. Figure 2 shows the position of several horizontal scrapers from the beam centerline, expressed in single beam (ideal) sigmas, which resulted in a 1 hour beam lifetime plotted as a function of beam current (electron and positron beam currents were within 10% of one another).

A second method of presenting scraper data is shown in Figure 3. For each position of the scraper (measured from the beam centerline) the beam lifetime is measured. If the transverse charge distribution were perfectly

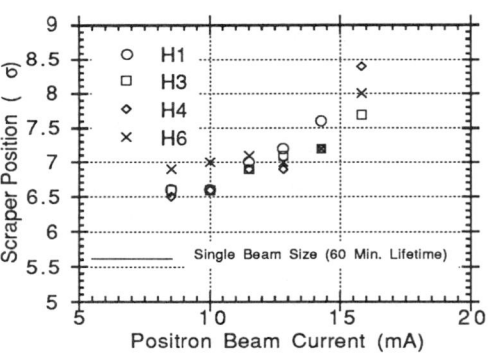

Fig. 2 - Horizontal scraper position normalized to local beam size (σ) which results in 60 min. lifetime vs. e^+ beam current.

gaussian, one could calculate the position (in units of σ) of the scraper from the beam center from this lifetime. This "ideal distribution" number is plotted on the ordinate. This presentation gives the "differential sigma" from the slope of a line drawn through the points, and the aperture required for a given life time may be read from the abscissa.

While the beam distribution is not nearly as distorted as is usually true for the vertical case, the increased aperture required may be critical in some situations. Multi-bunch operation at CESR is achieved by electrostatic separation of the beams in the horizontal plane. At the crossing points in the arcs sufficient separation must be

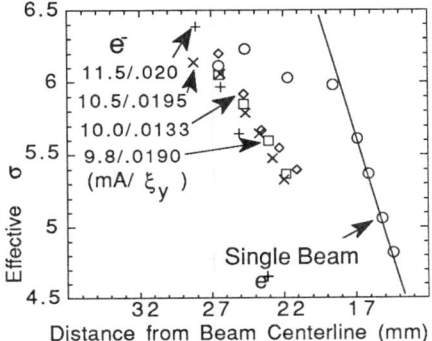
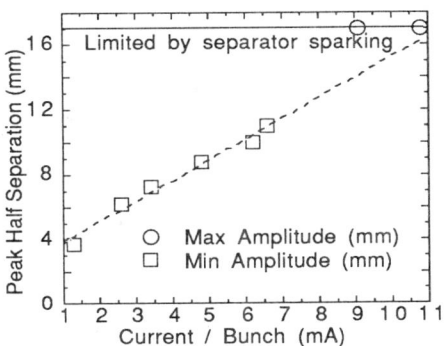

Fig. 3 - Beam lifetime as equivalent distance of a barrier from the center of a gaussian beam vs. scraper position.

Fig. 4 - 7 bunch separation requirements in CESR for > 1 hour beam lifetime.

provided to prevent particles out to at least 6 effective sigma amplitude from interacting with the passing bunch. Figure 4 shows the minimum peak orbit distortion (which is slightly more than half of the total separation between beams) required to maintain a 1 hour lifetime as a function of bunch current. These data are consistent with a 6 effective sigma separation requirement. In other words, the beams must be separated far enough so that particles one wishes to keep in the beam stay out of the center (±1 σ) of the opposing beam.

In order to minimize detector background from particle loss, it is a good design guideline never to have the smallest aperture (normalized by local beam sigma) near the interaction points. In this case, particles whose amplitude is increasing "slowly" (at least on the scale of a single turn) will initiate a shower far from the detector. Local gas scattering and synchrotron radiation would not be diminished, however.

Limits to β_y^* - The minimum value of β_y^* is usually determined by **aperture** or **chromaticity** becoming too large for satisfactory beam lifetime. The ß function grows as the beam moves in a field free region away from a minimum in ß (s=0) according to equation 5.

$$\beta(s) = \beta(0) + \frac{s^2}{\beta(0)} \qquad (5)$$

Where the dispersion function is zero, the beam size will scale as $\sqrt{\beta}$, resulting in loss of effective physical aperture in the first focussing quadrupole magnet. If an expensive increase in magnet bore is to be avoided, the first lens must be moved closer to the interaction point. Usually this means placing the magnet inside of the experiment detector, often in a longitudinal magnetic field of 1 tesla or greater. Both superconducting and permanent magnet[6] quadrupoles have been used in operating machines to minimize interaction point to quadrupole spacing.

Even if one were able to increase the physical aperture at the first quadrupole, the **chromatic errors** from the quad will scale as the β function at the quad divided by its focal length. Since this chromaticity must be compensated by increasing sextupole strengths and thereby reducing the dynamic aperture of the storage ring, a practical limit again is placed on β_y^*.

The continuous pressure on performance has stimulated clever engineering and cooperation between the detector and machine designers to reduce the distance from the interaction point to the first quadrupole. The limits to β_y^* imposed by the two effects above have been lowered in several machines until the beta to bunch length ratio approaches unity. Equation 5 shows that the rate of increase of β with distance from the interaction point increases as β* is lowered. If the bunch length, σ_s, is comparable with β*, a significant number of collisions will take place where β is larger than β*. The luminosity is lowered (equation 3) and the beam-beam parameter is increased (equation 2b).[7]   Performance may be lowered further by increasing tune modulation. An electron executing an energy oscillation will collide with the center of the opposite bunch at a a varying position depending on the phase of the energy oscillation at the collision point. A change in betatron phase advance between collisions results as the collision point moves from one interaction to the next. Since $\Delta\mu \approx \Delta s/\beta$, the modulation of betatron phase between interaction points will increase as β* is lowered, introducing coupling between synchrotron and betatron motion.[8,9] The resulting coupled resonances reduce the beam-beam limit. The approximate betatron tune modulation amplitude is given by:

$$\Delta\mu \approx \frac{2\pi}{k} \frac{\Delta s_{max}}{\beta^*} Q_s \sin\left(\frac{2\pi Q_s t}{k}\right) \quad (6)$$

There is some evidence from computer tracking[10] that a compensating effect is present for round beams when $\beta_y^*/\sigma_s \approx 1$. The effectiveness of the beam-beam kick is reduced since the force is exerted on an electron over a significant part of a betatron period. This averaging effect appears to reduce the strength of synchro-betatron resonances driven by the tune modulation mentioned above.

Measurements of beam-beam performance in the regime where $\beta_y^* \approx \sigma_s$ have been made at SPEAR and CESR.[11,12]   Figures 5 and 6 show the performance of CESR as β_y^* is varied with a constant σ_s of 2.2 cm. In Figure 5 the luminosity measured at each of the two interaction points is compared with the behavior expected at constant current with no "hourglass" effect from the small β, and with the purely geometric hourglass effect added. Figure 6 shows the beam-beam parameter (calculated from the luminosity) and maximum current (limited by beam lifetime or detector background) as β* is varied. While ξ_y and the maximum current decrease with β* in this regime, the luminosity achieved has a peak around $\beta_y^* \approx \sigma_s$.

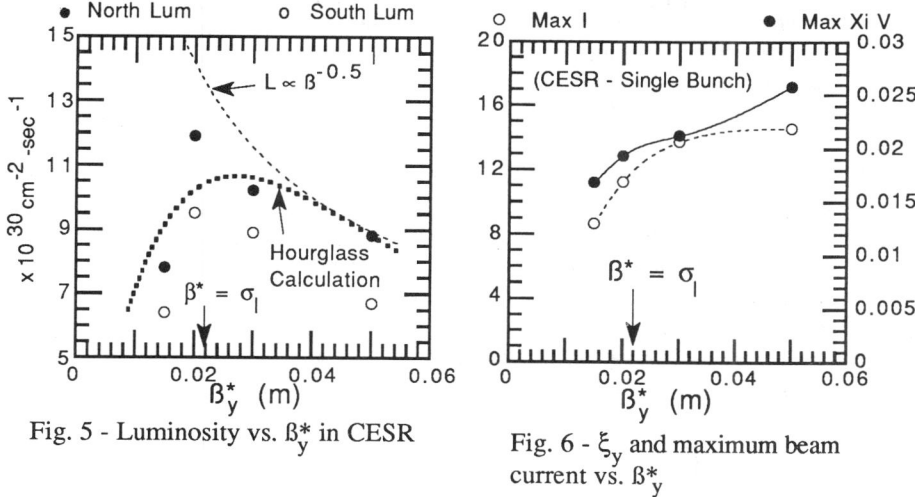

Fig. 5 - Luminosity vs. β_y^* in CESR

Fig. 6 - ξ_y and maximum beam current vs. β_y^*

PERFORMANCE LIMITATIONS - ξ

More so than the other parameters in equation 3, limits to the beam-beam space charge parameter, ξ, (often referred to as "tune shift") have proven difficult to predict by analytic or simulation approaches. While a few colliders have observed limits to ξ_x, in most machines ξ_x scales linearly with current, leaving the interesting beam dynamics to the vertical dimension. Both coupling and parametric[13] driving forces cause the

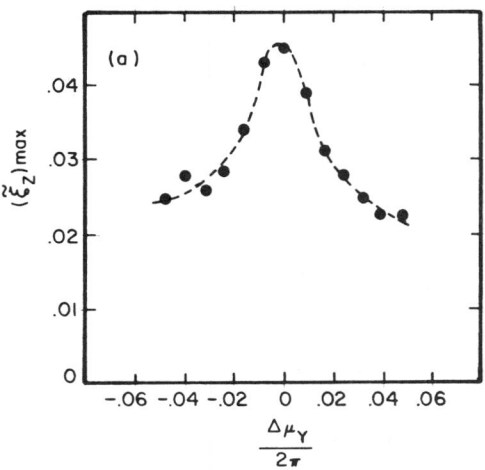

Fig. 7 - Maximum ξ_y vs. asymmetry in betatron phase advance. (From VEPP-2M, Ref. 28)

vertical emittance to grow significantly beyond the single beam value, reducing ξ_y below the value expected from the low current beam size. There have been several proposals for parameterization of observed values of ξ_y, two of the more often cited are those of Keil and Talman[14], and Seeman[4].

Caution must be used in the interpretation of experimental data since several well known effects can reduce the attainable value of ξ_y. Measurements on VEPP-2M show clearly the sensitivity to asymmetries in phase advance between interaction points (Figure 7). With perfect symmetry beam-beam driven resonances appear only according to the tune between interaction points.

When this symmetry is broken, resonance lines in the full machine tune plane become important, increasing opportunities for vertical emittance increase.

Dispersion at the interaction point, a finite crossing angle between electron and positron beams, and vertical dispersion or coupling in the arcs may increase vertical beam sizes and lower ξ_y. Optics non-linearities and small beam misalignment at the interaction point may increase sensitivity to particular classes of resonances. Figure 8 displays data from VEPP-4 showing sensitivity to horizontal misalignment of the beams at the interaction point. Although the luminosity is constant, beam loss when crossing the $7Q_x = 60$ resonance is very sensitive to small radial beam separation.

While attempting to parameterize experimental data, then, one should discard, or at least flag, data which is likely affected by "avoidable" limits to ξ_y, concentrating attention on those points representing fundamental limits.

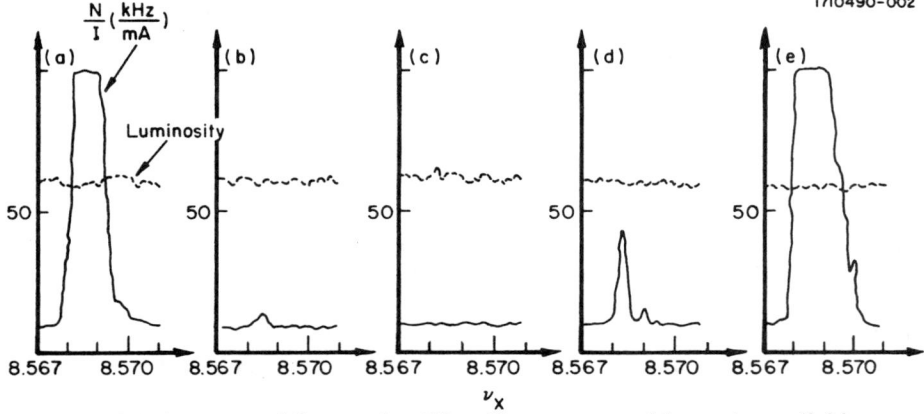

Fig. 8 - Particle loss rate while crossing $7Q_x=60$ resonance with varying radial beam separation in VEPP-4. a) $\Delta x=-0.1\ \sigma_x$; b) $\Delta x=-0.07\ \sigma_x$; c) $\Delta x=0.0\ \sigma_x$; d) $\Delta x=+0.07\ \sigma_x$; e) $\Delta x=+0.1\ \sigma_x$. (From Ref. 31)

Several scaling laws for ξ_y have been proposed. Some use true tune shift (ΔQ), others use the beam-beam parameter (ξ). Some possibilities are:

1. Constant ΔQ / I.R.

2. ξ_y scaling with damping decrement δ (Keil, Talman[14])

3. ΔQ scaling with $1/E_{beam}$, $1/\sqrt{k\rho}$ (Seeman[4])

Table 2 lists some parameters and measured luminosity and beam-beam parameters for several electron-positron storage rings. Several machines on this table (LEP, TRISTAN, PETRA>17 GeV) will not be discussed further since they are limited by single beam effects and any measured beam-beam parameters do not represent any beam-beam related limits. DORIS-2 will be included even though present operation is apparently not beam-beam limited.

Machine	(eff) rho	Circum	(k) #IR's	Energy GeV	Uo(MeV)	∂ (xE3)	H emit (μm-rad)	sigE/Eo (xE3)	B*(y)	B*(x)	Eta*(x)
VEPP-2M	1.22	18	2	0.5	0.0045	0.002	0.030	0.388	0.03	0.30	0.40
DCI	3.82	94.6	2	0.8	0.0095	0.003	0.068	0.351	2.00	2.00	0.70
ADONE	5	104.9	6	1.5	0.0896	0.005	0.371	0.575	3.00	9.00	2.20
SPEAR 1.2	12.7	234	2	1.2	0.0144	0.003	0.042	0.289	0.10	1.20	0.00
SPEAR 1.9	12.7	234	2	1.885	0.1130	0.012	0.435	0.454	0.10	1.20	0.01
SPEAR 2.1	12.7	234	2	2.08	0.1304	0.016	0.127	0.500	0.10	1.20	0.00
BEPC	10.345	240	2	1.6	0.0560	0.009	0.238	0.427	0.09	1.30	0.00
DORIS-2	12.22	288	2	5.3	5.7125	0.270	0.422	1.300	0.04	0.64	-0.39
VEPP-4	33	365	1	5	1.6756	0.168	0.104	0.746	0.12	3.00	0.00
KEK-AR 2 ip	23.2	375	2	5	2.3834	0.142	0.230	0.890	0.03	1.20	0.00
KEK-AR 1 ip	23.2	375	1	5	2.3834	0.284	0.170	0.890	0.03	1.40	0.00
CESR 1b 4.7	60	768	2	4.7	0.7195	0.038	0.180	0.520	0.03	1.00	1.10
CESR 1b 5.0	60	768	2	5	0.9216	0.046	0.200	0.554	0.03	2.00	1.10
CESR 1b 5.3	60	768	2	5.3	1.1635	0.055	0.225	0.587	0.03	1.00	1.50
CESR 1b 5.5	60	768	2	5.5	1.3493	0.061	0.240	0.609	0.03	1.03	1.00
CESR 7b 5.4	60	768	2	5.43	1.1432	0.059	0.200	0.601	0.02	1.00	0.55
PEP 6 ip	165.52	2200	6	14.5	23.65	0.136	0.120	0.966	0.11	2.95	0.01
PEP 2 ip	165.52	2200	2	14.5	23.65	0.408	0.117	0.966	0.14	2.90	0.02
PEP 1 ip	165.52	2200	1	13.69	18.72	0.686	0.100	0.912	0.05	1.34	0.00
PETRA 7	192	2304	4	7	0.1500	0.020	0.113	0.433	0.08	1.20	0.00
PETRA 11	192	2304	4	11	6.7463	0.077	0.150	0.681	0.08	1.20	0.00
PETRA 17	192	2304	4	17	38.48	0.283	0.200	1.052	0.08	1.20	0.00
TRISTAN	247	3018	4	30.4	318.30	1.297	0.136	1.659	0.10	2.00	0.00
LEP	3096.2	26658.9	4	46.5	133.59	0.359	0.012	0.717	0.07	1.75	0.00

Table 2 - Machine Characteristics (Numbers in italics indicate estimated or calculated values)

228 Beam-Beam Interaction: Experimental

Machine	Q(y)	Q(x)	Q(s)	bunch sig(l)	(xE10) Nb	xi x	xi y	ΔQy	By/By(0)	(xE30) Lpk	Ref#
VEPP-2M	3.08	3.06	0.010	0.030	0.64	0.015	0.050	0.035	0.546	3	9
DCI	1.80	3.80			11.80	0.044	0.041	0.055	2.11	0.36	15
ADONE	3.05	3.10	0.003	0.430	10.00	0.034	0.070	0.027	0.24	0.2	16
SPEAR 1.2	5.17	5.28					0.018	0.017	0.85	0.2	17
SPEAR 1.9	5.17	5.28		0.030	8.00	0.022	0.056	0.046	0.696	2	18
SPEAR 2.1	5.17	5.28					0.055	0.045	0.70	4	17
BEPC	6.80	5.80			8.50	0.051	0.035	0.044	1.70	2	19
DORIS-2	5.24	7.16			27.00	0.014	0.026	0.024	0.866	30	20
VEPP-4	9.57	8.55	0.024	0.060	7.60	0.017	0.060	0.046	0.638	5	9
KEK-AR 2 ip	9.20	9.13	0.042	0.015	17.10	0.034	0.030	0.027	0.821	24	21
KEK-AR 1 ip	10.27	10.13	0.037	0.014	17.10	0.046	0.045	0.047	1.086	33	21
CESR 1b 4.7	9.36	9.39	0.490	0.020	19.00	0.018	0.018	0.018	0.96	6	22
CESR 1b 5.0	9.36	9.39	0.490	0.021	21.40	0.025	0.022	0.022	0.95	6.5	22
CESR 1b 5.3	9.36	9.39	0.490	0.023	25.44	0.011	0.026	0.025	0.94	12.7	22
CESR 1b 5.5	9.37	9.40	0.490	0.024	22.40	0.016	0.028	0.027	0.94	15.5	22
CESR 7b 5.4	9.36	9.39	0.064	0.017	14.40	0.020	0.020	0.020	0.95	100	23
PEP 6 ip	18.19	21.25	0.035	0.020	37.00	0.049	0.045	0.031	0.518	32	18
PEP 2 ip	18.21	21.21	0.035	0.020	41.00	0.055	0.065	0.054	0.732	14	18
PEP 1 ip	18.21	21.30	0.024	0.021	32.00	0.054	0.050	0.049	0.961	55	18
PETRA 7	23.13	25.20	0.042	0.006	13.70	0.040	0.014	0.014	1.02	1.9	24
PETRA 11	23.12	25.19	0.054	0.007	13.70	0.019	0.024	0.024	1.043	8	24
PETRA 17	23.29	25.19	0.070	0.009	12.50	0.008	0.040	0.043	1.20	17	24
TRISTAN	38.72	36.60		0.012	20.00	0.011	0.034	0.033	0.93	10	25
LEP	77.29	71.38		0.018	17.40	0.072	0.023	0.024	1.09	4.45	26

Table 2 (cont.) Machine Characteristics

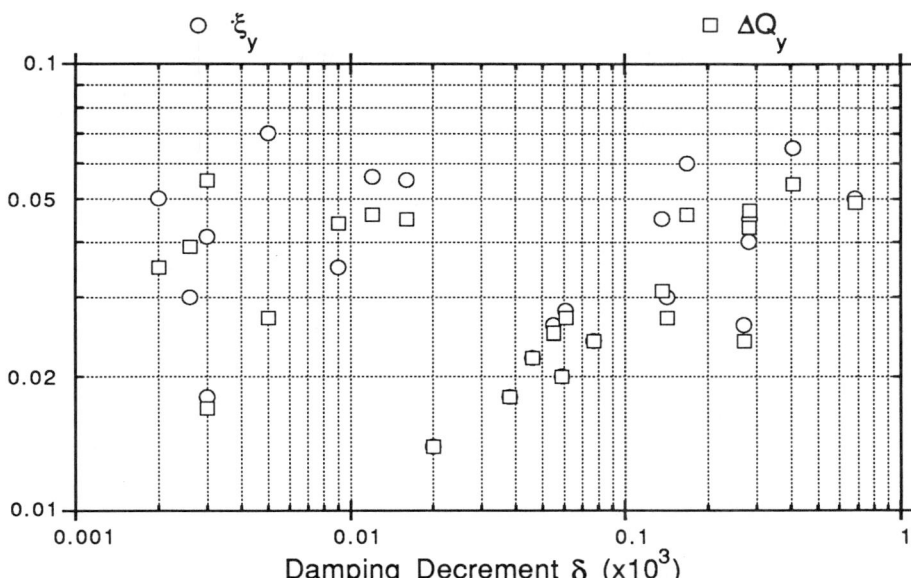

Fig. 9 - Vertical beam-beam parameter and coherent tune shift vs. damping decrement for several electron-positron colliders.

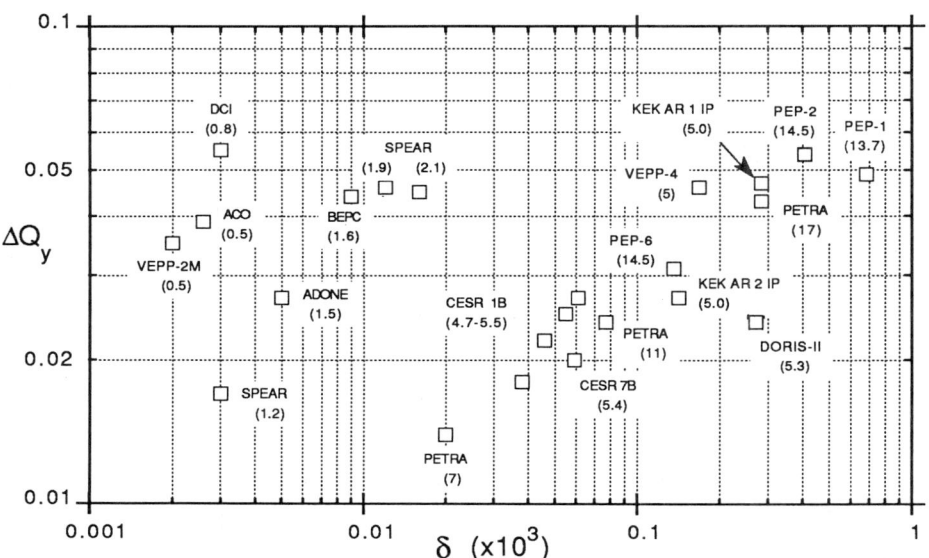

Fig. 10 - Vertical coherent tune shift vs. damping decrement for several electron-positron colliders. Beam energy (GeV) is shown in parenthesis.

Both ξ_y and ΔQ are plotted against damping decrement in Figure 9. Neither ξ_y nor ΔQ_y appear to be constant, nor is any strong functional dependence on δ obvious. The scatter in ΔQ_y does seem to be a bit less than ξ_y, and the ΔQ_y data have been plotted again in Figure 10 along with identification of the source of each datum. In this presentation there are many points within 15% of $\Delta Q_y = 0.045$ over more than 2 orders of magnitude change of δ (and a factor of 30 difference in beam energy). Furthermore, the data which fall below this group are mostly from machines operating below their maximum design energy.

One may ask how is the maximum design energy related to other parameters? An important operating parameter is the fractional energy spread[27], σ_δ, both for physics resolution and machine dynamic aperture. For iso-magnetic bend fields, $\sigma_\delta \propto E_0/\sqrt{\rho}$, suggesting that $E_{max} \propto \sqrt{\rho}$ would be appropriate. The maximum energy is compared with the bending radius of several machines in Figure 11. The scaling is somewhat steeper than $\rho^{-1/2}$, possibly due to increasing constraints of real estate as machine circumference grows. This factor in ΔQ scaling is consistent with Seeman's proposal.

Data points for a single machine operating at different energies tend to fall on a line in Figure 10. It would be appropriate to look at energy scaling of ΔQ_y for individual machines separate from overall energy scaling. Summaries from energy scaling data (ξ_y) for several machines are shown in Table 2. While it appears justified to pick any exponent between 0 and 3, there are 3 machines where $\xi_y \propto E_{beam}$, so an exponent of 1 will be used.

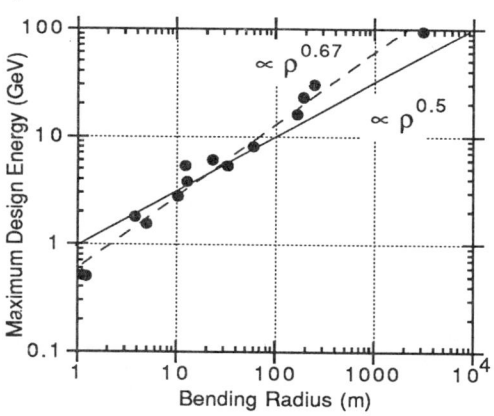

Fig. 11 - Maximum design energy of e^+e^- colliders as a function of bending radius

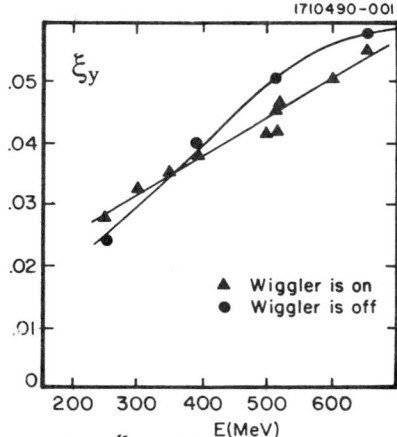

Fig. 12 - ξ_y vs Energy with and without 7.5 T wiggler in VEPP-2M (From Ref. 28)

We should ask if the scaling of ξ with energy for a machine can be altered. This is a critical question for asymmetric colliders with two equal circumference rings since the low energy ring is likely to be operating significantly below its "natural" maximum energy, or have many short, strong bending magnets. The addition of a wiggler to increase the damping (and energy spread) is a possible solution. Data from VEPP-2M[28] (Figure 12) suggest that this should be examined further. While

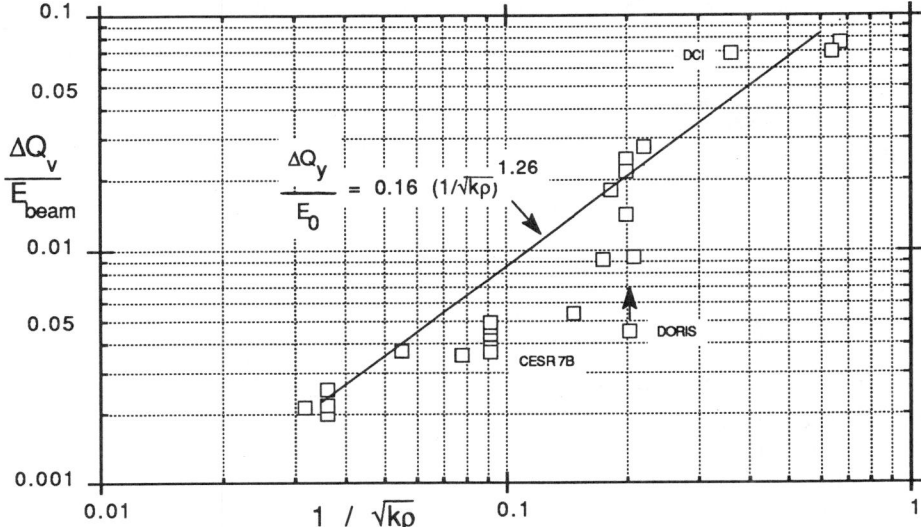

Fig. 13 - Vertical coherent tune shift divided by beam energy plotted as a function of one over the square root of the number of interaction regions times the bending radius. This is the same scaling used in Ref. 4.

luminosity was increased because of the larger emittance and higher current limit, ξ_y in fact decreased slightly with the wiggler on, possibly due to non-linearities introduced by the strong (7.5 T) wiggler. Additional experiments in this area would be appropriate.

The scaling of ΔQ with bending radius and beam energy favor Seeman's parameterization. The final factor, the number of interactions per ring (k), should enter as $k^{-1/2}$ according to the diffusion model[1] of the beam-beam interaction. Measurements at ADONE, CESR, and KEK AR[16,30,21] are consistent with this. Contrary experience comes from PEP[30] where maximum ξ_y is independent of the number of interaction regions, although reaching that maximum requires significantly more tuning as the number of interactions increases.

The data from Figure 10 are replotted in Figure 13 following Seeman's suggestion, with $\Delta Q_y/E_{beam}$ for the y value and $(k\rho)^{-1/2}$ on the x axis. The higher points are fit to an exponent of 1.26. Since the appearance of scaled data on a graph can be misleading, the rms spread of the scaled data from the fit line are compared with the spread in the unscaled data in Table 3.

Table 3 - R.M.S. Spread of Original Distributions and Fig. 13 Fit

Population RMS value of:	$\dfrac{\Delta Q_i}{\langle \Delta Q \rangle}$	$\dfrac{\xi_i}{\langle \xi \rangle}$	$\dfrac{(\Delta Q/E_0)_i}{0.16(1/\sqrt{k_i \rho_i})^{1.26}}$
Full	0.428	0.380	0.324
Less DORIS, DCI, CESR 7 bunch	0.440	0.364	0.255

The fit looks somewhat better if machines with unusual operating conditions are dropped from the group. (DORIS is single beam current limited, DCI had round beams, and CESR 7 bunch looses significant ξ from the separation scheme.) We might, with some justification, eliminate some of the other lower points, improving the fit of this parameterization. A more satisfying (but less realistic) approach would be to reconfigure the machines falling below the curve, eliminating the causes for low ξ_y, and make new measurements.

CONCLUSIONS

1) The performance of a collider should be beam-beam limited to make maximum use of available beam current and machine aperture. Single beam limits are usually avoidable by better engineering of RF systems, vacuum components, and feedback systems.

2) The interaction region should never include the limiting aperture of the machine. This may require building large bore (expensive) quadrupoles or clever schemes to move the i.r. quads close to the interaction point.

3) Best performance is obtained when $\sigma_s \approx \beta_y^*$ if optics do not limit β^*.

4) Asymmetries in phase advance between interaction points, dispersion at the interaction point, and other optics or misalignment errors should be avoided if possible.

5) Operating with the tune between interaction points above but close to a half integer multiple will provide the largest ξ_y (which appears in the luminosity formula) for a given ΔQ_y (which seems to be a more appropriate scaling parameter).

6) Damping decrement alone is not an appropriate scaling parameter for ξ_y. Computer simulations often show performance dependences on δ; however, they usually change radiation excitation as well as damping to keep the beam emittance constant. Referring to radiation effects rather than damping would be more accurate. Radiation excitation, (possibly the random kicks given to betatron motion) rather than damping, may be the more important effect in determinimg beam-beam limits.

7) Seeman's parameterization of ΔQ_y may be accurate if convincing evidence can be found that the data points which fall below the main trend do so because of

"avoidable" errors as discussed in section 3. There is a suggestion in the data that $\Delta Q_y \leq 0.05$ for all electron-positron colliders.

The author wishes to acknowledge the contributions of many people for both providing data and discussing the various models. M.H. Donald, J.M. Paterson, J. Seeman, M. Billing, D. Rubin, R. Siemann, and A.B. Temnykh have been particularly helpful.

APPENDIX - Symbol Definitions

E_0	beam energy
f^*	frequency of collisions (/second)
I	total current per beam
i_b	current per bunch
k	number of interaction points in the storage ring
L	Luminosity (/cm^2-/second)
$\dot{n}$	event rate per second
$N_b^{+/-}$	Number of particles/bunch (positrons/electrons)
$Q_{x,y,s}$	Betatron, synchrotron tune (number of periods/revolution)
r	beam aspect ratio σ_y/σ_x
r_e	classical electron radius (2.8177 x 10^{-15}m)
$R_{parasitic}$	longitudinal parasitic resistance
s	longitudinal coordinate
t	time
T_0	revolution period
x	horizontal (radial) coordinate
y	vertical coordinate
f	lens focal length
β	beta (amplitude) function
δ	damping decrement ($\delta = T_0/k\tau_y$)
γ	relativistic factor
μ	betatron phase advance (radians)
ρ	magnet bending radius (m)
σ_δ	beam rms energy spread (σ_E/E_0)
σ_T	process cross section (/cm^2)
$\sigma_{x,y,s}$	rms beam size (horizontal, vertical, longitudinal)
τ_y	damping time constant (vertical)
$\xi_{x,y}$	beam-beam space charge parameter
$*$	(star) indicates value at interaction point

REFERENCES

1. A. Chao, AIP Conf. Proc. 127, 201 (1983)
2. M. Zisman, "Instabilities", invited talk at this workshop
3. F. Pederson, "Feedback Systems", invited talk at this workshop
4. J. Seeman, Lecture Notes in Physics No. 247, 121 (Springer-Verlag, 1985)
5. S. Milton, CESR Operations Note 86-7 (1986)(internal note)
6. S. Herb, J. Kirchgessner, IEEE Conf. Proc. 87CH2387-9,130 (1987)
7. S. Milton, Cornell CBN 89-1 (1989)
8. S. Peggs, IEEE Trans. Nucl. Sci. NS-30, 2457 (1983)
9. G.M.Tumaikin, A.B.Temnykh, XIII Intl. Conf. H.E. Acc., 88 (1986)
10. S. Krishnagopal, R. Siemann, Phys. Rev. D, 41, 2312 (1990)
11. SPEAR Storage Ring Group, IX Intl. Conf. H.E. Acc., 37 (1974)
12. D. Rice, XIV Intl. Conf. H.E. Acc. (1989)
13. S. Peggs, R. Talman, XI Intl. Conf. H.E. Acc., 754 (1980)
14. E. Keil, R. Talman, CERN-ISR-TH/81-33 (1981)
15. R. Chehab et.al., XI Intl. Conf. H.E. Acc., 702 (1980)
16. F. Amman et.al., VIII Intl. Conf. H.E. Acc., 132 (1971)
17. H. Wiedemann, SLAC-PUB-2320 (1979)
18. M.H. Donald, J. Seeman, private communication (1989)
19. BEPC Storage Ring Commissioning Report (March, 1989)
20. H. Nesemann, B. Sarau, IEEE Trans. Nucl. Sci. NS-32, 1644 (1985)
21. E. Kikutani et.al., "Measurement of Luminosity in TRISTAN Accumulation Ring", XIV Intl. Conf. H.E. Part. Acc. (1989)
22. CESR run sheets for fills 82301.01, 82301.02, 83009.03, 83009.04, 82268.04, 82268.04, 83213.02, 83226.06
23. CESR run sheets for fills 88067.07, 88070.01
24. A. Piwinski, IEEE Trans. Nucl. Sci. NS-30, 2378 (1983)
25. Y. Funakoshi et.al., "Luminosity Tuning in TRISTAN Main Ring", XIV Intl. Conf. H.E. Part. Acc. (1989)
26. E. Blucher, private communication. (Data from Nov. 1989)
27. A. Chao, AIP Conf. Proc. 153, 103 (1985)
28. P. Ivanov, III ICFA Workshop on Advanced Beam Dynamics (1989)
29. D. Rubin, CESR Operations Note 88-6 (1988)
30. J.M. Paterson, J. Rees, private communciation
31. A.B. Temnykh, III ICFA Workshop on Advanced Beam Dynamics (1989)

RF Systems*

D.L. Rubin

Laboratory of Nuclear Studies, Cornell University, Ithaca, NY 14853

Introduction

Storage ring B factories are characterized by high current beams ($\sim 3A$) of short bunches ($\sigma_L \sim 1cm$) that radiate high power ($\sim 3MW$). The RF system is required to supply the power and sustain a voltage adequate for the strong longitudinal focusing. The beam cavity interaction limits the stability of single and multiple bunch beams. The optimized RF system therefore has low impedance and heavy damping of parasitic modes.

The task at hand is to develop an RF system that can meet these extraordinary requirements. To that end it is useful to review the character and especially the limitations of existing technology, including those imposed by input power window and cavity accelerating gradient. In view of these limits we find that single cell cavities are a logical alternative to the multicell structures common to operating storage rings. We then proceed to an optimization of the cell shape for such a single cell structure that minimizes higher order mode impedance and facilitates the coupling of HOM power out of the cell and into loads. The conceptual design of an associated fundamental power coupler and waveguide window suitable for a superconducting single cell cavity are described. The relative virtues of superconducting and room temperature structures are reviewed.

Single Cell Cavities

POWER WINDOW

The fundamental power that is coupled by the RF cavity from klystron to beam must pass through an RF window. The power through the window is limited by heating of the ceramic, sparking due to high fields associated with standing waves, multipacting, etc. Local field levels and vacuum conditions are all critical to the window performance but generally peculiar to the detailed design of the window and its mounting hardware. The power handling ability of a few cavity windows is summarized in Table I.[1][2] The tabulated power corresponds to the level at which the cavity was operated to accelerate beams.

Table I. RF Windows		
Machine	Power	Type
CESR	350kW	Cylindrical
TRISTAN	225kW	Coaxial disc
PETRA	120kW	Coaxial disc
PEP	$\leq$ 150kW	

---

* Work supported by the National Science Foundation.

Note that the PETRA window is adapted from a klystron design. As a cavity window it can not be operated reliably above 120kW. In the klystron it operates reliably up to 1MW [3]. Cavity and klystron window environments are very different. The klystron window is isolated from reflected power whereas the cavity window is exposed to full reflections at the driving frequency as well as at higher mode frequencies. In addition the klystron vacuum environment is generally superior to that of the cavity coupler.

It is clear that barring a major breakthrough in window cavity technology a B-Factory storage ring RF system will require at least 10 windows.

ACCELERATING GRADIENT

The accelerating gradient is limited by RF dissipation in a copper cavity and electron field emission in a superconducting structure at levels corresponding to $\leq 1.5 MV/m$ and $5 - 10 MV/m$ respectively. The gradients of several normal conducting RF systems are summarized in Table II. Note that in the design of all such systems the fundamental $(\frac{R}{Q})$ is maximized in order to minimize power consumption. We will find that high fundamental $(\frac{R}{Q})$ corresponds to high impedances in parasitic modes.

Table II. Normal Conducting RF Systems					
Machine	frequency	$(\frac{R}{Q})$	Gradient	P_{diss}	Number
	MHz	Ω/cell	MV/cell	kW/cell	cells
LEP	350			< 25	640
TRISTAN[1]	500	220	0.4	< 26	936
LEP SPS INJ	200	216	1	< 60	32
PEP[2]	350	291	0.37	17	120
CESR	500	256	0.35	14	28
HERA	500		0.33	< 14	~ 500

TRISTAN is the only storage ring with a significant complement of superconducting RF cavities. The gradients achieved in the operation of 16 5-cell, 500MHz cavities to accelerate beam average 4.4MV/m.[4] The TRISTAN superconducting cavities have $(\frac{R}{Q}) = 120\Omega/cell$.[5] In horizontal tests of 32, 5-cell TRISTAN cavities the average gradient is over 7MV/m.[6]

The peak accelerating voltage for a B factory storage ring is anticipated to be about 5 to 30 MV.[7] If the resonant frequency of the accelerating cavities is 500MHz then the active length of the cell is about 0.3m. Because the accelerating gradient of a copper system is limited to 1.5MV/m then 22 cells are required to establish a peak voltage of 10MV. A superconducting system with gradient of 7.5MV/m might be based on 5 to 15 cells. As noted above the minimum number of windows for the RF system is about ten. The numerology suggests for both copper and superconducting systems that the power to each cell be coupled by a dedicated window. The number of cells necessary to sustain the voltage is about the same as the number of windows required to deliver the power.

Typically the impedance of the storage ring is dominated by the RF cavities. The short range wake generated by the passage of the bunch through the cavities can result in bunch lengthening. The wake generated by the head of the bunch introduces a voltage that is destabilizing if its variation over the length of the bunch is comparable to that of the accelerating voltage. The instability threshold depends only on the loss parameter of each cell and the total number of cells. The optimized system has a minimum number of cells.

Multibunch stability thresholds are determined by the impedance of parasitic cavity modes and the decay time of energy stored in those modes. The impedances of the parasitic modes are the frequency domain decomposition of the loss parameter and so are a characteristic of the cell geometry. And the ring impedance is proportional to the total number of cells. Again the optimized system has as few cells as possible.

Damping is achieved by coupling higher order mode energy out of the cavities and into loads. The damping required to store high current beams in a B factory ($Q_{ext} <$ 100 [8]) can only be achieved if HOM couplers are located in each and every cell of the RF system. Since the multibunch threshold scales inversely with the mode shunt impedance, the loading in each cell necessarily increases with the total number of cells. The demand for heavy damping is conveniently addressed in single cells each with dedicated higher order mode loads.

Single cell cavities have some application in existing machines and are clearly practical devices. Thirty-two single cell copper cavities in the LEP injector operate at 200MHz and a gradient of 1.5MV/m. The power dissipated per cell is about 60kW. The shunt impedance is $R_s = 11.5 M\Omega$ at $Q_0 = 53000$, and $(\frac{R}{Q}) = 217\Omega/cell$.[9] A 500MHz single cell superconducting cavity tested in DORIS coupled 50kW to the electron beam.[3] We proceed with the notion of single cell cavities as the basis of the RF system, each cell with a complement of fundamental and HOM couplers.

Cavity Geometry

The optimization of the shape of the single cell is constrained by the requirements of fundamental and higher order mode shunt impedance and tolerable loss parameter. For superconducting cavities in which the effective accelerating gradient is limited by field emission, it is advantageous to choose a geometry that minimizes the ratio of peak surface to effective accelerating field. It is traditional to couple fundamental power into a superconducting cavity via an aperture in the beam tube. Good coupling then depends on a cell geometry in which a relatively large amount of energy is stored in the vicinity of the coupling hole.

Our strategy is to begin with a cavity geometry that is known to be free of multipacting when superconducting and that has a low ratio of peak surface to effective accelerating gradient. Geometries with reentrant noses are therefore excluded from consideration. We then proceed to optimize the geometry with respect to the defining parameters, namely the beam tube radius, the nose(iris) radius and the cell length.[10] While the excercise leads to solutions that differ in detail for superconducting and copper cavities, there are nevertheless many similarities, and in any event we are educated as to the tradeoffs in any design.

In terms of the cavity resonant modes we seek to maximize the $\left(\frac{R}{Q}\right)$ of the fundamental and minimize the $\left(\frac{R}{Q}\right)$ of all higher order modes. Furthermore it is attractive to locate couplers, both input and output along the beam tube where magnetic fields are relatively low, rather than inside the cell. (For a copper cavity it may be advantageous to couple to resonant modes through the main body of the cavity). For all modes we prefer to obtain strong coupling via a minimal discontinuity in the beam tube in order to minimize the overall loss parameter. It is therefore desireable to maximize the stored energy of the relevant modes at the coupling holes along the beam tube.

The $\left(\frac{R}{Q}\right)$ is computed directly with cavity codes URMEL[11] or SUPERFISH. The relative Q_{ext} of a mode from one geometry to the next is estimated by computing the fraction of the power dissipated in a band on the beam tube.[10] In summary we find that increasing the radius of the beam tube reduces the $\left(\frac{R}{Q}\right)$ of all longitudinal modes and increases the fraction of stored energy of all such modes in the beam tube. Upon reducing the cell length the impedance of the fundamental is increased and the higher order longitudinal modes begin to propogate into the beam tube. For a particular combination of cell length, iris radius and beam tube radius, the only remaining cavity modes with frequencies below the beam tube cutoff frequency are the fundamental accelerating mode, the TE111(644MHz) and TM110(680MHz) deflecting modes. For details of the optimization see H. Padamsee et. al.[10]

The resulting cell geometry that is the basis for further study is indicated in Figure 1. The cell length is 24cm, the beam tube radius is 12cm and the nose radius is 2cm. Its RF properties are summarized in Table III. $\left(\frac{R}{Q}\right)_{max}$ corresponds to the largest $\left(\frac{R}{Q}\right)$ of all of the many longitudinal modes that propogate out of the cell and into the beam tube but remain trapped wthin the boundaries of the tapered transition (see Figure 3.)

Table III. RF Properties of Optimized Cell	
frequency	500MHz
$\frac{R}{Q}$	89Ω/cell
E_{peak}/E_{acc}	2.5
H_{peak}/E_{acc}	55.8Oe/MV/m
$E_{surface}$ for 3MV/cell	25MV/m
$H_{surface}$ for 3MV/cell	558 Oersted
$k_L(\sigma_L = 1cm)$	0.11V/pico-C
$\left(\frac{R}{Q}\right)_{max}$(propogating HOM)	4Ω/cell

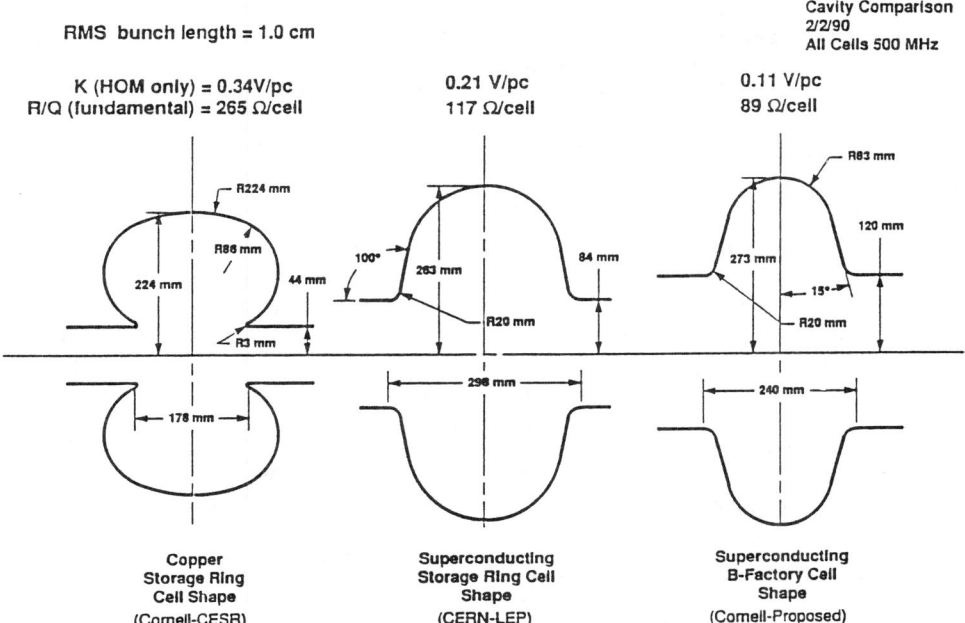

Fig. 1. At the left is a cross section of the CESR copper cell. The LEP superconducting cavity cell is shown in the center. The optimized B Factory superconducting cell is shown at the right. All three are scaled to 500 MHz. Loss parameter and fundamental $\left(\frac{R}{Q}\right)$ are indicated.

Couplers

Since we have supposed that the power window can transmit 500kW the fundamental coupler must be designed to match a 500kW load. The external Q associated with the coupler is chosen so that at the operating voltage, the full power is extracted by the beam and/or is dissipated in the cavity walls. Then the loaded $Q_L = \frac{P_{tot}}{\omega U}$ where $P_{tot} = P_{beam} + P_{walls}$ and the stored energy $U = \frac{V_{acc}^2}{(\frac{R}{Q})\omega}$. If the full forward power is matched into the beam and cavity walls then

$$Q_{ext} = Q_L = \frac{V_{acc}^2}{P_{tot}(\frac{R}{Q})}.$$

The total power and $(\frac{R}{Q})$ will not be very different for superconducting and normal conducting designs. But the accelerating voltage will be several times (3-7) greater for the superconducting cavity. Therefore, the copper cavity requires at least an order of magnitude stronger coupling and consequently a larger coupling aperture or more intrusive loop. If $V_{acc} = 3MV/cell (\sim 10MV/m)$, $P_{tot} = 500kW$, and $(\frac{R}{Q}) = 89\Omega/cell$ we find that $Q_{ext} \sim 2 \times 10^5$. Of course if $V_{acc} = 1MV/m$ then $Q_{ext} = 2 \times 10^4$.

Strong coupling is achieved in a copper cavity through an aperture or loop in the high magnetic field region of the cell near the equator. An advantage of such a configuration is that the discontinuities associated with the coupler are far from the beam. In superconducting operation departures from uniform curvature in the vicinity of the equator can induce multipacting and as a result in most SRF applications the hole is located on the beam tube. Because of the relative proximity to the beam, losses associated with the beam tube coupling iris can be significant and care must be taken to minimize the interference of the coupler with the wall currents.

A program of RF modeling, measurements, and computation has resulted in a conceptual design of a beam tube coupler for the B-factory single cell cavity described above. The coupling is achieved by way of a slot in the shorted end of a half height 500MHz waveguide that feeds power through a matching slot in the beam tube. The length (parallel to the beam axis) of the slot is equal to the height of the waveguide (10.2cm) and the width is 9.1cm. The waveguide is oriented so that electric fields propogating in the TE01 waveguide mode are parallel to the accelerating fields in the cavity. A tongue of width 3.7cm extends 8.3 cm into the slot. The tongue decreases the cutoff frequency of the simple rectangular opening and enhances the coupling. The geometry yields good coupling to the fundamental and the dipole modes trapped in the cell.

Table IV Fundamental Power Coupler

Mode	Frequency(MHz)	Q_{ext}
TM010	500	5×10^4
TE111	644	2×10^3
TM110	680	9×10^3

The waveguide coupling is expected to have high power handling capability. Unlike coaxial couplers there is no center conductor to be cooled or supported. In the case of the superconducting scenario, the waveguide can simply penetrate the cryostat via a transition similar to that of the beam tube penetrations. The transition to atmospheric waveguide is through a flat window that is matched to standard waveguide. The conceptual design of such a window and specifications are shown in Figure 2. The window is maintained at room temperature. The higher order modes are presumably directed to a load, also outside of the cryostat as shown in Figure 3.

Wall currents generate wake voltages due to the interaction with the break in the beam wall associated with the coupling hole. The losses are computed with the MAFIA[12] cavity code. The loss parameter for the cell not including the transition to a 5cm radius beam tube, is $k_L = 0.11 V/pico-C$ exclusive of the coupler and losses to the fundamental. The loss parameter increases to $k_L = 0.113 V/pico-C$ when the coupler (waveguide, slot, and tongue) is included (not quite 3%). The transverse wake voltage for the cell with coupler is given by $k_\perp = 0.37 V/pico-C/m$. The contribution to the wake voltages due to the coupler is quite small.

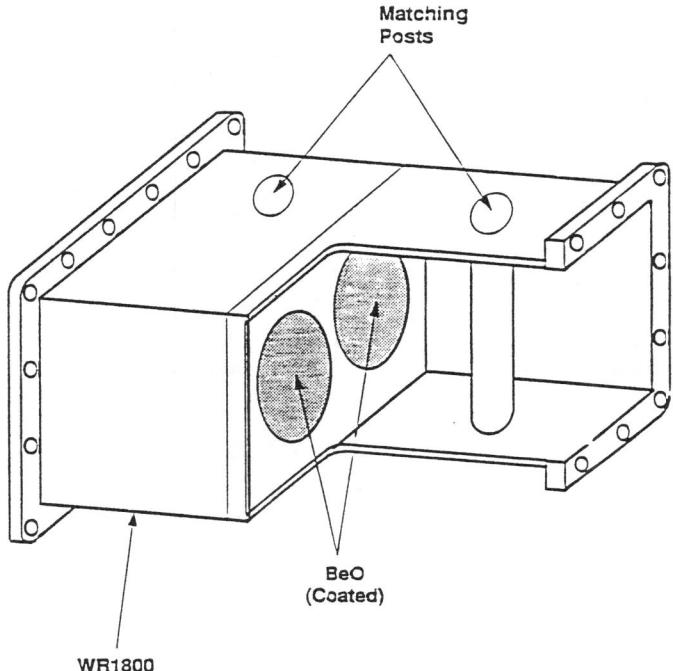

Fig. 2. 500 MHz high power window.

Single Bunch Stability

An estimate of the bunch lengthening threshold for a machine in which the longitudinal wake is dominated by the RF system follows from our knowledge of the longitudinal wake voltage and the peak accelerating voltage. The longitudinal wake generated by the passage of the bunch through the cell is decelerating. If the magnitude of the rate of change of the wake voltage over the length of the bunch approaches the time derivative of the externally applied accelerating voltage the bunch is likely to rearrange itself. The rate of change of the wake voltage over the length of the bunch is $\frac{1}{c}\frac{\partial V_{wake}}{\partial t} \sim \frac{qk_L}{\sigma_L}$, where k_L is the loss parameter per cell, q the bunch charge, and σ_L the bunch length. The rate of change of the accelerating voltage is $\frac{1}{c}\frac{\partial V_{acc}}{\partial t} = \omega V_{cell}/c$. If the ratio

$$\epsilon_w = \frac{\frac{\partial V_{wake}}{\partial t}}{\frac{\partial V_{acc}}{\partial t}} = \frac{qk_L c}{\sigma_L \omega_{RF} V_{cell}}$$

is greater than unity the charges are no longer an incoherent conglomeration. Then the charge threshold can be written as

$$q_{max} \leq \frac{\sigma_L \omega_{RF} V_{cell}}{k_L c}.$$

For $\sigma_L = 1cm$, $\omega_{RF} = 2\pi \times 500 MHz$, $V_{cell} = 1MV (\sim 3MV/m)$, and $k_L = 0.113 V/pico-C/cell$, $q_{max} \leq 9.1 \times 10^{-7}C = 5.6 \times 10^{12}$ electrons, at least an order of magnitude larger

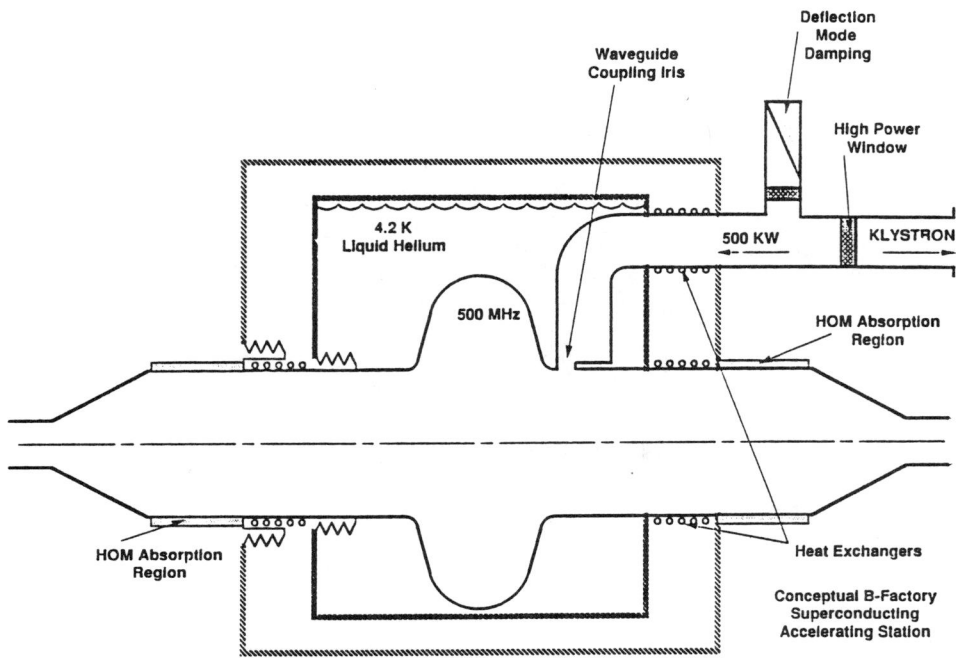

Fig. 3. Superconducting accelerating cavity.

than is likely in a storage ring. The threshold for the head tail instability due to the transverse wake is estimated[13] to be more than an order of magnitude higher than the bunch lengthening threshold. Of course the RF cavities with couplers are not the only source of longitudinal or transverse wakes. Indeed the taper transitions from large to small beam tubes in general contribute more to the loss than the cell itself. We can however be fairly confident that the proposed cell geometry does not constrain our choice of single bunch parameters.

Higher Order Mode Power

The beam power lost per cell to nonresonant wake fields is given by: $P = I_{avg}^2 k_L / f$ where I_{avg} is the average beam current and f the bunch frequency. For $I_{avg} = 3A$, $k_L = 0.13V/pico - C/cell$, and $f = 40MHz$, $P_{cell} = 26kW$. This substantial higher mode dissipation is a strong incentive to minimize the loss parameter of the cell and the total number of cells and to maximize the number of bunches in the beam. Note that the loss parameter of the cell in our conceptual design is relatively low (about 1/3 of that of the high shunt impedance version), due mostly to the large beam tube radius. The comparison with other geometries is indicated in Figure 1.

In addition to the losses associated with the single bunch passage there can be resonant buildup of stored energy in parasitic modes. Unless the modes are heavily damped, destabilizing voltages will rapidly accumulate.[10] For the highest impedance longitudinal mode in the optimized geometry, $(\frac{R}{Q}) \sim 4\Omega/cell$(including tapers). Then if $Q_{ext} \sim 100$, the power extracted from the beam and stored in the TM011 like parasitic mode is $P_{TM011} \sim 3kW/cell$ and for $Q_{ext} \sim 1000$, $P_{TM011} \sim 30kW/cell$ for $I_{avg} = 3A$. (For $Q_{ext} > 100$ the extracted power depends only on the average current.) The dense spectrum[10] of such modes essentially guarantees that some will interact resonantly with the beam. Heavy loading is therefore required simply to limit power consumption. Multibunch stability imposes additional constraints.

As noted above, all longitudinal modes propogate into the beam tube and outside the cooling vessel permitting relatively easy access to those modes. The possibility of lining the beam tube with an absorbing material such as a ferrite is under investigation. It offers the promise of broad banded loading of any modes with stored energy in the beam tube, which includes all longitudinal modes. Such a scheme precludes the need for additional coupling holes or loops to capture higher mode energy. There is a resistive wall loss associated with the absorber that for typical parameters is of the same magnitude as the wall loss of the entire remainder of the ring. The resistivity of the particular ferrite under study has a strong frequency dependence that peaks at about 2GHz and falls by an order of magnitude as the frequency increases to 12GHz as shown in Figure 4.

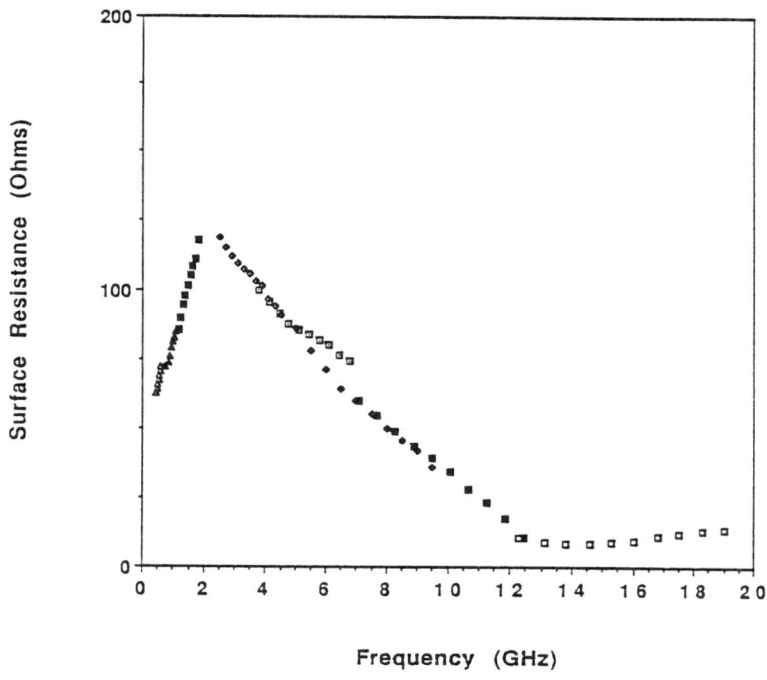

Fig. 4. Surface resistance of ferrite 50 as a function of frequency.

The general frequency dependence is a good match to machine requirements in so far as the modes trapped in the large radius beam tube are in the 1-3GHz range whereas a substantial fraction of the bunch power spectrum is above 12GHz. For the indicated absorption spectrum, the resistive wall heating of a 15cm long, 12cm radius ferrite 50 beam tube amounts to about 3.5kW for $I_{avg} = 3A, f = 40MHz$, and $\sigma_L = 1cm$. The effective resistance of the ferrite, averaged over the bunch power spectrum is about 800 times that of a copper tube of same dimensions.

The vacuum characteristics of the ferrite have been established by experiment to be adequate for the storage ring environment. Thermal properties and fabrication techniques are being investigated.

Superconducting or Copper RF Systems

Superconducting structures are attractive because they can sustain an accelerating voltage several times higher than the copper counterpart. Therefore fewer cells are required to provide the necessary longitudinal focusing. Higher mode losses increase linearly with the number of cells (of given geometry) and single and multibunch instability thresholds decrease inversely with the number of cells. Of course the relative merit of the superconducting system depends on the voltages that can be achieved. Table II summarizes the characteristics of some copper RF systems. Note that the peak gradient corresponds to about 1.3MV/m (at 500MHz the cell length is taken to be 0.3m). All of the cavities are designed with high fundamental $(\frac{R}{Q})$, about 2.5 times greater than that of the cell optimized for B-factory parameters. The existing systems are presumably dissipation limited (either by the economics of power or the technical difficulty of removing it). But similar gradients in the cell optimized for low HOM impedance and heavy HOM loading are attained only by a corresponding increase in dissipated power by a factor 2.5 to compensate for the deterioration of the $(\frac{R}{Q})$ in the fundamental.

The power dissipated in 4.2K Helium at $V_{acc} = 7MV/m, (\frac{R}{Q}) = 89\Omega/cell$, $Q_0 = 2 \times 10^9$ is 25W/cell corresponding to about 9.3 kW/cell of refrigerator power. The superconducting system is therefore capable of sustaining a much higher voltage at a fraction of the cost of the copper system both in terms of power and ring impedance. If on the other hand the number of cells (input windows) is determined by beam power rather than voltage requirements the advantage of superconducting RF is diminished.

Conclusions

The conceptual design of a B-Factory RF system consists of single cell cavities. The cell shape is optimized to:

1. Minimize $(\frac{R}{Q})$ of parasitic modes,
2. Force all longitudinal higher modes to propogate into the beam tube,
3. Yield strong coupling to the fundamental through an aperture in the beam tube,
4. Preserve fundamental $(\frac{R}{Q})$ at a level compatible with a superconducting system,
5. Minimize E_{peak}/E_{acc}.

Each cavity has a dedicated fundamental power feed. The waveguide coupler transmits power into the accelerating mode through a slot in the beam tube. The same slot couples to the only two parasitic modes that are trapped in the cell. The wake fields generated by the slot geometry are negligibly small. A ferrite absorber is perhaps the basis of a broad band higher order mode load if distributed along the large radius beam tube beyond the cutoff length of the fundamental. A 500kW RF window unit with impedance matched to standard waveguide is designed. The window is maintained at room temperature.

A system optimized for normal conducting operation is likely to have somewhat higher $(\frac{R}{Q})$ to reduce dissipation. Higher fundamental shunt impedance probably implies longitudinal parasitic modes that are trapped in the cell. Couplers are necessarily located near the equator of the cell. The ring HOM impedance and the AC power of the normal conducting system are higher than that of the superconducting system.

REFERENCES

1) Y.Kimura, XIII International Conference on High Energy Accelerators, August 7-11, 1986, ed. A.N.Skrinsky, p.47

2) PEP Proposal, SLAC 171

3) B. Dwerstag, Proceedings of the 4th Workshop on RF Superconductivity, August 14-18, 1989, KEK, Japan, p. 367

4) Shuichi Noguchi et. al., Particle Accelerators, Proceedings of the 14th International Conference on High Energy Accelerators, August 22-26, 1989, pp. 719-724.

5) Takaaki Furuya, et. al., Proceedings of the Third Workshop on RF Superconductivity, ed. K.W.Shepard, September 14-18, 1987, ANL-PHY-88-1, p.96

6) Y. Kojima, et. al., Proceedings of the 4th Workshop on RF Superconductivity, August 14-18,1989, KEK 89-21, ed. Y. Kojima, p. 95

7) CLNS 89/962

8) M. Zisman, "Current Limitations Summary", CALT-68-2552

9) P.E.Faugeras, et. al., 1987 IEEE Particle Accelerator Conference, March 16-19, 1987, p.1719

10) H. Padamsee et. al., CLNS 90/978, these proceedings

11) U.Laustroer,U. van Rienen, T. Weiland, DESY M-87-03

12) F.Ebeling et. al., Mafia User Guide, May 3, 1988

13) P. Wilson, Physics of High Energy Particle Accelerators, (Fermilab Summer School, 1981)

FEEDBACK SYSTEMS

F. Pedersen
CERN, Geneva Switzerland

ABSTRACT

Electronic feedback systems which actively damp longitudinal and transverse coherent coupled bunch instabilities are reviewed. Such feedback systems are essential to ensure stability of the high circulating currents required for B-factories[1,2].

INTRODUCTION

Limitations associated with the beam-beam effect imply that typical parameters for high luminosity (10^{33}-10^{34} cm^{-2}s^{-1}) B-factory collider rings require a high bunch frequency (typically 25-120 MHz), a large average beam current (typically 0.6-3.0 A), and a very short bunch length, which require a high rf voltage (typically 10-25 MV). This high rf voltage requires a significant number of cells.

The high Q transverse and longitudinal higher order modes (HOM's) in the rf cavities drive transverse and longitudinal coupled bunch modes with very fast growth rates, typically larger than the synchrotron frequency for undamped HOM's. Growth rates larger than the synchrotron frequency implies mode coupling between modes naturally separated by the synchrotron frequency, and the instability cannot generally be damped by feedback.

It is therefore essential to design the rf system so as to minimize the longitudinal and transverse HOM impedance. This means minimizing the number of cells (typically 8-24 cells), choosing an appropriate cavity shape which minimizes the number of HOM's and their R/Q (example: LEP superconducting cavity shape), and adding efficient HOM dampers. Efficient HOM dampers[3] are capable of reducing all HOM impedances by about two orders of magnitude (typically $Q_{HOM} = 5 \times 10^4 \rightarrow 500$, $R_L = 1$ M$\Omega \rightarrow 10$ kΩ per cell, $R_T = 10$ MΩ/m $\rightarrow 100$ kΩ/m per cell for a 500 MHz cavity).

In this case the growth rates are smaller than the synchrotron frequency and can be damped by electronic feedback, which still is needed since the fastest growth rates are larger than the synchrotron radiation damping rates by a significant factor. Reducing the parasitic coupling impedances to a level below which feedback is no longer needed would require damping by about three orders of magnitude, which seems extremely difficult or impossible.

A general classification of coherent beam modes is shown in Table I. In this paper we will mainly discuss feedback systems for longitudinal and transverse bunched beam modes.

LONGITUDINAL BUNCHED BEAM MODES

Longitudinal bunched beam modes and their electromagnetic interaction with the beam environment were first described by Sacherer[4,5,6]. For M equidistant bunches there are M coupled bunch modes characterized by the *integer number of waves n of the bunch motion around the ring* at any given instant:

$$0 \leq n \leq M - 1$$

The phase shift $\Delta\varphi$ of the perturbing bunch motion between two bunches is thus:

$$\Delta\varphi = 2\pi n/M \qquad (1)$$

Table I - Classification of coherent beam modes

	Coasting beam	Bunched beam
Longitudinal	n = azimuthal mode number = 1, 2, 3, ...	n = coupled bunch mode number = 0, 1, 2, ... (M-1) m = phase plane periodicity = 1 (dipole), 2 (quadrupole), 3 (sextupole), ... (q) = radial mode number)
		Mode coupling $\Rightarrow$ Single bunch "microwave" instability (turbulence)
Transverse	n = azimuthal mode number = ..., -1, 0, 1, 2, ... k = phase plane periodicity = 1 (dipole), 2 (quadrupole) 3 (sextupole), ...	n = coupled bunch mode number = 0, 1, 2, ... (M-1) m = head/tail mode number = ..., -2, -1, 0, 1, 2, ... k = phase plane periodicity = 1 (dipole), 2 (quadrupole), .. 3 (sextupole), ...
		Mode coupling $\Rightarrow$ Single bunch fast head-tail instability (turbulence)

The longitudinal motion within a bunch is characterized by the *within-bunch mode number m*, which describes the number of periods of density modulation per 2π phase advance in longitudinal phase plane.

The theory for longitudinal bunched beam mode interactions[7,8] contains in addition a *radial mode number q = m, m+2, ...* , which is describing an infinity of orthogonal radial modes with different density variations versus synchrotron amplitude (= radius). The first higher order radial mode q = 3 for the dipole mode (m = 1) has thus a line density pattern which looks like a sextupole mode (m = 3), but it oscillates with the synchrotron frequency and not 3 times this frequency.

Normally only the lowest radial mode is observed; probably due to the fact that higher radial modes have higher Landau damping thresholds[8].

As an example longitudinal modes in the PS Booster (M = 5) are shown in Fig. 1 for m = 1, 2, 3, 4, and n = 4.

It is often convenient to observe and calculate the interaction of these orthogonal modes in the frequency domain. They can be *described by a line spectrum* with frequencies:

$$f_p = |(n + pM)f_0 + mf_s|, \quad -\infty < p < +\infty \qquad (2)$$

where n and m are the mode numbers defined above, f_0 the revolution frequency, and f_s the synchrotron frequency. p is an integer, so a particular mode appears as many spectral lines, namely twice within every frequency span equal to the bunch frequency pMf_0, see Fig. 2.

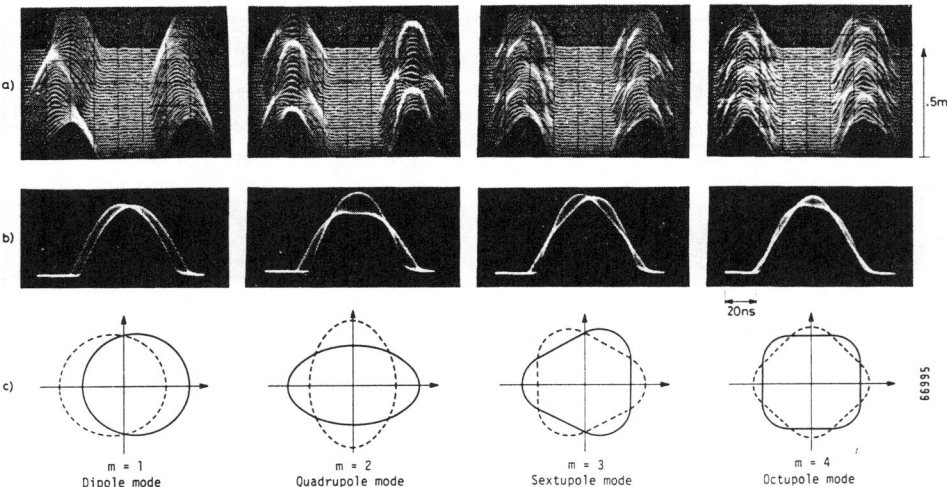

Figure 1 - Within bunch modes m = 1 to 4, coupled-bunch mode pattern n = 4.
a) Mountain-range display of one synchrotron period, b) Superimposed, c) Phase space.

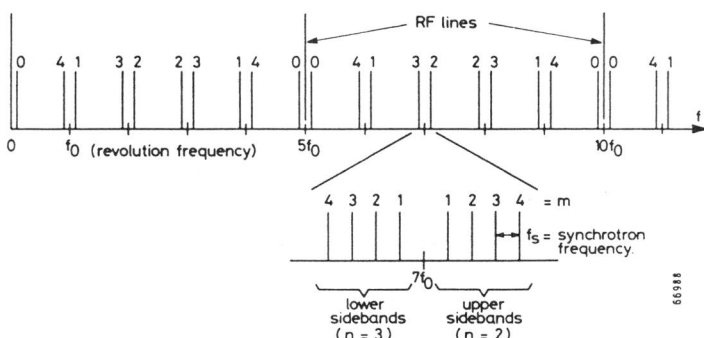

Figure 2 - Spectral lines of longitudinal modes.

The envelope or relative amplitude of all those lines belonging to a given mode depends on the mode numbers m (and q). The power spectrum envelope normalized as described in 6) is called the *form factor* $F_m(f)$ (Fig. 3). As an example of a measured amplitude envelope a coupled bunch octupole mode line spectrum is shown in Fig. 4. (PS Booster). The envelope peaks between the third and fourth harmonic of the bunch frequency.

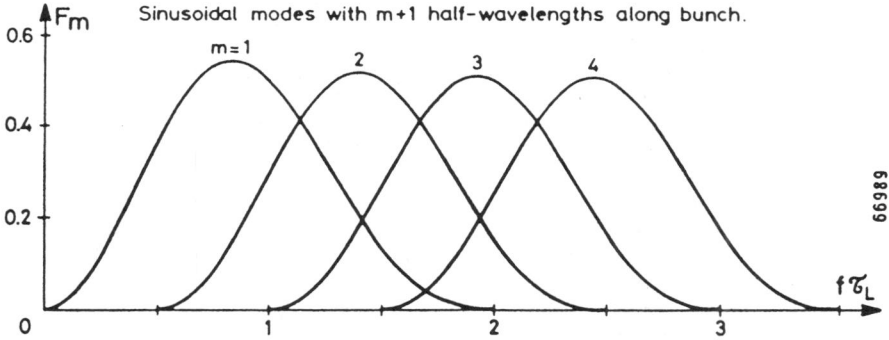

Figure 3 - Longitudinal form factors F

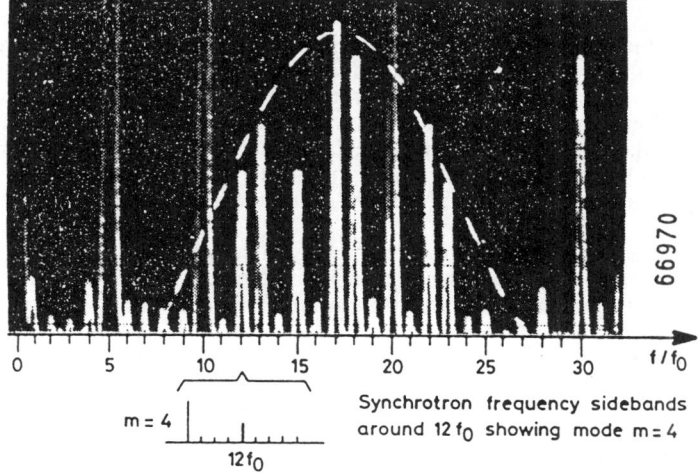

Figure 4 - Octupole mode line spectrum observed in the CERN PSB.

The coherent, complex frequency shift $\Delta\omega_{mn}$ is given by a weighted sum $(F_m(f_p\tau_L))$ of the longitudinal coupling impedance $Z_L(f_p)$ sampled at all those frequencies f_p where a spectral line belonging to mode (m,n) appear, Sacherer's[5,6] formulae:

$$\Delta\omega_{mn} = j \frac{m}{m+1} \frac{\omega_s I}{3B_0^2 V \cos\varphi_s} \sum_p F_m(f_p\tau_L) Z_L(f_p) / p \qquad (3)$$

where ω_s is the frequency, I = total beam current, B_0 = bunching factor, V is the rf voltage, and φ_s the stable phase angle. The mode growth rate is given by $-\text{Im}\{\Delta\omega_{mn}\}$ and the real frequency shift by $\text{Re}\{\Delta\omega_{mn}\} > 0$.

This is a very general formula from which several observations can be made.

Since p > 0 for the upper sidebands (Fig. 2) and $\cos\varphi_s < 0$ above transition, a passive resonant coupling impedance (Re{Z_L} > 0) will excite modes belonging to upper sidebands while modes belonging to lower sidebands are damped.

Modes n = M/2 (M even) and n = 0 are hard to excite due to upper/lower sideband cancellation, but asymmetry in the very large impedance of the fundamental rf resonance can damp or excite the n = 0 mode (Robinson damping).

For more than 3 bunches, the remaining modes can be strongly damped or excited by resonant coupling impedances with bandwidths less than the bunch frequency $\Delta f_{HOM} < Mf_0$.

All coupled bunch modes n = 0, 1, ..., M-1 appears twice in a band between two bunch frequency harmonics, once as upper sideband and once as lower sideband (Fig. 2). If $n = n_1$ is stable, the complementary mode $n = M-n_1$ is unstable with equal magnitude growth and damping rates.

The detuning of the fundamental rf resonance required to compensate the reactive component of the beam loading has such a sign ($f_{res} < f_{rf}$ above transition) that the n = 0 is damped (Robinson damping), but the n = M-1 is driven unstable by this detuning. This effect may become important in large rings with large average beam currents (low f_0, high I), if the required detuning approaches the revolution frequency.

Besides interacting with the parasitic HOM's and the fundamental rf resonance, the mode stability can also be influenced by an active damping feedback system. All modes can be damped by providing a large equivalent coupling impedance Z_{LFB} which has a dominating, stabilizing influence in the spectral line summation in Sacherer formulae. Since all coupled bunch modes n appear once in a band $Mf_0/2$ from one bunch harmonic to the symmetry point between two bunch harmonics, all modes can be damped with a feedback system bandwidth equal to half the bunch frequency.

This is a very fortunate fact of bunched beam instabilities: a powerful feedback system acting within a bandwidth equal to half the bunch frequency can cure instabilities resulting from interactions with parasitic impedances over a much larger bandwidth.

At low currents an active electronic damping system may not be required since other damping mechanisms may be adequate, namely:
- Landau damping by synchrotron frequency spread; can be enhanced by higher harmonic rf cavity,
- synchrotron radiation damping,
- amplitude limitation of dipole modes due to higher order phase modulation sidebands for large amplitudes[9],
- decoupling by spread in individual bunch frequencies[10].

LONGITUDINAL COUPLED BUNCH FEEDBACK SCHEMES

The requirements for the feedback system equivalent impedance is as follows:
- Bandwidth at least half the bunch frequency $\Delta f_{FB} \geq Mf_0/2$.
- Choose a center frequency such that the form factors $F_m(f)$ are sufficiently large for the modes m to be damped.
- The real part of the coupling impedance should change sign at each revolution harmonic: Re{Z_{LFB}} < 0 for upper sidebands and Re{Z_{LFB}} > 0 for lower sidebands (above transition).
- Only the perturbing part of the bunch spectrum should be transmitted to save power. The stationary spectrum like bunch frequency harmonics and inter-

mediate revolution frequency harmonics (due to slightly unequal bunch populations) should be rejected: $|Z_{LFB}(pf_0)| = 0$.

In the following, we will consider only modes with $n \neq 0$, since the $n = 0$ in phase modes generally interacts exclusively with the fundamental resonance of the rf cavity[11] and the feedback loops often associated with the rf system[12].

A possible feedback scheme using a *radial pick-up in a sector with high dispersion and a longitudinal momentum kicker*[10] is shown in .Fig. 5. The electrical delay from pick-up to kicker is equal to the beam transit time from pick-up to kicker.

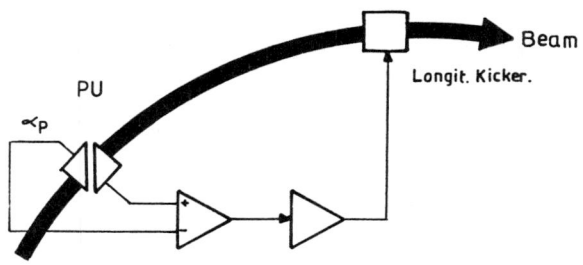

Figure 5 - Longitudinal feedback using radial pick-up

The sign reversal of the pick-up signal on either side of the central reference orbit defined by the electrical center of the radial pick-up provides the required change in sign of the $\text{Re}\{Z_{LFB}\}$ so both upper and lower sideband modes are damped.

Several practical problems associated with the use of this method makes it rather tricky and difficult. The stationary bunch spectrum is only suppressed if the beam is perfectly centered and the difference hybrid perfectly balanced. This may create the need for extra unnecessary power. In addition, the phase modulation sidebands of the common mode (= longitudinal) signal will antidamp either upper or lower mode sideband if the beam is displaced radially more than a certain critical distance, which diminish with frequency making it very difficult to damp higher order modes ($m \geq 2$).

A very high electronic gain is generally needed so noise is an important issue. When this method was tried in the PS[10], a comb filter with peaks at the revolution harmonics was needed as well as a suppression of the rf frequency.

The pick-up is also sensitive to radial betatron motion, so unless the dispersion and its derivative is zero at the location of the kicker radial betatron motion will be affected. This can, however, be used with advantage to damp radial betatron oscillations[13].

The scheme has also been applied experimentally to cure the coupled bunch instabilities in the FNAL Booster[14] but so far without much success.

A more successful scheme is the *bunch-by-bunch feedback scheme* shown in Fig. 6. Phases of individual bunches relative to the rf are detected and memorized bunch-by-bunch (dipole motion), as well as the peak detected bunch signal for the quadrupole motion.

After a phase shift of 90° (usually obtained by a high pass filter), the phase and amplitude signals are applied to fast phase and amplitude modulators inserted in the rf drive chain with appropriate timing so as to take into account electrical delays and pick-up to kicker transit time.

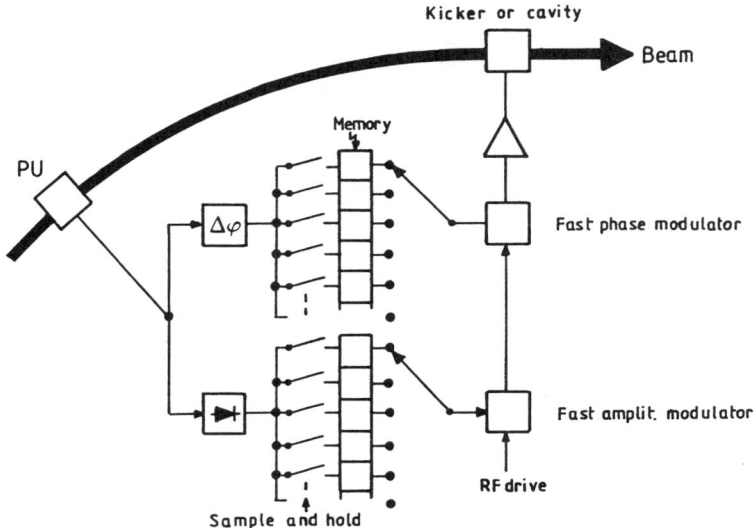

Figure 6 - Bunch-by-bunch longitudinal feedback for dipolar and quadrupolar modes

Since the sampling frequency equals the bunch frequency, it follows from the sampling theorem that information up to half the bunch frequency ($Mf_0/2$) can be transmitted. This is consistent with the preceding mode analysis, which requires that bandwidth to damp all modes. Phase and amplitude modulation generate two sidebands, each the modulation frequency ($\leq Mf_0/2$) above and below the carrier so a full bandwidth equal to the bunch frequency is required of the kicker or cavity. One of these sidebands could nevertheless be removed since the bunches would distinguish between dipole and quadrupole corrections through their frequencies.

The feedback loops for individual bunches are generally not independent, since they couple to each other through narrow band coupling impedances and possibly through the finite bandwidth of the kicker if the bunch frequency is high. This may make adjustment of the feedback system difficult.

Block diagrams of the systems implemented for the ISR[15] and PEP[16] are shown on Figs. 7 and 8. Examples of bunch-by-bunch longitudinal feedback systems are given in Table II.

Table II - Examples of bunch-by-bunch longitudinal feedback systems

Accelerator	h	M	m	Kicker	f	Power
ISR	30	20	1, 2	rf cavity	9.5M	-
PEP	2592	3 + 3	1	Spec. cavity	860M	55 kW
Sp$\bar{\text{p}}$S[17]	4620	6 + 6	1, 2	rf cavity	200M	-
LEP[18]	31320	4 + 4	1	rf cavity	352M	-

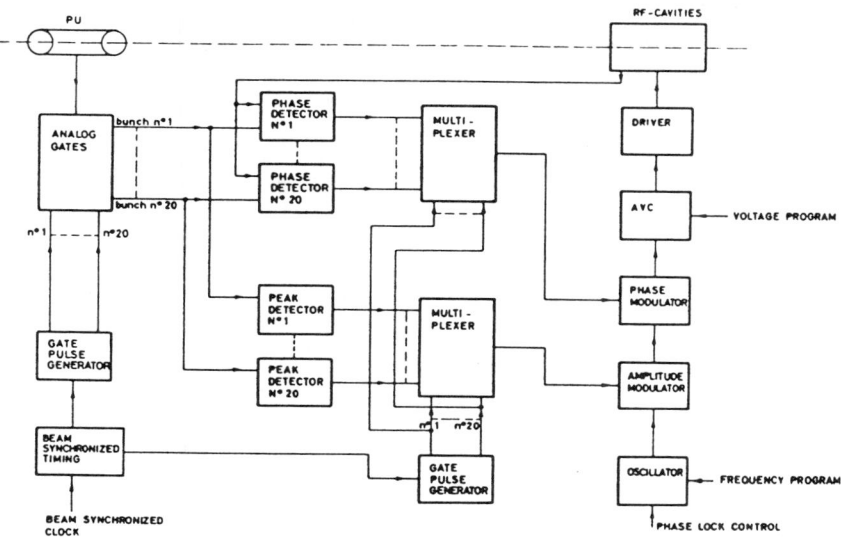

Figure 7 - ISR time domain system[15].

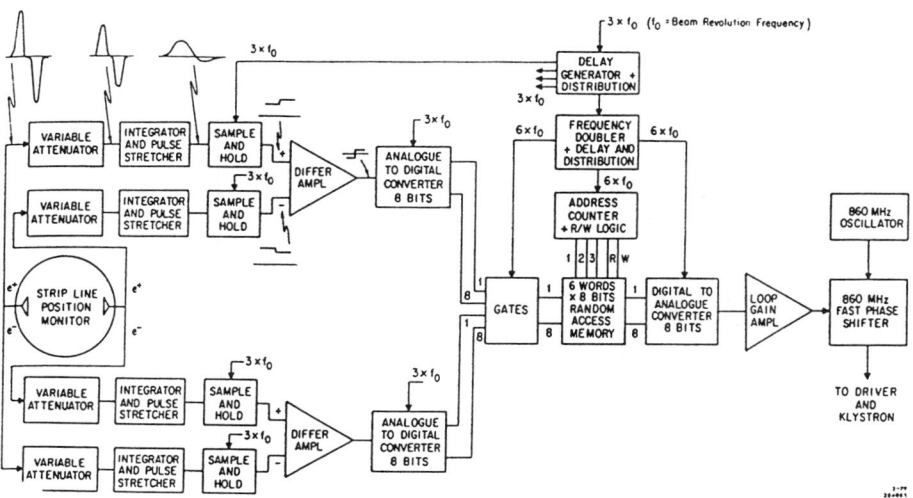

Figure 8 - PEP time domain system[16].

Another very powerful method for longitudinal coupled bunch mode damping is the *filter method* (Fig. 9) which uses a longitudinal pick-up and a longitudinal kicker together with specially shaped bandpass filters centered around (M-2)/2 (M even) or (M-1)/2 (M odd) consecutive revolution harmonics. The filters contain a notch centered exactly around the revolution harmonics. This notch serves the dual purpose of suppressing the unequal bunch line and providing a 180° phase shift such that both the upper and lower sideband modes are damped. The frequency band of the filters can be chosen such that several higher order modes can be damped simultaneously (m = 1, 2, 3). The roll-off of the bandpass filters can be tailored to the synchrotron frequency and the number of higher order modes to be damped so that the noise bandwidth is minimized.

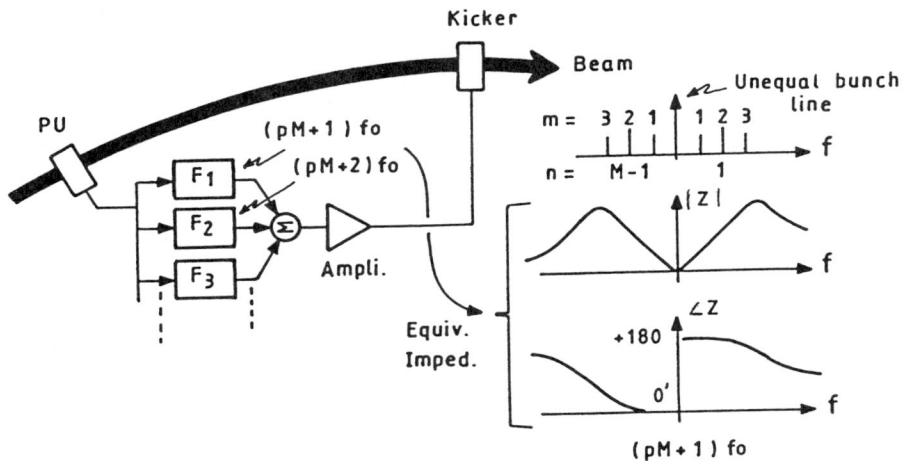

Figure 9 - Longitudinal coupled bunch feedback using the filter method.

The good signal to noise ratio and complete suppression of the stationary bunch spectrum mean low power requirements. Damping mode by mode rather than bunch-by-bunch has the advantage that modes are orthogonal, independent and decoupled for equidistant identical bunches, while bunch-by-bunch systems are not. This makes analysis and adjustment easier.

A block diagram of the longitudinal coupled bunch feedback system using the filter method for the PS Booster[6,19] is shown in Fig. 10. Since the revolution frequency varies by a factor 2.7 during the acceleration cycle, precise tracking and phase control of the filters are required. This is achieved by means of two-path filters (Fig. 11), where a low-pass to band-pass transformation takes place (Fig. 12), and the center frequency is determined by the frequency of the local oscillator signals applied to the mixers in quadrature. The local oscillator signals at each revolution harmonic is generated from the main rf signal by means of phase locked loops, which ensures automatic tracking.

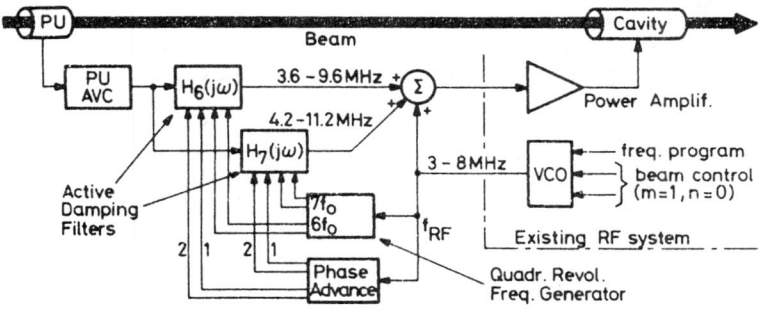

Figure 10 - CERN PSB longitudinal coupled bunch feedback system[19].

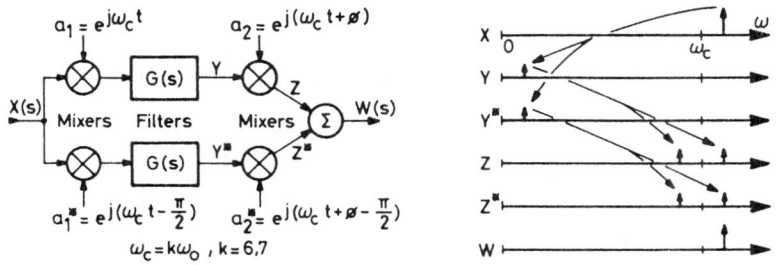

Figure 11 - Two path active filter with phase-shift control.

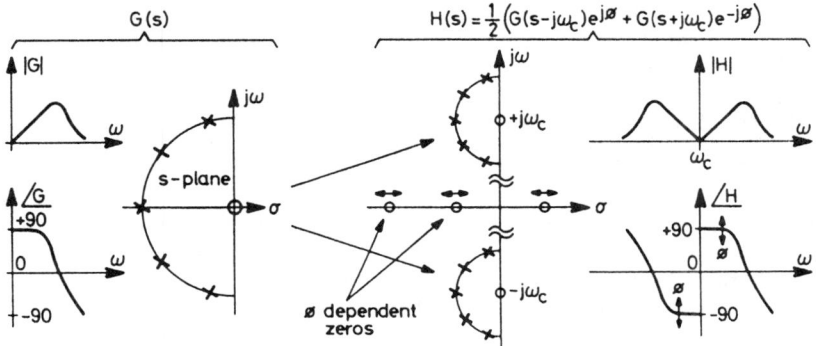

Figure 12 - Low-pass to band-pass transformation

This type of feedback makes it easy to measure the interaction of each individual mode with the longitudinal coupling impedance and with the feedback system itself. Each mode (m, n) controlled by the system can be selectively driven to a well defined initial amplitude and its growth or damping rate can be measured with the feedback on or off. By measuring the beam transfer function (BTF) in a band of frequencies in the neighbourhood of a mode frequency, a stability diagram can be measured by plotting the inverse of the BTF versus frequency[19]. From this Landau damping, real and imaginary frequency shifts can be derived. This type of measurements may help to identify the harmful impedances which are the source of the instability. Examples of longitudinal feedback systems using the filter method is shown in Table III. In some cases the main rf cavity could be used as kicker, in others a dedicated stripline or cavity kicker was used.

Table III - Examples of longitudinal feedback using filter method.

Accelerator	h	M	m	Kicker	Power	Frequency
PSB	5	4 × 5	1, 2, 3	rf cavity	-	3.6-11.2 MHz
PEP[20]	2592	3 + 3	1	spec. cavity	55 kW	860 MHz
PS	20	20	1	rf cavity	-	few modes only
NSLS,VUV	9	9	1, 2	stripline	100 W	400 MHz
EPA	8	8	1, 2	stripline	10 W	100 MHz
S-ACO	24	4	1	stripline	100 W	400 MHz

In some cases the fundamental resonance of the rf cavities is the main cause of coupled bunch instabilities. This happens often in medium to large synchrotron (low f_0) with cavities with moderate Q and with the cavities detuned, which may be required to compensate the reactive beam loading. While compensation or overcompensation of the reactive part of the beam load stabilizes mode n = 0 (Robinson) as well as n = 1, 2, ... ($\gamma > \gamma_t$), modes n = M-1, M-2, ... are driven unstable.

In such cases it may be simpler to *reduce the harmful impedance by rf feedback* rather than damp the resulting instabilities. Such an impedance reduction can be achieved by means of rf feedback around the final power amplifier. The impedance reduction is, however, limited by the achievable bandwidth, which is limited by the unavoidable delay in the feedback path. Also it should be noted that while the magnitude |Z| of the impedance is always reduced when Q is lowered, this is not the case with Re{Z}, which is the important quantity for growth rates of coupled bunch modes. When Q is lowered, the real part decreases inside the 3 dB bandwidth, while it actually increases for all frequencies outside.

If normal rf feedback around the power amplifier cannot provide adequate impedance reduction, a further reduction can be achieved by inserting a *comb filter in the rf feedback path*[21] with a periodicity equal to the revolution frequency such that an additional impedance reduction is obtained in bands near each revolution frequency. Such an impedance reduction scheme has the additional advantage that periodic beam loading transients in the rf system resulting from partial filling of the ring are reduced as well, since the impedance seen by the unequal bunch lines present with such a filling is reduced. This scheme has been very successful in the SPS, and is presently being implemented in the PS, where the periodic beam loading transients is the main issue.

GAIN, FREQUENCY, BANDWIDTH AND POWER CONSIDERATIONS FOR LONGITUDINAL FEEDBACK SYSTEMS FOR B-FACTORIES

The minimum *required bandwidth* is equal to half the bunch frequency $Mf_0/2$ provided the band extends from a bunch frequency harmonic pMf_0 to the symmetry point between two bunch frequency harmonics $(p + 1/2)Mf_0$.

The required gain $Z_{LFB} = GZ_{PU}$, where Z_{PU} is the pick-up coupling impedance and G the electronic gain from pick-up to integrated kicker voltage, is determined by the highest coupling impedance Z_L the system is designed to cure taking into account the weighting with the mode form factors F_m in Sacherer's formulae. Since the form factors all peaks above the vacuum pipe cut-off frequency, a lower gain is needed if the *center frequency* of the feedback band is chosen as high as possible just below the cut-off frequency.

The required *maximum feedback voltage* is determined by the need to maintain the required gain in presence of transient and undesired signals without saturation of the power amplifier. The *required power* is then determined by the kicker shunt impedance R_s, $P = V^2/2R_s$. The required power can thus be reduced by increasing the kicker impedance until the point where the kicker makes a similar contribution to the coupling impedance as the one it is designed to cure. Beyond this point the required power goes up since higher gain is needed.

Determining the magnitude of the transient and undesired signals may be difficult. Possible contributions may come from:
- Injection transients.
- Coherent random bunch motion, like rf phase noise mainly exciting the m=1, n=0 mode, but propagating to the other modes through mode to mode coupling, or filamentation from turbulent bunch lengthening.
- Stationary bunch spectrum if not completely suppressed.
- Intermodulation due to stationary bunch spectrum.
- Pick-up noise or dynamic range limitations.
- Digitizing noise of digital systems are used.
- Carrier feedthrough due to imperfect balance in the up/down mixers.

If fairly large charges are injected with a low repetition rate (synchrotron injection as in the CERN/PSI B-factory proposal), it is likely that the injection transients determine the required power. These can be broken down into the following components:
- time jitter,
- energy jitter,
- path length modulation of the circulating beam due to injection kick,
- bunch/bucket mismatch, which excites quadrupole (m = 2) injection oscillations.

If small charges are injected with a high repetition rate (Linac injection as in the SLAC/LBL B-factory proposal), it is less certain that injection transients dominate since the required dynamic range of the low level signal processing is very large.

LONGITUDINAL FEEDBACK SIGNAL PROCESSING FOR LARGE NUMBER OF BUNCHES

A solution to this problem is not only an issue for B-Factories. Many planned or existing synchrotrons have or will have the problem of damping longitudinal coupled bunch instabilities with a large number of bunches (for example, FNAL Booster, APS, ESRF, HERA, tau-charm factories).

The filter method provides excellent signal processing with a large dynamic range which results in minimum required power, but with one two-path filter with associated PLL synthesizer per revolution harmonic the amount of electronics becomes excessive if M is large.

For many bunches, two pairs of single side band quadrature mixers operated at a suitable harmonic of the bunch frequency can be used together with two identical periodic notch filters such that the complete assembly becomes a two-path filter with periodic notches (Fig. 13).

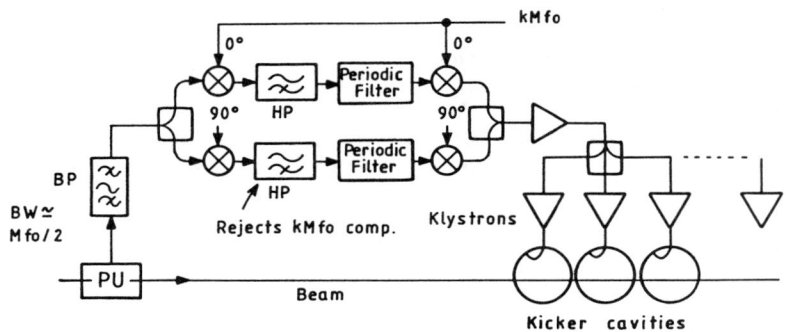

Figure 13 - Longitudinal feedback system for large number of bunches

The signal processing is similar to the scheme used by Boussard[21] for the comb filter rf cavity feedback. The feedback kicker can either be a number of stagger tuned UHV TV transmitter type klystrons with associated cavities if high power (tens of kW) is needed, or solid state amplifiers using stripline (loop) couplers if a lower power level (few kW) is sufficient.

Two well-tested means exist to make the periodic notch and comb filters. Both are used in stochastic cooling, where periodic filters are used for momentum cooling using the filter method. One method uses transmission line delay lines to obtain notch and peak filters, Fig. 14. The transmission lines can be low loss coax, superconducting coax, or an optical fiber.

Digitizing the input signal and converting the transmission line delays into shift register delays changes the analog notch/comb filter into a digital notch/comb filter (Fig. 15). Such digital filters have been built for the SPS rf cavity feedback[21] (comb filter, f_s = 4 MHz) and for stochastic cooling at low energy in LEAR[22] (notch filter, f_s = 200 MHz).

To leave room for the antialiazing filter preceding the ADC a useful bandwidth of 40% of the sampling frequency can normally be made available so that a sampling frequency of $f_s = 1.25$ Mf_0 is needed. An interesting possibility is to use the maximum bandwidth of 50% of the sampling frequency by reducing the sampling frequency to the bunch frequency $f_s = Mf_0$ and omit the antialiazing filter. This is possible since the mode spectrum is completely symmetric around the fold-over or Nyquist frequency $f_s/2 = Mf_0/2$.

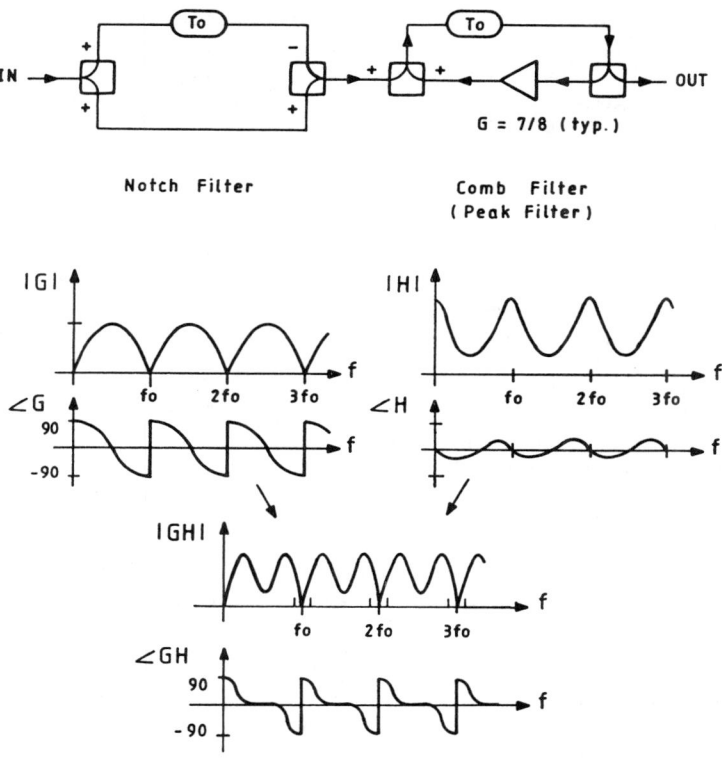

Figure 14 - Notch and comb filter using transmission line delays.

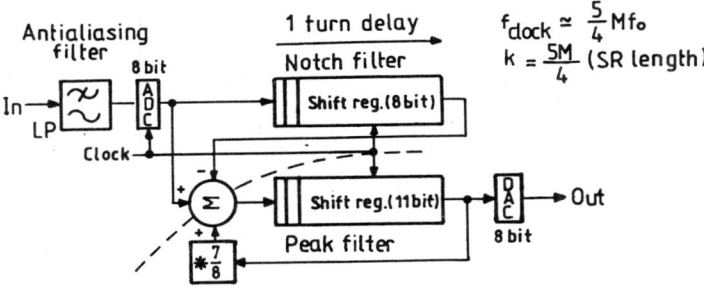

Figure 15 - Digital notch and comb filter.

The high sampling frequency required will, in most cases, limit ADC word length to 8 bits, which limits the dynamic range to about 46 dB. Careful pre-filtering makes best use of this dynamic range. The passive bandpass just after the pick-up (Fig. 13) rejects all bunch frequency harmonics except one, which limits the maximum signal level which can be applied to the first pair of mixers. That bunch frequency harmonic becomes a dc signal at the output of the mixers and is eliminated with high-pass filters. The strongest components remaining in the signal applied to the digital or cable filters is thus the unequal bunch lines which are eliminated by the notch filter, but limits the dynamic range. The unequal bunch lines could be as low as 40 dB below the bunch frequency harmonics if buckets are uniformly filled to within $\pm 1\%$, but could be as high as 20 dB below if a gap is needed in the electron beam to clear ions.

An interesting time domain interpretation of the feedback system in Fig. 13 can be made if the digital filters in Fig. 15 are used with a sampling frequency equal to the bunch frequency. In addition, we will choose the arbitrary phase of the local oscillator (Lo) at bunch frequency harmonic kMf_0 to be in phase with the corresponding bunch signal harmonic. The first mixer in the upper branch of the two-path filter can then be interpreted as a fast synchronous bunch-by-bunch amplitude detector (at that harmonic) and the mixer in the lower branch (90° LO) can be interpreted as a fast bunch-by-bunch phase detector. The shift registers in the two digital notch filters will then continuously store phases and amplitudes of all M bunches over the last turn. Subtracting previous turn phase or amplitude from present turn phase or amplitude provides a turn by turn differentiation of dipole and quadrupole motion, which produces the required 90° phase shift required to obtain damping.

The digital comb (peak) filter can be interpreted as M individual first-order recursive digital low-pass filters, which reduces the high-frequency noise amplified by the differentiation. The unequal bunch intensities will appear as a bunch-by-bunch amplitude variation in the amplitude branch, which again is eliminated by the turn-by-turn differentiation (or notch filter). This analysis shows that the system also can be analyzed as a bunch-by-bunch dipole and quadrupole damping system. The bunch-by-bunch time domain analysis and the mode-by-mode frequency domain analysis can thus in this case complement each other, although the frequency domain analysis in general is more powerful.

It is thus not obvious from the time domain analysis that the absolute phase of the local oscillator is unimportant, that sextupole and higher modes can be damped by the system if the frequency is chosen properly, and that one of the two amplitude and phase modulation sidebands generated by the two "up" mixers can be removed and thus reducing the required kicker bandwidth by one half.

Also mode-by-mode beam transfer functions across the feedback bandwidth is a powerful method to obtain correct delay of the system and to measure and correct phase non-linearities across the band. This method is commonly used in setting up stochastic cooling systems.

The choice between transmission line filters and digital filters is not obvious. Transmission line filters cannot be used in synchrotrons with variable frequency (FNAL Booster, TRIUMF Kaon Booster), while digital filters are easy to adapt for tracking. Digital filters are complex, while the almost passive transmission line filters are very simple in principle, but become complex if superconducting cables are required or servotuning or oven is required to maintain notch stability. The 8-bit limitation of very fast ADC's limits the dynamic range of digital filters, while cable filters have a very large dynamic range. On the other hand, digital filters have very precise notch frequencies, while the depth and frequency dispersion of the cable filter notches are limited by the quality of the transmission lines used. While transmission line filters can provide GHz bandwidths, present day digital ECL technology limits the digital filters to about 100 MHz bandwidth. This can be overcome though by stagger tuning several two-path filters like the one shown in Fig. 12. Most likely several systems of each type will appear in the near future.

TRANSVERSE BUNCHED BEAM MODES

The bunched beam transverse modes (Table I, Fig. 16), also called head-tail modes, and their electromagnetic interaction with the beam environment were first described in the most general form by Sacherer[23,24].

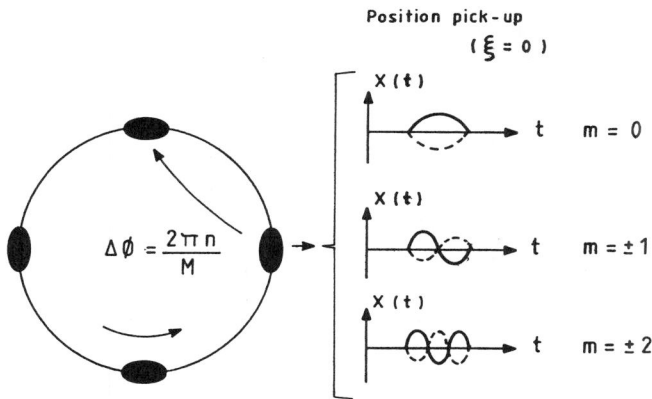

Figure 16 - Transverse bunch beam modes.

As in the longitudinal case there are M coupled bunch modes characterized by the *integer number of waves n of the bunch motion around the ring* at any given instant: $0 \leq n \leq M-1$. The phase shift $\Delta\varphi$ of the perturbing bunch motion between two bunches is thus

$$\Delta\varphi = 2\pi n / M \qquad (4)$$

The within-bunch mode number m (also called the *head-tail mode number*) is the net number of betatron periods per synchrotron period observed at any given instant as the synchrotron phase advance is followed over a complete period. At any given instant a closed pattern of $|m|$ betatron periods correspond to one synchrotron period with the phase either advancing ($m > 0$) or retarding ($m < 0$) as the synchrotron phase is advancing.

There will be $|m|$ positions along the bunch where particles above the synchronous energy have a betatron phase exactly opposite the phase of particles below the synchronous energy, so that no net betatron motion is observed at these positions, $|m|$ nodes appear at these positions (Fig. 17).

Non-zero chromaticity further complicates the motion as the energy modulations due to synchrotron motion modulates the tune. The betatron phase advance during the half of the synchrotron period above mean energy is compensated by an equally large betatron phase advance of opposite sign during the other half of the synchrotron period below mean energy, so that the closed pattern specified by m remains unchanged; only a temporary phase advance χ takes place between the head and the tail of the bunch.

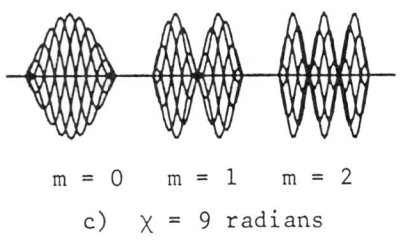

Figure 17 - Contortions of a single bunch on separate revolutions, and with six revolutions superimposed. Vertical axis is difference signal from position monitor, horizontal axis is time, and Q = 4.833. [23]

In addition, the coherent betatron motion along the bunch can be characterized by the betatron phase plane periodicity k. In general, only the dipole modes k = 1 with one period of density modulation in the transverse phase plane is observed, since the quadrupole modes k = 2 with two periods of density modulations corresponding to coherent beam width oscillations only interact very weakly with the vacuum chamber wall impedance. Transverse beam-ion quadrupolar instabilities have nevertheless been observed with coasting beams[25], and can be expected to occur for bunched beams as well.

The frequency spectrum as observed on a position pick-up (k = 1) or quadrupolar pick-up (k = 2) is a line spectrum (Fig. 18):

$$f_{m,n,k} = (n + pM + kQ)f_0 + mf_s \qquad -\infty < p < +\infty \qquad (5)$$

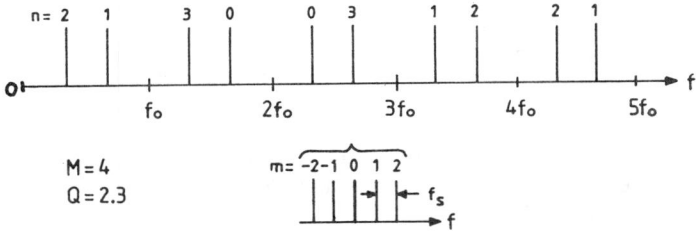

Figure 18 - Transverse coupled bunch head-tail mode line spectrum.

As in the longitudinal case, each mode m, n, k generates many lines in the line spectrum. For each mode two lines appear within a span equal to the bunch frequency Mf_0, and the mode lines repeat themselves every frequency interval equal to the bunch frequency.

The envelope or relative amplitude of the spectral lines depends on the mode number m. The power spectrum envelopes $h_m(\omega)$ are shown in Fig. 19. The width of the envelope is inversely proportional to the bunch length. The chromaticity $\xi = (\Delta Q/Q)/(\Delta p/p)$ will shift the center frequency of those envelopes to the chromatic frequency f_ξ given by:

$$f_\xi = \xi Q f_0 / \eta \tag{6}$$

The coherent, complex frequency shift is given by a weighted sum ($h_m(f-f_\xi)$) of the transverse coupling impedance $Z_T(f)$ sampled at all those frequencies f where a spectral line belonging to mode m,n appear, Sacherer's formulae[24]:

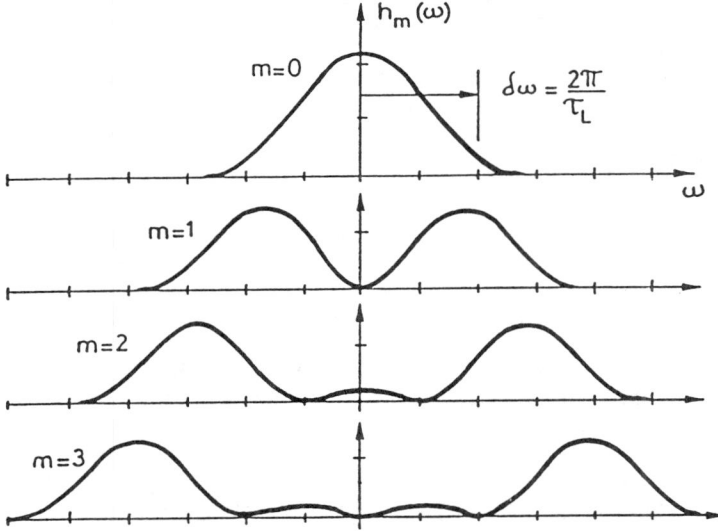

Figure 19 - Power spectrum envelopes for modes m = 0 to 3 for zero chromaticity.

$$\Delta\omega_{m,n} = \frac{1}{1+m} \frac{j}{2Q\omega_0} \frac{e\beta}{\gamma m_0} \frac{I_0}{L} \frac{\sum_p Z_T(f) h_m(f-f_\xi)}{\sum_p h_m(f-f_\xi)} \qquad (7)$$

where β and γ are the relativistic parameters, e the electron charge, m_0 the particle rest mass, I_0 the current per bunch, L the bunch length and Z_T the transverse coupling impedance in ohms/m.

The growth rate of a mode is given by $-\text{Im}\{\Delta\omega_{mn}\}$ and the real frequency shift by $\text{Re}\{\Delta\omega_{mn}\}$. The summation covers both negative and positive frequencies and since $\text{Re}\{Z_T(f)\} = -\text{Re}\{Z_T(-f)\} > 0$ for passive coupling impedances, the sidebands originating from negative frequencies (lower sidebands in Fig. 18) contribute to growth in the sum (7) while sidebands originating from positive frequencies (upper sidebands in Fig. 18) contribute to damping.

This is a very general formulae which is similar to the longitudinal case (3), although there are major differences, namely:
- the existence of an m = 0 mode in the transverse case which interacts strongly with low frequency impedances like the surface resistivity of the vacuum chamber,
- the asymmetry of the mode spectrum envelopes introduced by non-zero chromaticity. Due to this, single bunches can become unstable through interaction with broadband short-range impedances (so-called "head-tail" effect).

For a detailed discussion of the various scenarios involving interaction with narrow-band or broad-band impedances, and varying chromaticity refer to Refs. 23 and 24.

TRANSVERSE COUPLED BUNCH FEEDBACK SCHEMES

The requirements for the transverse feedback system equivalent impedance are similar to the longitudinal case, the signal processing is simpler since the betatron phase advance between pick-up and kicker can be used with advantage.
- Bandwidth at least half the bunch frequency, $\Delta f_{FB} \geq M f_0/2$ if all coupled bunch modes n must be damped. In some cases only few low frequency modes are driven unstable by the resistive wall impedance, and a lower bandwidth damper may be adequate.
- Choose a center frequency such that the form factors $h_m(f-f_\xi)$ are sufficiently for the modes m to be damped. Generally, low frequencies like 0 to $Mf_0/2$ are used since power at low frequencies is cheaper and the M = 0 mode normally is the fastest growing. To damp higher head-tail modes either the chromaticity must be made sufficiently large to obtain sufficient overlap with the higher order form factors or the frequency must be increased.
- The real part of the coupling impedance should change sign at each revolution harmonic: $\text{Re}\{Z_{TFB}\} > 0$ for upper sidebands, $\text{Re}\{Z_{TFB}\} < 0$ for lower sidebands. This can be achieved by simply arranging to have an odd number of a quarter betatron wavelength between pick-up and kicker.

Normally, the chromaticity ξ is controlled in such a way as to obtain a positive chromatic frequency f_ξ (Eq. 6) ($\xi > 0$ above transition, $\xi < 0$ below). A "passive" damping of all modes is obtained in this way by shifting the power spectrum envelopes h_m towards positive frequencies, where the interaction with the broadband, short-range component of the transverse coupling impedance has a stabilizing effect[23,24].

Often the singularity in $Z_T(f)$ due to the wall resistivity near zero frequency has a destabilizing effect on one or a few of the lowest frequency modes, which exceeds the stabilizing effects of positive f_ξ, Landau damping, or synchrotron radiation. These instabilities can be cured by a transverse damper of limited bandwidth[26,27] acting only on a few low frequency modes.

In other cases it may not be convenient or possible to apply the appropriate stabilizing chromaticity, so all modes may become unstable. Also resonant transverse coupling impedances at high frequency with bandwidths less than the bunch frequency may make modes with many different coupled bunch mode numbers n unstable. In these cases an all-mode damper (all M modes with m = 0) with a bandwidth of at least half the bunch frequency is needed.

The reactive part of Z_T gives rise to real frequency shifts of the coherent head-tail mode frequencies. Since $h_0(f)$ overlaps low frequencies with rather constant inductive reactance, while $h_1(f)$ overlaps much higher frequencies where the reactance may become capacitive, the coherent shift of the m = 0 modes is different from the coherent shift of the m = ±1 modes, in particular if the bunch length is short relative to the vacuum chamber. When the difference in frequency shift of the m = 0 modes and the m = -1 modes is equal to their separation by the synchrotron frequency, an instability known as the fast head-tail or the mode coupling instability occurs[28,29]. It is a single bunch instability since it is due to the broadband or short-range wake characteristics of the transverse coupling impedance. It is often an intensity limiting factor for large rings with few short bunches.

The threshold for this instability can be raised by providing a capacitive reactance component in the equivalent coupling impedance $\text{Im}\{Z_{TFB}\} < 0$ of an all-mode transverse feedback system[30,31] (BW $\geq Mf_0/2$). The gain of the reactive feedback system should be such that its capacitive contribution to the real coherent frequency shift of all M modes (with m = 0) cancels the mostly inductive contribution from the machines passive coupling impedance in the summation over all spectral lines in Sacherer's formulae (Eq. 7). Generally, a rather large capacitive $\text{Im}\{Z_{TFB}\}$ over a bandwidth about half the bunch frequency is used to cancel a smaller inductive $\text{Im}\{Z_T\}$ acting over a much larger bandwidth, namely the inverse of the bunch length. Such a reactive feedback coupling impedance can, for example, be obtained by having an even number of betatron wavelengths from pick-up to kicker.

Examples of a wide variety of transverse dampers are shown in Table IV. In spite of their differences, they can all be discussed using the typical block diagram in Fig. 20.

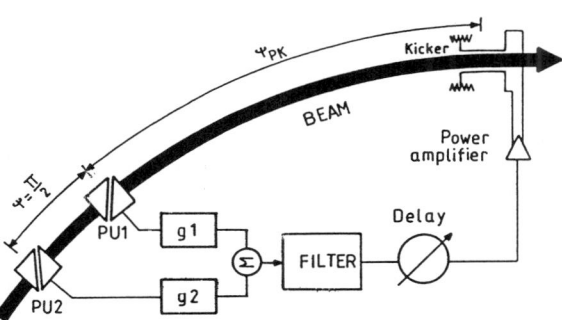

Figure 20 - Typical transverse system

Table IV - Transverse feedback systems

Machine	h	M	Mf_0 (Hz)	Bandwidth (Hz)	Kicker	V, I Power	Bunch by bunch	Delay track	Digital or Analog	Closed orbit suppr.	Notch filter	90° PU's
SPS[26]	4620	4620	200 M	0.003-1M	E	$2 kV_p$			A			+
ISR[32]	30	20	9.5 M	0.1-1.6M	E+B	2.5 kW			A		+	
PSB[33]	5	5	8M	0.005-50M	E+B	100 W		+	A	+		
FNAL Booster[34]	84	84	53M	0-27M	E+B	100W	+	+	A			
FNAL MR[35]	1113	1113	53M	0-27M	E+B	5 kW	+	+	D			
FNAL Tev.[36]	1113	1113	53M	0-27M	E+B	5kW	+	+	D	+		
SPEAR II[37]	280	1+1	1.3M	0-0.65M	E+B	2.5 kW	+		A	+		
NSLS Booster[38]	5	1	10.5M	10-250M	E+B	?			A		+	
PETRA[39]	400	80	10.4M	5M-10M	B	1 kW	+		D	+		
LEP[31]	31320	4+4	45 k	0-23k	B	$40 A_p$	+		D	+		+
AA[40]	1	1	1.85M	0.1-25M	E+B	10 W			A			
PEP[41]	2592	6+6	816 k	9.4-10.2M	B	$20 A_p$	+		A	+		

The first two (SPS and ISR) are examples of *low bandwidth resistive wall dampers* where only a few low frequency modes are damped. The low bandwidth often eliminates the need for precise delay tracking normally needed in variable frequency synchrotrons.

The remaining examples can be considered all-mode dampers since they all have a bandwidth of at least half the bunch frequency. A majority of these dampers are *bunch-by-bunch dampers* where the mean bunch positions are sampled bunch by bunch, delayed by the pick-up to kicker transit time and applied to the kicker. From the sampling theorem it follows that the bandwidth of such dampers are limited to half the sampling frequency.

Most bunch-by-bunch dampers use digital delays and processing. This simplifies *delay tracking* in variable frequency synchrotrons since the clock frequency varies as the bunch frequency. In other cases (FNAL Booster, PSB) switched delay lines are used.

The center of gravity detection followed by bunch-by-bunch sampling limits the action of the bunch-by-bunch dampers to the m = 0 mode except for sufficient non-zero chromaticity, which provides a center of gravity component of the higher order modes, so they can be damped by the feedback. This corresponds to overlap of the higher order form factors h_m with the damper bandwidth, normally dc to $Mf_0/2$.

Some analog, wide bandwidth transverse dampers (ex. PSB and AA) have bandwidths which exceed the inverse of the bunch length, which permits them to damp higher order head-tail modes, even for zero chromaticity.

Closed orbit errors at the pick-up may considerably increase the required power needed to maintain the required gain without saturation. This happens if the closed orbit error exceeds the amplitude of the injection oscillations, which normally should be determining for the required power. Two schemes are commonly used to reduce the stationary signals resulting from closed orbit errors. One is an automatic *closed orbit suppression* where the average common mode signal resulting from closed orbit errors is subtracted before further processing. This is implemented in an analog fashion in PSB, SPEAR, PEP and PETRA, in a digital fashion in LEP, and in a mixed digital/analog fashion in the Tevatron. Another scheme consists of using *notch filters* to suppress bunch frequency and revolution frequency harmonics (ISR and NSLS Booster). These can conveniently be implemented as correlator cable filters (NSLS Booster). Simple notch filters generate as correcting signal the difference in position of the last two turns of a given bunch in the time domain. In the frequency domain the phase shift of the notch filter at the lower sideband frequencies is different from the phase shift at the upper sideband frequencies. For this reason a phase advance that differs from an odd number of quarter wavelengths is needed from pick-up to kicker to obtain pure resistive feedback[38].

It is not convenient to move pick-up and kickers to compensate for changing tunes, notch filters, or ratio in resistive to reactive feedback. One simple scheme (Fig. 20) consists of providing *two pick-up signals 90° of betatron phase advance apart* (SPS and LEP) and use a suitable linear combination to compensate for changes in tune and desired ratio of resistive and reactive feedback (LEP).

Another scheme[39] uses a digital filter, which provides suitable linear combinations of the positions of the last three turns of given bunch to compensate for fractional tune changes in the range 0.1 to 0.4 using a single pick-up (PETRA and HERA). A similar scheme using an analog, passive correlator filter has been used in stochastic cooling systems to compensate for bad pick-up to kicker phase advance[42].

The pick-ups used are mostly electrostatic (plates, buttons), although electromagnetic and directional terminated striplines also have been used. Deflecting kickers are normally terminated striplines, which are electromagnetic and directional, although electrostatic (SPS) and magnetic (LEP, PETRA, PEP) deflectors also have been used at low frequencies.

The kicker frequency band used is normally from dc or a very low frequency to half the bunch frequency. This means that the power amplifier low frequency cut-off must be low enough to obtain a satisfactory phase at the lowest mode frequency qf_0 or $(1-q)f_0$. Often this excludes the use of commercially available power amplifiers and power combiners, and specially designed distributed tube amplifiers are often needed.

The extension of the pick-up frequency band to very low frequencies may present a similar problem. This can be obtained by using high impedance head amplifiers for electrostatic pick-ups (SPS, ISR, PSB, AA), by sample and hold (FNAL Booster, PEP), by integration and reset of stripline signals (SPEAR), by synchronous detection at the bunch frequency or a harmonic thereof (PETRA), or simply by the inherent sample and hold associated with digitizing bunch by bunch (all digital dampers).

The kicker low-frequency problem can be avoided by translation of the dc to $Mf_0/2$ band to $Mf_0/2$ to Mf_0 as has been done by a digital modulator implemented in PETRA. An obvious solution for bunched beam dampers is to move both pick-up band and kicker band to for example the band from $Mf_0/2$ to Mf_0 or higher, as demonstrated by the very simple damper used for the NSLS Booster.

A special problem occurs if counter-rotating beams of opposite polarity are circulating in the same ring (SPEAR, PEP, LEP, AA). In LEP, which uses non-directional pick-up and kickers, bunches of the two beams are separated by gating while directional kickers are used in the AA. Both gating and directional kickers and pick-ups are used in SPEAR.

Some transverse feedback systems normalize the pick-up signals to intensity (positions) such that damping rates are intensity independent (FNAL MR, FNAL Tevatron, PETRA, LEP), while others (SPS, ISR, PSB, AA) have no intensity normalization, which results in a constant equivalent coupling impedance with damping rates proportional to intensity as for a passive impedance.

TRANSVERSE DAMPERS FOR B-FACTORIES

Moving the feedback system frequency band from dc-$Mf_0/2$ to $Mf_0/2$ - Mf_0 eliminates any need for special power amplifiers and low frequency pick-ups or sample and hold circuits. Active closed orbit suppression combined with a simple notch filter can easily be made with band and suppress the revolution harmonics to the required level. The correct phase taking into account tunes and the notch filter can be obtained by using two pick-ups separated by 90° and combining their signals.

If the m = ±1 modes are unstable, chromaticity must be controlled to provide adequate overlap with the feedback system frequency band. If this is not possible, the frequency band could be moved to higher frequency, or a second system at higher frequency could be implemented in parallel. This point may need further work.

REFERENCES

1. Feasibility Study for an Asymmetric B-Factory Based on PEP, LBL PUB-5244, SLAC-352, CALT-68-1589, (1989).
2. T. Nakada (Editor), Feasibility Study for a B-Meson Factory in the ISR Tunnel, CERN 90-02, PSI PR 90-08, (1990).
3. P. Marchand, PSI Technical Note, TM-12-89-06, (1989).
4. F.J. Sacherer, IEEE Trans. Nuclear Sci., NS-20, 825 (1973).
5. F.J. Sacherer, IEEE Trans. Nuclear Sci., NS-24, 1393 (1977).
6. F. Pedersen and F.J. Sacherer, IEEE Trans. Nuclear Sci., NS-24, 1396 (1977).
7. F.J. Sacherer, CERN/SI-BR/72-5, (1972).
8. G. Besnier, Thesis, Université de Rennes, B-282, (1978).
9. S. Krinsky, IEEE Trans. Nuclear Sci., NS-32, 2320 (1985).
10. D. Boussard and J. Gareyte, Proc. 8th Int. Conf. on High Energy Accel. CERN 1971, p. 317.
11. K.W. Robinson, CEA Report CEAL-1010, (1964)
12. F. Pedersen, IEEE Trans. Nuclear Sci., NS-22, 1906 (1975).
13. A.W. Chao, P.L. Morton, and J.R. Rees, IEEE Trans. Nuclear Sci., NS-26, 3343 (1979).
14. V. Bharadwaj, private communication.
15. P. Bramham et al., Proc. 9th Int. Conf. on High Energy Accel, Stanford 1974, p. 359.
16. M.A. Allen, M. Cornacchia, and A. Millich, IEEE Trans. Nuclear Sci., NS-26, 3287 (1979).
17. D. Boussard, Antiprotons for Colliding Beam Facilities, CERN 84-15, 261 (1984).
18. J.C. Juillard and E. Peschardt, Proc. of European Particle Accelerator Conf., Nice 1990.
19. B. Kriegbaum and F. Pedersen, IEEE Trans. Nuclear Sci., NS-24, 1695 (1977).
20. M.A. Allen et al., IEEE Trans. Nuclear Sci., NS-28, 2317 (1981).
21. D. Boussard, G. Lambert, IEEE Trans. Nuclear Sci., NS-30, 2239 (1983).
22. M. Chanel, R. Maccaferri, and J.C. Perrier, Proc. of European Particle Accelerator Conf., Nice 1990.

23. F.J. Sacherer, Proc. 9th Int. Conf. on High Energy Accel. Stanford 1974, p. 347.
24. F. Sacherer, CERN 77-13, 198 (1977).
25. G. Carron et al., Proc. of the IEEE Particle Accelerator Conf., Chicago 1989, p. 803.
26. R. Bossart et al., IEEE Trans. Nuclear Sci., NS-26, 3284 (1979).
27. R. Stiening and E.J.N. Wilson, Nucl. Instr. and Methods 121, 283 (1974).
28. R.D. Kohaupt, Proc. 11th Int. Conf. on High Energy Accel., Geneva 1980, p. 562
29. B. Zotter, IEEE Trans. Nucl. Sci., NS-32, 2191 (1985).
30. S. Myers, Proc. of the IEEE Particle Accelerator Conf., Washington 1987, p.503
31. L. Arnaudon, S. Myers, R. Olsen, Proc. of the European Particle Accelerator Conf., Nice 1990.
32. L. Thorndahl and A. Vaughan, IEEE Trans. Nuclear Sci., NS-20, 807 (1973).
33. C. Carter et al., IEEE Trans. Nuclear Sci., NS-28, 2270 (1981).
34. C. Ankenbrandt et al., IEEE Trans. Nuclear Sci., NS-24, 1698 (1977).
35. E. Higgins et al., IEEE Trans. Nuclear Sci., NS-22, 1473 (1975).
36. J. Crisp et al., IEEE Trans. Nuclear Sci., NS-32, 2147 (1985).
37. J.L. Pellegrin, IEEE Trans. Nuclear Sci., NS-22, 1500 (1975).
38. J. Galayda, IEEE Trans. Nuclear Sci., NS-32, 2132 (1985).
39. D. Heins et al., DESY 89-157, (1989).
40. F. Pedersen, W. Pirkl, and K. Schindl, IEEE Trans. Nuclear Sci., NS-30, 2343 (1983).
41. C.W. Olson et al., IEEE Trans. Nuclear Sci., NS-28, 2296 (1981).
42. C. Taylor, Antiprotons for Colliding Beam FAcilities, CERN 84-15, 163 (1984).

COMPARISON OF ACCELERATOR PHYSICS ISSUES FOR SYMMETRIC AND ASYMMETRIC B-FACTORY RINGS

Maury Tigner
Cornell University, Ithaca, NY 14853

ABSTRACT

A systematic comparison of accelerator physics issues from the beam-beam interaction to single particle stability including ring and IR layout, synchrotron radiation and lost particle backgrounds, and single and multi-bunch instabilities is given. While some practical handicap probably accrues to the asymmetric design because of its extra constraints, the differences in the two approaches tend to be obscured by larger issues such as how to achieve the enormous increases in luminosity demanded of a b-factory.

INTRODUCTION

Many of the beam physics phenomena which govern the behavior of storage ring colliders are energy dependent. Thus we can expect to see some differences in behavior between machines with equal and unequal energies of the colliding beams. We shall examine the beam-beam effect as well as the questions of stability relating to the beam-environment interaction and the questions of single particle stability or dynamic aperture and geometry questions pertaining to the IR and ring layout as well. In each case we shall set down the things about the particular phenomenon that are the same for both kinds of machines and then discuss those that are different. There are indeed significant differences but lack of experimental experience makes judgement of their ultimate importance very difficult.

Some controversy still plagues the question of how much luminosity reduction is permitted if one utilizes b-mesons created in a center of mass system moving in the laboratory. For the purposes of this survey I shall assume that the design luminosity for an equal energy machine must be 10^{34} cm^{-2}sec^{-1} and that for a machine with significant asymmetry (energy ratio between 1.5 and 5) a design luminosity of 3×10^{33} will suffice.

BEAM-BEAM INTERACTION

Same
    As far as we know the achievable tune spread is limited by the same basic mechanisms that limit equal energy beam collisions. This means that the benefit or deficit in having round or flat profiles, equal or unequal phase shifts between crossings and bunch length close to or much less than beta star are the same.

Different

In configurations in which the two beams pass through common optical elements close to the IP it will be more difficult to get low beta star because the lens settings which would give low beta for the high energy beam (HEB) will overfocus the low energy beam (LEB). This condition will compromise the specific luminosity when the beams are of unequal energies. Of course, if very large crossing angles are used then the optics can, in principle, be completely independent. Whether the needed crabbing cavities are practical and do not compromise the rest of the design with too much impedance remains to be seen.

Additionally certain variables must now consciously be controlled. Obviously the two beams should have equal beam-beam tune spreads and transverse profiles to make optimum use of the beams. These conditions lead to a need to maintain the equality

$$N_2 \gamma_2 \beta*_{v1} = N_1 \gamma_1 \beta*_{v2}$$

Certain other restrictions have been suggested as additional assurances that collisions of unequal energy beams not give compromised performance by comparison with equal energies. One such set, used by the LBL/SLAC group[1] and checked with simulations, suggests equality of damping decrement between collisions for the two beams as well as equality of the synchrotron phase shift across the collisions or

$$\sigma_{s1} Q_{s1}/\beta*_{1,h,v} = \sigma_{s2} Q_{s2}/\beta*_{2h,v}$$

Collectively this set of conditions are referred to by their originators as "energy transparency".

Another set have been suggested on the basis of analytical theory and simulation[2]. In Hamiltonian perturbation theory one can derive the strength of the driving terms for the beam-beam resonances. If those terms are made the same for both beams then what space we have in the tune plane will be optimally used. This restriction leads to the further requirements

$$\beta*_1 = \beta*_2$$
$$\sigma_{s1} = \sigma_{s2}$$
$$\varepsilon_2 = \varepsilon_2$$
$$N_1 \gamma_1 = N_2 \gamma_2$$

An obvious question is whether the differing beam energies will influence the allowed beam-beam parameter for the lower (higher) energy beam. Unfortunately no firm answer can be given as this circumstance has never been confronted experimentally. What has been done is to explore the luminosity of a given machine as a function of energy. These data[3] indicate that in the regime in which the ξ value is saturated, the luminosity is proportional to E, the beam energy. One attempt to ameliorate this dependence by means of damping decrement wigglers showed no improvement. Taken at face value these data seem to indicate that in the assymetric case, other things being equal, one will achieve the same luminosity as one would have in the case of equal energy beams with the same center of mass energy i.e., higher tune spread for the higher energy beam cancels out lower tune spread for the lower energy beam, provided that the machine design is flexible enough to permit exploiting this possibility. It appears that uncertainty about the luminosity coefficient for unequal energy collisions will remain until it can be measured directly in an asymmetric collider configuration.

For configurations with relatively large assymetry, it is tempting to think of making the two rings of different radius. In such a case, the bunches in the smaller ring may collide with several of the bunches in the larger ring, giving opportunity for a rich spectrum of coherent beam-beam modes above and beyond those open to one bunch machines[4]. While not absolutely necessary in every case, it appears preferable to keep the rings of equal circumference to avoid blocking parts of the tune plane with these unwanted resonances.

IR LAYOUT AND LATTICE

Same

The basic options for equal or unequal energy beams are the same: head on collisions, beams crossing at an angle, each with either round or flat beams.

Different

In the case of unequal energies there is at least the apparent possibility of using simple magnetic separation to get the beams apart close enough to the IP to permit the needed close bunch spacing. Early studies indicate that this apparent advantage may be illusory because of the strong synchrotron radiation close to the IP which is a concommitant of such separation.

In cases where there are common optical elements, the differing energies will not only make achievement of the needed betas more complicated but may also have the potential for compromising the SR and lost particle engendered backgrounds due to the overfocusing mentioned before and due to the difficulty of keeping both beams small in the final quads and thereby minimizing synchrotron radiation there. To date no satisfactory optics for unequal beam energies with common optical elements have been found. Further design experience may lead us to some satisfactory solutions with adequate tuning flexibility.

SYNCHROTRON RADIATION BACKGROUNDS

<u>Same</u>
In any case it is crucially important to avoid rays of synchrotron radiation from striking parts of any mask from which scattered or fluorescence radiation can reach the detector pipe with large solid angle. This can be avoided in many examples by use of s-bends having opposite orbit curvature on opposite sides of the IP, throwing rays from the opposite directions to opposite sides of the pipe so they can be made to hit the mask side away from the detector while the bend serves to hide the detector from magnets further upstream.

<u>Different</u>
As mentioned in the previous section, magnetic separation is a tempting alternative in the unequal energy case. So far, no magnetic separation configuration with satisfactory SR background properties has been found.
In the "energy transparency" scheme of reference 1, the lower energy beam has a relatively larger emittance. This will give a relative disadvantage in SR background. Quantitative consequences remain to be understood.

LOST PARTICLE BACKGROUNDS

<u>Same</u>
The main defense against primary hits from coulomb scattered particles and particles that have lost energy through bremsstrahlung events upstream is to arrange the optics such that the majority are stopped before they reach the IP or are focussed through the IP without hitting. A significant help in accomplishing this is, in many examples, making the focus of the final lens in the plane of the bend nearest the IP.

<u>Different</u>
At this time there are no known differences of any significance.

SINGLE BUNCH STABILITY

<u>Same</u>
The basic phenomena affecting single bunch stability are expected to be the same in both equal and unequal beam energy cases. The primary concern will be bunch lengthening due to the fast head-tail instability. Transverse effects are not expected to play a role in single bunch stability in the parameter regimes now being studied. The prescription for avoiding bunch lengthening is the same, namely, keep momentum compacton factor up, bunch length up and number of particles per bunch down while minimizing the number of cavities and their beam coupling strengths.

Different
The various schemes for minimizing the potential sacrifice in beam-beam tune spread have different consequences for the ratio of beam currents in the two rings and thus for the number of cavities and consequent impedance per ring. The actual impedances will depend upon whether one uses normal or superconducting cavities and whether the rings have equal sizes. In principle it might thus be possible for one ring to be forced above the lengthening threshold. In examples seen to date such consequences seem avoidable. The same cannot be said for the multi-bunch instabilities.

MULTIBUNCH INSTABILITIES

Same
The beam currents needed for the enormous luminosities sought are ten times or more those now used in storage ring collider practice. In addition, the number of bunches per ring may be 100 times the present practice making the natural ringdown of even well damped cavity modes ineffective as a guard against multibunch instabilities. Thus, control of both longitudinal and transverse multibunch instabilities is one of the major engineering problems for b-factories whether of equal or unequal beam energies. Success will require extreme measures to cut down on the number of cavity cells required per ring as well as minimization of the R/Q's and Q's of all the modes, transverse and longitudinal. Naturally that holds also for other apparatus in the ring such as separators, septa, crab cavities, etc. which might have resonant modes that could drive multibunch instabilities.

Even with the best imaginable reduction in ring resonant impedances, it appears that feedback systems in transverse and longitudinal dimensions will be required. The challenge will be to keep the rf impedances low enough that the feedback systems are still manageable.

Different
The question here is whether the product of $I \cdot Z \cdot \tau$ is excessive for one of the rings in an asymmetric configuration due to the accelerator physics rules imposed to maximize beam-beam capability. Here I is the beam current, Z is the ring impedance, proportional to the number of cavity cells, and τ is the radiation damping time. One could make an analysis of the problem by setting forth the rf design and beam-beam design algorithms and studying their (non-linear) interplay. However we now have a number of examples to compare directly and that probably tells the story in a more direct fashion. If we use numbers of cavity cells for the impedance and use amps and seconds for the units of current and damping time we find that a "typical" symmetric machine example has the product about 0.8, whereas examples of asymmetric machines using either the "energy transparency" or "equal tuneprint" algorithms have the product 0.5 or less for both rings. Note that in these

examples the design luminosity of the symmetric machine is two to three times that of the asymmetric examples. In as much as these design examples are realistic, one can then say that for multibunch instabilities, the asymmetric design might have a slight advantage owing to the smaller currents required and that this obvious conclusion is not disturbed by the subtleties of the beam-beam optimization or energy ratios.

RING LAYOUT AND LATTICE

<u>Same</u>
In all cases we need to set the overall ring parameters to give a beam emittance of less than or equal to the operating emittance desired while permitting enough circumferential space, appropriately distributed, for auxiliary equipment. Items that may need to be evenly distributed are cavities and wigglers for emittance and damping enhancement. Particulary important in this instance is provision of sufficient room for flexible IR arrangements as that may need to be altered after initial running experience.

<u>Different</u>
In cases with rather larger energy ratios it is tempting to think of making the rings of different sizes which helps in keeping the rf power down while helping to keep the damping decrement for the lower energy beam up. In addition, separation of the beams at the IR could perhaps be more easily effected with kissing rings. However, the seemingly overwhelming burden of synchrotron radiation background in magnetic separation configurations together with the extra coherent beam-beam instabilities noted above make this approach seem too limited at present. Further developments may show ways around these problems.

SINGLE PARTICLE STABILITY

<u>Same</u>
Maintenance of sufficient dynamic aperture will be important in either case. As always, it is useful in this regard to keep the tunes and emittances as low as practical. It is commonly believed that in electron storage rings the dynamic aperture is set by the chromaticity and the distribution of its correction. Since at least half of the chromaticity comes from the IR normally and since we'll be wanting to use even stronger IP focusing than is now normal, it will be well to keep the number of IPs to a minimum. Also essential will be the production of working vacuums at least as good as in present practice even though the desorption may be ten times worse than now.

<u>Different</u>
In schemes having common optical elements for the two beams, the IR contribution to the chromaticity of the lower energy ring will be

larger than for the high energy ring. Special attention will have to be paid to maintaining sufficient dynamic aperture. Until more examples have been worked out it will not be known how much of a restriction on optics designs this is. It could be serious.

Depending upon the design algorithm that is used to minimize beam-beam problems for the low energy beam, the fractional depth of the rf potential will be different for the two beams and thus their lives due to beam-beam bremsstrahlung, the major life determining phenomenon, will also be different. In examples seen so far this does not appear to be a significant restriction.

R/D ISSUES

Same

In all cases the improvement of the luminosity coefficient, $\xi/\beta*$, in the hour glass regime, including the questions of best beam profile and tolerance for crossing at an angle, and the reduction in rf impedance per volt produced are of the utmost importance. Another prime focus for R/D must be the accelerator produced backgrounds from synchrotron radiation and lost particles. These issues can be explored by a combination of calculation, laboratory work and experiments on existing storage rings and some work is going forward on all these fronts.

Different

Unfortunately the crucial question of the influence of energy asymmetry on the luminosity cannot be tested experimentally in a direct way although the scaling of luminosity and saturated tune spread with energy in a single machine can shed light on this to some extent.

CONCLUSION

Collision of beams of unequal energies will no doubt bring with it more complications than were the energies equal. From what we know now, it appears that these differences may be small when compared with the challenge of increasing luminosities by two orders of magnitude to make a b-factory of any kind.

REFERENCES

1. LBL PUB-5263, Investigation of an Asymmetric B-Factory in the PEP Tunnel, March 1990.
2. S.Krishnagopal and R. Siemann, Beam Energy Inequality in the Beam-Beam Interaction, CLNS 89/967, Cornell, 1989.
3. D.H.Rice, these proceedings.
4. K.Hirata and E.Keil, Barycentre Motion of Beams due to Beam-Beam Interaction in Asymmetric Ring Colliders, CERN/LEP-TH/89-76.

Part III

CONTRIBUTED PAPERS

A COMPARISON OF FLAT BEAMS WITH ROUND

S.Krishnagopal & R.H.Siemann
Newman Laboratory, Cornell University, Ithaca, NY 14853

ABSTRACT

Beam-beam simulations that incorporate the phenomenon of phase-averaging are used to compare and contrast the dynamics of flat and round beams. Flat beams are found to require short bunches, and give a tune-shift limit of around 0.05. Round beams allow bunch-lengths of the order of β^*, and yield tune-shifts of 0.10 or more.

I: INTRODUCTION

Recently, the phenomenon of *phase-averaging*[1] has been shown to play an important part in reducing the strengths of synchrobetatron resonances. Even earlier, simulations that indirectly accounted for this by treating the beam-beam interaction as a thick lens, predicted large tune-shift limits (~ 0.10) for *round* beams[2]. This is higher than the experimentally achieved limits of up to 0.06 in present e^+e^- colliders running *flat* beams. There is thus a suggestion that round beams may allow for larger tune-shifts – and hence greater luminosity.

This motivated us to undertake a comparative simulation study of the beam-beam dynamics of flat and round beams. Our attempt has only been to compare the maximum tune-shifts obtained for the two; we have not, for example, tried to look at beam tails and lifetimes. Nor has any effort been made to consider effects other than the beam-beam interaction (such as RF requirements or optics design). Section II contains details of the simulation and the algorithm used for the comparison. Results are presented and discussed in Section III.

II: DESCRIPTION OF THE SIMULATIONS

<u>Simulation Details:</u> The simulation program was essentially the same as that in Ref. 1. It was three-dimensional and weak-strong. Radiation fluctuations and damping were included once every turn. Arc transport was linear and did not couple the horizontal and vertical. The beam-beam interaction was modelled as a thick lens, using nine beam-beam kicks per interaction. The amplitude function was allowed to vary quadratically about the interaction point (IP). Chromaticity and dispersion at the IP were zero.

The weak beam comprised of 1,000 test particles initially distributed in a random Gaussian in all six phase-space coordinates. The particles were tracked for five damping times, and beam-sizes averaged over the last 1,000 turns were used to calculate the tune-shift parameter.

The ring was based on the Cornell Electron Storage Ring (CESR), so that the revolution frequency $T_0 = 2.56$ μs, the beam energy $E_0 = 5.3$ GeV, and the R.F. cavity frequency $f_{rf} = 500$ MHz. Due to restrictions on computer time a rather high value of damping decrement (the fractional energy radiated per turn), $\delta = 10^{-3}$, was chosen, corresponding to a damping time of 1000 turns. The two transverse tunes were generally kept equal; the consequences of relaxing this constraint are discussed in Section III.

<u>Technique of Comparison:</u> We started with round beam parameters as chosen in Ref. 1. From this an equivalent flat beam was constructed by requiring that it have the same *nominal* tune-shift ξ and luminosity L (per unit current) as the round beam. Any discrepancy in the actual values could then be attributed to the different dynamics. From the formulæ for ξ and L it can easily be shown that given the amplitude function (β^*) and emittance (ϵ_0) of the round beam, those of the flat beam are completely determined provided the emittance ratio κ is known. We chose $\kappa = 0.15$; the other parameters are shown in Table 1.

TABLE I: Amplitude functions and emittances

ROUND	FLAT
$\beta_r^* = 3$ cm	$\beta_x^* = 1$ m $\beta_y^* = 1.5$ cm
$\epsilon_{0r} = 1\times 10^{-7}$ m	$\epsilon_{0x} = 2\times 10^{-7}$ m $\epsilon_{0y} = 3\times 10^{-9}$ m

A comment in passing: the luminosity is often parametrized as

$$L = \frac{I\gamma\xi}{2er_e\beta^*}(1+R_\sigma), \tag{1}$$

where β^* is the vertical amplitude function, R_σ is the ratio of the minor and major axes of the collision spot, and I is the beam current. It should be emphasized that the factor of two in favour of round beams (arising from the beam-profile) exactly compensates the factor of two in favour of flat beams (arising from the lower β^*), to give equal nominal luminosity. Consequently, differences in performance can only result from the beam-beam dynamics – that is from the tune-shift limit.

The remaining variables, with respect to which the tune-shift can be maximized, are the betatron and synchrotron tunes (Q_b and Q_s), the current (I), and the bunch-length (σ_l). A fair comparison cannot be obtained by simply making these equal for the two beams because the dynamics is inherently different, and what is optimal for the one beam may not be so for the other. We therefore decided to maximize the tune-shift limit *independently* for the two beams. For reasons of computer time, however, we restricted ourselves to two regions in tune space where good performance may *a priori* be expected – just above the fourth-order (3/4) and sixth-order (4/6) betatron resonances. For convenience they will be referred to as Case A and Case B, respectively.

Within these two regions then, the following algorithm was adopted, for both beams:
a) Perform a coarse betatron-tune scan to find the best operating-point (at a current of 10 mA).
b) At this operating-point increase the current to get the maximum tune-shift parameter, ξ_{max}.
c) Do a fine betatron-tune scan to optimize ξ_{max}. (In the first three steps, $Q_s = 0.11$, and $\sigma_l = \beta^*$.)
d) Optimize ξ_{max} w.r.t. Q_s.
e) Optimize ξ_{max} w.r.t. σ_l.

The results of these scans are presented in the next section in terms of the vertical tune-shift parameter. For round beams the horizontal ξ behaved the same. For flat beams, because of the large horizontal β^* the dynamics in that dimension is not expected to be synchrobetatron-resonance limited.

III: RESULTS AND DISCUSSION

Near the fourth-integer resonance (Case A), the initial betatron-tune scan gave the best ξ at different operating points for the two beams – at 0.75 for the round, and 0.78 for the flat beam. Above the sixth-integer resonance (Case B), both beams achieved the best ξ at the same tune – 0.67.

Figs. 1 show current scans at these operating points. In Case A, for flat beams the maximum is reached at 25mA, after which ξ decreases. For round beams this happens at 30mA. In Case B the corresponding currents are 20 & 40 mA respectively. In both cases ξ_{max} is greater for round beams by a factor of more than two and occurs at higher currents; both contribute to greater luminosity.

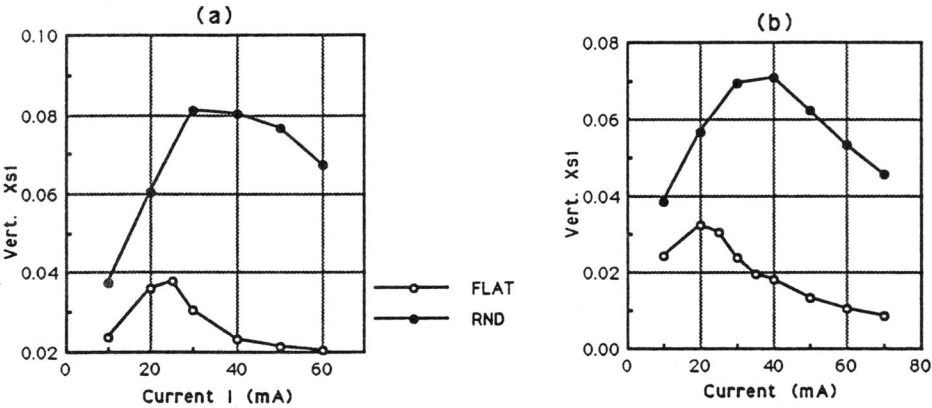

FIG. 1. Vertical tune-shift ξ as a function of current, above (a) the fourth-order, and (b) the sixth-order, betatron resonance. ($Q_s = 0.11$, $\sigma_l = \beta^*$).

At these currents a fine betatron-tune scan was performed to find the best tune. These are shown in Table 2 below.

Table II: Final Betatron-tunes

	ROUND	FLAT
Case A	0.742	0.775
Case B	0.665	0.670

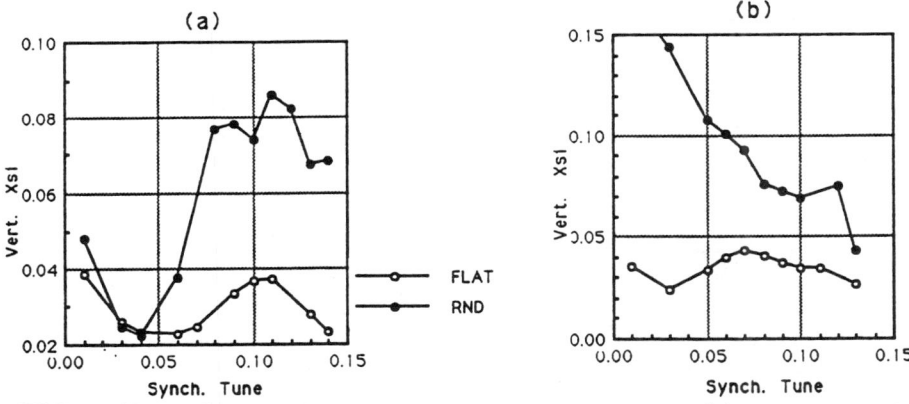

FIG. 2. Vertical ξ as a function of synchrotron tune Q_s. (a) I = 25mA, Q_b = 0.775 for flat beams, and I = 30 mA, Q_b = 0.742 for round beams. (b) I = 20mA, Q_b = 0.670 for flat beams, and I = 40mA, Q_b = 0.665 for round beams. ($\sigma_l = \beta^*$).

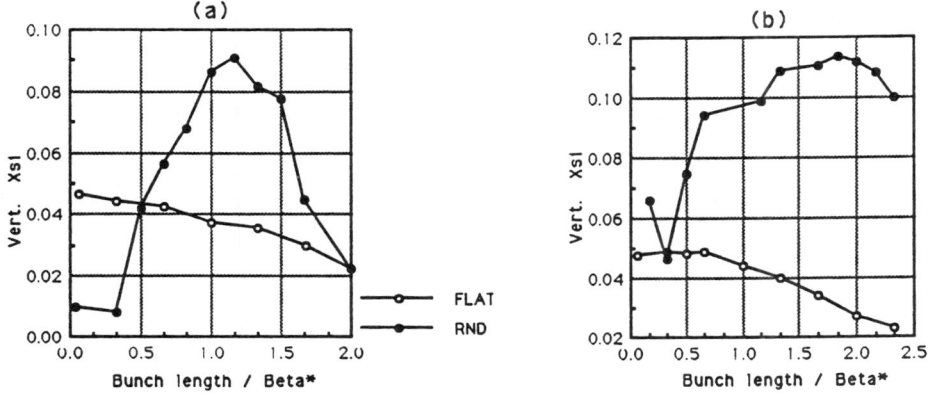

FIG. 3. Vertical ξ as a function of σ_l/β^*. (a) I = 25 mA, (0.775,0.11) for flat beams, and I = 30 mA, (0.742,0.11) for round beams. (b) I = 20 mA, (0.670,0.07) for flat beams, and I = 40mA, (0.665,0.06) for round beams.

The dependence on synchrotron tune is shown in Figs. 2, and warrants some attention. In Fig.(2a) note that for very small ($\leq .01$) synchrotron tunes ξ is large, because the tune-modulation is weak. As Q_s increases ξ falls, but not monotonically; both curves reach a maximum at $Q_s = 0.11$, before decreasing again beyond it. Thus, for Case A the optimal synchrotron tune is quite high for reasonable values of that parameter. In Fig.(2b) the behaviour is different. For round beams ξ falls steadily, so that it is best to run at the lowest practical Q_s. (For the next step in the algorithm $Q_s = 0.06$ was chosen as a reasonable value.) For flat beams $Q_s = 0.07$ is optimal.

Figs. 3 show the dependence on bunch-length (actually on σ_l/β^*). In both cases we see a qualitative difference between the behaviour of flat and round beams. For flat beams the decrease in ξ, though not as abrupt as in Ref. 3 because of phase-averaging, is still monotonic. So one is forced to go to as small bunch-lengths as possible to get large ξ. This is clearly limited by RF considerations.

For round beams, however, the curve is *not* monotonic – there is a maximum at $\sigma_l \geq \beta^*$. Thus one can reduce β^* as much as is feasible without having to worry about making σ_l even smaller. Too, from the figures it is clear that the ξ_{max} for flat beams is around 0.05, while for round beams it may reach as high as 0.10, or even more.

The consequences of going off the coupling resonance were also investigated at this point. Fig. 4 shows that for flat beams as the transverse tunes are split, with the horizontal tune Q_h tune increasing and the vertical tune Q_v decreasing (by equal amounts), ξ decreases for both cases (A & B). (The same thing happens if Q_h is fixed and Q_v is increased.) For round beams the influence of the coupling resonance is not an issue, and simulations confirm that for small distances away from it there is no substantial decrease in ξ.

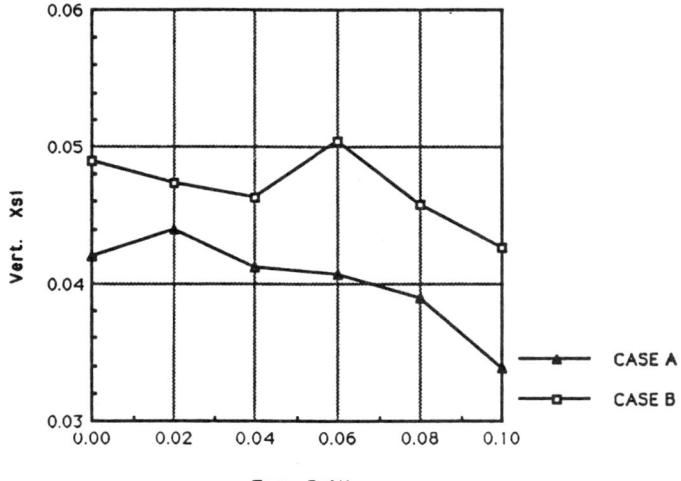

FIG. 4. Vertical ξ as a function of tune-split, $Q_h - Q_v$, for flat beams, as one goes above the coupling resonance; for the best conditions of Cases A and B (from Fig. 3). $\sigma_l = 1$ cm.

Discussion: The above simulations demonstrate one important point. The difference between flat and round beams is not just in the beam-profile and the consequent tune-shift limit (important though that is), but also in the details of the dynamics involved. Thus the picture that emerges is a complicated one, in which the interplay between the resonance structure around the operating point and the potential function defined by the beam-profile yields a rich dynamics, which must be thoroughly investigated before a decision can be made on the choice of various accelerator parameters.

In particular, we see that round beams not only give a larger tune-shift limit, but also allow us to run at smaller synchrotron tunes and larger bunch-lengths. This has important benefits for the RF system (higher single-bunch instability threshold and lower peak voltages). On the other hand, the need to have equal β^*s in both dimensions complicates IR optics design (separation and backgrounds) considerably.

IV: CONCLUSION

Our simulations of flat beams show bunch-length dependence and tune-shift limits consistent with experimental data from existing machines. Shorter σ_l are favoured, and $\xi_{max} \sim 0.05$.

Round beams with the same nominal tune-shift and luminosity show better actual performance (in simulations). The bunch-length dependence is more complicated, and $\sigma_l \sim \beta^*$ are favoured. Tune-shifts are greater than for flat beams, $\xi_{max} \sim 0.10$ or more, and occur at higher currents. The choice of operating point plays an important role here.

ACKNOWLEDGMENTS

This work was supported in part by the National Science Foundation.

REFERENCES

1. S.Krishnagopal & R.H.Siemann, Phys.Rev.D, **41**, 2312 (1990).

2. S.Krishnagopal & R.H.Siemann, in 'Proceedings of the 1989 IEEE Particle Accelerator Conference', Chicago, edited by F.Bennet & J.Kopta, p. 836.

3. S.Myers, Nucl. Instr. Methods **211**, 263 (1983).

The Advantages of Flat Beams in Head-on collisions

Andrew Hutton

This note extends some considerations of β_y^* (the value of the vertical beta function at the Interaction Point) in coupled and uncoupled machines due to E. Keil[Ref.1]. He showed that in two otherwise identical machines one of which is fully coupled, the other with a coupling ratio of r, then:

1. To obtain the same luminosity L with the same number of particles per bunch and the same number of bunches, the ratio of the β_y^* values must be

$$\beta_y^*/\beta_{yc}^* = (1+r)/2 \qquad \text{Eq. 1}$$

with

$$r = \sigma_y^*/\sigma_x^* = \beta_y^*/\beta_x^* = \varepsilon_y/\varepsilon_x \qquad \text{Eq. 2}$$

and this gives

$$\beta_x^*/\beta_{xc}^* = (1+r)/2r \qquad \text{Eq. 3}$$

2. To obtain the same luminosity under these conditions the sum of the vertical emittance ε_y and horizontal emittance ε_x is constant. This gives an emittance ratio in the two cases of

$$\varepsilon_y/\varepsilon_{yc} = 2r/(1+r) \qquad \text{Eq. 4}$$

and

$$\varepsilon_x/\varepsilon_{xc} = 2/(1+r) \qquad \text{Eq. 5}$$

In the context of the *B* Factory, the synchrotron radiation background is an extremely important factor in the design of the Interaction Region. It is therefore worthwhile to extend the comparison to the I.P. angular divergence, size of beams near the I.P. and the separation of closely spaced bunches.

Spot Size

The I.P. spot sizes are given by

$$\sigma_y^* = (\varepsilon_y \beta_y^*)^{1/2} \quad \text{Eq. 6}$$

and

$$\sigma_x^* = (\varepsilon_x \beta_x^*)^{1/2} \quad \text{Eq. 7}$$

The ratio of beam sizes for the flat and round beam cases is then

$$\sigma_y^*/\sigma_{yc}^* = (\varepsilon_y \beta_y^*)^{1/2}/(\varepsilon_{yc} \beta_{yc}^*)^{1/2} = r^{1/2} \quad \text{Eq. 8}$$

and

$$\sigma_x^*/\sigma_{xc}^* = (\varepsilon_x \beta_x^*)^{1/2}/(\varepsilon_{xc} \beta_{xc}^*)^{1/2} = 1/r^{1/2} \quad \text{Eq. 9}$$

Angular Divergence

The angular divergence is given by

$$\sigma'_y{}^* = \sigma'_x{}^* = (\varepsilon_y/\beta_y^*)^{1/2} = (\varepsilon_x/\beta_x^*)^{1/2} \quad \text{Eq.10}$$

The ratio of angular divergences in the coupled and flat cases is therefore

$$\sigma'_y{}^*/\sigma'_{yc}{}^* = (\varepsilon_y \beta_{yc}^*)^{1/2}/(\varepsilon_{yc}/\beta_y^*)^{1/2} = 2r^{1/2}/(1+r)$$
Eq.11

and $\sigma'_x{}^*/\sigma'_{xc}{}^* = (\varepsilon_x \beta_{xc}^*)^{1/2}/(\varepsilon_{xc}/\beta_x^*)^{1/2} = 2r^{1/2}/(1+r)$
Eq.12

which are identical as expected. As r is reduced, the angular divergences also decrease. For example, if r = 0.1 then the angular divergence is a factor 0.58 smaller than the round beam case. For a more extreme case of r = 0.02 (as in the crab crossing proposal in the Appendix of the Feasibility Study[Ref.2]) the angular divergence is then a factor 0.28 smaller than the round beam case.

Separation of closely spaced bunches.

The bunches must be separated close to the Interaction Point in order to avoid unwanted interaction between closely spaced bunches. The relevant parameter is the larger of the two transverse beam dimensions at the point where the succeeding bunches cross each other. In the case of very short bunch spacing, the nearest near collision occurs before the first quadrupole and therefore can be evaluated exactly. The general principle will also hold for near collisions beyond the first quadrupole but the exact values will depend on the details of the optics.

Let the first crossing be at a distance s from the I.P. then

$$\beta_y^s = \beta_y^* + s^2/\beta_y^* \qquad \text{Eq.13}$$

or in a form more suitable for the present discussion

$$\sigma_y^s = (\sigma_y^{*2} + s^2 \sigma'_y{}^{*2})^{1/2}$$
$$\text{and} \quad \sigma_x^s = (\sigma_x^{*2} + s^2 \sigma'_x{}^{*2})^{1/2} \qquad \text{Eq.14}$$

so in all cases σ_x^s is greater than or equal to σ_y^s as the angular divergences are equal in the two planes and σ_x^* is always greater than or equal to σ_y^*. So comparing the round and flat beam cases

$$\sigma_x^s / \sigma_{xc}^s = (\sigma_x^{*2} + s^2 \sigma'_x{}^{*2})^{1/2} / (\sigma_{xc}^{*2} + s^2 \sigma'_{xc}{}^{*2})^{1/2} \qquad \text{Eq.15}$$

For all practical designs $s \gg \beta_{xc}^*$ and so $\sigma_{xc}^* \ll s\sigma'_{xc}{}^*$. Using this approximation

$$\sigma_x^s / \sigma_{xc}^s = \left[\beta_{xc}^{*2}/s^2 r + 4r/(1+r)^2 \right]^{1/2} \qquad \text{Eq.16}$$

This implies that for the same separation criteria for the near collisions, the bending strength required to produce the separation is significantly less for flat beams than for round beams. Numerically, with $s/\beta_{xc}^* = 30$, the reduction is a factor 0.58 for $r = 0.1$ and 0.36 for $r = 0.02$. With the same bend length, the bending radius is reduced by the same factor giving a reduction in the number and energy of the synchrotron radiation photons.

Synchrotron Radiation from the I.R. quadruples

The beam sizes in the I.R. quadrupoles will also be smaller by a similar factor, so the synchrotron radiation is more easily collimated. If the quadrupole bore is not determined by the beam stay-clear, but is fixed by other constraints, then the smaller beam sizes in the quadrupoles will also lead to significant reductions in the number and energy of the synchrotron radiation photons. If the scaling rules for β_y^* given above are used, the total amount of focusing strength required to produce flat beams is about the same as is required to produce round beams Ref. 3 so there is no penalty to having a smaller β_y^* in the flat beam case. Thus, all of the effects lead to a reduction in the synchrotron radiation background and to an easier collimation scheme in the case of flat beams.

Factors against Flat Beams

The minimum β_y^* is determined by the bunch length σ_s, the usual criteria being that $\beta_y^* = (1-2)\, \sigma_s$. This means that for a flat beam, this criteria is approximately a factor two more stringent than for round beams. This is by far the most important problem to be evaluated and will lead to a push for the highest reasonable R.F. frequency (probably in the region of 500 MHz).

With the two rings stacked vertically in the PEP tunnel, it is inevitable that there will be vertical bending to bring the beams into collision. It will be the vertical quantum excitation of the beam in these bends that will determine the minimum value of r and hence how much advantage can be obtained from the flat beams. Clearly the weaker these bends the better for all effects.

There are some indications from simulations that round beams may permit a larger value of the acceptable tune shift for the beam-beam interaction[Ref.4]. Since there is no experimental verification of this effect at present, it is difficult to evaluate the probability of luminosity gains from this effect in a real machine.

Note that the usual argument in favour of round beams, a factor two in luminosity, is only valid at equal β_y^*. The scaling law used here leads to equal luminosities in the round and flat beam cases.

Conclusions

The present situation of the *B* Factory studies is that the problems of the synchrotron radiation background in the detector appear almost insurmountable for head-on collisions with round beams. It would seem reasonable to explore the possibility of colliding flat beams in the head-on configuration and to evaluate the synchrotron radiation background in the detector. All of the formulae derived in this paper would indicate that this configuration should lead to a significant improvement in the detector environment.

Acknowledgements

I would like to thank Witold Kozanecki for pointing out an error in the first draft of this paper.

References

1. ...E Keil. "The Choice of β_y^* in Fully Coupled Machines" LEP Note 113 (1978)

2. "Feasibility Study for an Asymmetric *B* Factory Based on PEP" LBL PUB-5244, SLAC-352, CALT-68-1589 (1989)

3. H. Kaiser and K. Steffan. "Notes relating to Combined Function and Strongly Coupled Electron Storage Rings", DESY PET-78/10 (1978)

4. R.H. Siemann, "Simulations of Electron-Positron Storage Rings", CBN-89-4 (1984)

COMPARISON OF LUMINOSITIES FOR ROUND AND FLAT BEAMS

V.A.Lebedev
Institute of nuclear physics, Novosibirsk, 630090, USSR

ABSTRACT

The paper compares of the ultimate luminosities which can be achieved for an asymmetric B-factory (4*7 GeV) in cases of the flat and round beams.

INTRODUCTION

The luminosity for colliding beams is defined by the well-known expression:

$$L = \frac{N_1 N_2 f}{4\pi \sigma_x \sigma_z} . \qquad (1)$$

Here N_1 and N_2 are the number of particles per bunch in beams 1 and 2 respectively, f is the collision frequency, and σ_x and σ_z is transverse beam size in the x and z directions at the interaction point. We assume that both beams have the same size. The principal limit to increasing the luminosity of colliding beams is the electromagnetic interaction. It is characterized by the tune shifts:

$$\xi_{1x,z} = \frac{e^2 N_2 \beta^*_{1x,z}}{2\pi E_1 \sigma_{x,z} (\sigma_x + \sigma_z)} , \quad 1 \leftrightarrow 2, \qquad (2)$$

where E_1 and E_2 are the beam energies, and where β^*_{1x}, β^*_{1z}, β^*_{2x} and β^*_{2z} are β-functions at the interaction point. By combining Eqs. (1) and (2) we get:

$$L = \frac{I_1 E_1 (1+r)}{2 e^3 \beta^*_{1z}} \xi_{1z} = \frac{I_2 E_2 (1+r)}{2 e^3 \beta^*_{2z}} \xi_{2z} , \qquad (3)$$

where $r = \sigma_z/\sigma_x$. One can see from Eq. (2) that the higher luminosity is achieved for a round beam, rather than for a flat one. The gain can be as high as a factor of 4. One part of this gain is due to the factor $(1 + r) = 2$, and another part is due to the tune shift which increases for round beams [1]. Now we need to compare the minimal possible β-function values which can be obtained in the cases of round and flat beams since this will determine the resulting luminosities.

DEFINITION OF PARAMETERS OF THE BEAMS

We suppose that the B-factory consists of two intersecting storage rings with two interaction points. To choose the beam parameters we need to use the following requirements:

1. It is better from the point of view of the accelerator design to have a small difference in beam energies; however, the resolution of CP-violation experiments increase with the difference in beam energies. Here we choose beam energies of 4 and 7 GeV. At this working point the experimental resolution decrease rapidly with smaller beam energy differences, but increases rather slowly with larger energy difference [2].

2. We assume the following parameters to be the same for both beams: the transverse size at the interaction region -- $\sigma_x = \sigma_{1x} = \sigma_{2z}$, $\sigma_z = \sigma_{1z} = \sigma_{2z}$ and the tune shifts - $\xi_x = \xi_{1x} = \xi_{2x}$, $\xi_z = \xi_{1z} = \xi_{2z}$.

3. The minimal longitudinal distance between bunches is determined by the requirement that the two beams should be separated by a few times of the beam size at the parasitic interaction points.

4. Bunch length has to be smaller than β-function at interaction point.

5. The number of particles per bunch is confined by the broad band longitudinal impedance $|Z_n/n| \simeq 0.2 \, \Omega$ (for given bunch length).

6. The decrease of the β-function at the interaction point is confined by the chromaticity in the final focus (FF).

Table 1

The main parameters of the beams

		Round		Flat	
Energy	[GeV]	4	7	4	7
Beam current	[A]	2.09	1.96	0.92	0.91
Particles per bunch	[10^{11}]	3.85	3.62	1.41	1.35
Distance between bunches	[m]	8.85		7.08	
Emittance ε_z	[nm·rad]	148	90	3.87	2.3
ε_x	[nm·rad]	148	90	387	230
R.M.S. momentum spread	$\Delta p/p$ 10^{-4}	9	7.1	9	7.1
Bunch length	[cm]	1.67	1.42	0.7	0.7
In the interaction point:					
β-function β_z	[cm]	3.4	5.6	0.88	1.48
β_x	[cm]	3.4	5.6	14.	23.5
dispersion function		0	0	0	0
transverse sizes σ_z	[μm]	71		5.84	
σ_x	[μm]	71		232	
beam-beam tune shift ξ_z		0.07	0.07	0.05	0.05
ξ_x		0.07	0.07	0.02	0.02
FF chromaticity (for one IP) ν'_z		-24	-24	-25	-26
ν'_x		-24	-24	-22	-15
Circumference	[m]	685		685	
Betatron tunes ν_z		≈15		≈18	
ν_x		≈15		≈12	
Synchrotron tune ν_s	10^{-2}	2.8	2.6	4.1	3.6
Momentum compaction α	10^{-3}	4.75	4.75	3.4	3.4
Accelerating voltage	[MV]	4.2	5.92	11.2	15
Synchrotron radiation losses	[MV]	1.82	3.2	1.82	3.2
Synchrotron radiation power	[MW]	3.8	6.27	1.67	2.93
Power of coherent loss	[MW]	0.8	1.2	0.15	0.20
Number of cavities		12	16	6	8
Frequency of RF system	[MHz]	511		511	
Confinement for longitudinal impedance $|Z_n/n|$	[Ω]	180	235	0.15	0.20
Luminosity	[cm^{-2}s^{-1}]	$7.35 \cdot 10^{33}$		$4.7 \cdot 10^{33}$	

We have assumed $(p\, d\nu/dp)_{max} \simeq 25$ for one interaction point.

7. The maximum gradient of FF quadrupoles is confined by value of magnetic field on the coils and also synchrotron radiation from the lens. We selected the magnetic field at aperture 15σ not higher than 20 kG.

8. We have assumed that the radial emittance is enlarged by the wigler-magnets. This allows independent control of the beam emittance and the ring momentum compaction parameter α.

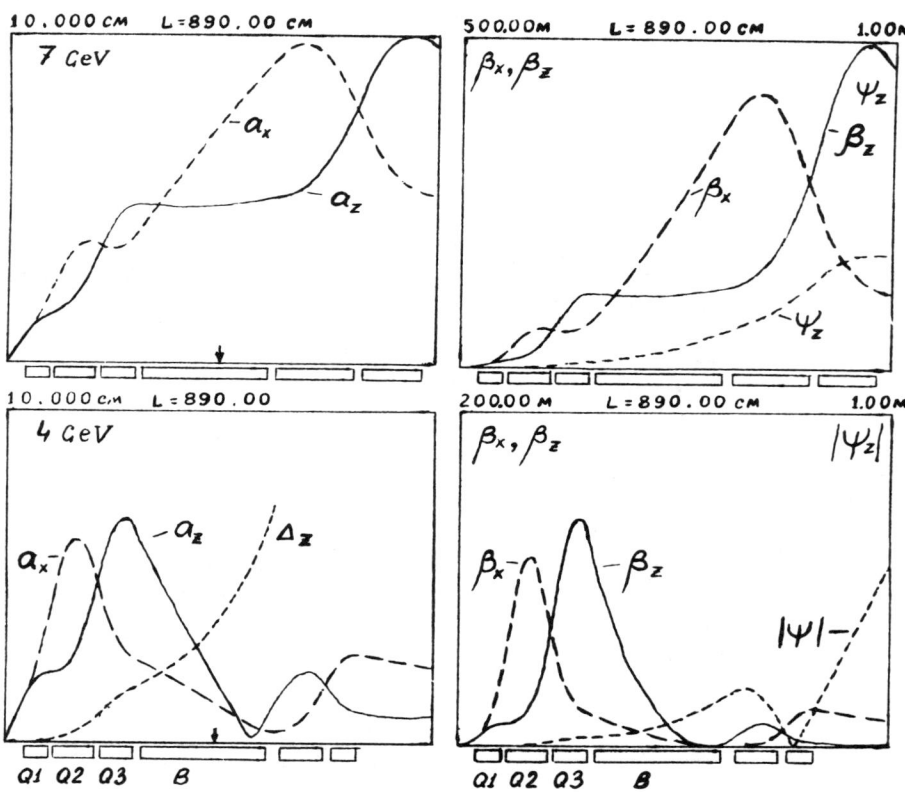

Fig.1. Round beam. IR optic functions for the high and low energy rings, beam sizes for 15σ and value of the transverse beam separation. The arrow shows the position of the parasitic interaction point.

It is necessary to note that the strong focusing of beams at the interaction region causes both chromaticity of the betatron frequencies and β-functions. It is suggested that the β-function chromaticity be set equal to zero at the interaction point (it seems right for a collider with ultimate luminosity from the point of view the beam-beam effects). There is a simple connection between the chromaticity ν' and the amplitude of the

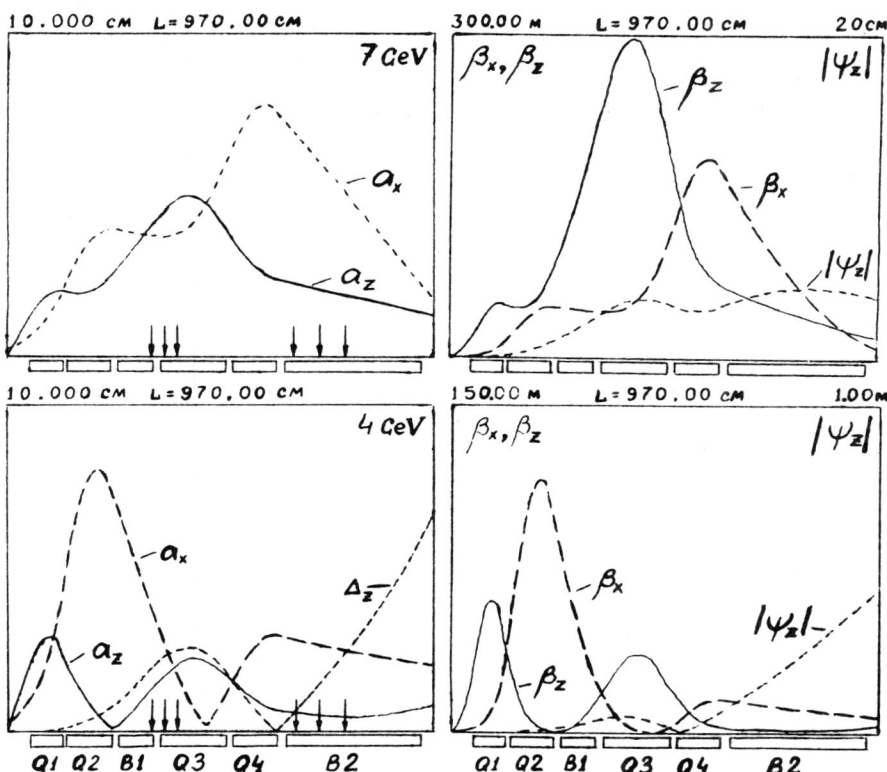

Fig.2. Flat beam. IR optic functions for the high and low energy rings, beam sizes in vertical plane for $6\sigma_x$ and in horizontal plane for $12\sigma_x$, value of the transverse beam separation. The arrow shows the position of the parasitic interaction point for bunch spacings of 649, 708 and 767 cm.

β-functions oscillations:

$$\left(\frac{\Delta\beta}{\beta}\right)_{max} = 2\pi\nu'\frac{\Delta p}{p} \quad , \quad \nu' \equiv p\frac{d\nu}{dp} . \tag{4}$$

For $\Delta p/p = 0.005$ (7σ) and $p\, d\nu/dp = 25$, Eq.4 gives $(\Delta\beta/\beta)_{max} = 0.78$. This limits any further decrease in the β-functions at the interaction points.

Beam parameters for round and flat beams are shown in Table 1. As one can see from the Table, the vertical β-function for the flat beam is 5 times smaller than β-function for the round beam. As result the flat beam has a higher luminosity.

FINAL FOCUS DESIGN

Now we describe the interaction point optics and the beam separation. The β-functions, dispersion function, beam sizes and beam-beam separation for the round and flat beams are shown in Figs. 1 and 2 and the parameters of optical elements are given in Tables 2 and 3. We use a quadrupole triplet for initial focusing of the round beams in order to get equal chromaticities in the x and z directions. This is better than a quadrupole doublet which roughly doubles the chromaticity in one plane while setting it close to zero in the other plane. The beam separation begins at the first quadrupole which is vertically displaced from the beam axes. The separation is completed after the separation magnet where beams are directed into different channels. The first two lenses in the channels are close spaced in transverse direction. To allow the independent control of the lenses gradient the quadrupoles with neutral pole are used.

In the case of the flat beams a quadrupole doublet is used for initial focusing, and, as with the round beams, the beam separation also begins at the first quadrupole and is in the vertical plane. A smaller vertical emittance allows the use of a smaller separation magnetic

Table 2
Parameters of the optical elements near
the interaction point for round beams

Name	Length(cm)	Gradient(kG/cm)	Field(kG)*
O	40		
Q1	50	-4.57	1.5
O	10		
Q2	85	3.49	2.5
O	10		
Q3	70	-2.78	2.5
O	15		
B	260	0.06	3.5

Table 3
Parameters of the optical elements near
the interaction point for flat beams

Name	Length(cm)	Gradient(kG/cm)	Field(kG)*
O	50		
Q1	75	-4.75	1.3
O	10		
Q2	100	2.20	1.3
O	15		
B1	80		1.3
O	15		
Q3	150	-1.30	
O	15		
Q4	100	1.60	
O	20		
B	310		-1.0

*Values of the magnetic field are given for the high energy beam.

field. This is important since it reduces the amount of synchrotron radiation heating of vacuum chamber and synchrotron radiation background inside detector. The vacuum chamber which is common to both beams is longer than in the case for the round beams. This leads to a larger beam separation at the entrance to the separate beam channels. The beam separation at the two parasitic interaction points is equal to 30 σ_z and to 45 σ_z for the beam of larger size (7 GeV).

To decrease the synchrotron radiation background the transverse beam position has odd symmetry about the interaction point. A preliminary design of the beam pipe near interaction point is shown in Fig.3. This scheme gives good suppression of the synchrotron radiation background. Photons can hit the vertex detector only after three reflections which leads to background of $\simeq 10^8$ photons/s. To simplify the optics, the relative vertical positions of the two rings alternate after each interaction point. For example, the high energy storage ring goes from the top position to the bottom one.

As one can see in Table 1, the larger luminosity for round beam is connected with the larger beam current. The main beam current limitation in the case of the flat beam is the bunch lengthening which is more intensive due to

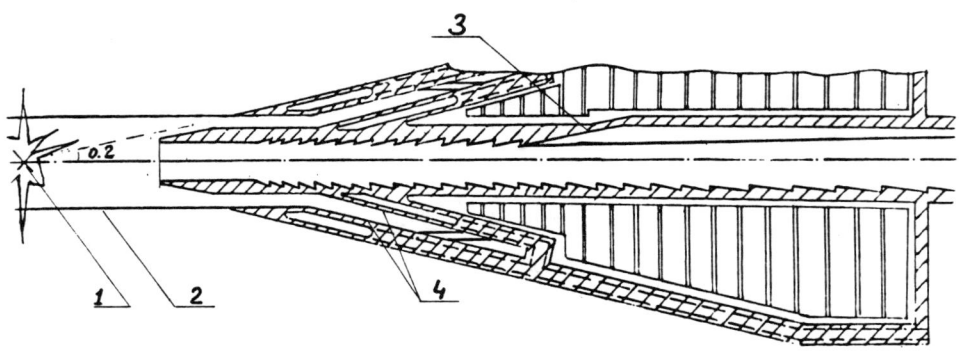

Fig.3. Shape of beam pipe near interaction point.
  1 - interaction point, 2 - beryllium beam pipe,
  3 - SR receiver, 4 - absorber for high order modes, 5 - vacuum pump.

the shorter bunch length. However, in reality, other limitations (for example, multi bunch instabilities or RF power limitation) might determine the beam current. In this case, the luminosity for the flat beams might be larger than that for round beams as the parameter L/I is larger for the flat beams.

The author extends their sincere gratitude to N.S. Dikansky, A.A.Zholents, and V.P.Yakovlev as well as to our other colleagues working on the B-factory project for useful discussions and T.Ellison for help in the English editing of this paper.

REFERENCES

1. R.H.Siemann, CLNS 88/865 (Cornell 1988).
2. Report of B-factory working group:1.Physics and techniques. SLAC-PUB-4838;CLNS 89/884;LBL-26790, 1989.

ROUND BEAMS GENERATED BY VERTICAL DISPERSION*

P. Bagley[†]

Wilson Laboratory, Cornell University, Ithaca, New York, 14853

ABSTRACT

Simulations[1] suggest that in e^+e^- storage rings collisions of round beams (equal emittances and equal β^*) can produce very large tune shifts and luminosities. We understand how to make equal β^*'s, but generating equal emittances is more difficult. We describe an equal emittance scheme that uses several skew quads to couple horizontal dispersion into vertical dispersion. These skew quads also produce a coupling bump. At the interaction point and at all other points outside the coupling bump, the motion is not coupled, so that the 'A' normal mode corresponds to horizontal motion and the 'B' normal mode corresponds to vertical motion. We present a round beam lattice for CESR that incorporates this scheme.

INTRODUCTION

Simulations[1] suggest that in e^+e^- storage rings collisions of round beams (equal emittances and equal β^*) can produce very large tune shifts and luminosities. We understand how to make equal β^*'s, but generating equal emittances is more difficult. At present there are several schemes for this.

- **Rotate all quads by 45 degrees** [2] - This effectively makes all the quads into skew quads. Unfortunately it requires a dedicated machine. It is not practical for a test of round beams on an existing machine.
- **Exchange of phase plane** [3]
- **Operate on the coupling resonance** - This is the simplest method, but it has several problems. The motion is strongly coupled through the entire ring, including the interaction point. This may make orbit and β correction difficult. Also the strong coupling at the interaction point may adversely affect the beam beam interaction. A more serious problem is that the operating point is restricted to a small subspace of the tune plane by the requirement that the tunes be equal.
- **Generate with vertical dispersion**

---

* Work supported by the National Science Foundation.
† Supported by an AT&T fellowship.

THEORY

Normal Mode Decomposition[4-6]

A full turn coupled transfer matrix T is decomposed into normal modes as follows:

$$T = \begin{pmatrix} M & n \\ m & N \end{pmatrix} = VUV^{-1} \qquad (1)$$

where

$$U = \begin{pmatrix} A & 0 \\ 0 & B \end{pmatrix} \; ; \; V = \begin{pmatrix} \gamma I & C \\ -C^\dagger & \gamma I \end{pmatrix} \text{ and } \gamma^2 + |C| = 1 \qquad (2)$$

A is the full turn transfer matrix for one of the normal modes and has the form

$$A = \begin{pmatrix} \cos 2\pi\nu_A + \alpha_A \sin 2\pi\nu_A & \beta_A \sin 2\pi\nu_A \\ -\gamma_A \sin 2\pi\nu_A & \cos 2\pi\nu_A - \alpha_A \sin 2\pi\nu_A \end{pmatrix} \qquad (3)$$

and similarly for B. $T, U,$ and V are 4x4 matrices. $A, B,$ and C are 2x2 matrices and I is the 2x2 identity matrix. The laboratory phase space coordinates X are related to the normal mode coordinates W by $X = VW$. The same relation, $X = VW$, holds for energy displacements so the normal mode dispersions may be calculated from

$$(\eta_A, \eta'_A, \eta_B, \eta'_B)^T = V^{-1}(\eta_x, \eta'_x, \eta_y, \eta'_y)^T \qquad (4)$$

Following Peggs[5], we transform the normal mode coordinates to remove the α and β dependence

$$\overline{W} = GW = \begin{pmatrix} G_A & 0 \\ 0 & G_B \end{pmatrix} W \; ; \text{ where } G_A = \begin{pmatrix} \frac{1}{\sqrt{\beta_A}} & 0 \\ \frac{\alpha_A}{\sqrt{\beta_A}} & \sqrt{\beta_A} \end{pmatrix} \qquad (5)$$

and similarly for G_B. Applying the same transformation to U and V yields

$$\overline{V} \equiv GVG^{-1} = \begin{pmatrix} \gamma I & G_A C G_B^{-1} \\ -G_B C^\dagger G_A^{-1} & \gamma I \end{pmatrix} = \begin{pmatrix} \gamma I & \overline{C} \\ -\overline{C}^\dagger & \gamma I \end{pmatrix} \qquad (6)$$

$$\overline{U} \equiv GUG^{-1} = \begin{pmatrix} G_A A G_A^{-1} & 0 \\ 0 & G_B B G_B^{-1} \end{pmatrix} = \begin{pmatrix} R(2\pi\nu_A) & 0 \\ 0 & R(2\pi\nu_B) \end{pmatrix} \qquad (7)$$

where $R(\phi)$ is the rotation matrix for an angle ϕ

$$R(\phi) = \begin{pmatrix} \cos\phi & \sin\phi \\ -\sin\phi & \cos\phi \end{pmatrix} \qquad (8)$$

Physical Meaning of $\overline{C}$

Consider the motion in the x-y coordinate system as a consequence of the excitation of only the A mode. Figure 1 shows the beam ellipse (i.e. this is what would be seen on a screen in the beamline) for this case.

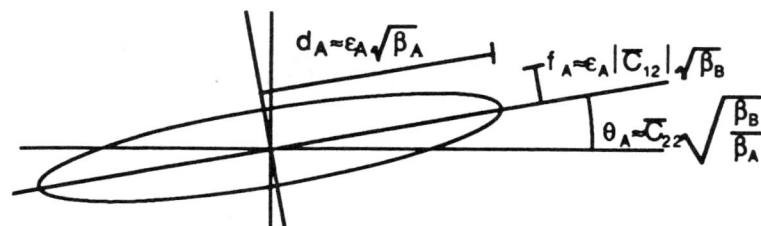

Figure 1 : Beam ellipse for weak coupling and the 'A' mode excited.

For weak coupling, where $|\overline{C}_{ij}|^2 \ll 1$ and $\gamma \approx 1$, we find $\theta_A \approx \overline{C}_{22}\sqrt{\beta_B/\beta_A}$ $d_A \approx \epsilon_A\sqrt{\beta_A}$; $f_A \approx \epsilon_A|\overline{C}_{12}|\sqrt{\beta_B}$. In this approximation, after scaling by the ratios of the root betas, $\overline{C}_{22}$ is the angle through which the ellipse has been rotated and $\overline{C}_{12}$ is the ratio of the lengths of the minor to major axes. The results for the B mode excitation are easily obtained by switching the horizontal and vertical axes in Figure 1 and $\beta_A \longleftrightarrow \beta_B$; $\epsilon_A \longrightarrow \epsilon_B$; $\overline{C}_{22} \longrightarrow \overline{C}_{11}$.

Three of the four elements of $\overline{C}$, the $\overline{C}_{11}$, the $\overline{C}_{12}$ and the $\overline{C}_{22}$ can be measured [7].

Propagation of Normal Modes and Coupling

We now calculate how the normal modes and the coupling propagate through a normal (upright) quadrupole. In the limit of zero strength, a quad becomes a drift space, so this calculation also applies to propagation through a drift space.

Let point 2 come after point 1, with only a horizontally focusing quad between them. Let the full turn transfer matrix at point 1 be T_1 and the transfer matrix for the quad be Q. Then,

$$T_1 = \begin{pmatrix} M_1 & n_1 \\ m_1 & N_1 \end{pmatrix} \quad ; \quad Q = \begin{pmatrix} F & 0 \\ 0 & D \end{pmatrix} \tag{9}$$

where F and D describe transport through a quad with length l, strength K and focal length $f = 1/Kl$.

$$F = \begin{pmatrix} \cos\sqrt{K}l & \frac{1}{\sqrt{K}}\sin\sqrt{K}l \\ -\sqrt{K}\sin\sqrt{K}l & \cos\sqrt{K}l \end{pmatrix} \tag{10}$$

$$D = \begin{pmatrix} \cosh\sqrt{K}l & \frac{1}{\sqrt{K}}\sinh\sqrt{K}l \\ \sqrt{K}\sinh\sqrt{K}l & \cosh\sqrt{K}l \end{pmatrix} \qquad (11)$$

The full turn transfer matrix at point 2 is

$$T_2 \equiv \begin{pmatrix} M_2 & n_2 \\ m_2 & N_2 \end{pmatrix} = QT_1Q^{-1} = \begin{pmatrix} FM_1F^{-1} & Fn_1D^{-1} \\ Dm_1F^{-1} & DN_1D^{-1} \end{pmatrix} \qquad (12)$$

Decomposing T_2 into normal modes yields

$$A_2 = FA_1F^{-1} \quad ; \quad B_2 = DB_1D^{-1} \quad ; \quad C_2 = FC_1D^{-1} \qquad (13)$$

So the normal modes propagate through this quad in exactly the same way as the horizontal and vertical modes would if there were no couplers in the lattice. Specifically the twiss parameters become

$$\alpha_{A2} = \alpha_{A1}\left(\cos^2\sqrt{K}l - \sin^2\sqrt{K}l\right) + \frac{1}{\sqrt{K}}\sin\sqrt{K}l\cos\sqrt{K}l\,(K\beta_{A1} - \gamma_{A1}) \qquad (14)$$

$$\beta_{A2} = \beta_{A1}\cos^2\sqrt{K}l - \frac{2\alpha_{A1}}{\sqrt{K}}\cos\sqrt{K}l\sin\sqrt{K}l + \frac{\gamma_{A1}}{K}\sin^2\sqrt{K}l \qquad (15)$$

$$\gamma_{A2} = \gamma_{A1}\cos^2\sqrt{K}l + 2\sqrt{K}\alpha_{A1}\cos\sqrt{K}l\sin\sqrt{K}l + K\beta_{A1}\sin^2\sqrt{K}l \qquad (16)$$

$$\nu_{A2} = \nu_{A1} \quad ; \quad \phi_{A21} = \int_{s_1}^{s_2}\frac{ds}{\beta_A(s)} \qquad (17)$$

where ϕ_{A21} is the A mode phase advance from point 1 to point 2. Similarly for the B mode,

$$\alpha_{B2} = \alpha_{B1}\left(\cosh^2\sqrt{K}l + \sinh^2\sqrt{K}l\right) - \frac{1}{\sqrt{K}}\sinh\sqrt{K}l\cosh\sqrt{K}l\,(K\beta_{B1} + \gamma_{B1}) \qquad (18)$$

$$\beta_{B2} = \beta_{B1}\cosh^2\sqrt{K}l - \frac{2}{\sqrt{K}}\alpha_{B1}\cosh\sqrt{K}l\sinh\sqrt{K}l + \frac{\gamma_{B1}}{K}\sinh^2\sqrt{K}l \qquad (19)$$

$$\gamma_{B2} = \gamma_{B1}\cosh^2\sqrt{K}l - 2\alpha_{B1}\sqrt{K}\cosh\sqrt{K}l\sinh\sqrt{K}l + K\beta_{B1}\sinh^2\sqrt{K}l \qquad (20)$$

$$\nu_{B2} = \nu_{B1} \quad ; \quad \phi_{B21} = \int_{s_1}^{s_2}\frac{ds}{\beta_B(s)} \qquad (21)$$

where ϕ_{B21} is the B mode phase advance from point 1 to point 2.

Recall from eqn. (13) that $A_2 = FA_1F^{-1}$ and $B_2 = DB_1D^{-1}$. These imply that

$$U_{21} = \begin{pmatrix} F & 0 \\ 0 & D \end{pmatrix} = T_{21} \tag{22}$$

i.e. if there are no couplers between points 1 and 2, the normal mode transfer matrix from point 1 to point 2 is equal to the transfer matrix from point 1 to point 2. This implies that between points 1 and 2, the vertical motion propagates with the phase advance of the B mode. So between couplers, the vertical dispersion also propagates with the B mode phase advance.

Recall from eqn. (13) that $C_2 = FC_1D^{-1}$ and recall from eqn.s (14) to (20) the expressions for the twiss parameters at point 2. These can be used to express $\overline{C}_2$ in terms of $\overline{C}_1$.

$$\overline{C}_2 = R(\phi_{A21})\,\overline{C}_1\,R(-\phi_{B21}) \tag{23}$$

$$\begin{pmatrix} \overline{C}_{11} + \overline{C}_{22} \\ \overline{C}_{12} - \overline{C}_{21} \end{pmatrix}_2 = R(\phi_{B21} - \phi_{A21}) \begin{pmatrix} \overline{C}_{11} + \overline{C}_{22} \\ \overline{C}_{12} - \overline{C}_{21} \end{pmatrix}_1 \tag{24}$$

$$\begin{pmatrix} \overline{C}_{11} - \overline{C}_{22} \\ \overline{C}_{12} + \overline{C}_{21} \end{pmatrix}_2 = R(\phi_{A21} + \phi_{B21}) \begin{pmatrix} \overline{C}_{11} - \overline{C}_{22} \\ \overline{C}_{12} + \overline{C}_{21} \end{pmatrix}_1 \tag{25}$$

where the R's are rotation matrices. This clearly shows that the elements of $\overline{C}$ have components that propagate as the sum and the difference of the normal mode phase advances. This is an exact expression, weak coupling is not assumed.

Generation of Closed Coupling Bump and a Vertical Dispersion Wave

We will ignore the vertical dispersion from the vertical steerings used to center the orbit vertically in the quads. In CESR we typically use about 60 vertical correction coils with rms kick of about 0.2 mrad. For $\sqrt{\beta} \approx 5$ meters and vertical tune about 9.35, these correctors generate about 3 mm of vertical dispersion. Since we will generate about 3 m of vertical dispersion, this may be neglected.

We generate this vertical dispersion using 8 skew quads at locations of nonzero horizontal dispersion. (To get about 3 m of vertical dispersion with the typical β's and tune above, and typical horizontal dispersion of 2 m, the 8 skew quads will have focal lengths of about 4 m. These are very strong skew quads. In practice we use skew quads at locations with much higher β and reduce the strength.) These very strong skew quads will generally create a large coupling throughout the machine. We want to avoid this, so we will constrain the skew quad settings so they produce a "closed" coupling bump, but a large vertical dispersion wave throughout the entire machine. This vertical dispersion wave will generate the "vertical" emittance.

Is this possible? We have shown that between couplers, the elements of the $\overline{C}$'s have components that propagate as the sum and as the difference of the normal mode phase advances. Also between couplers the vertical dispersion propagates as the phase advance of normal mode B. Since the coupling and the vertical dispersion propagate differently it is possible to close the coupling bump and leave a large vertical dispersion wave around the ring.

EXAMPLE

We have designed a lattice for CESR with the parameters in Table I. To simplify installation of this lattice these conditions have the CLEO solenoid on, and the REC quads rotated to their usual angles. We adjusted the angles of quads 1 and 2 and the strength of the skew quad at 6. Generally it requires 2 degrees of freedom to close the coupling bump; we use the third degree of freedom to adjust the size of the η_B wave.

Graphs of the $\sqrt{\beta}$'s, η's and $\overline{C}$'s are shown in Figures 2 to 7. As promised there is a large η_B wave through the machine and no coupling in the North Interaction Region (NIR). Where the $\overline{C} = 0$ (most of the machine), the A mode is purely horizontal and the B mode is purely vertical. The emittances shown in Table I are calculated for the normal modes (see Appendix).

CONCLUSIONS

We have described a scheme for generating equal emittances in both normal modes and shown an example. This scheme uses a closed coupling bump and a large vertical dispersion wave. At the interaction point there is no coupling and zero dispersion.

We have installed this lattice into CESR and collided beams. Unfortunately there is a problem with the lattice. To prevent collisions in the South Interaction Region (SIR), we separated the beams horizontally. Thus, the beams are off center in the SIR's final focus quads. This produced large differences in the damping partition numbers for the electrons and positrons and consequently large differences in their beam sizes at the collision point in the North. This problem is due to the horizontal separation and this specific set of optics. It is not intrinsic to this scheme for generating equal emittances. We are currently redesigning the optics to avoid this problem.

Table I : Parameters for CESR Round Beam Lattice

	'A' mode	'B' mode
β^* (cm)	31	20
η^* (cm)	-2.8	0
η'^*	0	1.0
ν	9.753	9.774
ϵ (10^{-7} m·rad)	1.4	1.4

$$\overline{C}^* = \begin{pmatrix} 0 & 0 \\ 0 & 0 \end{pmatrix} \quad ; \quad E = 5.18 \text{ GeV}$$

$\sigma_l =$ bunch length $= 2.2$ cm ; $\nu_s = 0.046$

$\dfrac{\sigma_\epsilon}{E} =$ fractional energy spread $= 6.0 \times 10^{-4}$

1 collision point at the NIR at arc position $= 384.214$ m
beams are separated at the SIR at arc postition $= 0.$ m
1 bunch/beam

	arc position	1/(skew focal length)
South West REC quad (rotated)	1.25 m	0.150/m
South East REC quad (rotated)	767.18 m	-0.150/m
Q1 West (rotated)	2.54 m	-0.041/m
Q1 East (rotated)	765.89 m	0.041/m
Q2 West (rotated)	3.88 m	0.011/m
Q2 East (rotated)	764.55 m	-0.011/m
Skew Quad at 6 West	29.05 m	-0.035/m
Skew Quad at 6 East	739.38 m	0.035/m

CLEO solenoid has length $= 3.44$ m, a magnetic field of 1.5 Tesla and is centered at arc position$=0$.

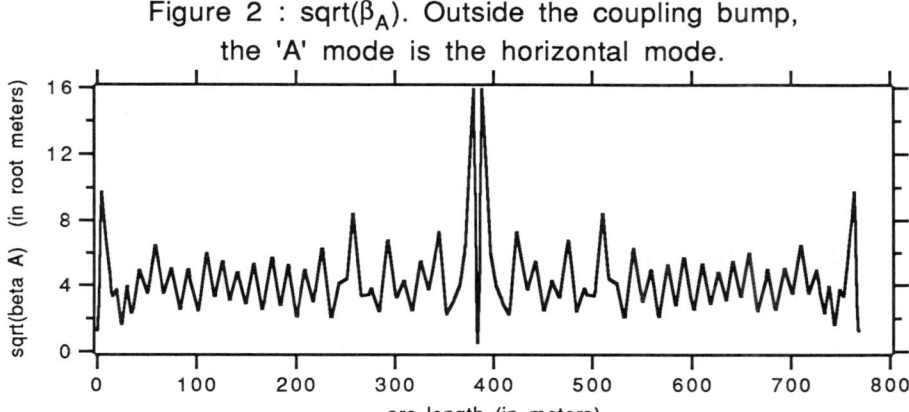

Figure 2 : sqrt(β_A). Outside the coupling bump, the 'A' mode is the horizontal mode.

Figure 3 : sqrt(β_B). Outside the coupling bump, the 'B' mode is the vertical mode.

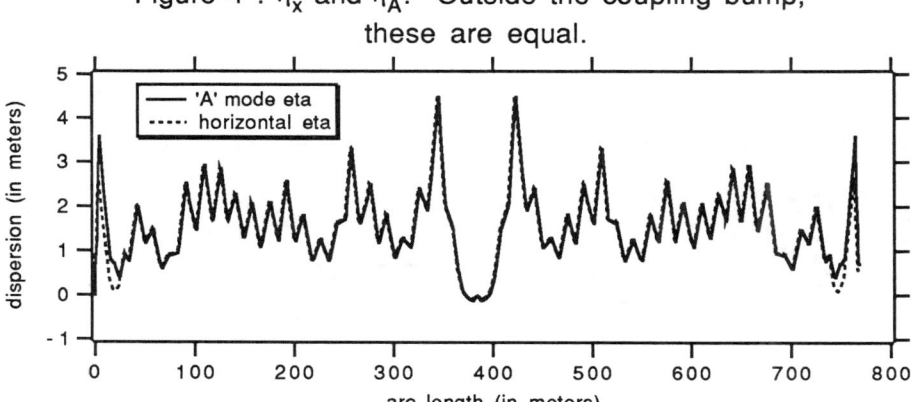

Figure 4 : η_x and η_A. Outside the coupling bump, these are equal.

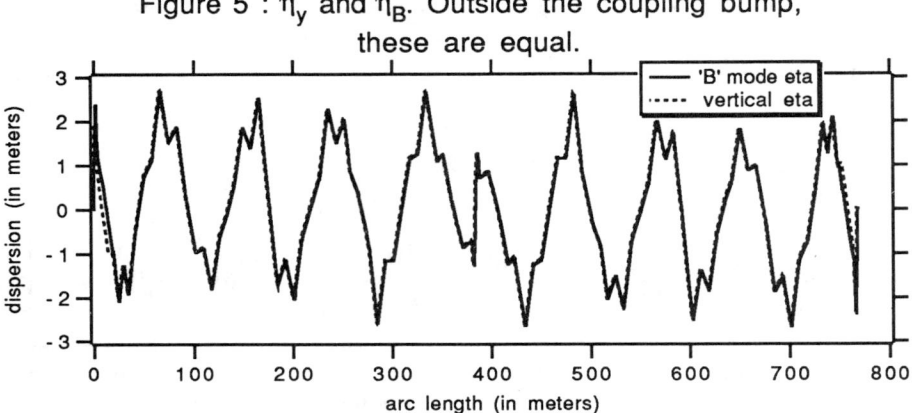

Figure 5 : η_y and η_B. Outside the coupling bump, these are equal.

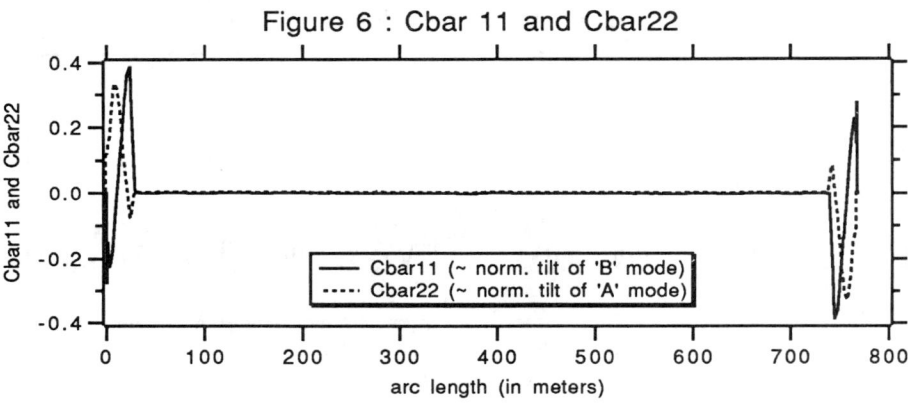

Figure 6 : Cbar 11 and Cbar22

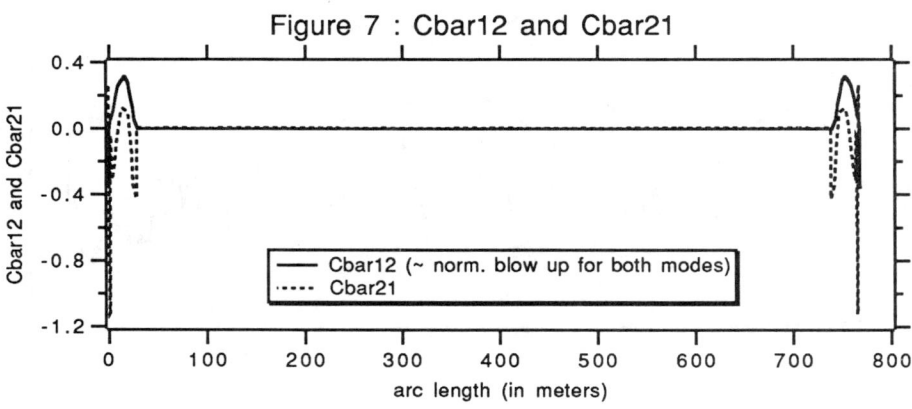

Figure 7 : Cbar12 and Cbar21

APPENDIX

When we calculate the normal mode emittances, we ignore sextupoles and the path length effects of vertical bends.[8] . We use the same notation as Sands[9] .

The normal mode emittances and the energy spread are given by

$$\epsilon_z = \frac{1}{4}\tau_z Q_z \quad ; \quad \epsilon_\epsilon = \frac{1}{4}\tau_E Q_E \tag{26}$$

where ϵ_ϵ is the energy spread, z can be either A or B and

$$Q_z = \frac{1}{LE^2}\oint N\langle u^2\rangle H_z\, ds \quad ; \quad Q_\epsilon = \frac{1}{L}\oint N\langle u^2\rangle\, ds \tag{27}$$

$$H_z = \frac{1}{\beta_z}\left[\eta_z^2 + (\alpha_z\eta_z + \beta_z\eta_z')^2\right] \tag{28}$$

$$N\langle u^2\rangle = \frac{55}{24\sqrt{3}}r_e\hbar mc^4\left(\frac{E}{mc^2}\right)^7\frac{1}{\rho^3} \tag{29}$$

$$\tau_z = \frac{2E_0 T_0}{U_0(1-D_z)} \quad ; \quad \tau_\epsilon = \frac{2E_0 T_0}{U_0(2+D_\epsilon)} \tag{30}$$

where

$$D_\epsilon = \frac{\oint \eta_x G(G^2 + 2K_1)\, ds}{\oint G^2\, ds} \tag{31}$$

$$D_A = \frac{\oint \eta_{x|A} G(G^2 + 2K_1)\, ds}{\oint G^2\, ds} = \frac{\oint \gamma\eta_A G(G^2 + 2K_1)\, ds}{\oint G^2\, ds} \tag{32}$$

$$D_B = \frac{\oint \eta_{x|B} G(G^2 + 2K_1)\, ds}{\oint G^2\, ds} = \frac{\oint (C_{11}\eta_B + C_{12}\eta_B') G(G^2 + 2K_1)\, ds}{\oint G^2\, ds} \tag{33}$$

$\eta_{x|z}$ is the projection of the normal mode dispersion and its derivative onto the horizontal dispersion and $G = 1/\rho$ where ρ is the bend radius. These D's are usually small compared to 1, so they only have a small effect on the damping rates and the emittances.

ACKNOWLEDGMENTS

I thank the CESR operations group and the accelerator operators. I especially thank D.L. Rubin and M. Billing for many helpful discussions.

REFERENCES

1) S. Krishnagopal and R. Siemann, "Simulation of Round Beams" Proc. of the 1989 IEEE Particle Accel. Conf., March 1989, p.836

2) J. Byrd, D. Sagan, and R. Talman, "An Uncoupled, Round Beam, Electron Accelerator Lattice", CESR note CBN89-5

3) M. Donald and R. Servrankx, "Round Beams Generated by Exchange of Phase Plane", in these proceedings

4) D. Edwards and L. Teng, "Parameterization of Linear Coupled Motion in Periodic Systems", IEEE Transactions on Nuclear Studies, **NS-20**, No.3, June 1973, p.885

5) S. Peggs, "Coupling and Decoupling in Storage Rings", IEEE Transactions on Nuclear Science, **NS-30**, No.4, August 1983, p.2460

6) M. Billing, "Theory of Weakly Coupled Transverse Motion in Storage Rings", CBN 85-2

7) P. Bagley and D.L. Rubin, "Correction of Transverse Coupling in a Storage Ring", Proc. of the 1989 IEEE Particle Accel. Conf., March 1989, p.874

8) For a more complete calculation see T. Raubenheimer, "A Formalism and Computer Program for Coupled Lattices", Proc. of the 1989 IEEE Particle Accel. Conf., March 1989, p.1313

9) M. Sands, "The Physics of Electron Storage Rings", SLAC-121, 1970

BEAM SEPARATION IN AN ASYMMETRIC ENERGY COLLIDER USING A TILTED EXPERIMENTAL SOLENOID

T. Risselada
CERN, Geneva, Switzerland

ABSTRACT

The CERN - PSI feasibility study of a B-Meson Factory in the ISR tunnel has resulted in the proposal of a 3.5 on 8.0 GeV machine with a bunch spacing of 12 m, to be reduced eventually to 3 m. The beams will be vertically separated by the radial field component of the 1.5 T experimental solenoids, rotated for this purpose by 4 degrees, and by the fields acting on the off-centered beams in the central interaction region quadrupoles. Vertical separations of 6 σ_x are obtained at 3 m from IP, and of 2.5 σ_x at 1.5 m. The solenoid tilt and the amount of synchrotron radiation arriving in the central detector are compatible with the requirements of the experiments. The geometry and the performance of the proposed separation scheme are presented.

INTRODUCTION

A joint working group at CERN and PSI has recently studied the feasibility of a B-Meson Factory in the CERN ISR tunnel (circumference 960 m), using the existing CERN lepton injectors[1]. The proposed machine will have 3.5 on 8 GeV beams colliding at 2 interaction points, with an initial luminosity of 10^{33} cm^{-2} s^{-1} and a potential luminosity increase to 10^{34} cm^{-2} s^{-1}. If this machine fails to produce the expected luminosity in asymmetric mode it will have to be converted to symmetric operation at the expense of a modification to the interaction region layout.

The symmetric machine proposed by PSI in 1989 to the Swiss authorities[2] has been used as a starting point, but the energy asymmetry, the geometry of the existing ISR tunnel and the necessity of lowering the energy loss per turn required several modifications to the lattice. The proposed interaction region optics is close to the design of a 4 on 7 GeV variant of the PSI proposal[3]. The CERN-PSI reference machine (L = 10^{33} cm^{-2} s^{-1}) will have 80 bunches, and the ultimate asymmetric machine (L = 10^{34} cm^{-2} s^{-1}) is assumed to have 320 bunches. The beams collide in the center of two 50 m long straight sections and then traverse four common quadrupoles before being separated by a vertical septum. Five matching quadrupoles in each ring match the interaction region lattice functions to those of the arcs, which are superimposed with a distance of 0.8 m. Two additional 50 m long dispersionless straight sections are foreseen for the r.f.system. The interaction region optics and the arc lattice are described in separate reports[5,6].

SEPARATION USING A TILTED SOLENOID

The large number of circulating bunches requires a scheme separating the beams as close as 6 m to IP initially, and eventually at 1.5 m. This can only be obtained at the expense of adding transverse magnetic fields inside the central detector between IP and the first quadrupoles. The longitudinal field of 1.5 T, required for particle detection, is produced by a 3 m long superconducting solenoid. Rather than inserting additional dipole magnets we have studied the possibility of rotating the solenoid around the vertical axis by an angle of 4 degrees, which yields a radial field of 0.1 T[4]. This solution has been shown to be compatible with the experimental physics requirements[1].

In the study the solenoid has been provisionally modeled as a pure radial dipole field. The beams enter the first quadrupole with y = 2.5 mm, y' = 6 mrad (3.5 GeV) and y = 1.1 mm, y' = 2.6 mrad (8 GeV). If it turns out that the separation is too much reduced by the solenoid end effects, the rotation angle may be further increased without inconvenience for the detector.

QUADRUPOLE POSITION OFFSET

After crossing at IP the beams traverse 4 common quadrupoles. Q1 and Q2 are superconducting quadrupoles mounted in the same cryostat, Q3 and Q4 are normal magnets. The strengths of these quadrupoles have been established[5] by matching the lattice parameters of the two beams simultaneously (table I), imposing $\beta_x = 1.00$ m and $\beta_y = 0.03$ m at IP for both beams. The radial field of the tilted solenoid is present in the drift space between IP and Q1, and inside Q1.

Table I. Parameters of the common central low-beta insertion

element	length [m]	gradient [T/m]
IP	0.00	
drift	0.85	
Q1	0.60	- 46.8
drift	0.30	
Q2	0.60	+ 35.5
drift	0.50	
Q3	1.00	- 16.2
drift	1.24	
Q4	1.00	+ 8.1

In addition to the energy difference the two beams have separate trajectories at the entrance of the first quadrupole and therefore at least one of the two beams is bent in each quadrupole. As the solenoid field extends up to the exit of Q1 the trajectories of both beams are bent in this quadrupole. In order to minimize the amount of synchrotron radiation arriving at the central detector the vertical positions of the 4 quadrupoles have to be determined with great care.

In a first iteration it was attempted to center all quadrupoles around the incoming beam. In this case the majority of the photons would have been emitted towards the arcs, and very few towards IP. Unfortunately this solution did not provide any separation at 3 m from IP, a parasitic bunch crossing point in the case of 160 or 320 bunches.

A second solution where all quadrupoles are centered around the low energy beam yields both adequate separation and acceptable synchrotron radiation levels. The vertical geometry of this solution is shown in figure 1.

VERTICAL BENDING MAGNETS

In the above mentioned quadrupole geometry the beams leave the last of the 4 common quadrupoles at y = + 81.4 mm (3.5 GeV) and y = + 56.5 mm (8 GeV), which allows them to enter the separate channels of a double vertical septum magnet VB1.

The septum bending angles must be at least 30 mrad so as to obtain sufficient separation at the next (separate) quadrupole, and to reach the 0.8 m arc separation before the end of the straight section, limited to a length of 50 m because of the circular ISR tunnel geometry. Synchrotron radiation considerations limit their field, which implies a length of about 3 m. The resulting sagitta of 45 mm does not allow the beams to cross over at the exit of the septum, and therefore the low energy beam must always run on top.

A second vertical bend VB2 brings the beams back into the horizontal planes of the arcs. Since the betatron phase advance between VB1 and VB2 is small the two bends must have comparable amplitudes (and opposite signs), so as not to create a too large vertical dispersion. For this reason bending angles of + 33 and - 51 mrad at 3.5 GeV and - 32 and + 22 mrad at 8 GeV, as shown in figure 1, will have to be used for VB1 and VB2 respectively, rather than the geometry presented in the Feasibility Study Report [1].

RESULTS AND CONCLUSION.

With the above described solution a vertical separation of at least 6 σ_x is obtained beyond 3 m from IP. The values at the first 4 possible parasitic bunch crossings are presented in table II. The width of the wider of the two beams is taken for σ_x. The separation criterion of 5 σ_x is satisfied for bunch numbers up to 160. With 320 bunches the solenoid tilt will have to be increased by a factor of 2.

Table II. obtained vertical beam separation in numbers of σ_x

Distance from IP [m]	Vertical separation [σ_x]
1.5	2.5
3.0	9.5
4.5	6.0
6.0	6.0

The vertical dispersion with the above presented solution is shown in figure 2. Its maximum values in the interaction region straight sections are 0.12 m (3.5 GeV) and 0.18 m (8 GeV). This will have to be compensated at the entrance of the arc, for example in the horizontal dispersion suppressor cell.

ACKNOWLEDGEMENTS

This report presents the result of contributions by several subgroups of the CERN - PSI working group. The author would like to acknowledge particularly the constant support of working group chairman Y. Baconnier who initiated this study, and of T. Taylor who proposed the idea of a tilted solenoid.

REFERENCES

1. Feasibility Study for a B-Meson Factory in the CERN-ISR tunnel, CERN 90-02, and PSI 90-08, April 1990. See also L. Rivkin, these Proceedings and references 4, 5 and 6 mentioned below.

2. Proposal for an Electron Positron Collider for Heavy Flavour Particle Physics and Synchrotron Radiation, PSI PR-88-09, July 1988.

3. K. Wille, Proposal for a High Luminosity Electron Positron Collider at PSI with an option for Symmetric and Asymmetric Collider Mode, XIV International Conference on High Energy Accelerators, August 1989, Tsukuba, Japan.

4. T. Risselada and T. Taylor, Interaction Region Optics and Separation for an Asymmetric Energy Beauty Factory in the ISR Tunnel, CERN-PS/PA Note 90-06.

5. T. Risselada, Asymmetric BFI Low-Beta matching, CERN-PS/PA Note 90-10.

6. T. Risselada and L. Rivkin, Ring Layout and Lattice Design for a B-Factory in the ISR Tunnel, CERN-PS/PA Note 90-07.

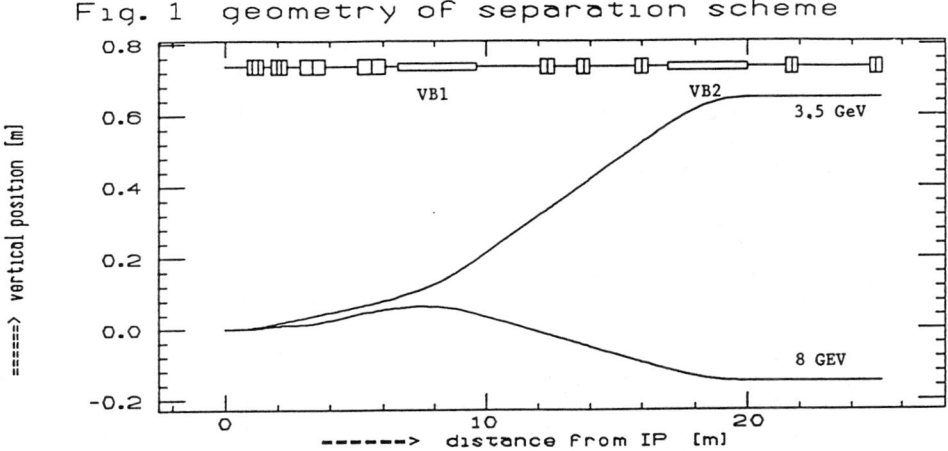

Fig. 1 geometry of separation scheme

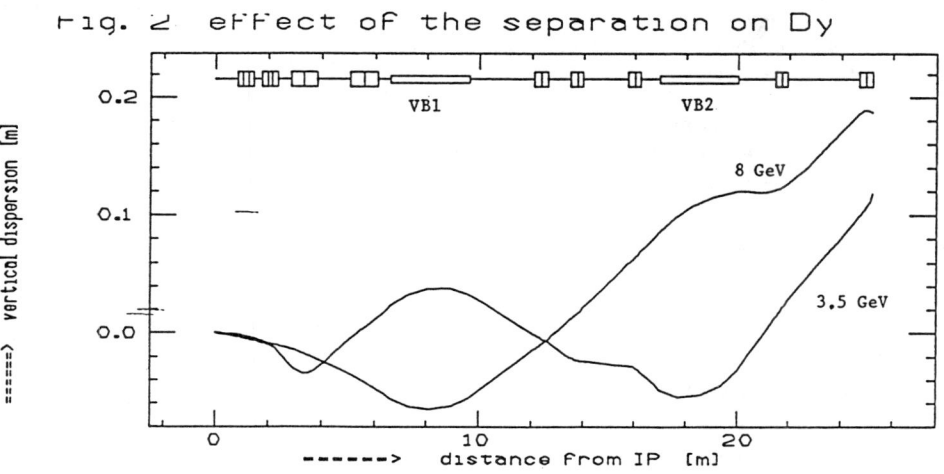

Fig. 2 effect of the separation on Dy

A high Luminosity e^+e^- collider with a crossing angle

Rüdiger Schmidt
CERN, CH-1211 Geneva 23, Switzerland

Abstract

In this paper an e^+e^- collider with a crossing angle is discussed. It is shown, that for flat beams a crossing in the horizontal plane is preferred. The basic parameters for a machine with a luminosity of $10^{34}s^{-1}cm^{-2}$ are given.

1 Introduction

For the next generation colliders high energy physics requires a luminosity two orders of magnitude higher than presently achieved, for e^+e^- machines as well as for hadron colliders.

In order to achieve a luminosity above $10^{34}s^{-1}cm^{-2}$ for the future hadron colliders such as the LHC and the SSC it is intended to bring two beams with a large number of bunches into collision. To avoid crossings outside the physics experiments the beams collide with a small angle [1,2,3]. In the design for the future high luminosity e^+e^- colliders a crossing angle has been excluded because of the painful experience with DORIS, where the beams were colliding with a vertical angle [4]. The existence of strong synchro-betatron resonances was observed with a limiting effect on the performance of DORIS.

Since then no e^+e^- storage ring operated with a crossing angle. It is not yet understood which maximum crossing angle can be tolerated. Not much work has been done on problems related to a crossing angle for e^+e^- colliders since the DORIS experiments, whereas different groups are working on the maximum value for the crossing angle which can be tolerated for future hadron colliders [5,6,7]. For e^+e^- collider one suggestion is the crab crossing geometry, which can suppress synchobetatron resonances [8].

In this paper we discuss an e^+e^- collider with a crossing angle. It is shown that for flat beams a horizontal crossing angle is better than a vertical one. The crossing angle can be as small as about 1 mrad. It is not yet clear if such an angle is tolerable, but at least it can be said that this angle is small compared to the vertical crossing angle at DORIS.

The basic design principles for an e^+e^- collider with a crossing angle are the following:

- Both beams have many bunches, circulating in two distinct vacuum chambers. The distance between two bunches is in the order of 0.5 m.
- The emittances of the beams are very low.
- The intensity per bunch is very small.
- The total current is high.
- The beams collide with a 'small' crossing angle in order to avoid parasitic bunch crossings.

- Because of the small emittance a small diameter of the vacuum chamber in the physics experiments is feasible.
- Because of the small emittance the dynamic aperture does not need to be large.

Here we only consider a collider with both beams having the same energy. First we show how the basic parameters for such a collider are derived. Then we discuss some aspects of the interaction with a crossing angle. In an example the basic parameters for a B-factory with a luminosity of $10^{34} s^{-1} cm^{-2}$ are given.

2 The basic equations

The luminosity of a collider is given by :

$$L = \frac{N^2 \times f}{4\pi \times \sigma_x \times \sigma_z} \qquad (1)$$

with :
N...Number of particles per bunch.
f...Number of crossings per second.
σ_x, σ_z...beam dimensions at the interaction point in both planes.

This equation is valid for head-on collision. It is approximately true if the crossing angle is small, which is assumed in the following.

With the definition $\kappa = \sigma_z/\sigma_x$ the luminosity and the tune shifts are :

$$L = \frac{N^2 \times f}{4\pi \times \sigma_x^2 \times \kappa} \qquad (2)$$

$$\delta Q_x = \frac{r_e \times N \times \beta_x}{(2\pi \times \sigma_x^2) \times (1 + \kappa) \times \gamma} \qquad (3)$$

$$\delta Q_z = \frac{r_e \times N \times \beta_z}{(2\pi \times \sigma_x^2) \times \kappa \times (1 + \kappa) \times \gamma} \qquad (4)$$

with γ the relativistic factor and r_e the classical electron radius. The beam-beam effect limits the tune shift to a value between 0.03 and 0.05. The optimal luminosity is achieved with the same tune shift in both planes : $\delta Q_x = \delta Q_z$ which requires for the ratio of the beta functions at the IP : $\beta_z/\beta_x = \kappa$.
The beam size at the interaction point for a given tune shift can be expressed in the following way :

$$\sigma_x = \sqrt{\frac{r_e \times N \times \beta_x}{2\pi \times (1 + \kappa) \times \delta Q_x \times \gamma}} \qquad (5)$$

In conventional e^+e^- storage rings with a small number of bunches the emittance of the beam is large. This allows to have strong bunches without reaching the beam-beam limit. The equations below show how a collider with many bunches is optimized : the total current is limited by the power requirements because of the energy loss due to the synchrotron radiation. The current is distributed between many bunches. The emittance is reduced to keep a high luminosity and to keep the beam-beam tune shift close to the maximum possible value.

The required number of particles per bunch for a given luminosity is (from eq.2 and eq.5) :

$$N = \frac{2 \times L \times r_e \times \beta_y}{f \times \gamma \times (1+\kappa) \times \delta Q_x} \tag{6}$$

and the required horizontal emittance :

$$\epsilon_x = \frac{r_e^2 \times \beta_x \times \kappa \times L}{\pi \times \gamma^2 \times (1+\kappa)^2 \times \delta Q_x^2 \times f} \tag{7}$$

The last equation shows, that the emittance is inversely proportional to the bunch frequency.

3 An interaction region with a crossing angle

The geometry around the interaction point is defined in fig.1. The distance between two bunches is given by s_b, the first parasitic crossing point is $0.5 \times s_b$ away from the interaction point. For a drift space (no quadrupoles) the β-function is given by: $\beta(L) = \beta^* + L^2/\beta^*$, β^* is taken at the interaction point.

The minimum crossing angle is determined by the requirement to have a large enough separation at the parasitic crossing points. The distance between the beams at the parasitic crossing is given by $d = L \times \alpha$. This distance expressed in units of the beam size σ should be larger than a certain minimum value, i.e. $d/\sigma(L) \geq (d/\sigma(L))_{min}$.

For $L \gg \beta^*$ the beam size increases linearly with the distance from the interaction point : $\sigma(L) = (L \times \sigma^*)/\beta^*$ and the separation d/σ_L becomes constant. A reasonable assumption for the separation is $(d/\sigma(L))_{min} = 10$ for the interaction points at a distance $L \gg \beta*$. For one or two parasitic crossing points a smaller separation can be tolerated.

One finds for the minimum crossing angle respecting the condition for the separation :

$$\alpha_{min} = (d/\sigma(L))_{min} \times \frac{\sigma^*}{\beta^*} \tag{8}$$

with σ^* the beam size and β^* the value of the β function at the interaction point. For the required separation of 10 σ the crossing angle is $\alpha = 10\sigma^*/\beta^*$.

A large crossing angle reduces the strength of the parasitic beam-beam crossings, but increases the strength of the synchro-betatron resonances. In first order this strength is proportional to [4] :

$$R = \frac{\alpha}{2} \times \frac{\sigma_l}{\sigma^*} \tag{9}$$

with σ_l the r.m.s. bunch length.

R gives the separation of a small amplitude particle at a longitudinal displacement of σ_l when the bunch centres collide (see fig. 2). In the case of a zero crossing angle the parameter R vanishes.

More recent studies show that in the case of very flat beams and a large crossing angle in the vertical plane the resonance strength saturates [9]. In the following we continue the discussion with the approximate formulas.

We assume that a certain crossing angle can be tolerated, corresponding to the parameter $R = R_{max}$. Then a combination of (8) and (9) yields a condition for the ratio of the bunch length to the beta function at the interaction point :

$$\frac{\sigma_l}{\beta^*} = \frac{2 \times R_{max}}{(d/\sigma(L))_{min}} \qquad (10)$$

In order to evaluate this expression numerically, some assumptions about the value of R_{max} and $(d/\sigma(L))_{min}$ have to be made. We assume for R_{max} a value of 0.1, about five times smaller than at DORIS. For the separation at the parasitic crossing points we require $(d/\sigma(L))_{min} = 10$. This yields for the ratio of the bunch length and the beta function a value of 0.02. With a bunch length of 1 cm, the minimum beta function is 50 cm. In a geometry with a vertical crossing this condition is impossible to fulfill : for flat beams the vertical beta function is only about 1 cm. Therefore a horizontal crossing is mandatory.

Even if synchro-betatron resonances are not considered, a horizontal crossing is preferable. With nominal coupling the required crossing angles in both planes are identical. It is well known that the beam-beam effect leads to an increase of the vertical emittance and to non Gaussian tails in the vertical particle distribution. If a geometry with a vertical crossing angle is chosen, the angle must be larger than the angle calculated with nominal coupling to have a certain security margin.

4 Example for a B-factory with a luminosity of $10^{34}s^{-1}cm^{-2}$

We assume the following parameters :
Beam energy : E = 5.3 GeV for both beams.
Gamma factor : $\gamma = 10372$.
Bunch distance : $s_b = 1.2$ m.
Frequency of the collisions : f = 500 MHz.
Beta functions at the interaction point : $\beta_x = 50cm$ and $\beta_y = 1cm$.
Coupling : $\kappa = 0.02$.
Tune shifts : $\delta Q_x = 0.05, \delta Q_z = 0.05$.
Bunch length : $\sigma_l = 1cm$.
Classical electon radius : $r_e = 2.8179 \times 10^{-15}m$.

Using these parameters we find :
Number of particles per bunch : $N_b = 2.13 \times 10^{10}$.
Horizontal emittance : $\epsilon_x = 1.805 \times 10^{-8}m \times rad$.
Bunch sizes at the interaction point : $\sigma_x = 95\mu m, \sigma_z = 1.9\mu m$.
Crossing angle in the horizontal plane : $\alpha = 1.90 mrad$.
Separation at the first parasitic crossing point : d = 5.1 σ.
Separation at the next crossing point : d = 7.7 σ.

Parameter for the strength of the synchro-betratron resonances : P = 0.1
Current in one beam : I = 1.7 A.

The separation at the first two parasitic crossing points is 5.1 and 7.7 σ. These values are probably sufficient for a small number of crossing points. Further away the separation reaches the required value of 10 σ.

5 Conclusions

For an example design of a B-factory with two beams of the same energy the crossing angle is 1.9 mrad. A collider with beams of different energies can be designed following the same principles. The angle is substantially smaller than the angle used at DORIS, and the synchro-betatron resonances are expected to be weak. Nevertheless, simulations and experiments are needed to study if an angle of this order is acceptable. It is also of interest to compare the strengths of the synchro-betatron resonances generated by a crossing angle with those generated by nonvanishing dispersion at the interaction point. If detailed studies show that this crossing angle is not acceptable, a crab crossing geometry can be employed. The fields in the crab cavities are very moderate due to the small value of the crossing angle.

An angle of 1.9 mrad leads to a beam separation of 9.5 mm after a distance of $5m$. This is too small to install elements to separate the two beams into different vacuum chambers. Therefore an additional separation is needed to increase the distance between the two beams. This could be done with electrostatic separators, either in the horizontal or in the vertical plane.

The masking of the physics detectors against synchroton radiation is probably easier in an interaction region with a crossing angle, as it was found at Cornell studying the synchrotron radiation background for different interaction region geometries [10].

In many ways such a collider resembles more to a synchrotron light source than to a conventional storage ring : small emittance, many bunches, high total current. It has to be studied, if beams with so many bunches are stable, and if a feedback system can be designed to damp coupled bunch instabilities. A detailed layout of the interaction region is required to calculate the minimum possible β function.

6 Acknowledgements

The author would like to thank the following collegues for some enlightening discussions : K. Cornelis, J.Gareyte and W.Herr.

References

[1] G.Brianti and K.Hübner, The Large Hadron Collider in the LEP tunnel, CERN 87-05 (1987)
[2] SSC - Conceptual design, SSC-SR-2020 (1986).
[3] D.Neuffer and S.Peggs, SSC-63 (1986)
[4] A.Piwinski,Limitation of the Luminosity by Satellite resonances, DESY 77/18, (1977).

[5] W.Herr, Tune shifts and spreads due to short and long range beam-beam interactions in the LHC, CERN SL/90-06 (AP).
[6] W.Herr, Synchro-Betatron resonances induced by a non-zero crossing angle in the LHC, to be published.
[7] A.Piwinski, Computer simulation of Satellite Resonances caused by the Beam-Beam Interaction at a crossing Angle in the SSC, SSC-57 (1986).
[8] K.Oide, K.Yokoya, SLAC-PUB-4832 (1989)
[9] S.A.Kheifets, G.A. Voss, A kick experienced by a particle obliquely traversing a bunch, SLAC-PUB 4994, June 1989
[10] P.Bagley, Cornell, private communication.

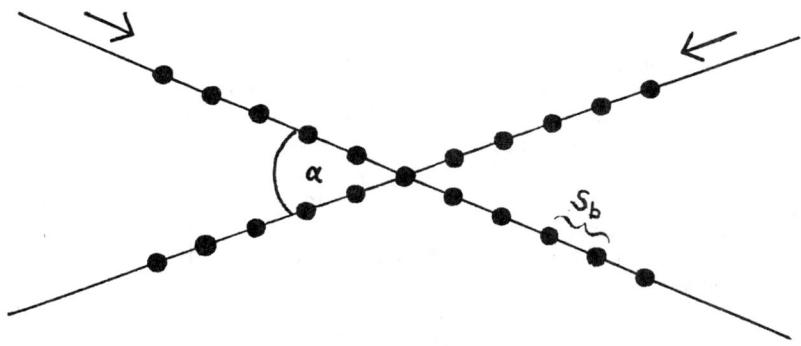

Fig.1 Geometry around the interaction region with a crossing angle.

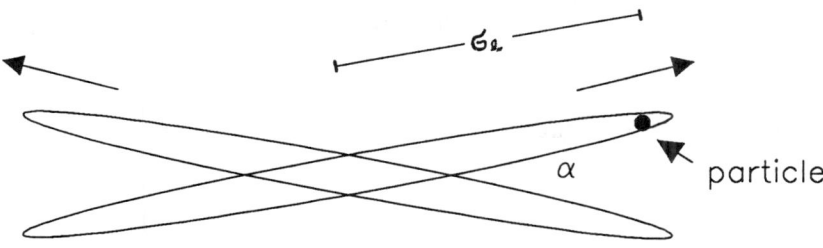

Fig.2 Geometry of the interaction point when the bunch centres collide

B FACTORY WITH A FINITE CROSSING ANGLE[*]

B. Autin, Y. Baconnier, K. Hirata, A. Hoffmann, J. Jowett,
H. Lengeler, D. Möhl, H. Moshammer
T. Risselada, H.H. Umstätter, T. Taylor, A. Verdier, T. Wang

Abstract

The R&D working group for the design of a B factory in the ISR tunnel has defined the characteristics of a very high luminosity machine based on a geometric separation of the beams at the interaction point (IP). Such a machine has the ISR configuration, namely two rings crossing horizontally. Each beam is made of many bunches with a relatively small number of particles per bunch and is flat at the IP. The orbits being straight in the interaction region, the synchrotron light background is maintained at a low level. In order to avoid synchro-betatron excitation through the beam-beam interaction and to preserve the luminosity, the bunches are rotated horizontally upstream to the IP and re-aligned after the crossing using two deflecting rf cavities per ring (*crab crossing* scheme); this way, the bunches fully overlap when they collide. The commisonning of the machine would begin with symmetric energies where the beam-beam effect is known and be progressively extended to asymmetric energies, a regime for which no experience exists yet.

Geneva, Switzerland

26 February 1990

[*]CERN PS/AR/90-06 European Organisation for Nuclear Research

Introduction

Given all the constraints imposed by the beam-beam interaction, the synchrotron radiation and the problems associated with short dense bunches, the only parameter which is, to some extent, free to reach extreme luminosities is the number of bunches. These bunches must collide at the interaction point only and be quickly separated to avoid extra interactions. Magnetic separation assumes a large asymmetry in beam energies and creates synchrotron radiation in the collision area. Geometric separation with a finite crossing angle is free of these limitations. One may thus contemplate a scenario of machine commissioning where the high luminosity regime would first be tested for beams of equal energies, then progressively extended to beams of unequal energies for which there is presently no experience.

Problems associated with finite crossing angle

The concept of crossing lepton beams at a finite angle has been tested on DORIS and turned out to be unsuccessful but for reasons which have been understood [1]. In the DORIS configuration, flat beams collide in the vertical plane and are thus submitted to strong transverse fields which vary along the bunch length and drive exceedingly large synchro-betatron resonances. Furthermore, if the crossing angle is calculated to make the optics independent for the two beams, its value is such that the luminosity is severely affected by a lack of overlapping of the incoming bunches.

Crab crossing

The solution to these problems lies in a method nicknamed *crab crossing* [2,3]. The bunches are rotated in the real space so that they fully overlap and, within some tolerances, synchro - betatron resonances are no longer excited [4,5].

The bunch rotation is produced by radio frequency transverse fields in a special cavity located at an odd multiple of quarter betatron wavelengths from the interaction point. A second identical cavity is located downstream to the interaction point to realign the bunch along the closed orbit. The spatial dependence of the transverse field is

$$E = E_0 \sin\left(2\pi \frac{z}{\lambda}\right)$$

where $2\sigma_z$ is the bunch length and λ the rf wavelength; the peak field necessary to achieve the bunch rotation is thus

$$E_0 = \frac{E}{\sin\left(2\pi \frac{\sigma_z}{\lambda}\right)}$$

The bunch rotation angle θ at the interaction point is achieved with an integrated field

$$E_0 \, l = \frac{p\,c}{e} \frac{\sigma_z \tan\theta}{\sin\left(\frac{2\pi \sigma_z}{\lambda}\right) \sqrt{\beta \beta^*}}$$

or, in the approximation of a large rf wavelength

B Factory with a Finite Crossing Angle

$$E_0 \, l = \frac{p\,c}{e} \frac{\lambda \tan\theta}{2\pi \sqrt{\beta\,\beta^*}}$$

β and $\beta*$ are the values of the β-function at the cavity and at the interaction point. For a given rotation θ, the peak field is roughly proportional to λ; there is however a tendency to choose a rather low frequency to reduce, as we shall see, the higher mode losses.

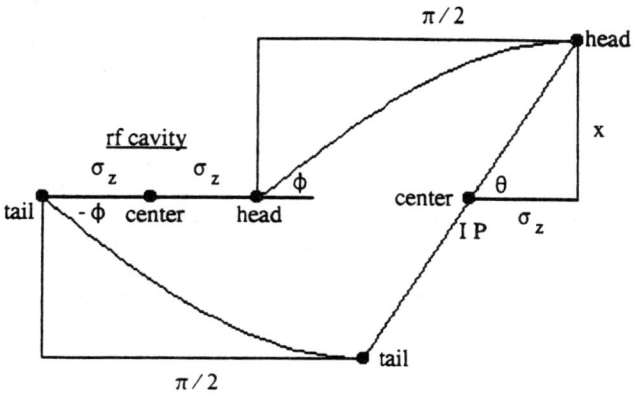

Principle of crab crossing. In an rf cavity where the transverse modes are excited, the particles at the head and at the tail of the bunch are deflected in opposite directions so that they oscillate in the magnetic field of the insertion quadrupoles and reach their peak amplitude at an odd multiple of quarter betatron wavelengths corresponding to the distance from the cavity to the interaction point (IP).

The β values at the cavity and at the interaction point have to be as large as possible to keep the rf deflection within reasonable limits. For this single reason, crossing must take place in the horizontal plane and the beams have to be flat. Any increase of luminosity related to round beams is thus discarded but, anyhow, it seems that round beams create too much complication for the reward that can be expected [6].

Super-conducting cavities are chosen because they can withstand a transverse field of 5 MV/m against .25 MV/m for normal cavities [7]. As for the accelerating cavities, a major concern is the evacuation of the higher order modes power. The loss factor k is estimated to be .15 V/pC/m at 350 MHz and varies like the square of the frequency. The rf power generated by the higher order modes is given by

$$P_{hom} = k\,f_b\,q^2$$

where f_b is the bunch frequency and q the bunch charge. Provisional estimates of the cavity impedance [8] cope with the tolerances imposed for bunch stability.

Determination of the crossing angle

The crossing angle is determined by the separation of the two adjacent quadrupoles belonging to each beam next to the interaction point. This way, there are no quadrupoles common to the two beams, the optics design is simplified [9] and the synchrotron radiation is reduced to its lowest level.

The pole tip radius a of the quadrupole is equal to 10 Max[σ_x, σ_y] augmented of 5 mm for the vacuum chamber. The beam sizes are calculated at the quadrupole location assuming the quadrupole infinitely thin. The distance between the quadrupole axes is $4a$. In this configuration, the final focussing is achieved with permanent magnets [10].

Parameter list

The parameter list has been established for symmetric energies. The bunch population n is determined from the luminosity expression

$$L = \frac{\gamma \Delta Q}{2 r_e} \frac{n f_b}{\beta_y}$$

for given high bunch frequency, low vertical β-value and moderate beam-beam tune shift ΔQ. r_e is the classical electron radius and γ the ratio of the total particle energy to its rest energy. The horizontal beam emittance ε_x is then inferred from the expression of the beam-beam tune shift

$$\Delta Q_y = \frac{r_e}{2 \pi \gamma \varepsilon_x} n$$

The vertical emittance corresponds to 2% coupling between the horizontal and vertical betatron oscillations. The horizontal β-value is such that the horizontal and vertical beam-beam tune shifts are the same

$$\frac{\beta_x}{\beta_y} = \frac{\varepsilon_x}{\varepsilon_y}$$

For both the higher order modes power and the synchrotron power density

$$P_{sy} = \frac{2 r_e E_0}{3 e} \frac{I \gamma^4}{\rho^2}$$

the beam current I has to be as low as possible. This is the reason why the vertical β-value at the crossing point is only 1cm at the cost of an accordingly small bunch length. Furthermore, as the geometry of the interaction region imposes two rings with a common horizontal median plane of symmetry, the ISR configuration becomes attractive and ρ has been deliberately increased with respect to the basic design in order to stimulate the study of a combined function lattice reminding that the ISR bending radius was 79 m.

B Factory with a Finite Crossing Angle

..........General characteristics..........................

Energy [GeV]	5.3
Revolution frequency [Hz]	318310.
Number of bunches	800
Bunch spacing [m]	1.18
Bunch population	$7.11 \cdot 10^{10}$
Beam current [A]	2.9
Half standard bunch length [m]	0.005
Bending radius [m]	75
Synchrotron power density [kW/m]	5.72
Horizontal emittance [mm.mrad]	0.102527
Vertical emittance [mm.mrad]	0.00205055

..............Interaction Region..........................

Luminosity [cm^{-2} s^{-1}]	$1. \cdot 10^{34}$
Half crossing angle [mrad]	20.13
Beam-beam tune shift	0.03
Beam disruption	0.188
Horizontal ß at IP [m]	0.5
Vertical ß at IP [m]	0.01
Distance from IP to first quadrupole [m]	1
Detector half integrated field [T m]	1

(over)

..............Crab crossing..............................

Number of cavities per crossing	2
Horizontal ß at the cavity [m]	30
Integrated electric field in a cavity [V]	$2.58 \cdot 10^{6}$
Deflecting rf frequency [GHz]	0.509296
Number of super-conducting cells	2
Length of a cavity [m]	0.707
Radius of a cavity [m]	0.373
Loss factor at 350 Mhz [V pC^{-1} m^{-1}]	0.15
Higher order mode power per cell [kW]	3.085

..............Beam separation..............................

Vertical deflection at detector exit [mrad]	1.139
First quadrupole half aperture [cm]	1.006
Beam separation at first quadrupole [cm]	4.025

Parameter list for equal energies

■ Final remarks

A B-factory with a finite crossing angle has very promising features related to the high bunch frequency which paves the way towards ultimate luminosities and independent focussing for the two rings which almost eliminates the synchrotron light background in the interaction region. These advantages are at the cost of an extra system: the *crab-crossing* scheme. The preliminary studies do not reveal special difficulties when the needed deflecting cavities are compared with the accelerating cavities. More work has clearly to be done in the domain of the choice of parameters, cavity design, special magnets, coupled bunch instabilities, etc. but it is likely that the machine can be evaluated in detail by the end of 1990. Last, let us note that the interaction areas of a B-factory based on a crab-crossing scheme and of high energy linear colliders are quite similar. More generally, any high luminosity lepton colliders would benefit from the experience developed in bunch rotation schemes. It is therefore our recommendation that the present study be pursued beyond the present mandate of the working group.

Working group

B. Autin (coordinator), Y. Baconnier, K. Hirata, A. Hofmann, J. Jowett, H. Lengeler, D. Möhl, H. Moshammer, T. Risselada, H.H. Umstätter, T. Taylor, A. Verdier, T. Wang.

References

[1] A. Piwinski, IEEE Trans. Nucl. Sci., NS-24, No 3 (1977).
[2] R. Palmer, SLAC-PUB-4707 (1988).
[3] G. A. Voss, J. M. Paterson, S. A. Kheifets, SLAC-PUB-5011 (1989).
[4] K. Oide, K. Yokoya, SLAC-PUB-4832 (1989).
[5] H. Moshammer, private communication.
[6] K . Hirata, H. Moshammer, F. Ruggiero, CERN SL-AP/90-02.
[7] H. Lengeler, private communication.
[8] J. Jowett, private communication.
[9] T. Risselada, private communication.
[10] H. H. Umstätter, private communication.

DISPERSIVE CRAB CROSSING:
AN ALTERNATIVE CROSSING ANGLE SCHEME

G. Jackson
Fermilab, PO Box 500, Batavia, IL 60510, USA

ABSTRACT

In order to maximize luminosity, many colliding beam storage ring designs are employing two rings, each carrying many bunches. Since the distance between bunches is small, beam-beam interactions other than the desired one at the interaction point occur unless the beams are quickly separated once outside of the high energy physics detector. One possible interaction region geometry is to have the beams collide at an angle. For luminosity degradation and beam stability reasons, this scheme has been augmented with transverse deflecting RF cavities which generate a correlation between horizontal position and time within the bunch, causing the beams to collide in a head-on fashion in their center of momentum inertial frame. The required transverse impedance, in addition to the development costs, of the transverse deflection cavities make this idea unattractive. In this paper an alternate scheme is proposed which makes use of the already existing accelerating RF cavities in the rings.

INTRODUCTION

The beam-beam interaction has the effect of limiting the product of the single bunch current and the number of interactions per bunch per turn. Therefore, to attain the high luminosities required by colliding beam storage rings such as B-meson factories, multiple bunches per beam with only one beam-beam crossing per turn is desirable. Due to limitations in magnet strengths and physical space near an interaction region, a useful optics configuration is to bring the beams into collision with a relative crossing angle.

There are two disadvantages with this scheme. First, when the crossing angle times the bunch length is comparable to the beam width a decrease in luminosity is suffered as compared to head-on collision geometries.[1] Second, when beams cross at an angle synchrobetatron resonances are excited.[2] Since the synchrotron tune of electron-positron storage rings is on the order of 0.05, excitation of synchrotron sidebands modulating major coupling resonances fills the tune plane with "forbidden" regions, reducing the beam-beam limit and hence the beam current.

In the case of linear colliders the concept of "crab crossing" was invented to circumvent the luminosity loss issue.[3] Later, crab crossing was proposed to eliminate synchrobetatron resonance excitation in storage rings.[4] In the remainder of this paper the original scheme discussed in refs. 3 and 4 will be referred to as "transverse" crab crossing.

TRANSVERSE CRAB CROSSING

Transverse crab crossing involves correlating the horizontal and longitudinal beam emittance envelopes such that in their center of momentum frame the bunches appear to collide head-on. The correlation is accomplished through the use of a set of horizontal deflecting RF cavities set 90° in betatron phase advance from the interaction point. A diagram of this scheme appears in figure 1.

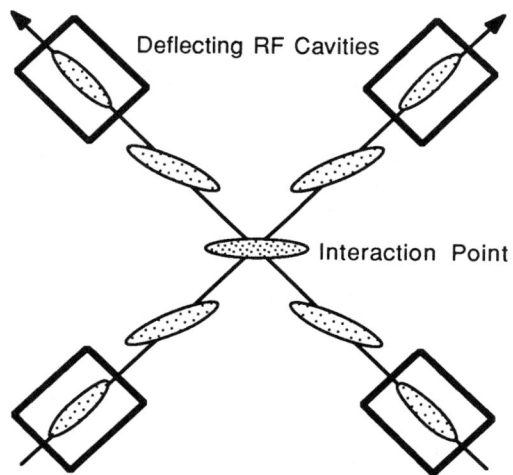

Fig. 1. Sketch of an interaction region in which an exaggerated crossing angle is assumed. The relative positions of the four horizontal deflection cavities and their effect on the beam envelopes are included.

Assume that a horizontal deflecting RF cavity of wavelength λ has an equivalent voltage amplitude V_x. The cavity is placed in a storage ring of energy E_o at a lattice location 90° upstream of the interaction point. Figure 2 contains a sketch of this interaction region, indicating the deflection applied to a bunch and the resultant beam envelope trajectory. If β^* and β_c are the values of the β-function[10] at the interaction point and the RF cavity respectively, then the crossing angle exhibited by the beam at the interaction point is given by

$$\theta = \frac{2\pi}{\lambda} \frac{e V_x}{E_o} \sqrt{\beta^* \beta_c} \qquad (1)$$

The assumption in this equation is that the bunch length is small compared to λ, such that the RF wave is approximated by a straight line. Note that the crab angle, which should equal the crossing angle of the beam, is proportional to the square root of β^*. In flat beam e+e- interaction region designs, this functional dependence restricts crab crossing to the horizontal plane.

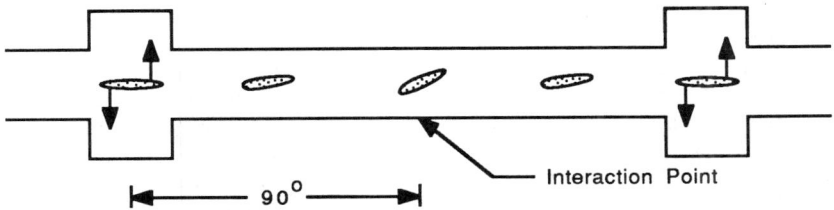

Fig. 2. Sketch of the evolution of one bunch in a single ring undergoing crab crossing. At the interaction point the deflection in the upstream deflecting RF cavity is converted into an azimuth dependent transverse offset.

Looking at figure 2, it is straightforward to visualize crab crossing by treating the head and tail of the bunch as individual particles. These particles receive opposing kicks in the upstream RF cavity, which generate free betatron oscillations. At the interaction point these oscillations reach their maximum amplitudes, and then proceed to advance another 90° at the crossing of the downstream cavity. At this point the betatron oscillation is removed.

Table I. Values of a PEP based asymmetric B-meson factory transverse crab crossing design.

Parameter	Units	Ring #1	Ring #2
E_o	GeV	3	9
eV_x	MeV	1.4	2.5
λ	m	0.42	0.42
β^*	m	0.5	0.5
β_c	m	10	10
θ	mrad	15.6	9.3

As an instructive example, take the parameters (see table I) of a recent asymmetric B-meson factory design.[5] A total of almost 8 MV of transverse voltage must be produced to produce the required crossing angle of 25 mrad.

A number of problems exist with the concept of transverse crab crossing when it is applied to storage rings. These include tolerance issues such as RF voltage and phase regulation necessary to insure beam collisions.[6,7] In addition, the implications of transverse transient beam loading[8], especially in the presence of an azimuthal bunch distribution which contains a gap to prevent ion trapping[9], has not been seriously studied yet. The addition of a transverse deflecting cavity will partial negate the great care taken to minimize the transverse impedance of the storage ring, making transverse instabilities more likely to occur. Finally, the cost and time required to research and development such deflecting cavities is substantial.

DISPERSIVE CRAB CROSSING

With the above arguments in mind, it is desirable to ponder whether any alternatives exist. Clearly, the fundamental concept of crab crossing the beams to attain functional, high luminosity beam collisions is worth preserving. In fact, an alternative method for generating the crab crossing angle is possible. The basic idea is to replace the transverse deflecting RF cavities with the acceleration RF cavities in which the horizontal dispersion is nonzero. Since these cavities already exist, many of the above objections are eliminated. This alternative scheme is called "dispersive" crab crossing.

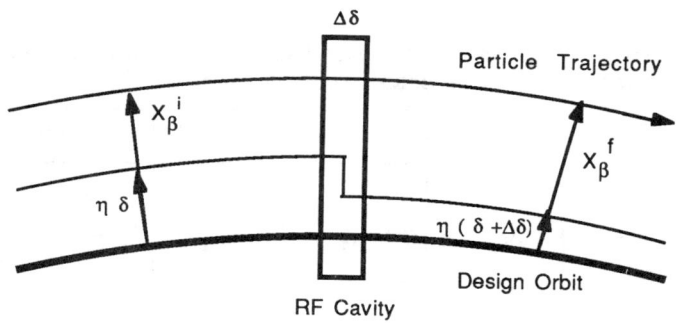

Fig. 3. Change in particle coordinates caused by an instantaneous fractional energy change $\Delta\delta$.

Relative to the design orbit, the horizontal position of a particle is the sum of its betatron position[10] (x_β) and its equilibrium orbit position[1], which is equal to the local horizontal dispersion (η_x) times the particle's fractional energy offset ($\delta = (E-E_o)/E_o$). Upon passage through a RF cavity, assume that the particle is decelerated relative to a synchronous particle. Since the particle cannot change its physical location instantaneously, the change in equilibrium orbit position is compensated by a matching change in betatron amplitude. Figure 3 contains a sketch of this situation.

The change in betatron position, and for generality sake betatron angle, produced by a fractional energy deflection relative to a synchronous particle, are given by

$$\Delta x_\beta = - \eta_x \Delta \delta \quad , \quad \Delta x'_\beta = - \eta'_x \Delta \delta \quad . \quad (2)$$

Assuming for the time being that η'_x is zero, the change in betatron position for a small deviation away from the RF synchronous angle (ϕ_s) is

$$\frac{\Delta x_\beta}{s} = \frac{2 \pi e V_o \eta_c}{\lambda \; E_o} \cos \phi_s \quad , \quad (3)$$

where V_o is the amplitude of the RF voltage and η_c is the value of the dispersion at the cavity. Note that the expression in equation 3 is also the instantaneous betatron angle the beam envelope develops at the RF cavity. Propagating the betatron oscillation of the beam envelope 180° downstream to the interaction point, the equation for the crab angle of the beam is

$$\theta = \frac{2 \pi e V_o \eta_c}{\lambda \; E_o} \sqrt{\frac{\beta^*}{\beta_c}} \cos \phi_s \quad . \quad (4)$$

Just as in the case of transverse crab crossing, the crab angle scales with the square root of β^*. On the other hand, the angle now depends on the ratio of η_c and the square root of β_c. For the same example B-meson factory used in the discussion of transverse crab crossing, table II contains parameter values for a dispersive crab design.

In the design scenario outlined in Table II, the available RF voltages are 34.2 and 21.3 MV for rings #1 and #2, respectively. Since the synchronous phase angles in the two rings are on the order of a few degrees, they have been neglected. Note that the RF voltages required to generate the same crab angles are small compared to the total amount of RF available. Therefore, only 5 to 7 cavities need to be placed on either side of the interaction region in both rings. The majority of the cavities can be placed at any other convenient azimuth around the rings.

Table II. Values of a PEP based asymmetric B-meson factory dispersive crab crossing design.

Parameter	Units	Ring #1	Ring #2
E_o	GeV	3	9
eV_o	MeV	4.7	8.3
λ	m	0.42	0.42
β^*	m	0.5	0.5
β_c	m	10	10
η_c	m	3	3
θ	mrad	15.6	9.3

The next issue to contemplate is the placement of the cavities near the interaction region. Since the already existing cavities are being used, phase stability must be a consideration. The RF cavities should be placed such that the beam is always at the stable, synchronous phase angle. This requires that if the upstream cavity is $180° + n \cdot 360°$ away from the interaction point, the downstream cavity must be $m \cdot 360°$ away in order to close the crab induced free betatron oscillations (where m and n are arbitrary integers). Figure 4 contains a sketch of such an interaction region geometry.

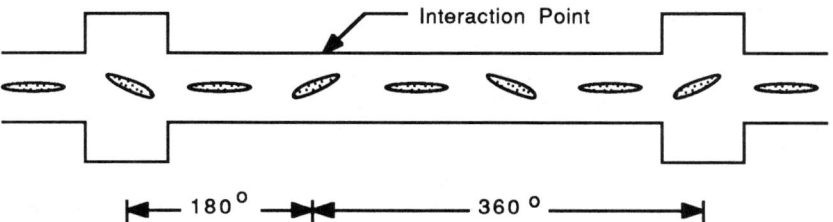

Fig. 4. Example of a dispersive crab crossing interaction region design for a single beam line. Note the asymmetric placement of the RF cavities to insure that the beam always crosses at the stable synchronous phase angle.

SYNCHROBETATRON COUPLING

Potentially the most serious drawback to dispersive crab crossing is the excitation of synchrobetatron coupling due to the existence of horizontal dispersion in the RF cavities.[11] As discussed in detail above, passage through an RF cavity in which dispersion is nonzero causes a horizontal betatron kick. The resultant free betatron oscillation around the ring generally has a different path length as compared to a particle with the same energy but not undergoing betatron oscillations. This causes the particle to arrive at the cavity the next turn with a time delay, determining the next RF/betatron kick to be received. The betatron oscillation dependence of the path length is due to the fact that particles to the outside of the design orbit have a longer path length inside dipole magnets. If the synchrotron (Q_s) and betatron (Q_x) tunes have the relationship

$$i\, Q_x + j\, Q_y = k \quad , \quad (5)$$

where i, j, and k are integers, the particle motion is unstable.

The widths of these synchrobetatron resonances are beam current independent, and scale as η_c divided by the square root of β_c (exactly the same dependence as the crab angle). But in the dispersive crab crossing scheme the betatron oscillations induced by the RF cavities are purposely cancelled for the majority of the accelerator. Therefore, if the interaction region dipoles between the crab cavities are placed such that the path length through the region is independent of betatron amplitude (a task made significantly easier by the asymmetric placement of the cavities on either side of the interaction region), synchrobetatron resonances will not be excited.

Even if the path length modulation is not compensated, one usually finds that other mechanisms[11], usually beam current dependent, are the dominant sources of synchrobetatron coupling.[12] For example, at CESR the ratio of η_c to the square root of β_c is roughly 3/6 $m^{1/2}$. And yet, attempts to find the source of synchrobetatron resonance excitation have always led away from the RF cavities.[13]

COHERENT INSTABILITIES

Because existing RF cavities are used in the dispersive crab crossing scheme, the introduction of longitudinal impedance by the cavities must already be dealt with. As a result, dispersive crab crossing does not induce a cost with respect to longitudinal instabilities.

In fact, the same can be said with respect to transverse instabilities. In order to minimize the impact of RF cavity generated transverse wake fields, the value of β_c should be as small as possible. But this is exactly the direction required to maximize the crab crossing angle.

TRANSIENT BEAM LOADING

Transient beam loading occurs when beams with high currents are accelerated through RF cavities with a very large shunt impedance, and is a potential problem for at least two reasons. First, potential well distortion[8] disrupts the linear dependence of RF voltage on arrival time at the cavity, destroying the center of momentum frame head-on collision geometry which crab crossing is intended to generated. Second, the gap in the azimuthal bunch distribution around the ring designed to prevent ion trapping will produce a bunch dependent beam loading voltage which is out of phase with the RF drive. The result is that without active compensation, each bunch will see a unique RF phase. Each bunch would then receive a different average energy kick, leading to bunch dependent interaction point offsets in the horizontal plane.

For higher order mode reasons, the RF cavities considered for B-meson factories have recently been designed with very small shunt impedances.[15] As a result, the above two concerns should be greatly diminished. In addition, feed-forward transient beam loading compensation systems are quite common and are well understood.

CONCLUSIONS

Though much work still needs to be done, it is already clear that dispersive crab crossing has many advantages, and no substantial disadvantages, when compared to transverse crab crossing. As in the case of transverse crab crossing, systematic simulation studies of tolerances and impedance effects (such as transient beam loading) need to be done. Such simulation studies are presently underway by the author (dispersive) and A.Piwinski[14] (transverse). Comparisons of these results with experimental results and with one another should be a future priority.

ACKNOWLEDGEMENTS

This work was supported by the Universities Research Association, Inc., under contract DE-AC02-76CH03000 from the United States Department of Energy. Suggestions and references provided by G.Dugan and P.Morton enhanced the scope and completeness of this work.

REFERENCES

1. M.Sands, SLAC-121 (1970).
2. A.Piwinski, IEEE Trans. on Nucl. Sci. NS-24, 1408 (1977).
3. R.Palmer, SLAC-PUB-4707 (1988).
4. K.Oide and K.Yokoya, SLAC-PUB-4832 (1989).
5. "Feasibility Study for an Asymmetric B-Factory Based on PEP", LBL PUB-5244, SLAC-352, CALT-68-1589 (1989).
6. S.A.Kheifets, J.M.Paterson, G.A.Voss, SLAC-PUB-5011 (1989).
7. A.Piwinski, DESY-HERA 90-04 (1990).
8. P.B.Wilson, AIP Conf. Proc. No. 87, Fermilab, 450 (1981).
9. Y.Baconnier and G. Brianti, CERN/SPS/80-2 (DI) (1980).
10. E.D.Courant and H.S.Snyder, Ann. Phys. 3, 1 (1958).
11. A.Piwinski, CERN 87-03 (1987).
12. SPEAR Group, 1975 Part. Acc. Conf., IEEE Trans. Nucl. Sci. NS-22, 1366 (1975).
13. G.Jackson and R.Siemann, Nucl. Instr. Meth. in Phys. Research A286, 17 (1990).
14. A.Piwinski, DESY-HERA 90-04 (1990).
15. D.Rubin, this proceedings.

Bunch Length Compression Using Crab Cavities

Y. Orlov, Cornell University

1. Introduction

The ratio σ_L^*/β^* of the bunch length σ_L to the beta-function β in the interaction point IP must be small, given both the hourglass effect and synchrobetatron couplings. But, simultaneously we need as small a β^* as possible for high luminosity, and a rather big σ_L outside of the IP, because of problems with RF power, high order modes losses, and single bunch instabilities. Therefore, it would be better to have a small σ_L^* locally, in the IP only.

The proposed structure SS^{-1}, where S is the structure between some initial point 0 and the IP (and the corresponding transition matrix), and S^{-1} is the inverse structure (and the inverse matrix) between the IP and some final point 1 (fig. 1), compresses longitudinal coordinates s in the interaction region IR only. The features of this structure are quite different for cases $k \neq 0$ and $k = 0$; $k = \frac{\partial s^*}{\partial s_0}$, where s^* and s_0 are longitudinal coordinates at the IP and at point 0, respectively.

In the case of $k \neq 0$ we can manage $\eta^* = 0$ (i.e., zero dispersion function at the IP), and in this case the compression of the bunch length is accompanied by the decompression of the energy $\Delta \mathcal{E}/\mathcal{E}$ (momentum $\Delta p/p$) spread at the IP. In the case of $k = 0$ we have inevitably $\eta^* \neq 0$, but the energy spread at the IP can be almost the same as at point 0. And it is much more curious and perhaps important that we can manage the full insensitivity of the vertical (or horizontal, but not both) oscillations to the beam-beam interactions. Namely, we have at point 1, compared with point 0, after the beam-beam kick between them,

and
$$\begin{aligned} y_1 &= y_0 \\ y_1' &= y_0' \\ \left(\tfrac{\Delta p}{p}\right)_1 &= \left(\tfrac{\Delta p}{p}\right)_0 \end{aligned} \qquad (1)$$

$$s_1 = s_0 + \delta s_1 \;\; ; \;\; \delta s_1 = f\left(\frac{\Delta p}{p}\right)_1 \; , \qquad (2)$$

where f is some nonlinear function dependent on the beam structure at the IP. More accurately, $\delta s_1 = f(\frac{\Delta p}{p}, x)$, where x is the other (horizontal, if y is vertical) transverse coordinate of the particle.

It creates the possibility of designing the collider without tune shift ξ_y but with the usual tune shift ξ_x, and with an unusual longitudinal tune shift ξ_L.

2. The Basic Ideas

The combination of an RF deflector ("crab cavity") in which a transverse angle kick $\Delta y'$ depends on the longitudinal coordinate s (fig. 2),

$$\Delta y' = bs \; , \qquad (3)$$

together with some magnetic lattice M containing bending magnets can work as a longitudinally focusing system (fig. 3). (Really, I have placed a lense, F, just before the "crab"; see fig. 1.) After passing through such a system, the final longitudinal coordinate of a particle s can be compressed with a factor $k < 1$ and will depend only on the initial angle y_0' in the case $k = 0$:

$$s^* = ks_0 + M_{32} y_0' \qquad (4)$$

(M_{ik} = a matrix element of the matrix M corresponding to the lattice M). The focusing lattice m which focuses the particles into the IP does not change this result.

Let us consider the case $k = 0$. If the IP is the longitudinal focus, then the initial bunch length σ_{L0} is transformed into some angle spread,

$$y'^* = S_{23} s_0 + \ldots \qquad (5)$$

(S_{23} is the matrix element of the matrix S describing the transition from point 0 to the IP). Therefore, the inverse structure S^{-1} transforms any angle kick produced at the IP, $(\delta y')^*$, into a longitudinal kick at point 1:

$$\delta s_1 = S_{32}^{-1}(\delta y')^* \qquad (6)$$

Other ideas can be derived from the analysis of 4x4 matrices. It is impossible to have a 2x2 matrix S with $S_{12} = S_{22} = 0$. But for a 4x4 matrix, i.e., in a system with synchrobetatron coupling, it is possible, and after proper choice of the structure parameters we can get, for example,

$$S^{-1} = \begin{pmatrix} -\frac{abD(e^\varphi-1)}{m_{11}} & 0 & \frac{bl}{2}-q & aD(e^\varphi-1) \\ 0 & 0 & -\frac{1}{aD(e^\varphi-1)} & 0 \\ \frac{m_{21}}{b} - \frac{e^\varphi+1}{m_{11}\sqrt{n}} & -\frac{m_{11}}{b} & 1 & 0 \\ b/m_{11} & 0 & 0 & 0 \end{pmatrix} \qquad (7)$$

$$D = 1 + \frac{l}{a\sqrt{n}} \frac{e^\varphi+1}{e^\varphi-1} \qquad (8)$$

The meaning of these matrix elements will be explained later. The most important thing is that $S_{12}^{-1} = S_{22}^{-1} = 0$. As a result, the betatron coordinates y, y' at point 1 do not depend on the particle-beam kick in the IP. Together with $S_{42}^{-1} = 0$ it gives the equations (1).

But it is not enough for full insensitivity of the y-oscillation to beam-beam kicks. For this insensitivity the final longitudinal kick δs_1 must be independent of the y, y' coordinates. It will depend on $\Delta p/p$ only if in the S matrix (0 $\to$ IP transition) we can manage $S_{11} = S_{12} = S_{13} = 0$. It is impossible to have $S_{11} = S_{12} = 0$ for a 2x2 matrix, but possible for 4x4 matrixes. For example, in our case, when $k = 0$,

$$S = \begin{pmatrix} 0 & 0 & 0 & \frac{m_{11}}{b} \\ 0 & -\frac{aD(e^\varphi-1)l}{m_{11}} & -\frac{b}{m_{11}} & \frac{m_{21}}{b} - \frac{e^\varphi+1}{m_{11}\sqrt{n}} \\ 0 & -aD(e^\varphi-1) & 0 & 0 \\ \frac{1}{aD(e^\varphi-1)} & \frac{bl}{2}-q & 0 & 1 \end{pmatrix} \qquad (9)$$

It means that

$$\eta^* = \frac{m_{11}}{b}, \quad k = 0, \qquad (10)$$

and the y-size at the IP depends only on the initial momentum spread:

$$\sigma_y^* = \eta^*(\frac{\Delta p}{p})_0 \tag{11}$$

As a result we have the equation (2). Therefore, the structure SS^{-1} does not change the transverse y, y' coordinates *at all* despite the presence of an additional (undescribed by SS^{-1}) effect of the beam-beam kick at the IP between S and S^{-1}; and this structure does not produce any additional $y - s$ coupling *outside* itself despite that kick. Now the beam-beam drama is being played not in the (x, y) space, but in the (x, p) space. The y-oscillation has become free.

3. The Main Conditions Placed on the Parameters of the Structure

The compression factor produced by the structure S is

$$k = 1 - ab(e^\varphi - 1) - (\frac{bl}{2} + q)\frac{e^\varphi + 1}{\sqrt{n}} \tag{12}$$

where b is the deflector "crab" strength (3) and l is its length. I have added one more independent parameter, q, which represents another possibility of an RF transverse deflection. Together with the first "crab", q produces the displacement

$$\Delta y = (\frac{bl}{2} + q)s \tag{13}$$

The amplifying factor $e^\varphi = sh\varphi + ch\varphi$ describes the strength of the defocusing lense D_0 (fig. 1), which serves as an amplifier of the $\Delta y, \Delta y'$ deflections.

$$\varphi = \sqrt{\frac{\partial B/\partial y}{BR}} l_D \tag{14}$$

where l_D is the length of the D_0 lense and $(\frac{\partial B}{\partial y})$ is its field gradient. The amplifying factor e^φ is limited only by demands placed on the β_y, β_x beta-functions (i.e., on the physical sizes of the lenses and magnets) *inside* the M lattice. A moderate number is, for example, $\varphi = \pi/2, (\beta_y)_{\max} \approx (\beta_x)_{\max} \sim 100, e^\varphi = 4.8$. When φ is higher, $(\beta)_{\max}$ will grow $\sim e^{2\varphi}$.

If the radius of a bending magnet $\Phi_\pm$ is R (fig. 1), then in (7), (9), (12)

$$\frac{1}{a} = R\left[\frac{1}{R^2} + \frac{\partial B/\partial y}{BR}\right], n = \frac{R}{a} \tag{15}$$

The bigger the a, the less deflector strength b is needed to have the given compression factor k.

The focusing force $1/\sqrt{aR}$ of the combined functions magnets $\Phi_\pm$ must be exactly the same as the defocusing force $\sqrt{\frac{\partial B/\partial y}{BR}}$ of the D_0. This means that the gradient $\frac{\partial B}{\partial y}$ in the D_0 and in the $\Phi_\pm$ are not exactly equal:

$$\left(\frac{\partial B/\partial y}{BR}\right)_{D_0} = \left[\frac{1}{R^2} + \frac{\partial B/\partial y}{BR}\right]_\pm \equiv \frac{1}{aR} \tag{16}$$

The length $l_\pm$ of the $\Phi_\pm$ magnet of the S structure is defined by the condition

$$\phi_\pm \equiv l_\pm/\sqrt{aR} = \sqrt{n}\,\frac{l_\pm}{R} = \frac{\pi}{2} \tag{17}$$

The field index n is not a completely free parameter: in order to have the simplest S structure, and simultaneously to cancel the important part of the momentum $(\Delta p/p)_0$ dependence of the s^*,

$$\sqrt{n} = \frac{e^\varphi}{\frac{l_\pm}{\sqrt{aR}} - \sin\frac{l_\pm}{\sqrt{aR}}} = \frac{e^\varphi}{\frac{\pi}{2} - 1} \tag{18}$$

For $\varphi = \pi/2$, $\sqrt{n} = 8.4208$, and $n = 70.91$; these are reasonable numbers. If condition (17) is fulfilled, then $S_{34} = 0$. In order to cancel also the dependence of the s^* on the initial transverse coordinate y_0, i.e., to have $S_{31} = 0$, we need a thin focusing lense F with the strength

$$\frac{1}{f} = \frac{1}{a} \cdot \frac{(e^\varphi + 1)}{\sqrt{n}(e^\varphi - 1)} \tag{19}$$

For $\varphi = \pi/2$ and n from (18), $f/a = 5.4645$. If the derivative of the dispersion function $\eta_0' = 0$ and conditions (18) and (19) are met, the longitudinal coordinate s^* at the IP does not depend on $(\frac{\Delta p}{p})_0$ at all.

Other conditions are placed on the parameter q and on the final focusing system m (fig. 1). In general there is less freedom in choosing parameters when $k \neq 0$. If the main purpose is to compress the bunch, then in the case of $k = 0$, we need to use a flat beam; the structure SS^{-1} will be placed in the horizontal plane, i.e., $B = B_y$; and in the equations (1) there will be $x_1 = x_0$, $x_1' = x_0'$ instead of $y_1 = y_0$, $y_1' = y_0'$, i.e., the horizontal oscillations, not vertical ones, will be insensitive to the beam-beam kicks. The horizontal dispersion function $\eta^* \neq 0$, but η^* can be sufficiently small, and focusing system m can be no stronger than usual. If the main purpose is to cancel the vertical tune shift ξ_y, then in order to have simultaneously a small $\sigma_y^* = \eta_y^*(\Delta p/p)_0$, focusing system m must be stronger than in a "normal" collider.

The only additional condition in the case $k = 0$ is

$$m_{12} = m_{11}\left(\frac{l}{2} + \frac{q}{b}\right) \tag{20}$$

where m_{ik} are the matrix elements of the matrix m.

We can use the parameter q to fulfill condition (20). This gives us enough parameters for focusing the particles onto the IP in both directions, y and x.

4. The Calculation of the Matrices

The matrices are acting on the vectors

$$\vec{V} = \begin{pmatrix} y \\ y' \\ s \\ \frac{\Delta p}{p} \end{pmatrix} \tag{21}$$

where y, y' are betatron variables, and s is the longitudinal coordinate with $s = 0$ in the center of the bunch.

It follows from fig. 1 that

$$S = mMCF \tag{22}$$

$$M = \Phi_- D_0 \Phi_+ D_0 \tag{23}$$

For all lenses without magnetic field B on the axis, the 4x4 matrices have the obvious shape

$$(\text{Lense}) = \begin{pmatrix} 2 \times & 2 & 0 & 0 \\ m\ a\ t & r\ i\ x & 0 & 0 \\ 0 & 0 & 1 & 0 \\ 0 & 0 & 0 & 1 \end{pmatrix} \qquad (24)$$

Using the sympecticity condition and equations (definitions) (3) and (13) for a crab cavity (there is really a set of cavities), we have

$$\Delta\left(\frac{\Delta p}{p}\right) = by + \left(\frac{bl}{2} - q\right)y' \qquad (25)$$

Therefore,

$$C = \begin{pmatrix} 1 & l & \frac{bl}{2}+q & 0 \\ 0 & 1 & b & 0 \\ 0 & 0 & 1 & 0 \\ b & \frac{bl}{2}-q & 0 & 1 \end{pmatrix} \qquad (26)$$

To calculate the matrix $\Phi_\pm$ we can use the equation of the motion

$$y'' + \left(\frac{1}{R^2} + \frac{\partial B/\partial y}{BR}\right) y = \frac{1}{R}\frac{\Delta p}{p} \qquad (27)$$

It gives

$$\begin{array}{rl} y = & y_0 \cos\phi + \sqrt{aR}\ y_0' \sin\phi \pm a(1-\cos\phi)\frac{\Delta p}{p} \\ y' = & -\frac{1}{\sqrt{aR}} y_0 \sin\phi + y_0' \cos\phi \pm \frac{1}{\sqrt{n}}\sin\phi\frac{\Delta p}{p} \end{array} \qquad (28)$$

where

$$y_0 = \eta_0\left(\frac{\Delta p}{p}\right)_0 + (y_0)_{\text{betatron}}, \quad y_0' = \eta_0'\left(\frac{\Delta p}{p}\right)_0 + (y_0')_{\text{betatron}}, \qquad (29)$$

$$y' \equiv dy/dz\ ,\ \phi = z/\sqrt{aR} \qquad (30)$$

At the end of the magnet $z = l_\pm$. In our structure $\frac{l_\pm}{\sqrt{aR}} = \frac{\pi}{2}$.

When passing through a bending magnet with a radius R, the particle longitudinally shifts,

$$s - s_0 = -\int_{z_0}^{z} \frac{y(z)dz}{R} \qquad (31)$$

Putting (28) here, we have

$$s = \mp \frac{1}{\sqrt{n}} \sin\phi \cdot y_0 \mp a(1 - \cos\phi)y_0' + s_0 - \frac{a}{n}(\phi - \sin\phi)\left(\frac{\Delta p}{p}\right)_0 \qquad (32)$$

Therefore

$$\Phi_\pm = \begin{pmatrix} \cos\phi & \sqrt{aR}\sin\phi & 0 & \pm a(1-\cos\phi) \\ -\frac{1}{\sqrt{aR}}\sin\phi & \cos\phi & 0 & \pm\frac{1}{\sqrt{n}}\sin\phi \\ \mp\frac{1}{\sqrt{n}}\sin\phi & \mp a(1-\cos\phi) & 1 & -\frac{a}{n}(\phi - \sin\phi) \\ 0 & 0 & 0 & 1 \end{pmatrix} \qquad (33)$$

In our structure $\sin\phi = 1, \cos\phi = 0, \phi = \frac{\pi}{2}$, and under condition (18) we have

$$M = \begin{pmatrix} -1 & 0 & 0 & a(e^\varphi - 1) \\ 0 & -1 & 0 & -\frac{e^\varphi + 1}{\sqrt{n}} \\ -\frac{e^\varphi + 1}{\sqrt{n}} & -a(e^\varphi - 1) & 1 & 0 \\ 0 & 0 & 0 & 1 \end{pmatrix} \qquad (34)$$

$$M^{-1} = M \qquad (35)$$

The product (22) gives us (9) when $k = 0$.

5. One Example

Deflector C contains N crab cavities, with $\lambda_{RF}/2$ of every length,

$$l = N\lambda_{RF}/2, \quad b \approx 1.7 \frac{E_{\max}}{\gamma mc^2} N, \qquad (36)$$

where $E =$ an electric field of one cavity in Mv/m, mc^2 in Mev, $\gamma = E/mc^2$. Assuming the superconducting cavities with $E_{\max} = 25$Mv/m, we will have (for $\gamma = 10^4$) $b \approx 0.83 \times 10^{-4} N\,cm^{-1}$. If $R = 100m$, then $a = R/n = 141cm$. In the case $\lambda_{RF} = 60cm, k = 0$ we need $N = 15$ (if $q = 0$); $l = 4.5m$, $b = 1.25 \times 10^{-3} cm^{-1}$.

The length of every element of the structure M, D_0 and $\Phi_{\pm}$, is equal to $R/\sqrt{n} = 11.9m$, so the length of the S structure $L_s > 4 \times 11.9 + 4.5 = 52m$.
In (9) $D = 1 + \frac{450 \times 5.8}{141 \times 3.8 \times 8.42} = 1.58$, and $s^* = -141 \times 1.58 \times 3.8 y_0' = -0.85 \times 10^3 y_0'(cm)$. In order to have $s^* < 1.5mm$ we need $|y_0'| < 0.1 mrad$.
$\eta^* = m_{11}/b$. If $s^*/s_0 \ll 1$ and $(\frac{\Delta p}{p})_0$ is sufficiently small, then

$$(y')^* \approx s_0/\eta_y^* \tag{37}$$

For the angles of the particles, let us assume a technical limit, $(y')^*_{\max} \sim 25 mrad$. Supposing $s_0 \sim 0.5cm$, we have $\eta_y^* \sim 20cm$. Since

$$\sigma_y^* = \eta_y^* \sigma_{e0} \tag{38}$$

then $\sigma_y^* \sim 200\sigma_{e0} mm$. If $\sigma_{e0} = 0.3 \times 10^{-3}$ then $\sigma_y^* = 60\mu m$. This is not a small size. However, the independence of the vertical oscillations of the beam-beam kicks in the case $k = 0$ permits us to hope that the luminosity $\mathcal{L} = N_B^2 f_c/A$ will depend only on horizontal parameters, $\mathcal{L} \propto \xi_x I(1 + \sigma_x^*/\sigma_y^*)/\beta_x^*$, and that by playing with $\sigma_x^*/\sigma_y^* \approx \eta_x^*/\eta_y^*$ and β_x^* we can achieve a big $\mathcal{L}$ even with a big A.

For bunch compression only, we can consider the SS^{-1} system in the horizontal plane; with a less abnormal limit, $(x')^*_{\max} \sim 5 mrad$, we will have more or less normal $\eta_x^* \sim 100cm$, $\sigma_x^* \sim 100\sigma_{e0} cm$, i.e., σ_x^* close to CESR's horizontal size. The horizontal oscillations will be independent of beam-beam kicks.

The momentum spread $(\frac{\Delta p}{p})$ in the case $k = 0, \eta_0 = 0$, is almost unchanged, $(\frac{\Delta p}{p})^* \approx (\frac{\Delta p}{p})_0$.

In equation (6) $S_{32}^{-1} = \eta^*$, i.e., $\delta s_1 = \eta^*(\delta y')^*$.

F = thin focusing lense
C = "crab" cavities
D_0 = thick defocusing lense, $B=0$, $\frac{\partial B}{\partial x}<0$, ="amplifier"
φ_+ = focusing magnet (combined functions), $B>0$, $\frac{\partial B}{\partial x}>0$
φ_- = focusing magnet, $B<0$, $\frac{\partial B}{\partial x}>0$.
m = focusing system, $B=0$.
m^{-1}, C^{-1}, F^{-1} = reverse matrixes

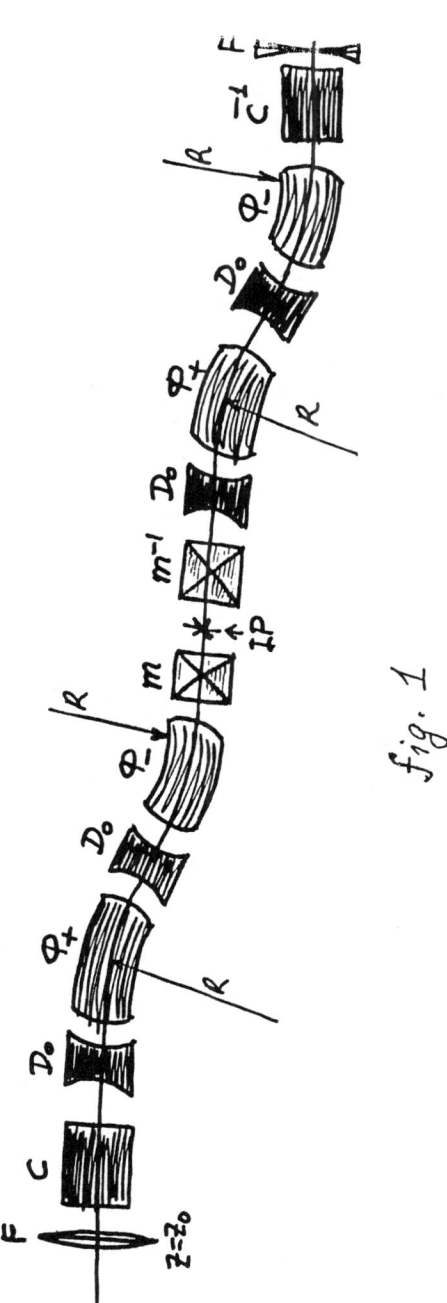

fig. 1

Bunch Length Compression

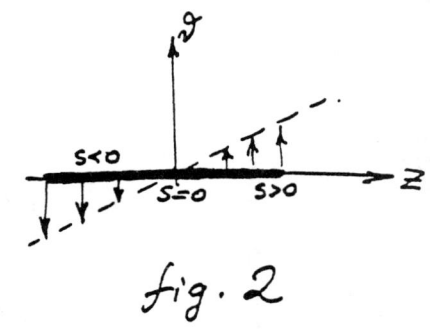

fig. 2

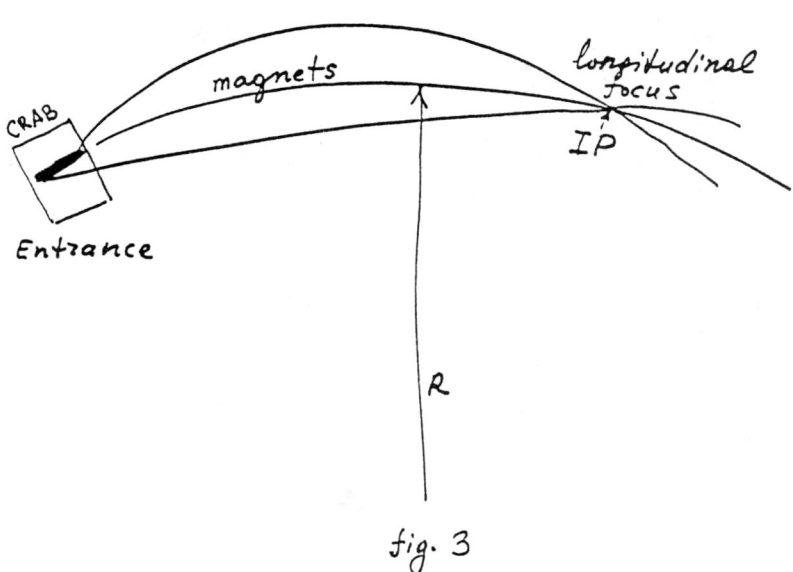

fig. 3

MONOCHROMATIC DESIGN FOR THE APIARY B-FACTORY

A. Dubrovin, A. Zholents

Institute of Nuclear Physics, Novosibirsk, USSR

A. Garren*

Lawrence Berkeley Laboratory, Berkeley, California, USA

Abstract

There are several ways to design e^+e^- circular colliders to maximize luminosity. Here we present a design that achieves two principal goals: high luminosity and very small center-of-mass energy spread. The design possesses several attractive features, especially for an asymmetric B-factory.

1. Introduction

Here we shall discuss the logical constraints of the monochromatic approach to the asymmetric B-factory design. The first issue to consider is the means of obtaining monochromatic collisions. The idea is that if the two colliding beams have equal and opposite dispersion functions $D^{\pm}$ at the interaction point (IP), large enough to make the dispersive horizontal beam size σ_{xs} large compared to the betatron size σ_{xb}, then all pairs of colliding particles will have nearly equal center-of-mass (c.m.) energies, see Figure 1. Specifically, the rms c.m. energy spread σ_W is given by[1]

$$\sigma_W = \frac{W}{2}\sqrt{\frac{\sigma_{\varepsilon_1}^2 + \sigma_{\varepsilon_2}^2}{1 + (D^*)^2(\sigma_{\varepsilon_1}^2 + \sigma_{\varepsilon_2}^2)/(\sigma_{xb_1}^2 + \sigma_{xb_2}^2)}} \tag{1}$$

which is smaller by the factor

$$\lambda = \sqrt{\frac{1 + (D^*)^2(\sigma_{\varepsilon_1}^2 + \sigma_{\varepsilon_2}^2)}{(\sigma_{xb_1}^2 + \sigma_{xb_2}^2)}} \tag{2}$$

---

* Prepared for the U.S. Department of Energy under contract numbers DE-AC03-76SF00098, DE-AC03-76SF00515, and DE-AC03-81-ER40050

than the energy spread obtained with zero dispersion at the IP. Here W is the nominal c.m. energy, D^* is the dispersion at the IP and $\sigma_{\varepsilon_{1,2}}$ are the relative energy spreads of the two beams.

For $\lambda \gg 1$ and for the case of equal horizontal beta-function values at the IP, β_x^*, Eq.(1) can be further reduced to

$$\sigma_W \simeq \frac{W}{2}\sqrt{\frac{\epsilon_{x_1} + \epsilon_{x_2}}{H}} \tag{3}$$

where $H = (D^*)^2/\beta_x^*$ and $\epsilon_{x_{1,2}}$ are the horizontal beam emittances.

One can see from the above equations that in the monochromatic case σ_W depends only on the beam emittances and the value of H, which may be called the 'dispersive invariant' because it is constant in bend-free parts of the lattice. Therefore, to have σ_W small, one needs very small emittances and large H.

The second issue is maximizing the luminosity. The usual tools for increasing the luminosity are reducing the beta-function values at the IP, increasing the bunch crossing frequency and blowing up the transverse beam size at the IP. The latter allows adjustment of the particle density at the IP so that reasonable values of the linear beam-beam tune shifts can be achieved. One usually controls the beam emittances to provide this density adjustment. In the monochromatic case we can adjust the density simply by choosing the proper value of D^*. This second approach has the advantages that the beam size can be kept small all around the machine except in the vicinity of the IP, and that wigglers for increasing the emittance are unnecessary.

For the case that the two beams have equal emittances and equal beta-function values at the IP, equal and opposite dispersions, and equal energy spreads, the luminosity for the monochromatic case is given by

$$L = f\frac{\pi}{4}\left(\frac{W}{e^2}\right)\frac{\sigma_\varepsilon^2}{\beta_z^*}H\xi_x\xi_z \tag{4}$$

where e is the electron charge, f is the bunch crossing frequency, β_z^* is the vertical beta-function at the IP and $\xi_{x,z}$ are the saturated beam-beam tune-shift parameters.

In deriving this formula we also assumed the existence of horizontal and vertical tune-shift limits and flat beams, with the horizontal beam size composed mainly of contributions from synchrotron oscillations:

$$\sigma_x = \sqrt{\epsilon_x \beta_x^* + (D^*)^2 \sigma_\epsilon^2} \simeq |D| \sigma_\epsilon = \sigma_{xs} \qquad (5)$$

One can see from Eq.4 that a large H is useful not only for improving the energy resolution, but also for producing high luminosity.

We now wish to point out some further useful consequences that follow the choice of the monochromatic concept for B-factory design. A salient feature of this design is that all of the most important machine parameters are very closely connected with each other: the choice of any particular parameter immediately follows from the choice of the previous one and vice versa. Thus, every new step lies not along a logical line but along an internal logical circle (see Fig.2), which we will now discuss.

We have shown that monochromatization requires large dispersion and small emittance. To create small emittances one must provide strong focusing in the arc lattice, which leads to a low momentum compaction factor. The latter simultaneously diminishes the accelerating voltage needed for short bunch production and achieves lower synchrotron tunes, which might be very useful for beam-beam effects. The short bunches allow low beta-function values at the IP, but to have these one needs to find a solution to the chromaticity correction problem, which arises from the extremely high contributions to the total chromaticity from the final-focusing quadrupoles when the beta- functions are very small at the IP. Fortunately this solution is at hand, since the large D^* provides a large dispersive invariant H in the final-focus quadrupoles, making this the ideal place to install sextupoles capable of canceling almost all chromatic effects from these quadrupoles.

Having closed the logical circle, we proceed to discuss a concrete example which, in our opinion, could well be appropriate for the Apiary B-factory.

2. Separation scheme and final focus

For an asymmetric B-factory, the separation scheme and final-focus schemes are so closely connected that it is best to discuss them together. We choose 8.5 GeV for the high energy beam (HEB) and 3.3 GeV for the low energy beam (LEB). Magnetic fields are used to produce the separation, which is possible due to the energy asymmetry.

The beams collide head on at the IP, and the separation begins in the first bending magnet B1 placed 30 cm away from the IP (see Fig.3). Then both beam trajectories pass through the common vertically focusing quadrupole QD1 installed in such a way that the LEB passes through an almost constant 3.8 kG magnetic field, while the HEB is shifted towards the lens center and the magnetic field along its trajectory is gradually reduced from 2.3 kG at the entrance to -0.6 kG at the exit. This lens continues the orbit separation and begins the focussing. There is a strong overfocusing effect by QD1 on the LEB and underfocussing on the HEB. The overfocusing makes it difficult to obtain low beta-function values for both beams at the IP without exciting very high values in the interaction region (IR) quadrupoles.

To solve this problem, a double quadrupole lens (see Fig.4) has been proposed as the second lens of the final-focus scheme (see Ref.2). It has two centers with equal but opposite gradients and provides focusing in the horizontal plane for the LEB and in the vertical plane for the HEB (see Fig.5 where the plots of the dispersion and beta-functions of both beams are presented). This lens must provide a high quality magnetic field with a sufficient aperture for both beams. Hence the distance between the two lens centers cannot be small; it should be approximately equal to twenty times the horizontal rms radius of the largest beam plus the value of the gap height between the central poles of the lens, which cannot be less than several centimeters. Therefore, to produce sufficient orbit separation with comparatively small magnetic fields, we need some drift space (magnetic fields must be small for effective synchrotron radiation masking and to avoid large emittance excitation in the separation magnets). Beyond the double-focusing quadrupole the lattices for both beams become totally independent.

The main problem for the optics after the double quadrupole is to bring the dispersion to zero in the simplest way possible. For the HEB this can be done with the help of two additional bending magnets B4, B5 and of three quadrupoles QF3, QD4, QF5. Magnet B4 must be installed immediately after the horizontal focusing quadrupole QF3 (exactly at the $\pi/2$ betatron phase-advance position from the IP), because β_x is large there. This decreases the H most effectively due to the third term in the exact expression for H:

$$H = \gamma_x D^2 + 2\alpha_x DD' + \beta_x D'^2 \qquad (6)$$

where $\alpha_x, \beta_x, \gamma_x$ are Twiss parameters.

The magnet B5 is the last to influence the dispersion. It must have a negative magnetic field in order to cancel both the dispersion and the total bending angle of the trajectory. The last two quadrupoles QD6 and QF7 create waists of β_x and β_z.

Now one can see the important roles played by the bending magnets, both in the orbit separation and in the beam optics. The possibility of combining these roles is the direct consequence of the asymmetric-energy approach to collider design.

The optics of the LEB is a little more complicated due to the negative sign of the dispersion at the IP. To apply a similar approach here as that used for the HEB, we need first to reverse the sign of the dispersion. Therefore, magnet B3 in the LEB lattice is placed at the $3\pi/2$ betatron phase-advance position from the IP. In order to have such a phase advance we produce a waist of β_x just in front of this magnet. Two additional quadrupoles are used for this purpose, as well as for control of β_z and for the production of a large β_x in the region of B3 (which is needed for the same reasons as in the HEB lattice at the B4 location).

The remainder of the LEB lattice is very similar to the HEB lattice. Finally, at a point about 40 m from the IP, we have zero dispersion D, waists in β_x and β_z, and the HEB and LEB beam lines are parallel, separated horizontally by 2.6 m.

To bring the beam lines to suitable positions in the arc tunnel, directly above and below each other, we must shift the LEB in both the horizontal and vertical directions. The easiest way to do this is to use two bending magnets with opposite polarity, with six quadrupoles between them. These quadrupoles must have the same median plane as the bending magnets and produce unity transport matrices between the bending magnets in both transverse directions, in order to cancel the dispersion produced by the bends. This array of bend and quadrupole magnets lie in a tilted plane.

Finally, the IR insertion is completed with a set of quadrupoles that match the orbit functions to the proper values of the arc cells.

3. Beam-beam effects

A detailed discussion of the beam-beam effects in the monochromatic case is beyond the scope of this article, but we cannot totally avoid this discussion due to a rather controversial aspect of the monochromatic design, namely the finite dispersion at the IP, which is well known to be a source of synchro-betatron resonances. Are these resonances strong enough to produce significant effects on beam blowup and lifetime? The answer is not very clear at present, particularly in the case of large dispersion values. These questions are currently under intensive study in Novosibirsk at INP. Preliminary results of these studies presented at this workshop (see Ref.3) show quite good behavior of the vertical beam size when the dispersion is large. One plot from this work is reproduced in Fig.6, which shows the vertical beamsize blowup at a fixed synchrotron tune as a function of the betatron tunes. Good regions, where the beam blowup is less than 10%, are shown here for three particular cases. Case one has comparatively small dispersion, with the parameter λ (ratio of the synchrotron size at the IP to the betatron size) having the relativlely small value of 2.5. In the second case the dispersion was increased and λ became equal to 5. In the third case λ was equal to 7.5. The different shadings correspond to the three λ values. The areas for larger λ include those for smaller λ. One observes that larger good-region areas correspond to larger λ values. The reason for this is the diminishing strength of all coupling and pure horizontal synchrobetatron resonances as well as betatron resonances one with increasing values of λ. There is a tentative phenomenological explanation which agrees with the analytical calculations of the resonance widths. In the monochromatic regime the amplitudes of the horizontal betatron oscillations of the particles at the IP are much smaller than the total horizontal beam size and the positions of the particles at the IP from one turn to another turn do not change very much. Therefore, the stability of their motion might be less sensitive to the non-linearity of the space charge field of the counterrotating bunch.

4. Example of Possible Parameters for a B-factory based on PEP

First, we present the list of B-factory parameters in the conservative case with values of the nominal beam-beam vertical and horizontal tune shifts equal to 0.03 and 0.01, respectively, and a bunch spacing of 3.78 m. (See Table I). Then we show a further scenario for luminosity improvements.

Among these parameters, the longitudinal threshold impedance for the LEB seems to be the most troublesome. However, we believe that its value is acceptable for two reasons. One is the recent impedance measurements in LEP, which gave a very small impedance value for short bunches [Ref.4]. The other is the center-of-mass energy spread insensitivity to the beam energy spread in the monochromatic case. This allows us, in principle, to exceed the longitudinal threshold impedance. By increasing the accelerating voltage the bunchlength can be kept constant.

After having realized the luminosity given in Table I, further improvements might be made as follows. The coupling must be diminished from 22.7 % to 8.2 % to increase the beam density to a value corresponding with an increase of the vertical beam-beam tune shift to 0.05. This step will allow a luminosity 8×10^{33} cm^{-2} s^{-1} to be achieved without any changes in beam currents or number of particles per bunch. Then, diminishing the bunch spacing to 2.52 m while increasing the HEB current to 1.3 A and the LEB current to 3.3 A, without changing the number of particles per bunch, gives an additional factor 1.5 in the luminosity gain. This will allow us finally to exceed the 10^{34} cm^{-2}s^{-1} luminosity barrier.

5. Summary

We believe that the study presented above confirms our initial attraction to the monochromatic approach in asymmetric B-factory design. The collider promises to be a low current, low emittance machine, insensitive to beam energy-spread blowup due to the microwave instability. It also should possess a large dynamic aperture due to the small transverse dimensions of the beam and the possibility of using sextupoles near the final-focus quadrupoles. Finally, it should provide an energy resolution of 0.9 MeV at $\Upsilon(4S)$ and 0.7 MeV at $\Upsilon(1S)$.

At present, the dispersion at the IP does not look as dangerous as it seemed to be in the past. Moreover, if further studies of beam-beam effects should give additional evidence in favor of using large dispersion values in head-on collisions, it also might suggest the possiblility of having collisions with a small crossing angle $\sim \sigma_{xb}/\sigma_{xs}$. The latter could so radically reduce the complexity of the synchrotron radiation masking problem as to more than balance any disadvantage of dispersion.

Acknowledgements: One of the authors, A. Zholents, thanks G. Travish and J. Bengtsson for their help in preparing this article. He also thanks all the staff of the LBL Exploratory Studies Group for their hospitality and the very friendly atmosphere he enjoyed during his stay in Berkeley.

References

[1] I.Ya. Protopopov, A.N. Skrinsky, A.A. Zholents, Monochromatization of Colliding Beam Energy in Storage Rings, Proc. of the VI All-union Conf. on Charged Particles Acc., Dubna, 1979, p.17. Preprint INP 79-6, Novosibirsk 1979.

[2] A.N. Dubrovin, A.M. Vlasov, A.A. Zholents, Interaction Region of 4x7 GeV Asymmetric B-Factory, Preprint INP 89-97, Novosibirsk 1989.

[3] A.L. Gerasimov, D.N. Shatilov, A.A. Zholents, Study of Beam-Beam Effects in the Presence of Horizontal Dispersion at Interaction Point, Report on the Asymmetric B-Factory Workshop, Berkeley, 1990.

[4] B. Brandt, I. Fieguth, et al., Further Measurements of the Impedance of LEP, LEP Commissioning Note 21, December 1989.

Table I: Monochromatic APIARY Parameters

	LEB	HEB
Energy, E[GeV]	3.3	8.5
Circumference, C[m]	2200	2200
Number of bunches, k_B	582	582
Particles per bunch, $N_b[10^{11}]$	1.76	0.68
Total current, I[A]	2.23	0.86
Longitudinal threshold impedance, $|Z/n|$ [Ω]	0.046	0.15
Emittances [nm-rad]:		
horizontal, ϵ_x	6.6	6.6
vertical, ϵ_z	1.5	1.5
Bunch length, σ_ℓ [mm]	7.5	7.5
Energy spread, σ_ϵ [10^{-3}]	1	1
Momentum compaction, α	0.001	0.0005
Synchrotron tune, ν_s	0.047	0.023
RF parameters:		
frequency, f_{rf} [MHz]	476	476
voltage, V_{rf} [MV]	13	17
Beta function values at IP [cm]:		
horizontal, β_x	35	35
vertical, β_z	1	1
Horizontal dispersion function at IP, D [cm]	-40	40
Beam sizes at IP [mm]:		
horizontal, σ_x	0.4	0.4
vertical, σ_z	0.004	0.004
Nominal beam-beam tune shifts:		
horizontal, ξ_x	0.01	0.01
vertical, ξ_z	0.03	0.03
Center-of-mass energy spread, σ_W[MeV]	0.9	
Luminosity, L [cm^{-2}s^{-1}]	4.8×10^{33}	

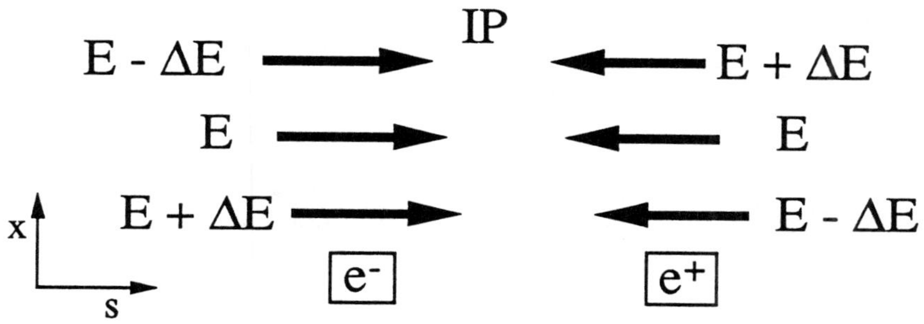

$$
\begin{array}{ccc}
E - \Delta E \longrightarrow & \text{IP} & \longleftarrow E + \Delta E \\
E \longrightarrow & & \longleftarrow E \\
E + \Delta E \longrightarrow & & \longleftarrow E - \Delta E \\
\boxed{e^-} & & \boxed{e^+}
\end{array}
$$

Figure 1. Monochromatic c.m. collisions produced by large dispersion values at the IP.

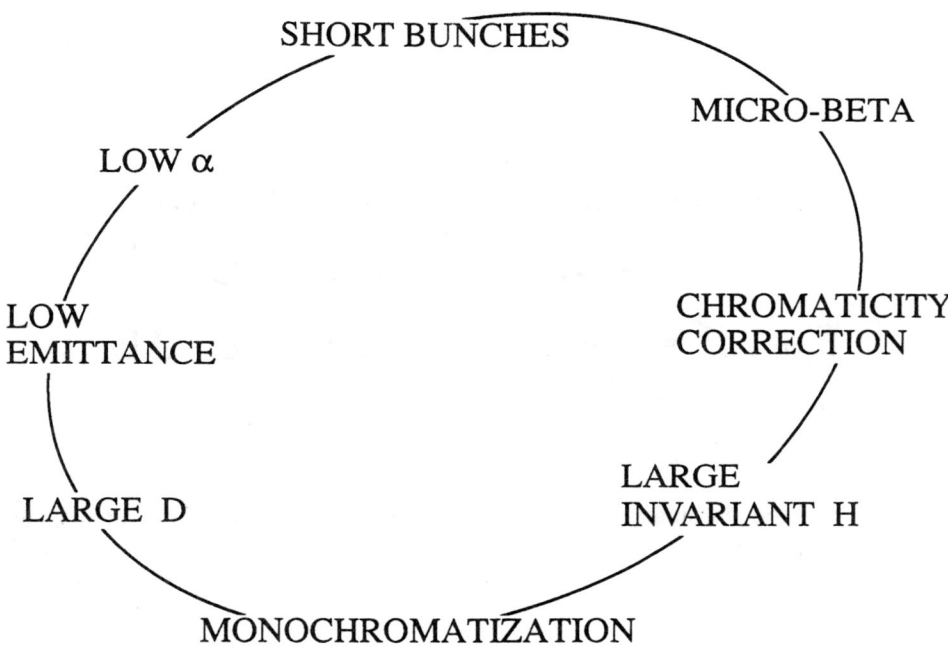

Figure 2. The logical circle of monochromatic e⁺e⁻ collider design.

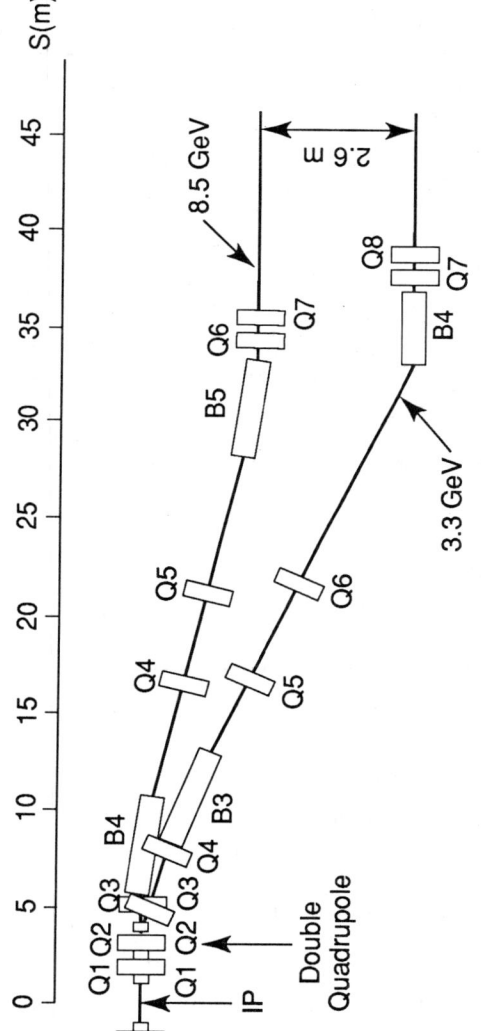

Figure 3. Schematic plan view of interaction region. The layout reflects about the IP.

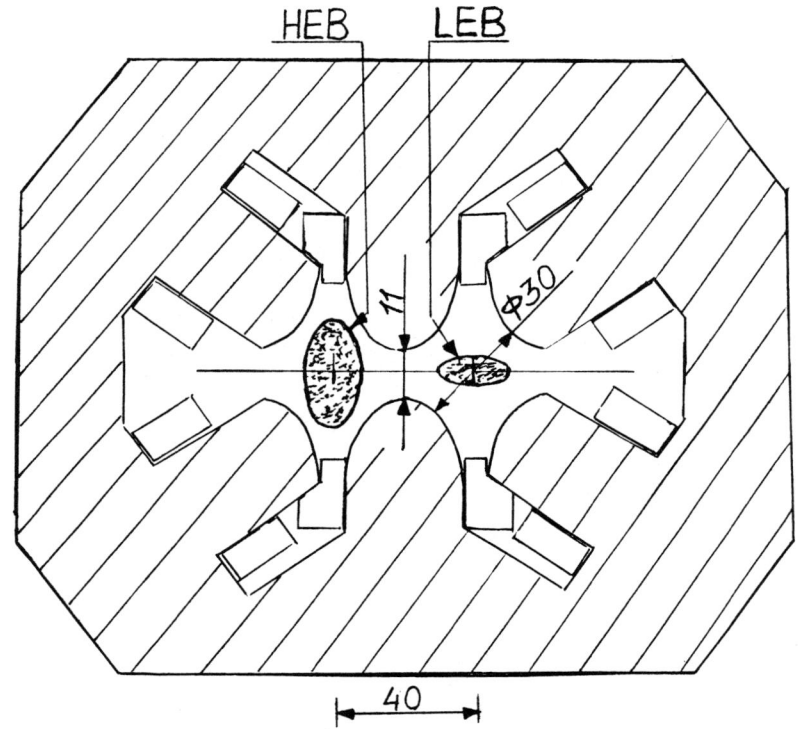

APIARY double quadrupole

length   50 cm
gradient 3.25 kG/cm
Amp turn 3 µA
(all dimensions in mm)

Figure 4. Cross Section of Q2, the double quadrupole lens. The gradients on the two beams are opposite.

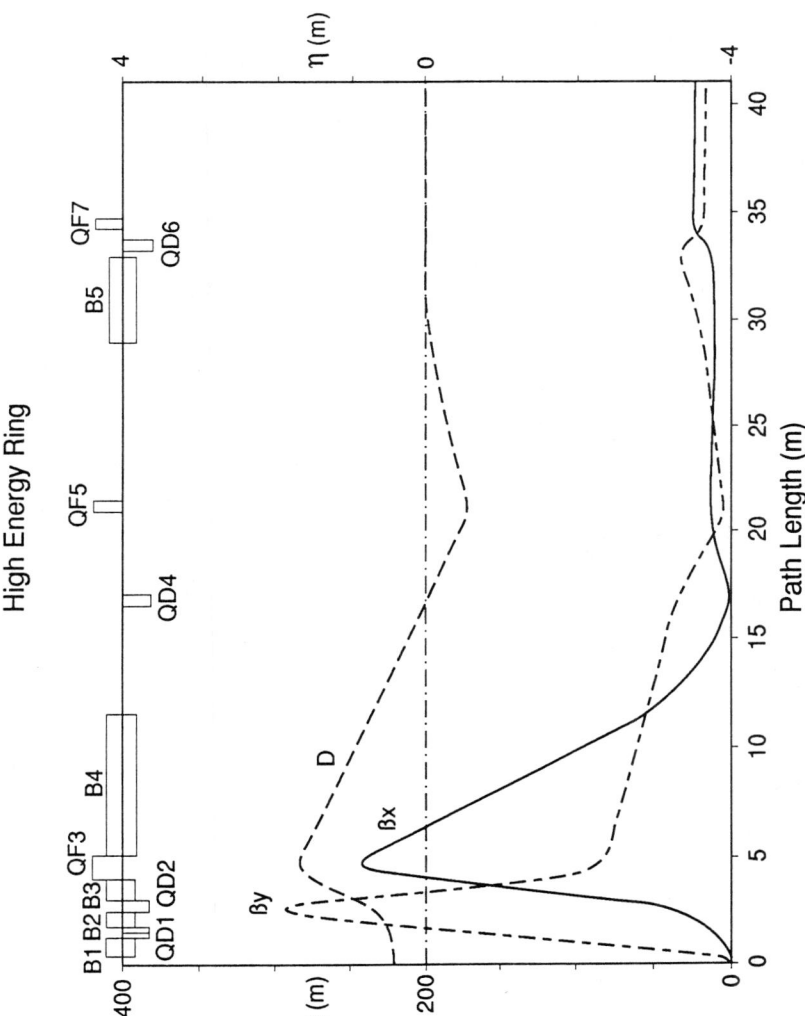

Figure 5a. Lattice and Orbit functions of one half of the interaction region for the high-energy ring.

362   Monochromatic Design

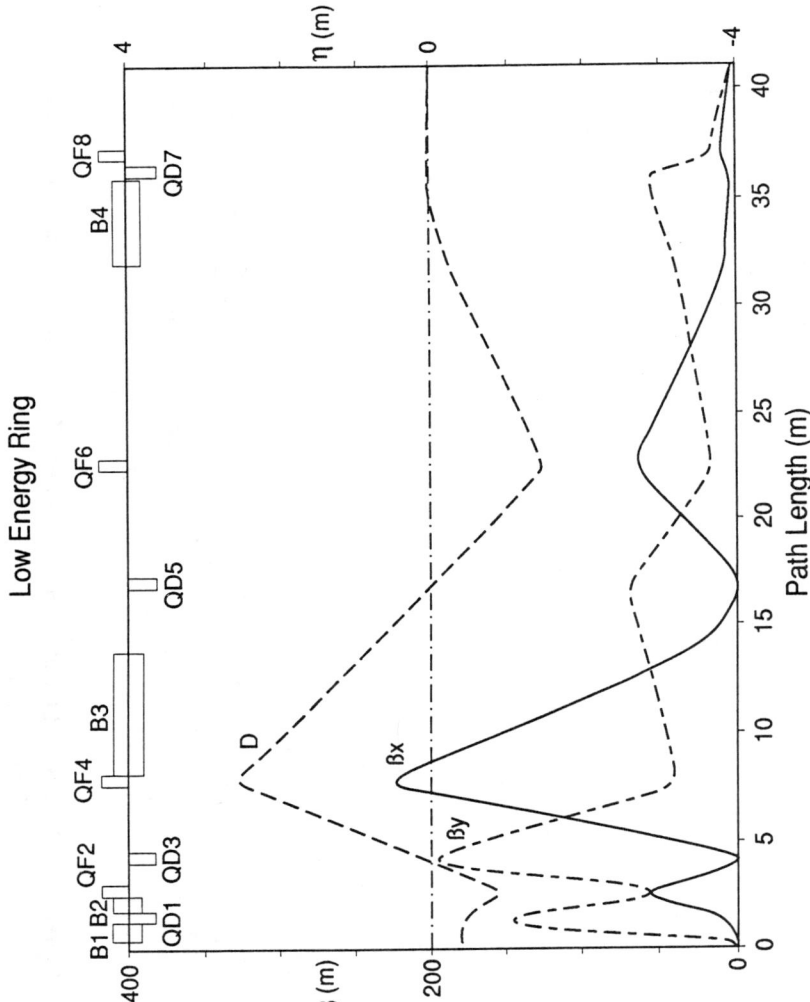

Figure 5b.   Lattice and Orbit functions of one half of the interaction region for the low-energy ring.

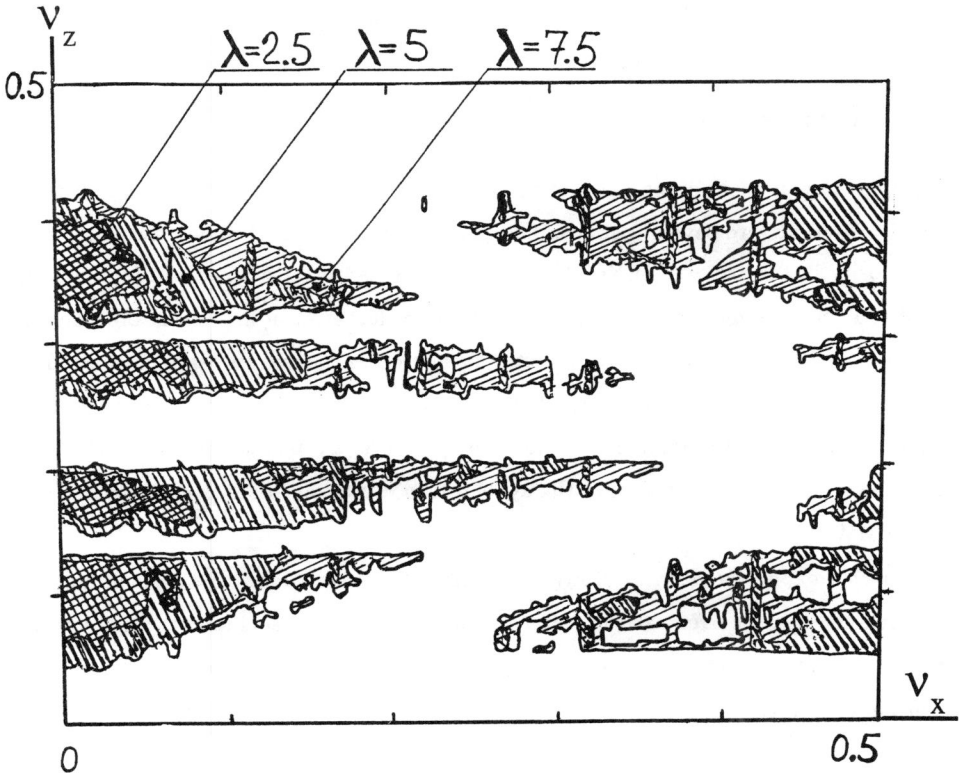

Figure 6. Vertical beam size blowup dependence on $\lambda = \sigma_{xs}/\sigma_{xb}$. Shaded regions have blowup less than 10%.

Isochronous Storage Rings and High Luminosity Electron-Positron Colliders

Claudio Pellegrini
and
David Robin

University of California at Los Angeles

1 Introduction

The interest in studying CP and possibly CPT violations in B and ϕ meson decay has lead recently to several proposals for the construction of B and ϕ factories, with luminosity in the range of $10^{33} cm^{-2} s^{-1}$ to $10^{34} cm^{-2} s^{-1}$. With a conventional storage ring collider, this high luminosity, 10 to 100 times larger than maximum obtained up to now, is obtained by increasing the stored electron and positron beam currents from the 10-100 mA range to the several Ampere level. This very large beam current raises questions of collective instabilities, and vacuum and RF system design. In addition the RF power needed to compensate the synchrotron radiation losses is of the order of 5 to 10 MW.

In this paper we want to study the possibility of obtaining this high luminosity keeping the beam current at the $100 mA$ level, by using an "isochronous ring" and reducing the bunch length and the beta function at the interaction point. The luminosity can be written as

$$\mathcal{L} = \frac{\mathcal{P}_1 \mathcal{D}_1}{4\pi r_e mc^2 \sigma_{L2}} \qquad (or\ with\ 2 \leftrightarrow 1) \qquad (1)$$

where $\mathcal{P}_1$ and $\mathcal{D}_1$ are the power and the disruption parameter of beam 1, and σ_{L2} is the bunch length of beam 2. The luminosity can alternatively be written as

$$\mathcal{L} = \frac{fN^2}{4\pi\sigma_T^2} = \frac{fN^2}{4\pi\varepsilon\beta^*} \qquad (2)$$

where the assumption has been made that the beams are symmetric and round. N is the number of particles in each bunch, f is the frequency of collisions, σ_T is the width of the beam, ε is the emitance of the beam and β^* is the beta function at the interaction region. There is an effective limit to how small one can make β^* so as to increase $\mathcal{L}$.

$$\beta^*_{min} \approx \sigma_L \qquad (3)$$

Making $\beta^* < \beta^*_{min}$ would not result in an increased Luminosity. If one wants to increase $\mathcal{L}$ without increasing the power, one would have to either increase $\mathcal{D}$ and/or decrease σ_L.

In this paper, we propose to use an "isochronous storage ring", having the particle revolution frequency independent of energy, to achieve a high luminosity with a small beam current. Isochronous rings have been discussed before as possible damping rings for linear colliders and for Free Electron Lasers and some of their properties have already been discussed[1,2]. In this paper, we will make a more detailed study of the beam dynamics in such a ring. The advantages of an isochronous ring are: the possibility of reducing the bunch length from the present centimeter range to the millimeter range; the elimination of one class of resonances, the synchro-betatron resonances, limiting the beam density at the collision point and the interaction region geometry. By reducing σ_L by 10 and increasing $\mathcal{D}$ by a factor of 2, we can reduce the current to $\sim 100 mA$ for a luminosity of $10^{34} cm^{-2} s^{-1}$.

We will first discuss the basic concept and some of the main properties of an isochronous ring. This discussion will define the conditions for a stable operation of an isochronous ring. We will then consider the possibility of using an isochronous ring as a collider, along with its advantages and disadvantages. As an example, we will then give the main parameters of a B-factory based on this idea.

2 Isochronous Storage Rings

This discussion of isochronous storage rings will be preceded by a short summary of the general equations of motion for the longitudinal degree of freedom of a storage ring. Then the differences between conventional rings and isochronous rings will be more clearly illustrated. The main difference between a conventional and and isochronous storage ring lies in the longitudinal beam dynamics; the transverse beam dynamics are not influenced, except for the synchro-betatron coupling effects, and will not be discussed here. In particular the synchrotron oscillation frequency is assumed to be zero or very small. Defining what we mean by very small is one of the key questions to be discussed here.

Using as variables the relative energy deviation, $\delta = (E - E_s)/E_s$, and the angular distance $\psi = \phi - \phi_s$, from a reference particle of energy E_s and angle ϕ_s, we can write the equations of motion as[3]

$$\psi' = \alpha(\delta)\delta \tag{4}$$
$$\delta' = -\kappa(\psi + \phi_s) - \frac{U_0}{E_s}(1 + J_\epsilon \delta) + \text{fluctuations} \tag{5}$$

where U_0 is the energy radiated per turn from the reference particle, J_ϵ is the radiation damping partition number and with fluctuations we indicate the term arising from fluctuations in the emission of synchrotron radiation. The superscript $'$ implies a derivative with respect to $\omega_0 t$, where t is time and ω_0 is the revolution frequency around the ring of the reference particle. For a storage ring with an energy dispersion function, η, and a bending radius, ρ, we have[3]

$$J_\epsilon = 2 + \frac{<(1-2n)\frac{\eta}{\rho^3}>}{<\frac{1}{\rho^2}>} \tag{6}$$

n being the bending magnet field index.

The quantity $\kappa(\psi + \phi_s)$ represents the combined effect of the RF field, and of the field due to the interaction with the ring's longitudinal coupling impedance, $\kappa = \kappa_{RF} + \kappa_C$. These effects have both been linearized. This condition can be realized for the RF term, while the coupling impedance terms will also contain nonlinear terms, which we initially neglect. We define also the synchrotron phase, ϕ_s, so that $\kappa\phi_s = \frac{U_0}{E_s}$. For an isochronous ring we want to make the quantity κ small, to reduce the coupling between energy and angle change.

The momentum compaction term, $\alpha(\delta)$, is dependent upon two quantities; the difference in velocity between the test particle and the ideal particle, and the difference in path length between the test particle and the ideal particle as they travel around the ring. The faster the test particle moves, the farther it moves tending to decrease ψ; however, the longer the path length, the longer it will take to move around the ring tending to increase ψ. The momentum compaction term α incorporates all this information, and is defined as

$$\alpha = \frac{(\frac{\Delta\psi}{2\pi})}{(\frac{\Delta E}{E})} \qquad (7)$$

where $\Delta\psi$ is the change in ψ per turn for given ΔE. In the simplest approximation

$$\alpha \approx <\frac{\eta}{\rho}> -\frac{1}{\gamma^2} = \alpha_1 \qquad (8)$$

where γ is the ratio of the particles total energy to that of its rest energy[3].

For highly relativistic particles, the first term is dominant; it is usually positive but it can be made nearly zero or negative by having in the ring regions of inverted bending, $\rho < 0$, or of negative dispersion, $\eta < 0$. The full expression for α that we should use in (4), is

$$\alpha = \frac{1}{L\delta} \oint ds [\sqrt{(1+\frac{x}{\rho_s})^2 + (x')^2 + (z')^2} - 1] - \frac{1}{\gamma^2} \qquad (9)$$

To simplify this initial discussion of an "isochronous ring", we will expand α to first order in delta and neglect the contributions from the the betatron oscillations:

$$\alpha = \alpha_1 + \alpha_2 \delta \qquad (10)$$

with α_1 given by equation (8). The α_2 term is given by

$$\alpha_2 = \frac{1}{L} \oint ds \frac{(\eta')^2}{2} \tag{11}$$

The value of α_1 can be adjusted to zero or negative but, as one can see from (11), α_2 is always positive. Therefore one can never completely eliminate the effect of α_2. As a first step in understanding the behavior of an isochronous ring, we will study the system of equation (4) and (5) with α given by (10).

The simplest case is that when $\alpha_1 = 0$, and in addition we neglect damping. We are now reduced to the equation:

$$\delta'' + 2\pi\omega_0\kappa\alpha_2\delta^2 = 0 \tag{12}$$

This equation leads to unstable solutions, described in phase space by the curve, $(\delta')^2 + (\frac{4\pi}{3})\omega_0\kappa\alpha_2\delta^3 =$ constant , and shown in Figure 1 for a particular case[1].

There are two ways to make the longitudinal motion stable. The first is to make $\alpha_1 \neq 0$ and dominant over α_2; this introduces an elastic-like focusing force, and thus provides a stable oscillation region near the origin ($\delta = \psi = 0$). The area of the stable region depends on the relative magnitude of α_1 and α_2, and also on κ, and can be usually made large enough for a convenient accelerator operation. All existing proton and electron synchrotrons and storage rings operate in this mode. The second way of providing stability, when α_1 is made small, is to increase the rate at which the electrons lose energy through the emission of synchrotron light; in other words, increase the damping of the system. If $\alpha_1 = 0$ there can be no absolute stability in the system no matter how large the damping is. However, damping will slow down the rate of the instability growth. Indirectly, damping does provide stability. Without damping one would require a larger value of α_1 for stability then with damping. The more damping, the smaller α_1 needs to be. In the limit of severe damping, one could get by with $\alpha_1 = 0$ for some time. A more detailed discussion on the limits of stability will now follow.

3 Limits of Stability

This discussion of stability limits for α will involve the second order equations of motion including damping with α given by (10). We will continue now to neglect the fluctuation term. The question which we try to answer is: given a specific value for α_2, what is the smallest value of α_1 necessary to make the equations of motion stable? The answer is that this value of α_1 is dependent upon both the initial value of ψ and δ and also on the amount of damping.

Initially we chose the following values for ψ_0, δ_0 and κT_0: $\psi_0 = 0.0001$, $\delta_0 = 0.001$ and $\kappa T_0 = -0.01$, where T_0 is the period of oscillation. The choice of ψ_0 corresponds to an initial displacement of about $1 cm$ for a ring with a circumference of $760 m$, as given in Table 1. The choice of δ_0 is made due to a restriction on the allowable energy spread which is required by B meson physics. The choice of κT_0 is consistent with possible parameters for an isochronous ring.

In the case of zero damping we find that for each value of α_2, there exists a certain value of α_1 which just makes the motion stable. Plotting δ versus ψ one finds that a closed curve is mapped out over time as can be seen in Figure 2. This curve is typical of all limiting cases. All choices of ψ_0 and δ_0 lying inside the curve will result in stable states. All choices of ψ_0 and δ_0 lying outside the curve will result in unstable states. A choice of α_1 greater than the limiting value will also produce a closed curve. This curve will lie inside this separatrix and will be smoother than that for the limiting case. But for smaller values of α_1, the curve will be open and the trajectory will be unstable, similar to the curve in Figure 1.

It is interesting to see what happens when ψ and δ are plotted against time using the case of zero damping for illustrative purposes. In Figure 3 there are four plots. The first is a plot made for a choice of α_1 which lies far within the stability region. It appears regular and sinusoidal for both ψ and δ. This is a direct result of the fact that the first order term, α_1, completely dominates the second order term, α_2, and the motion is therefore almost that of a simple harmonic oscillator. The second plot is for a smaller choice of α_1 which is also within the stability limits but is closer to the limiting value. The period of oscillation for ψ and δ is increasing but the shape still basically resembles that of sine waves. The third plot is for a choice of α_1 which is very close to the stability limit. In this plot the motion is oscillatory; however, the period has increased dramatically and the curve is strongly distorted from a sinusoid. During portions of the oscillation, ψ remains close to zero and δ is nearly constant. If α_1 was chosen to lie right on the limit of stability, the period of motion would increase to infinity. Finally there is a plot for a value of α_1 which lies just outside the limit of stability. The motion appears to be stable for a while but at some time both ψ and δ diverge. The farther outside the limit of stability, the shorter time it will take for ψ and δ to diverge.

Now the case with damping is considered. Again we find that for every value of α_2 there exists a specific choice of α_1 such that all larger values are stable and all smaller values are unstable. Plotting δ versus ψ as we did before for this limiting case, we do not find a closed separatrix but instead a spiral which is sharp at the bottom and slowly converges to the point (0,0) as seen in Figure 4. In the case in which α_1 is smaller than this critical value, the motion is unbounded as in the case with zero damping. As the value of α_1 increases, moving the system farther into the stability region, the curve becomes smoother and rotates quicker. And in a plot of ψ and δ versus time, one finds that the motion is that of a damped oscillator which is converging on the reference particle's coordinates.

Let us now take a further look at the limits of stability. We have found that on the boundary between stability and instability there exists one value of α_1 for each value of α_2. For three different cases of damping we plot these values of α_2 against α_1 as seen in Figure 5. One finds that α_1 increases steadily with α_2. Also one finds that as the damping increases the values of α_1 decrease, and the curve is lowered. So for stable operation of a storage ring, α_1 must be chosen to lie above the curves.

Finally, to get an indication of how these requirements of α_1 depend upon the initial values of ψ and δ, we choose the values $\psi_0 = 0.0003$ and $\delta_0 = 0.003$. Again making a plot of α_1 versus α_2 we find similar curves to that of the first choice of ψ_0 and δ_0 as can be seen in Figure 6. But the corresponding values of α_1 are almost a factor of three bigger. So the "tighter" the initial conditions for the bunch, the smaller one can make α_1.

4 Future Investigations

We have shown that one can produce stability in a ring with values of α_1 which are many orders of magnitude smaller than α_2 and much smaller than used in existing rings. Since the size of the beam bunch is related to the momentum compaction, one should be able to decrease the bunch length by decreasing the momentum compaction factor. In fact, one can make a estimate of how much a decrease in α_1 would produce a subsequent decrease in the beam length, σ_L. In the case where $\alpha_2 = 0$, σ_L is proportional to $\sqrt{\alpha_1}$. In the case where $\alpha_2 \neq 0$, one can still use this relationship to obtain an estimate of the bunch length, the reason being that for stable motion, the α_1 term $(\alpha_1 \delta)$ is larger than the α_2 term $(\alpha_2 \delta^2)$. A simulation was done including the effect of fluctuations in the photon emission spectrum. The results give us some confidence in this statement. For a "normal" ring α_1 is $10^{-3} \sim 10^{-4}$ giving $\sigma_L \sim 1cm$. In our case, we can have $\alpha_1 \sim 10^{-6}$ which means a bunch length, $\sigma_L \sim 1mm$. Also our first look at the effects of other many particle instabilities seems to indicate threshold limits for the small momentum rings are no worse then that of conventional rings.

A small σ_L allows us to use a small β^*, increasing the beam at the collision point and the luminosity. It might therefore be possible to build high luminosity colliders using smaller beam currents. But before we can make definitive statements, more work has to be done. A more general form of α will have to be used, including the effect of betatron oscillations and higher order terms in δ. We also will have to take into account other many particle collective effects such as the longitudinal microwave instability and the head-tail effect. Finally a "real" machine needs to be designed to see if these parameters for α_1 and α_2 are possible. We expect to report in the near future the results of this additional work.

5 Example of an "Isochronous Ring B-Factory"

As was mentioned earlier, one of the possible uses of an isochronous ring would be for a B-Factory. A tentative list of parameters for a B-Factory using an isochronous ring is given in Table 1; we also give a parameter list for a conventional ring, whose values were taken from an example given by R. H. Siemann[4].

The important difference between the two colliders lies in the bunch length, σ_L, which differs by a factor of 7.5. Our choice of α_1 was made by choosing an appropriate value from Figure 5 which was consistent with our other isochronous collider parameters. As a result of this smaller bunch length, the beta function of the isochronous ring is smaller, which in turn provides a bunch width considerable smaller than that of the conventional ring. The total current can then be reduced.

Our justification of this difference lies in the fact that the ratio of the two values of the momentum compaction is about 650. By the relationship given in the previous section, the bunch length could be reduced by a factor on the order of 25. Thus a reduction in σ_L of 7.5 would be very reasonable.

References

[1] S. Chattopadhyay et al., Proceedings of the ICFA Workshop on Low Emittance $e^- - e^+$ Beams, Brookhaven National Laboratory, March 20 - 25, 1987

[2] D. A. G. Deacon, "Basic Theory of the Isochronous Storage Ring Laser", *Physics Reports* Vol. 76 No. 5, pp. 349-391 (Oct. 1981)

[3] M. Sands, SLAC-121 (1979)

[4] R. H. Siemann,"Very High Luminosity Electron Colliders", Lecture given at the US Particle Accelerator School at Brookhaven National Laboratory, July 24 - Aug. 4, 1989

[5] S.Y. Lee and J.M. Wang,"Microwave Instability Across the Transition Energy", IEEE Transactions on Nuclear Science, Vol. NS-32, No. 5, pp 2323 - 2325, October 1985

[6] S.Y. Lee and J. Wei, "Nonlinear Synchrotron Motion Near Transition Energy in RHIC" European Particle Accelerator Conference, Vol. 1, pp. 764 - 765, 1989

[7] C. Pellegrini, "Single Beam Coherent Instabilities in Circular Accelerators and Storage Rings", *Physics of High Energy Particle Accelerators* (Fermilab Summer School, 1981), AIP Conference Proceedings No.87, pp 77 - 147, 1982

List of Figures

Figure 1: An example of unstable motion for the case of $\alpha_1 = 0$ and zero damping.

Figure 2: An example which lies on the stability limit for the case of zero damping.

Figure 3: $\frac{\delta}{\delta_0}$ and $\frac{\psi}{\psi_0}$ versus number of revolutions around the ring: Three examples within the stability region and one outside for the case of zero damping.

Figure 4: An example which lies on the limit of stability for the case of finite damping.

Figure 5: Three curves which define the stability limits for α_1 shown for three different values of damping with $\delta_0 = 0.001$ and $\phi_0 = 0.0001$.

Figure 6: Three curves which define the stability limits for α_1 shown for three different values of damping with $\delta_0 = 0.003$ and $\phi_0 = 0.0003$.

Figure 7: Two curves which define the threshold number of particles which can be circulated in the ring as a function of inverse damping time. Ring paramaters used in the calculation are given in Table 1.

Table 1

B-factory Parameters	"Isochronous"	Conventional
Circumference [m]	760	628
Energy, E [GeV]	5	5
Luminosity, $\mathcal{L}$ [$cm^{-2}s^{-1}$]	10^{34}	10^{34}
Disruption, $\mathcal{D}$	1	0.47
Tune Shift, ξ	0.1	0.05
Bunch Length, σ_L [m]	0.001	0.0075
Current, I [mA]	60	1800
Number of Bunches	20	42
Current/Bunch [mA]	3	42
Electrons/Bunch	4.5×10^{10}	5.6×10^{11}
Energy Loss/Revolution, U_0 [MeV]	3.5	1
Damping Period, τ_D [ms]	3.6	10.4
Synchrotron Radation Power [MW]	0.2	1.8
Transverse Emittance, ε_T [$m-rad$]	10^{-8}	5.1×10^{-7}
β^* [mm]	1.3	10
Bunch Width, σ^* [m]	3.6×10^{-6}	7.1×10^{-5}
Momentum Compaction (first order), α_1	3×10^{-5}	1.86×10^{-2}

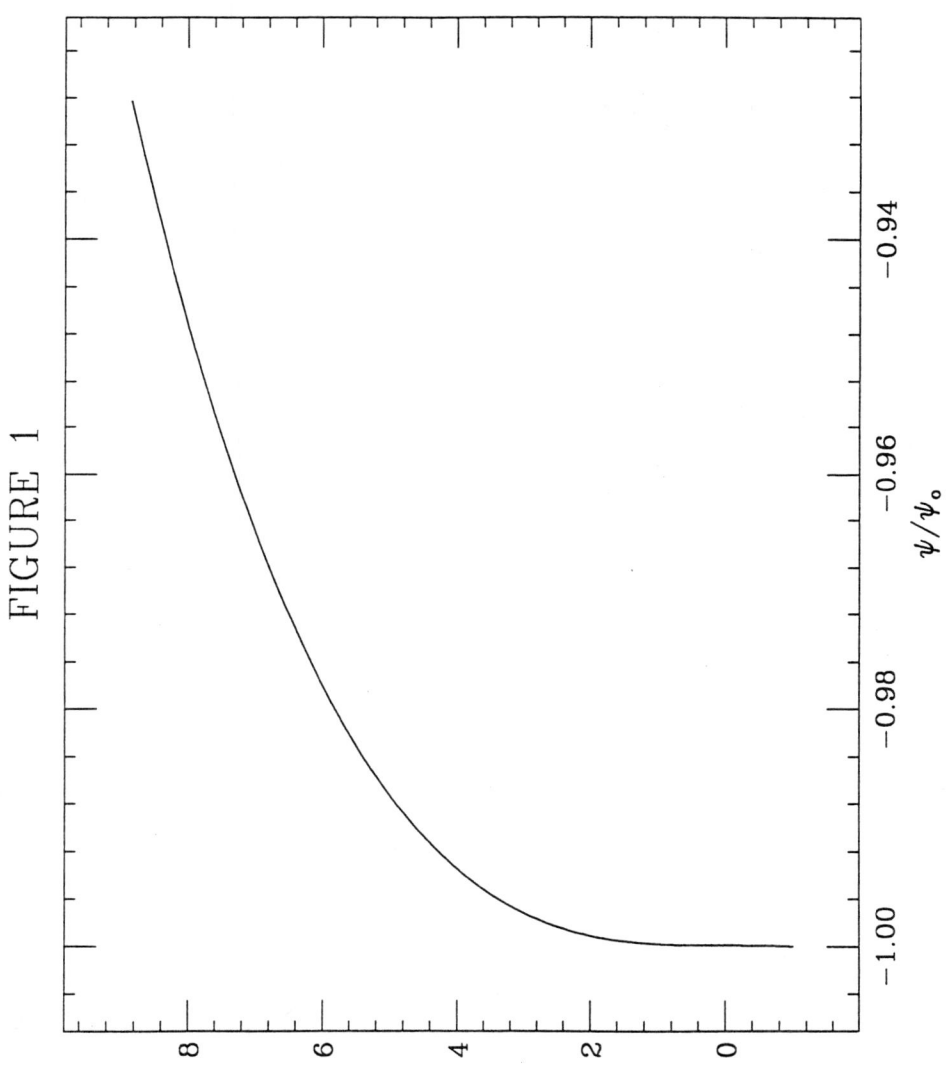

FIGURE 1

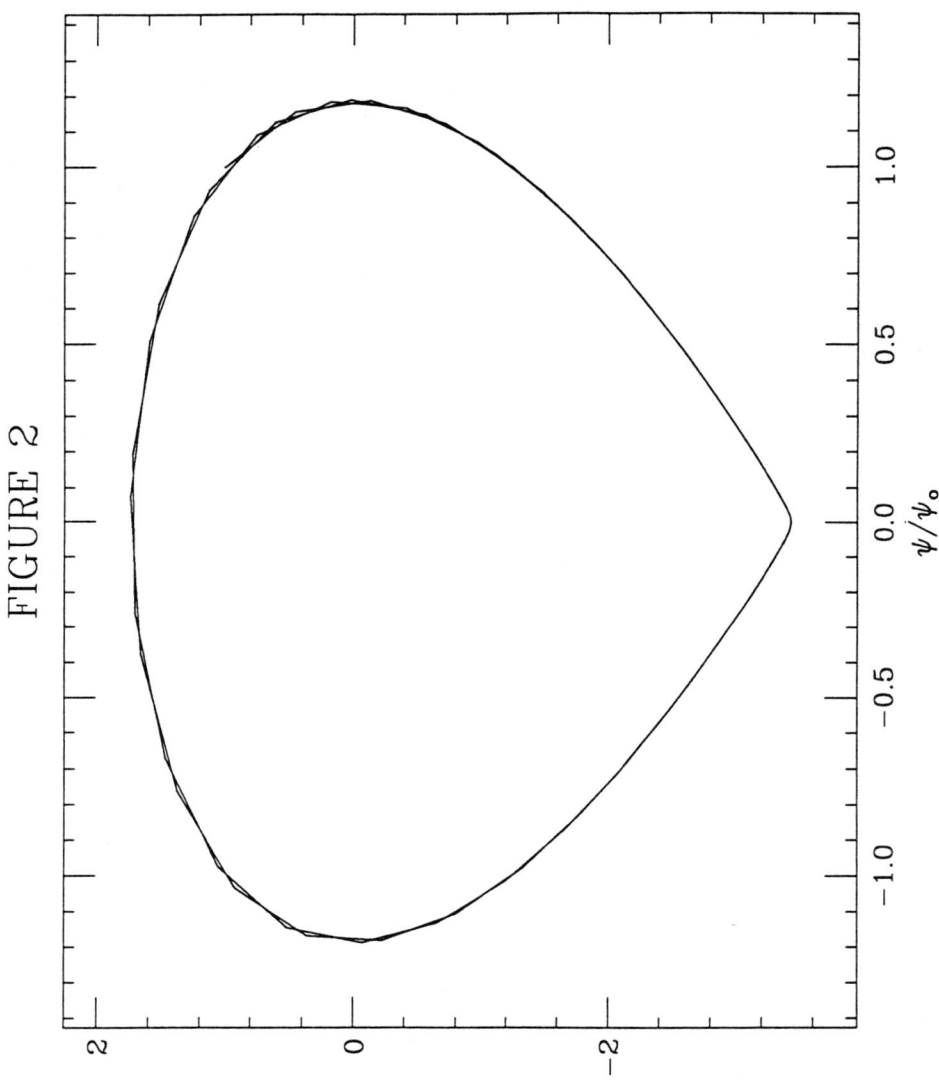

FIGURE 2

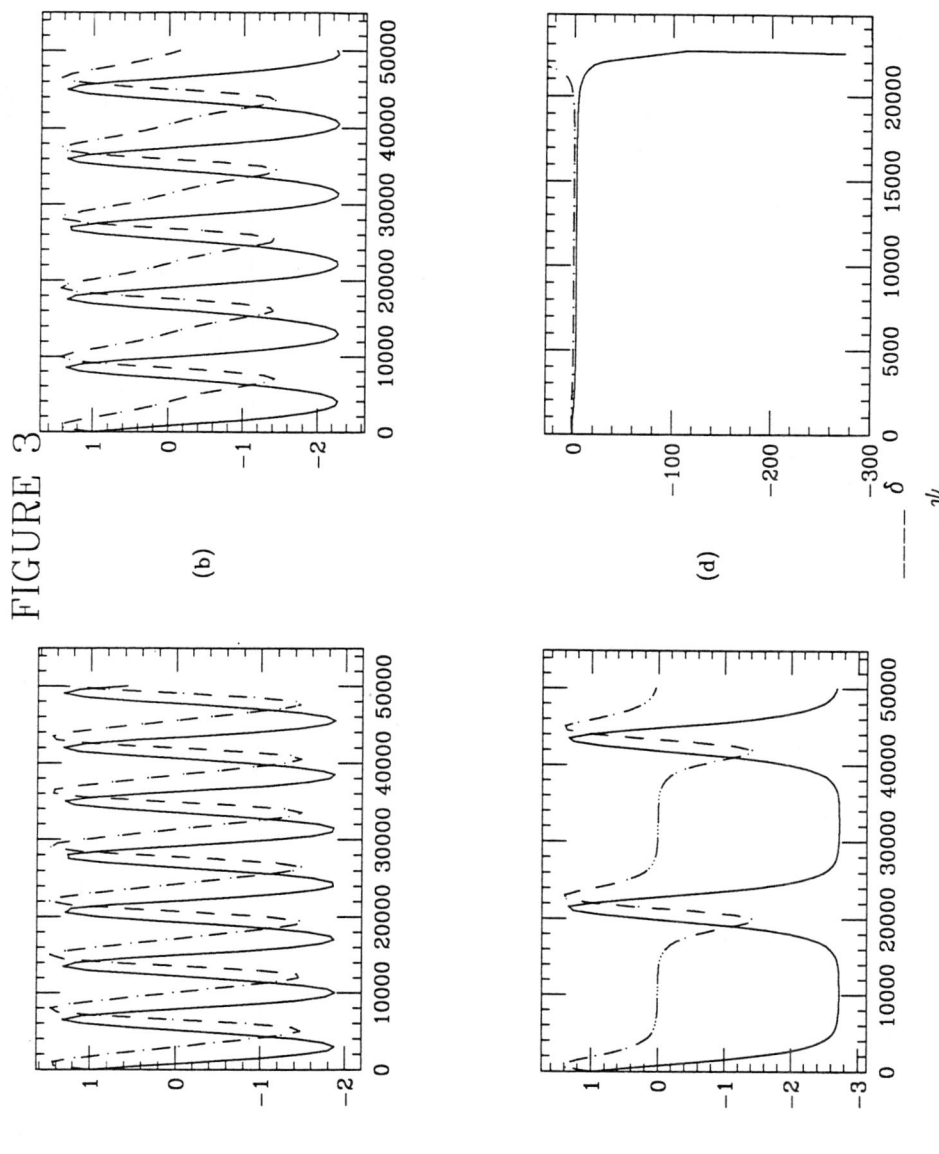

FIGURE 3

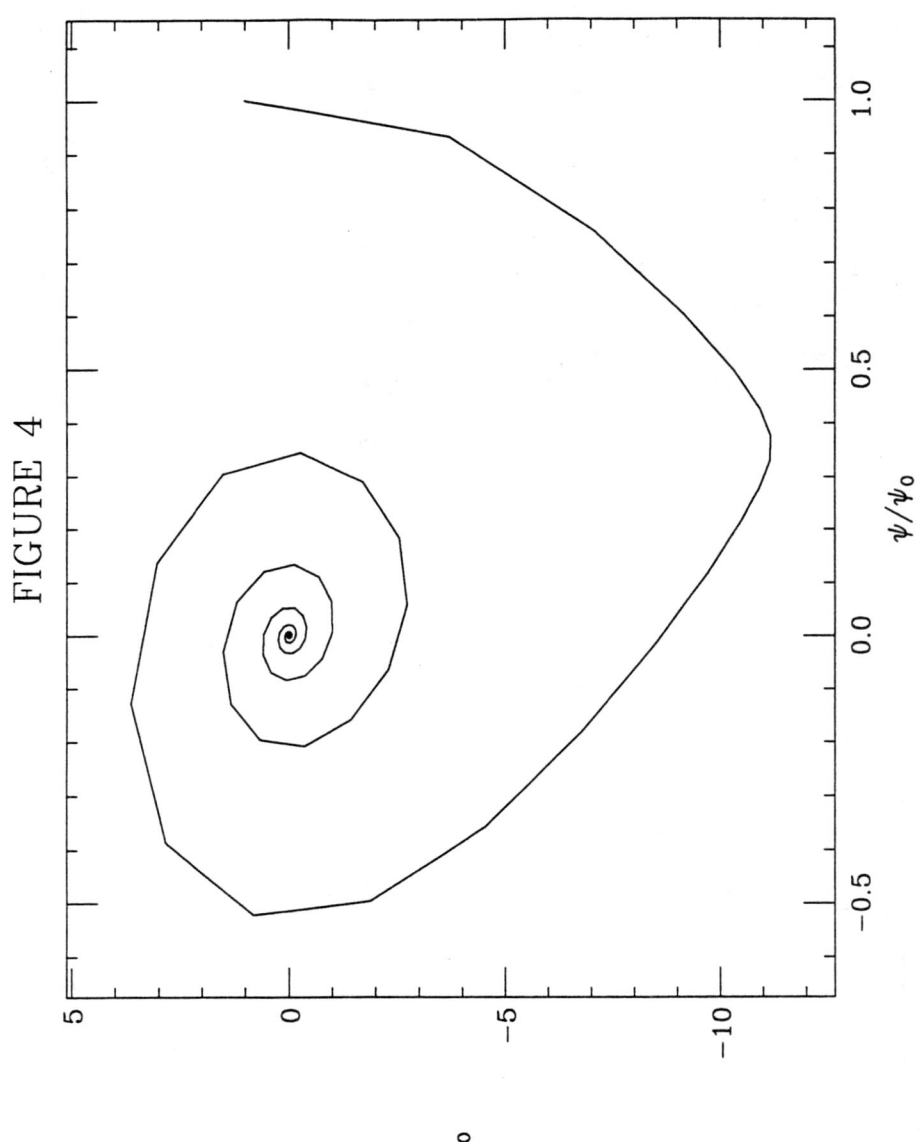

FIGURE 4

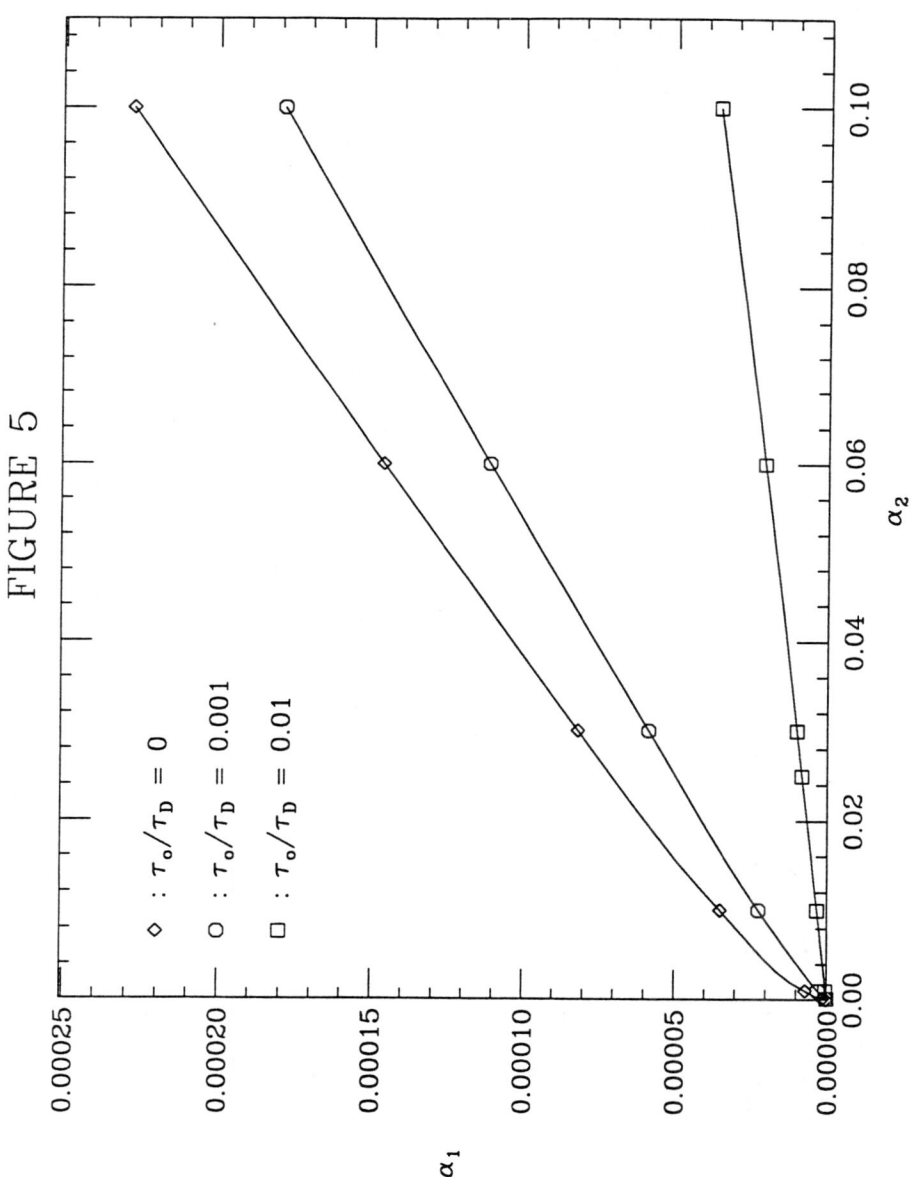

FIGURE 5

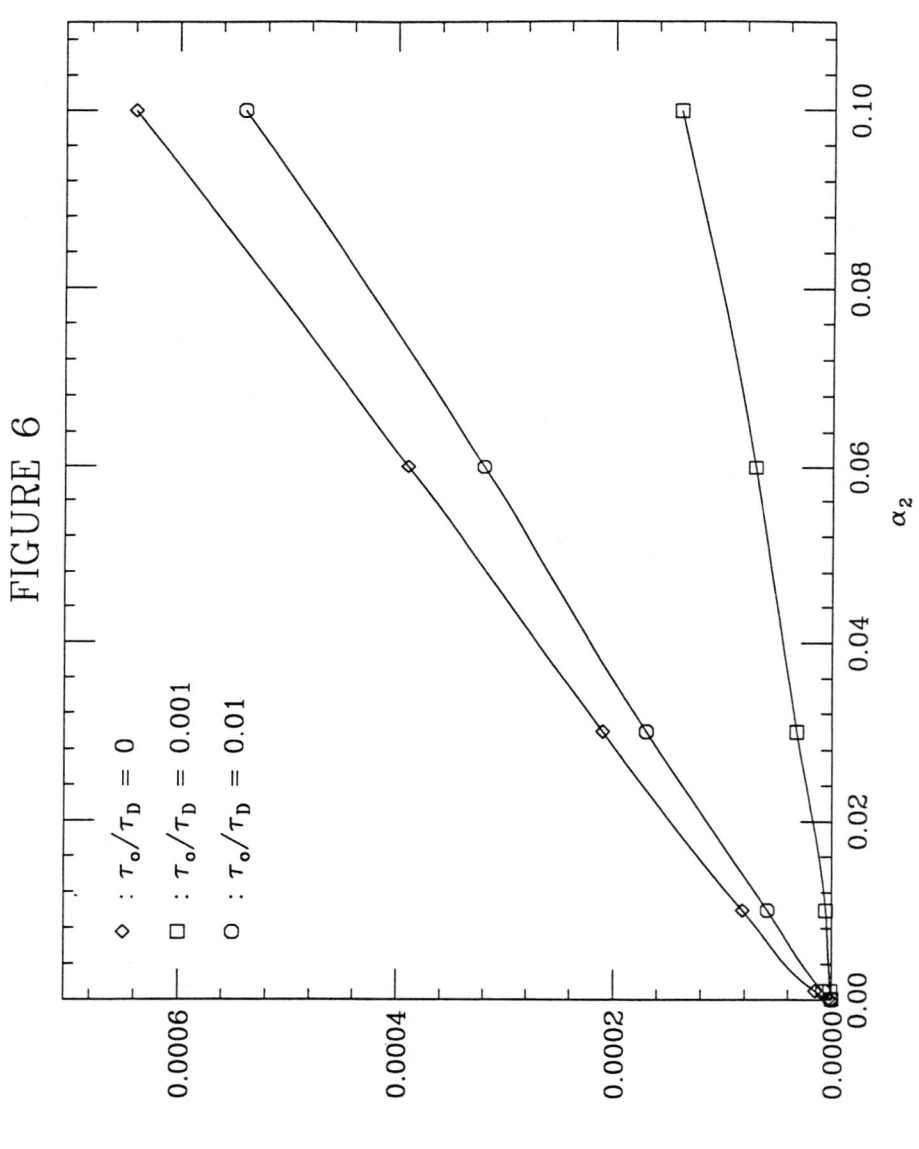

FIGURE 6

R&D of damped cavities at KEK

E. KIKUTANI

National Laboratory for High Energy Physics

1. Introduction

Now many high-energy physics laboratories around the world have their plans to construct B-meson factories, which mainly aim at observe CP-violation effects in the B-meson sector. Luminosity demanded by this physics program is order of 10^{33} through $10^{34} cm^{-2} s^{-1}$. From many experiences of operation of $e^+ e^-$ colliding machines we know that it is reasonable to operate a collider in a multibunch mode to get such a high luminosity. The number of bunches is so large that the spacing of bunches is only a few times of RF wave length. In this situation, as is well known, wake fields which left by the preceding bunches may be sources of instabilities (coupled bunch instabilities).

On the other hand, in the study of linear colliders, coupled bunch instabilities are also serious because linear colliders also must use the multibunch operation to increase the luminosity. The problems are not only for linear accelerators themselves but for their damping rings.

To solve the problem, one of the solutions is to use the damped cavities proposed by R. Palmer[1]. That is, a cavity couples with wave guides and quality factors of unwanted modes are decreased to unharmful values. From this point of view, we are now studying possible schemes of damped cavities at KEK.

2. KEK type damped cavities

Calculation method

In order to calculate the external Q of higher modes, T. Kageyama[2] applied a formula which described in the text by J.C.Slater[3], i.e.,

$$\frac{dL}{d\lambda_g} = \frac{2n+1}{4} + \frac{Q_{ext} \cdot v_g}{\pi v_p},\qquad(1)$$

where L is the distance from the cavity to the short-circuited end of the wave guide, λ_g the wave length in the wave guide, v_g the group velocity, v_p the phase velocity and n is an integer. Actuall calculation to get resonant frequencies is done with the computer code MAFIA[4].

Type 1

At present, two types of damped cavities are under study at KEK. The first one is developed by T. Kageyama. A schematic view of this cavity is shown in Fig. 1. Four wave guides (the cross section is 4cm × 11cm) are attached to both end-boards of the cavity.

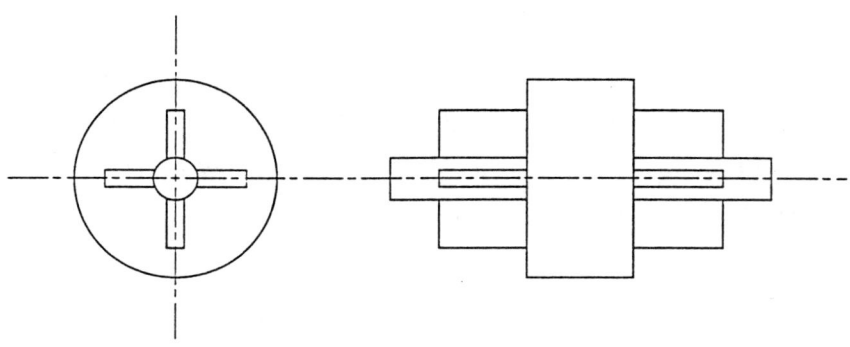

Fig.1 Schematic view of damped cavity type 1

Varying L (length of the wave guide) step by step, tuning curves are obtained. Figure 2 shows the curve for TM110 mode. From the fitted value of $\frac{dL}{d\lambda_g}$ the external Q is calculated to be ∼ 10 by expression (1).

At present S-band model of this cavity is under construction and actual measurement of the external Q will start in coming summer.

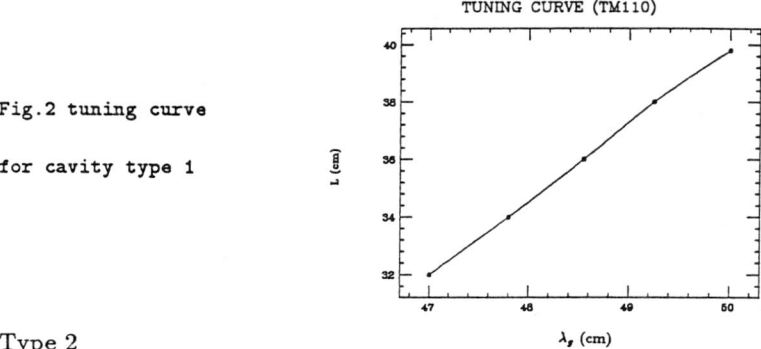

Fig.2 tuning curve for cavity type 1

Type 2

The second type of the damped cavity is developed by M. Suetake[5]. In this type, a disc which makes a boundary of the cells is slotted as shown in Fig. 3. The external Q for some higher modes are calculated with the same method as the type 1 using a model cavity shown in Fig. 4. Fig.5 shows a typical result for TM110 mode(plotted with a symbol "o"). The external Q varies as a function of window width to the wave guide, W1. You can see that the external Q is decreased to the order of 10 by optimizing the width.

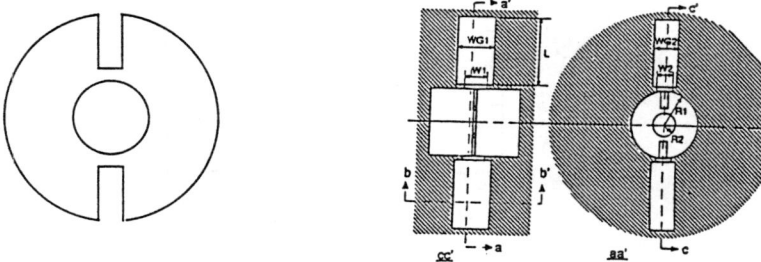

Fig.3 The slotted disc   Fig.4 Damped cavity type 2

On the other hand, the quality factor of the fundamental mode (TM010) is not significantly affected by making slots and by opening windows, as shown in the table below.

	no window, no slot	no window	no slot	with slots and window
π mode	14470	14340	12370	12850
0 mode	-	-	-	14400

For this type, a prototype (C-band 2/3 π mode caivity[6]) was made and the external Q was measured. Result is also shown in Fig. 5 with a symbol "×". The measured external Q is decreased to the order of 10 and the values agrees with the calculated ones.

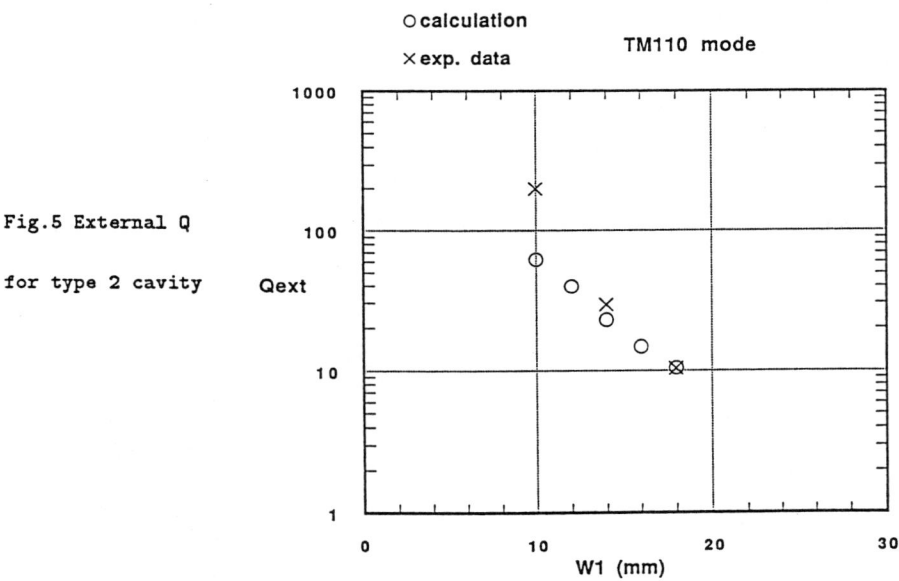

Fig.5 External Q for type 2 cavity

3. Summary

Two types of KEK damped cavities are now under study. With both types, the calculated external Q is decreased to order of 10. For the second type, a model cavity was constructed and the measured external Q agrees with the calculated one.

References

1. R. Palmer, SLAC-PUB-4542.

2. T. Kageyama, KEK-89-4.

3. J.C.Slater, Micro Wave Electronics, Van Nostrand, 1950

4. R. Klatt et al., Proceedings of the 1986 Linear Accelerator Conference, SLAC-Report-303, p. 276

5. M. Suetake et al., KEK Preprint 89-164.

6. This is designed as one candidate cavity for Japan Linear Collider.

REDUCTION OF BEAM-BEAM SYNCHROBETATRON RESONANCES USING COMPENSATING INTERACTION REGIONS[*]

John T. Seeman
Stanford Linear Accelerator Center,
Stanford University, Stanford, CA 94309

INTRODUCTION

The next generation of high luminosity circular electron-positron colliders requires many intense bunches in each beam, of order 200 to 1100. The interbunch spacing dictates that two separated rings are necessary. The ensuing problem is how to design the interaction regions so that the two beams can be made to collide and separate using physically realizable magnets, providing low detector backgrounds, having a practical vacuum system design, and obeying the constrains of the beam-beam interaction.[1,2] One of the solutions involves the use of a finite crossing angle between the trajectories of the two beams at the interaction point (IP). The problem with a crossing angle is the well-known problem of synchro-betatron resonances driven by the beam-beam force.[3]

A solution to the synchro-betatron problem is to make the beams effectively collide head-on by adding a tilt to each bunch at the IP, a so-called crab crossing. There are several methods to produce these tilted bunches.[4,5] However, all of these solutions involve placing RF cavities near the interaction region where either the betatron or dispersion functions are large. These crab cavities in turn generate additional design problems. Some of the problems are: (1) a significantly larger impedance of the ring vacuum system, (2) the large cavity voltages (of order 10 MV), (3) the rotation of both beams, and (4) the control and stability of these cavities.

An alternative solution discussed here has two interaction regions physically adjacent to each other where the beams collide at a finite crossing angle without tilted bunches. To ameliorate the synchro-betatron resonances, there is a specially chosen betatron phase advance between the two IRs so that the coupling effects from the beam-beam forces cancel. The consequence of this arrangement is that the interaction region design is now straightforward.

COMPENSATING INTERACTION REGIONS

A schematic layout of the interaction region is shown in Fig. 1. The beams cross each interaction region at a horizontal angle of ϕ for a total crossing angle of 2ϕ. As a result of the beam-beam force of a single collision, the head of each bunch is deflected in the opposite direction from that of the tail. The second

---

[*] Work supported by Department of Energy contract DE-AC03-76SF00515.

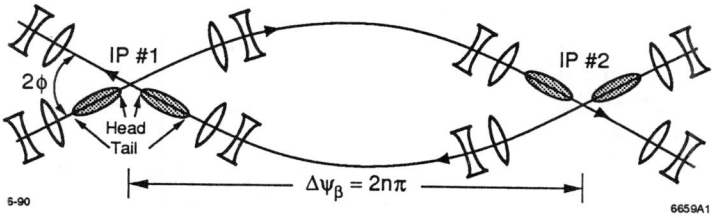

Fig. 1: Layout of compensating interaction regions.

interaction region is placed at a betatron phase advance of exactly $2n\pi$ from the first IP, with n an integer. Furthermore, the crossing angle is arranged to change sign between IPs. With this placement, the deflection angles of the head and the tail resulting from the first interaction region are, to first order, exactly cancelled by that of the second. The position offsets are also zero. Thus, the design goal is satisfied.

The horizontal incoherent beam-beam effects add linearly from the two collisions because of the integer phase separation. This addition will cause the horizontal beam-beam tune shifts to add directly from the two collisions and will somewhat reduce the maximum horizontal tune shift observed in either IP. This should not strongly affect the maximum luminosity since the beam-beam limit is most often determined by the limiting vertical tune shift (for flat beams). The linear addition of the vertical tune shifts from the two interaction regions can be broken by adjusting the vertical phase advance between interaction regions so it is not $2n\pi$ but has a sizable fractional part, e.g., $2n\pi + 0.19$. The design of an interaction region satisfying these two phase advance constraints should be straight forward. Some horizontal bending of the trajectories between the two IPs is needed to provide the proper geometry, requiring a dispersion matched lattice to be made.

The basic reason this scheme is worth investigation is that the direct coupling of the longitudinal position and the beam-beam force of the core of the beam is eliminated. The resonant condition is then removed. Unfortunately, a particle which is executing both a synchrotron oscillation and a betatron oscillation will see some coupling due to the nonlinear nature of the beam-beam force. Therefore, nonlinear effects influencing the sparse tails of the beams may likely appear upon detailed simulations. A detailed simulation of the beam tails will be conducted soon to investigate the strength of these possible nonlinear effects and how important they are.

In this design with compensating IPs, each individual bunch will collide with two opposing bunches thus interacting through the beam-beam force to all other bunches. This situation requires either that the intensity of all the bunches be the same to a small tolerance or that the collider operates just below the bunch intensity where the tune shift saturates and the vertical beam size starts to increase. This second scenario is the most advantageous and can likely be achieved by actively controlling the vertical-horizontal coupling with magnets throughout the course of a physics run as the currents decay.

The design of the interaction region for compensating IPs is easier than either a crab-crossing IP or a head-on IP. The first advantage is that no crab cavities need be planned for. Secondly, there are no bending magnets within the IP, significantly reducing the synchrotron radiation background problem. In this new scheme, the two IPs should be sufficiently close to each other so that radiation mixing of phase space is minimized. However, the added complexities are: (1) the phase advances must be carefully planned, (2) the bending needed between the IPs must be designed to provide for dispersion matching, and (3) the trajectories for the two beams between IPs need to have the same path length.

CONCLUSIONS

The concept of mutually compensating IPs to reduce the effects of synchrobetatron resonances associated with the beam-beam effect appears sufficiently interesting to pursue further. With success, the interaction region design will be simplified significantly. A detailed beam-beam simulation of this proposed concept is needed next to study possible nonlinear effects. A realistic IP layout and design are also needed.

REFERENCES

1. R. Siemann, Proceedings of the 14th International Conference on High Energy Accelerators, Tsukuba, Japan, 1989, p. 1305.
2. S. Chattopadhyay, Proceedings of the 14th International Conference on High Energy Accelerators, Tsukuba, Japan, 1989, p. 1329.
3. A. Piwinski, "Synchrobetatron Resonances," Nonlinear Dynamics Aspects of Particle Accelerators, (Springer-Verlag, 1985) p. 104.
4. G. Voss, J. Paterson, and S. Kheifets, "Crab-Crossing in a Tau-Charm Facility," Tau-Charm Factory Workshop, SLAC, May 1989; SLAC–PUB–5011 (1989).
5. G. Jackson, Proceedings of the Workshop on Beam Dynamics Issues in High-Luminosity Asymmetric Collider Rings, LBL, February 1990.

SYNCHRO-BEAM INTERACTION

K. Hirata
KEK, Tsukuba, Ibaraki 305, Japan

H. Moshammer and F. Ruggiero
CERN, CH-1211 Geneva 23, Switzerland

M. Bassetti
Laboratori Nazionali di Frascati dell' INFN, 00044 Frascati, Italy

ABSTRACT

A symplectic mapping is presented, describing the effect of the beam-beam interaction in presence of synchrotron motion. In addition to the usual transverse kick, this mapping includes the longitudinal displacement of the collision point and the energy variation caused by the electric field of the opposite bunch. These two effects are shown to play a concurrent role in preserving symplecticity. The case of weak-strong interaction in e^+e^- colliding rings is investigated by simulation, assuming only one betatron degree of freedom.

INTRODUCTION

Many years ago, Augustin remarked that a particle can lose or gain energy as a consequence of the beam-beam interaction [1]. Indeed, in the extreme relativistic limit, half of the force felt by the particle is due to the electric field of the opposite bunch and thus can have a component in the direction of the particle velocity. Later on, Piwinski developed a theory for synchro-betatron resonances driven by this effect in the case of collisions with a crossing angle [2]. Then one of the authors (M.B.) remarked that a variation of energy is possible even when the (average) crossing angle is zero [3]. Recently, he proposed a heuristic derivation of a new luminosity limit for e^+e^- storage rings, based on the comparison between the typical energy variation due to beam-beam collisions and the quantum-induced energy fluctuations over one machine revolution [4].

In this paper, we discuss the beam-beam interaction in weak-strong approximation. The evolution of a test particle in the weak bunch depends on the electromagnetic field created by all the particles in the strong bunch and the latter can be considered as an external *prescribed* field, since we neglect the reaction of the weak bunch on the particles of the strong one. Therefore the beam-beam interaction (in weak-strong approximation) and the interaction with bending and focusing magnets or RF-cavities of the storage ring must be treated on an equal footing. As a consequence, apart from radiation effects, the evolution of a test particle in the *six-dimensional* phase space is derivable from a Hamiltonian and the corresponding single-turn mapping must be *symplectic*.

This condition is important both in particle tracking and for theoretical considerations; it implies the conservation of the six-dimensional phase-space volume element and any deviation from symplecticity can lead to unphysical effects, such as damping or antidamping of the particle oscillations as well as undue modifications of the synchro-betatron coupling.

We consider the case of a single interaction point (IP) and treat the strong bunch in *thin lens* approximation. Betatron and synchrotron transfer matrices allow us to describe the symplectic single-turn evolution, from IP to IP, in the absence of beam-beam interaction. Then, if the beam-beam collisions took always place at the IP, the associated map should also be symplectic. However, as a consequence of synchrotron oscillations, the point where a test particle collides with the strong bunch does not coincide with the IP and the corresponding longitudinal displacement changes from turn to turn. Therefore, we should require the symplecticity of the transformation consisting of a drift from the IP to the collision point, followed by the beam-beam map (i.e. transverse kick plus energy variation) and by a backward drift to the IP. This transformation, to be applied at the IP, will be called the 'synchro-beam mapping' and is discussed in the next section for a round beam with Gaussian radial distribution. There, we also consider the possible variation of transverse beam dimensions over the interaction region.

In the last section, we describe the complete set of equations (including radiation effects) used in the simulation of the weak-strong beam-beam interaction for e^+e^- storage rings and present some preliminary simulation results. They seem to confirm the importance of our symplecticity condition, at least when the r.m.s. value of the transverse betatron slopes is not negligible with respect to the relative energy spread: this is more likely the case for round beams and becomes generally true when approaching the beam-beam limit in the strong-strong interaction. Finally, we draw some conclusion about possible scaling laws and implications for future high-luminosity storage rings.

A SYMPLECTIC MAP FOR SYNCHRO-BEAM INTERACTION

To make the discussion as simple as possible, we assume that the interaction region is dispersion-free and consider the case of a round beam with Gaussian radial distribution. We also assume that the motion of a test particle in the weak beam remains in a plane, so that we can reduce the number of transverse degrees of freedom of our problem from two to one. For the sake of definiteness, we suppose that the strong beam consists of positrons (with positive charge e) while the test particle is an electron (with negative charge $-e$).

In order to discuss symplecticity, we use as independent variable the curvilinear abscissa s along the reference orbit of the storage ring ($s = 0$ at the IP) and fix the set of canonical variables by choosing the betatron and synchrotron variables for the ideal linear machine, without beam-beam interaction. In the dispersion-free interaction region, these variables coincide with the following physical variables:

- x: [coordinate] transverse displacement.
- $x' \equiv dx/ds$: [momentum conjugate to x] transverse slope.
- $z \equiv s - ct$: [coordinate] longitudinal displacement with respect to the center of the weak bunch (synchronous particle). If $z(s) > 0$, a particle of the weak bunch arrives at s ahead of the synchronous particle.
- $\varepsilon \equiv (E - E_0)/E_0$: [momentum conjugate to z] relative energy deviation from the nominal energy $E_0 = \gamma mc^2$.

We denote all four variables by the vector $\mathbf{x} = (x, x', z, \varepsilon)^t$. These variables are defined for a test particle in the weak beam.

Imagine a particle arriving at the IP with positive z. This implies that the center of the other bunch is not yet at the IP: it is still at $s = z$. Thus, the real collision between this particle and the strong bunch takes place at $s = z/2$, to be called the collision point (CP). This effect can be described by a set of three subsequent transformations, namely

$$\mathbf{x}_{IP} \xrightarrow{D(z/2)} \mathbf{x}_{CP} \xrightarrow{B(z/2)} \mathbf{x}_{CP}^{new} \xrightarrow{D(-z/2)} \mathbf{x}_{IP}^{new}.$$

First the particle goes to $s = z/2$ by drift, D, then it undergoes the beam-beam interaction B at $s = z/2$ and finally it is brought back to the IP, again by drift. In the following, we consider the mapping corresponding to the concatenation of these three transformations and show that it is symplectic. Since the single-turn evolution along the arc of the storage ring, from IP to IP, can be assumed to be known and corresponds to betatron and synchrotron oscillations plus radiation effects, it is useful to include the back-drift in the 'synchro-beam' interaction.

The mapping corresponding to a drift of length $z/2$ can be written

$$\begin{pmatrix} x \\ x' \end{pmatrix}_{CP} = \begin{pmatrix} 1 & z/2 \\ 0 & 1 \end{pmatrix} \begin{pmatrix} x \\ x' \end{pmatrix}_{IP}. \qquad (1)$$

The longitudinal variables z and ε remain unchanged. However, since the length of the drift depends on the dynamical variable z, it is easy to see that the corresponding mapping in the four-dimensional phase space $\mathbf{x}$ is *not* symplectic. Indeed the Jacobian matrix

$$M(CP; IP) \equiv \frac{\partial(\mathbf{x}_{CP})}{\partial(\mathbf{x}_{IP})}$$

is not a symplectic matrix, i.e.

$$M(CP; IP)\, J\, M(CP; IP)^t \neq J, \qquad (2)$$

where M^t is the transpose of M and J denotes the (antisymmetric) 4×4 unit symplectic matrix.

In the vicinity of the CP, the evolution of a particle of the weak beam can be described by the following Hamiltonian, valid in the extreme relativistic limit and to first order in the betatron slopes:

$$H(x, x', z, \varepsilon; s) = \frac{x'^2}{2} + U(x, \sigma(s)) \frac{\theta(s - \frac{z-\Delta}{2}) - \theta(s - \frac{z+\Delta}{2})}{\Delta}, \qquad (3)$$

where $U(x, \sigma)$ is the beam-beam potential, $\sigma(s)$ is the transverse size of the strong bunch and $\theta(s)$ is a step function, defined by

$$\theta(s) = \begin{cases} -1/2 & \text{for } s < 0, \\ 1/2 & \text{for } s > 0. \end{cases}$$

The thin lens approximation for the strong bunch corresponds to the limit $\Delta \to 0$ (see Fig. 1) and the interaction potential in the previous Hamiltonian is then multiplied by a Dirac delta-function $\delta(s - z/2)$

$$\delta(s - \frac{z}{2}) = \lim_{\Delta \to 0} \frac{\theta(s - \frac{z-\Delta}{2}) - \theta(s - \frac{z+\Delta}{2})}{\Delta}.$$

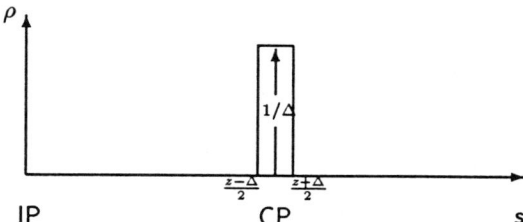

Figure 1: Longitudinal density $\rho(s, z)$ of the strong bunch.

The potential U is defined in such a way that its gradient with respect to x gives the transverse beam-beam kick. This leaves the possibility of adding an arbitrary function of σ, usually chosen so that U vanishes at $x = 0$ or at large amplitudes. However, as we will see, the derivative of U with respect to s (and thus with respect to σ) can be interpreted as the contribution of the longitudinal electric field of the strong beam to the energy variation of a particle in the weak beam. This is possible only if the derivative of U with respect to σ vanishes at large amplitudes and so removes the arbitrariness in the choice of U (up to an inessential constant).

The Hamilton equations read

$$\frac{dx}{ds} = \frac{\partial H}{\partial x'} = x',$$

$$\frac{dx'}{ds} = -\frac{\partial H}{\partial x} = -\frac{\partial U(x,\sigma)}{\partial x}\frac{\theta(s-\frac{z-\Delta}{2}) - \theta(s-\frac{z+\Delta}{2})}{\Delta},$$

$$\frac{dz}{ds} = \frac{\partial H}{\partial \varepsilon} = 0,$$

$$\frac{d\varepsilon}{ds} = -\frac{\partial H}{\partial z} = -U(x,\sigma)\frac{\delta(s-\frac{z+\Delta}{2}) - \delta(s-\frac{z-\Delta}{2})}{2\Delta}.$$

We start by integrating the equations of motion for x' and ε to lowest order in Δ, thus obtaining

$$x'(\frac{z+\Delta}{2}) - x'(\frac{z-\Delta}{2}) = -\left[\frac{\partial U(x,\sigma)}{\partial x}\right]_{\frac{z}{2}},$$

$$\varepsilon(\frac{z+\Delta}{2}) - \varepsilon(\frac{z-\Delta}{2}) = -\frac{1}{2\Delta}\left\{[U(x,\sigma)]_{\frac{z+\Delta}{2}} - [U(x,\sigma)]_{\frac{z-\Delta}{2}}\right\}.$$

As shown by the last equation, the energy change is proportional to the total variation of the beam-beam potential U and the latter can be computed by a first order Taylor expansion with respect to all of its arguments. Therefore we get

$$\varepsilon(\frac{z+\Delta}{2}) - \varepsilon(\frac{z-\Delta}{2}) = -\frac{1}{2}\left\{\left[\frac{\partial U(x,\sigma)}{\partial x}\right]_{\frac{z}{2}}\frac{x(\frac{z+\Delta}{2}) - x(\frac{z-\Delta}{2})}{\Delta} + \left[\frac{\partial U(x,\sigma)}{\partial \sigma}\frac{d\sigma}{ds}\right]_{\frac{z}{2}}\right\}.$$

The factor 1/2 in front of this expression reminds us that only the electric field can do work. We see that the energy variation has a contribution coming from the transverse electric field (associated with $\partial U/\partial x$) and another contribution associated with the dependence of the transverse beam size on s, i.e. with $d\sigma/ds$; this last contribution can be identified with the effect of the longitudinal electric field of the strong bunch [5].

Since the slope x' has a discontinuity at the CP, we assume that the effect of the beam-beam kick be constant over the length Δ of the interaction. Thus the variation of x corresponds to 'uniform acceleration' and reads

$$\frac{x(\frac{z+\Delta}{2}) - x(\frac{z-\Delta}{2})}{\Delta} = x'(\frac{z-\Delta}{2}) + \frac{1}{2}\left[\frac{x'(\frac{z+\Delta}{2}) - x'(\frac{z-\Delta}{2})}{\Delta}\right]\Delta = \frac{x'(\frac{z+\Delta}{2}) + x'(\frac{z-\Delta}{2})}{2}.$$

This regularization procedure is therefore equivalent to replacing the slope at the CP by the average of its values just before and just after the beam-beam collision. Using the notation

$$f(x,\sigma) = \frac{\partial U(x,\sigma)}{\partial x}, \qquad (4)$$

$$g(x,\sigma) = \frac{1}{2}\frac{d\sigma}{ds}\frac{\partial U(x,\sigma)}{\partial \sigma}, \qquad (5)$$

the beam-beam mapping at the collision point can be written as follows:

$$\left.\begin{aligned}
x_{CP}^{new} &= x_{CP}, \\
x'^{new}_{CP} &= x'_{CP} - f(x_{CP}, \sigma_{CP}), \\
z_{CP}^{new} &= z_{CP}, \\
\varepsilon_{CP}^{new} &= \varepsilon_{CP} - \tfrac{1}{2} f(x_{CP}, \sigma_{CP}) \left[x'_{CP} - \tfrac{1}{2} f(x_{CP}, \sigma_{CP}) \right] - g(x_{CP}, \sigma_{CP}),
\end{aligned}\right\} \quad (6)$$

where $\sigma_{CP} \equiv \sigma(z/2)$.

In the case of a round beam with Gaussian distribution, the radial kick in the extreme relativistic limit reads

$$f(x, \sigma) = -\frac{2 N r_e}{\gamma} \frac{1}{x} \left[\exp\left(\frac{-x^2}{2\sigma^2}\right) - 1 \right], \quad (7)$$

where N is the number of particles in the strong beam, r_e denotes the classical electron (proton) radius and γ is the Lorentz factor for the weak beam. This kick can be derived from the potential [6]

$$U(x, \sigma) = -\frac{N r_e}{\gamma} \int_0^\infty \frac{\exp(\frac{-x^2}{2\sigma^2 + t})}{2\sigma^2 + t} dt.$$

An alternative potential [7] would lead to an unphysical longitudinal electric field, which does not vanish at $x \to \infty$. Hence, from Eq. (5), we have

$$g(x, \sigma) = \frac{N r_e}{\gamma} \frac{1}{\sigma} \frac{d\sigma}{ds} \exp\left(\frac{-x^2}{2\sigma^2}\right). \quad (8)$$

The transverse beam size $\sigma(s)$ can be expressed in terms of the betatron function $\beta(s)$ and of the transverse emittance ϵ_t as $\sigma(s) = \sqrt{\epsilon_t \beta(s)}$. Thus the logarithmic derivative of the beam size at the CP ($s = z/2$) becomes

$$\left[\frac{1}{\sigma} \frac{d\sigma}{ds} \right]_{CP} = \frac{1}{\beta_{IP}} \frac{(z/2\beta_{IP})}{\sqrt{1 + (z/2\beta_{IP})^2}}, \quad (9)$$

where β_{IP} is the betatron function at $s = 0$.

Similarly to the drift, already discussed, the beam-beam map at CP is *not* symplectic. This can be easily seen by testing the symplecticity condition Eq. (2) for the Jacobian matrix

$$M(CP^{new}; CP) \equiv \frac{\partial (\mathbf{x})_{CP}^{new}}{\partial (\mathbf{x})_{CP}}$$

and follows from the fact that we have integrated the Hamilton equations across the discontinuity at $s = z/2$, instead of fixing the initial and final values of s independently of the dynamical variable z. However, if we concatenate the drift

$D(z/2)$ from IP to CP with the beam-beam mapping at the CP and the back-drift $D(-z/2)$ from CP to IP, we get the following symplectic 'synchro-beam' mapping, from $\mathbf{x}_{IP}$ to $\mathbf{x}_{IP}^{new}$:

$$\left.\begin{array}{rcl}x^{new} & = & x + \frac{1}{2}zf(x_{CP}, \sigma_{CP}), \\ x'^{new} & = & x' - f(x_{CP}, \sigma_{CP}), \\ z^{new} & = & z, \\ \varepsilon^{new} & = & \varepsilon - \frac{1}{2}f(x_{CP}, \sigma_{CP})[x' - \frac{1}{2}f(x_{CP}, \sigma_{CP})] - g(x_{CP}, \sigma_{CP}),\end{array}\right\} \quad (10)$$

where
$$x_{CP} = x + \frac{1}{2}z\,x'.$$

The symplecticity of this mapping can be easily checked by examining the Jacobian matrix
$$\frac{\partial(\mathbf{x})_{IP}^{new}}{\partial(\mathbf{x})_{IP}}.$$

This is also clear from the fact that Eqs. (10) are derivable from the following effective Hamiltonian H_{eff}:

$$H_{eff} = U(x_{CP}, \sigma_{CP})\,\delta(s).$$

This result is general and not restricted to the case of a round beam with Gaussian distribution.

We can say that, from the point of view of symplecticity, the energy variation due to the beam-beam interaction is complementary to the drift $D(z/2)$ and the back-drift $D(-z/2)$. It is now clear that if one considers the fact that the collision point is changing at every turn, one should also consider the energy variation and vice versa. Otherwise the symplecticity is lost.

WEAK-STRONG SIMULATION

Let us describe the implementation of the synchro-beam mapping discussed above for particle tracking in weak-strong approximation. The code includes betatron and synchrotron oscillations with damping and noise, associated with synchrotron radiation in the arcs of an e^+e^- storage ring. Thanks to the back-drift, $D(-z/2)$, we can apply the revolution matrix representing betatron and synchrotron oscillations just from IP to IP. This may be an important point, since the phase advance is defined for a fixed path length (s) and not for a fixed time interval (t). Had we used t as independent variable, the description would have been completely different (although equivalent).

THE PROGRAM

In order to work with essential parameters only, we introduce normalized variables by means of the following (non-canonical) transformation:

$$Q = \frac{x}{\sigma_x^0}, \quad P = \beta_{IP} \frac{x'}{\sigma_x^0},$$

$$Z = \frac{z}{\sigma_z^0}, \quad \Sigma = \frac{\varepsilon}{\sigma_\varepsilon^0},$$

where σ_x^0, σ_z^0 and σ_ε^0 are the nominal values at IP of the transverse beam size, bunch length and relative energy spread, respectively. For simplicity, we assume that the betatron function and the nominal emittances are the same for the weak and for the strong beam.

The program tracks each particle as follows:

Oscillation The transformations from IP to IP are just rotations with damping and quantum diffusion.

Thus we first apply the mapping

$$\begin{pmatrix} Q \\ P \end{pmatrix} \longrightarrow U(2\pi\nu_x) \begin{pmatrix} Q \\ P \end{pmatrix},$$

$$\begin{pmatrix} Z \\ \Sigma \end{pmatrix} \longrightarrow U(-2\pi\nu_s) \begin{pmatrix} Z \\ \Sigma \end{pmatrix},$$

where

$$U(\alpha) = \begin{pmatrix} \cos\alpha & \sin\alpha \\ -\sin\alpha & \cos\alpha \end{pmatrix}.$$

Radiation For transverse diffusion, the symmetric prescription discussed in [8] is used. The longitudinal diffusion is applied only to Σ. Thus,

$$Q \longrightarrow \lambda_x Q + \sqrt{1 - \lambda_x^2}\, \hat{r}_1,$$

$$P \longrightarrow \lambda_x P + \sqrt{1 - \lambda_x^2}\, \hat{r}_2,$$

$$\Sigma \longrightarrow \lambda_s^2 \Sigma + \sqrt{1 - \lambda_s^4}\, \hat{r}_3,$$

where the $\hat{r}$'s are independent, random Gaussian variables with unit standard deviation. Here the λ's are damping factors defined by

$$\lambda_x = e^{-\delta_x}, \quad \text{and} \quad \lambda_s = e^{-\delta_s},$$

where the δ's denote the corresponding damping decrements [9]. Their inverse values $\tau_x = 1/\delta_x$ and $\tau_s = 1/\delta_s$ represent the number of beam-beam collisions per betatron or synchrotron damping time, respectively.

Synchro-Beam Interaction The mapping (10), representing the synchro-beam interaction for a round beam becomes

$$\begin{aligned}
Q &\longrightarrow Q - R_1 Z \, \delta P, \\
P &\longrightarrow P + \delta P, \\
Z &\longrightarrow Z, \\
\Sigma &\longrightarrow \Sigma + R_2 \, \delta P \, (P + \delta P/2) - G,
\end{aligned} \qquad (11)$$

where

$$\delta P \equiv \kappa_r \frac{1}{Q + R_1 Z P} [\exp(-A_{col}^2) - 1],$$

$$G \equiv \kappa_r R_1 R_2 \frac{Z}{1 + R_1^2 Z^2} \exp(-A_{col}^2),$$

$$A_{col} = \frac{(Q + R_1 Z P)}{\sqrt{2(1 + R_1^2 Z^2)}},$$

$$R_1 = \frac{\sigma_z^0}{2\beta_{IP}}, \quad R_2 = \frac{\epsilon_t}{2\sigma_\epsilon^0 \beta_{IP}}$$

and

$$\kappa_r = 8\pi\eta.$$

Here η is the nominal beam-beam parameter, which is proportional to the current of the strong beam

$$\eta = \frac{N r_e}{4\pi\gamma\epsilon_t},$$

The denominator of A_{col} represents the variation of the beam size (of the strong beam) due to the motion of the CP.

Let us note that the only dimensionless free parameters are the following:

- nominal beam-beam parameter η,
- transverse and synchrotron tunes ν_x and ν_s,
- transverse and synchrotron damping decrements δ_x and δ_s,
- ratios R_1 and R_2.

In the case of flat beam, all the betatron parameters refer to the vertical plane (denoted by a subscript y) and the synchro-beam mapping should be modified as follows [5]:

$$\delta P \equiv -\kappa_f \, \text{erf}(A_{col}),$$

$$G \equiv \kappa_f R_1 R_2 \sqrt{\frac{2}{\pi}} \frac{Z}{\sqrt{1 + R_1^2 Z^2}} \exp(-A_{col}^2),$$

$$\kappa_f = 2\pi^{3/2}\eta.$$

Simulation Results

Due to CPU time restrictions, no extensive survey of the parameter space has been done yet and we only present some preliminary results. We have mainly studied the case of flat beam, using the following standard set of parameters: $\eta = 0.05$ $\nu_y = 0.15$, $\nu_s = 0.085$, $\delta_y = 1/800$, $\delta_s = 2\delta_y$, $R_1 = 0.5$, $R_2 = 0.005$.

Dependence on ν We vary the vertical tune ν_y for a fixed value of the beam-beam parameter η (equal to 0.01). The equilibrium vertical beam size is plotted as a function of ν_y in Fig. 2, where synchro-betatron side-bands appear clearly. This effect is still present for vanishing values of the parameter R_2, expressing

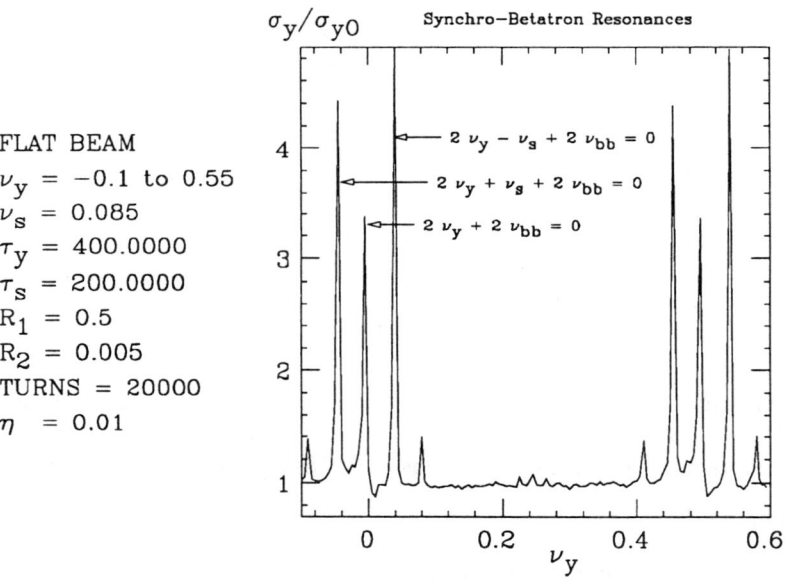

Figure 2: Tune dependence of the vertical beam size σ_y, normalized to its nominal value σ_{y0}. The parameters are the standard ones except for η, which is 0.01.

the relative magnitude of the energy variation due to the beam-beam interaction. Therefore it is mainly a consequence of the longitudinal modulation experienced by the collision point and of our thin lens approximation for the strong bunch. However, for finite values of R_2, we have some indication that the synchro-betatron side-bands *do not disappear* even when the strong bunch is longitudinally distributed and consists of a sequence of many thin slices: this preliminary result seems to contradict the conclusions of Ref. 10.

Dependence on η If we consider only transverse motion, ignoring the energy variation and the modulation of the CP, it is already known that the transverse beam size grows up smoothly when increasing η. This means that there is no remarkable threshold value for η.

With the synchro-beam interaction, when we increase η, the beam size grows up slowly at the beginning but suddenly blows up by orders of magnitude at a certain point. For practical purposes, we define η_∞ as the value of η corresponding to an increase of the transverse beam size by a factor 10 over its nominal, unperturbed value. In Fig. 3 we plot the η-dependence of the beam size, corresponding to different values of the parameter R_1, for a vertical damping decrement $\delta_y = 800$. For a given value of η, the equilibrium beam size is

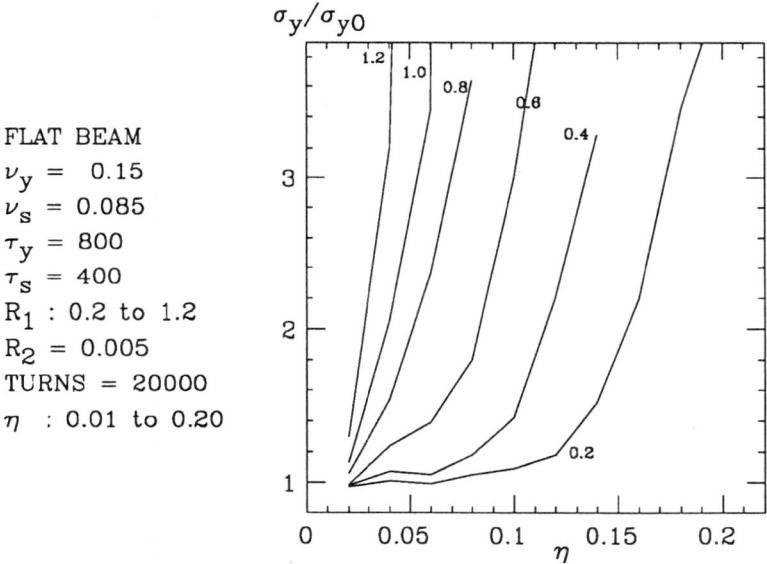

Figure 3: The η dependence of σ_y for $\delta_y = 1/800$ and different values of R_1.

smaller in the case of rapid damping and, correspondingly, the threshold value η_∞ is larger.

We now consider the dependence of the threshold value η_∞ on R_1 and R_2. Its dependence on ν_y and δ_y can be qualitatively deduced from the previous discussion.

Dependence on R_1 In Fig. 4, we show the dependence of η_∞ on R_1 for different values of δ_y. It appears that the product $\eta_\infty R_1$ is approximately constant (for fixed values of the δ's, the ν's and R_2). *For realistic values of the parameters, the dependence of this product on the δ's, the ν's and R_2 is rather weak.*

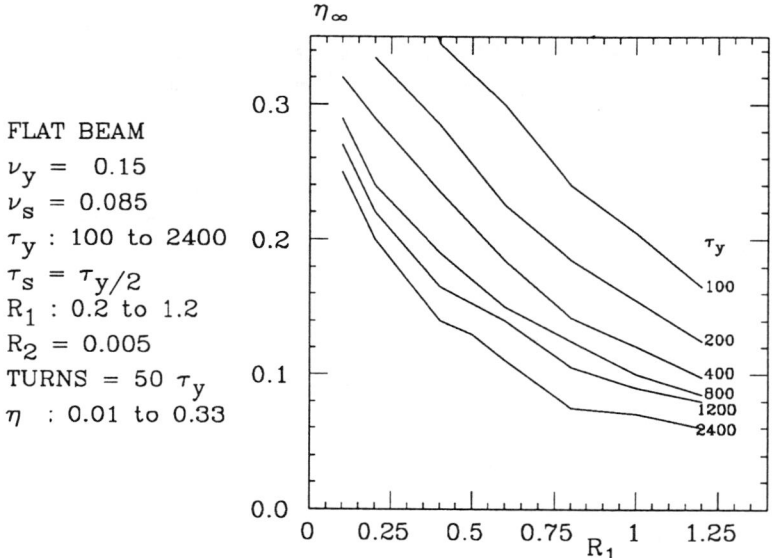

Figure 4: The R_1 dependence of η_∞ for different values of $\delta_y = 1/\tau_y$.

Note that the disruption parameter D is given by

$$D = 8\pi\eta R_1.$$

This implies that the limiting value D_∞ is almost universal.

Dependence on R_2 In Fig. 5, we plot the threshold value η_∞ versus R_2 for different values of δ_y. It is remarkable that, at the beginning, the beam is more stable when increasing R_2. However, beyond a certain value of R_2 corresponding to a maximum for η_∞, the stability of the beam decreases and the beam-beam limit is reduced by almost a factor two for R_2 close to unity.

These results are a consequence of the symplecticity condition (satisfied by our synchro-beam mapping) on the synchro-betatron coupling. Indeed, as shown in Fig. 6, the bunch length at the beam-beam limit is an increasing function of R_2. This means that the betatron oscillations can 'release' part of their energy to the synchrotron oscillations and, for small values of R_2, this mechanism helps stabilizing the beam. Beyond a certain value of R_2, the enhanced longitudinal modulation of the CP, associated with larger synchrotron oscillations, becomes the dominant effect and reduces the threshold value η_∞. Let us notice that, for values of R_2 comparable to unity, the bunch lengthening tends to saturate. The corresponding decrease of the transverse beam size before blow up is then a consequence of the reduction of η_∞.

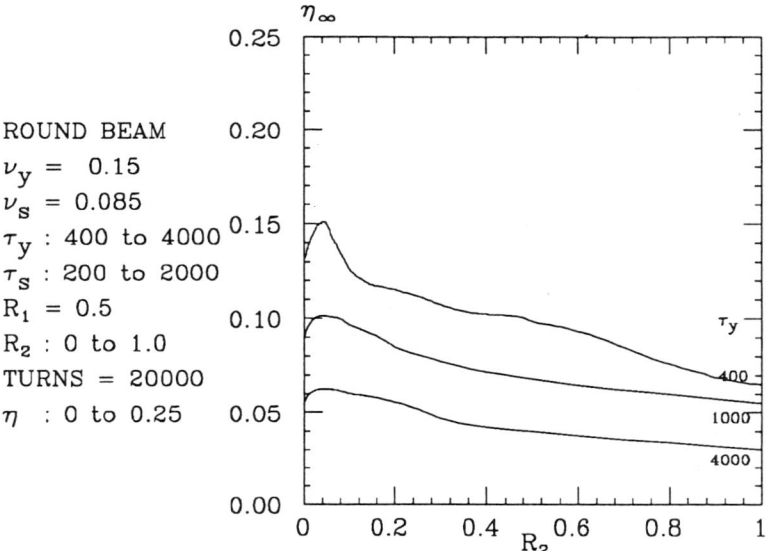

Figure 5: The R_2 dependence of η_∞ for different values of $\delta_y = 1/\tau_y$.

Summary of Simulation Results Our preliminary simulation results can be summarized as follows:

- For a fixed beam-beam parameter η, we observe clear synchro-betatron side-bands.

- For η small enough, the equilibrium beam size remains almost constant. It grows up suddenly at $\eta = \eta_\infty$.

- The threshold value η_∞ is smaller for slower damping.

- The threshold is lower for larger values of R_1. The product $\eta_\infty R_1$, proportional to the disruption parameter D, is nearly constant and of the order of 0.1, for $R_2 = 0.005$.

- The threshold η_∞ has a peak corresponding to a value of R_2 roughly given by $R_1/10$. For larger values of R_2 it is considerably reduced. Meanwhile we observe a substantial increase of bunch length.

- For round beams we usually get a larger beam size than for flat beams.

Note that we use relatively fast damping. For more realistic values of the δ's, the threshold seems to be smaller.

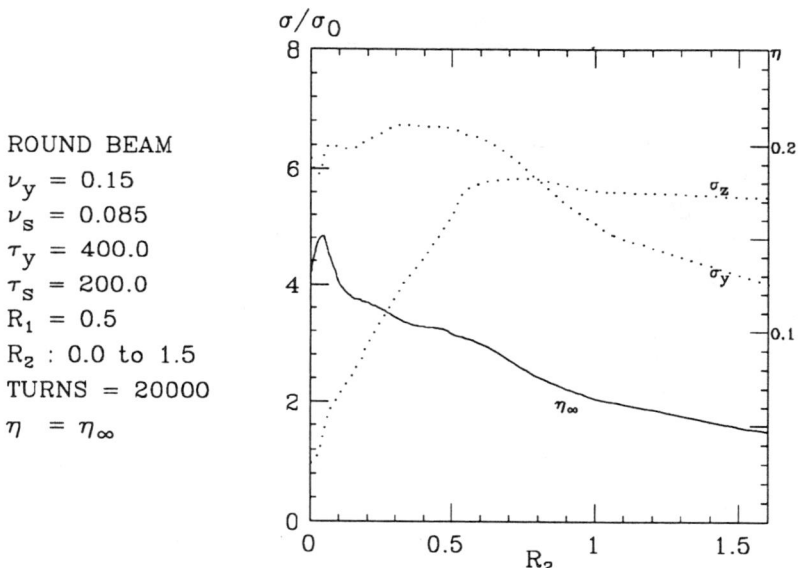

Figure 6: The R_2 dependence of η_∞, σ_y and σ_z (just before blow up) for $\delta_y = 1/400$ and $R_1 = 0.5$.

DISCUSSION

We have proposed a symplectic 'synchro-beam' mapping, which describes the weak-strong beam-beam interaction in presence of synchrotron oscillations. This mapping is valid in the extreme relativistic approximation and to first order in the betatron slopes of the particles in the weak and in the strong bunches.

The symplecticity condition can be fulfilled only including the energy variation of the particles in the weak beam due to the electric field generated by the strong beam. When the bunch length is of the order of the betatron function at interaction ($R_1 \sim 1$) and the beam-beam parameter is close to its threshold value ($\eta \sim 0.05$), the contribution of the longitudinal component of the electric field to the energy variation is comparable to that of the transverse component: from the mapping (11), we see that they are both proportional to the parameter R_2.

In our weak-strong simulation, that includes synchrotron oscillations plus only one degree of freedom for betatron oscillations, we used fast damping and rather large values for R_2. In existing low energy e^+e^- storage rings, damping decrements in the range 10^{-5}–10^{-4} are rather frequent. This is not incompatible with the threshold values η_∞ we have found, since they tend to decrease for slower damping (see Fig. 5). As for the values of R_2 in existing or future machines, they are typically in the range 10^{-4}–10^{-3}, although values as large as 10^{-2} may be expected for round beams. This is a consequence of the fact that, in flat beam operation, the vertical emittance is usually made as small as possible. Indeed no storage ring has ever been operated at optimal coupling and the reason might be that optimal coupling corresponds to larger values of R_2.

By a generalization of our mapping (that will be discussed in a forthcoming paper), it is possible to perform six-dimensional weak-strong simulation using the Bassetti-Erskine formula [11]. Since our mapping can be applied at the nominal IP, it may be interesting to include it in some six-dimensional tracking code, such as MAD or SAD, to study pp or $\bar{p}p$ colliders, where symplecticity is more important than in the case of e^+e^- rings.

As pointed out before, one should consider both the energy variation due to the beam-beam interaction and the modulation of the CP. Some existing codes include only the latter. When R_2 is large, as may be the case for a round beam, these codes are not reliable. It could appear that we can totally ignore the energy variation and still preserve symplecticity by continuously decreasing R_2. However, $R_2 = 0$ represents a singular point for the transformation to normalized variables and we cannot go back to the original symplectic mapping for the physical variables. Indeed the energy change can be totally neglected only when σ_ϵ^0 is infinitely large. For example, in Fig. 7 we show the evolution of

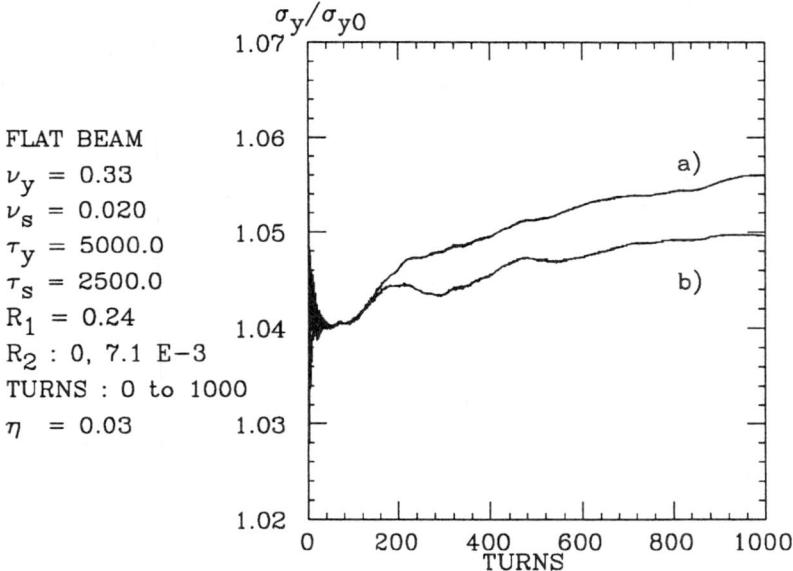

Figure 7: Evolution of the vertical beam size for $R_2 = 0$ (a) and for $R_2 = 7.1 \times 10^{-3}$ (b).

the vertical beam size as a function of the number of turns, over a time interval shorter than one damping time. The corresponding divergence between the case of vanishing R_2 (a) and of small but finite R_2 (b) is a direct consequence of the lack of symplecticity for $R_2 = 0$ and suggests the possibility of important long term effects in hadron colliders.

Our mapping can be used also in the strong-strong case. One should slice the two bunches longitudinally (i.e. in the z-direction) and evaluate the effect of the collision between two slices [10]. The number of slices should be such that the changes of x' and y' remain small enough, so that one can still neglect terms of order $(x')^2$ and $(y')^2$. As shown in Ref. 5, the conservation of energy and momentum after the collision between two slices is automatically insured by our mapping.

ACKNOWLEDGEMENTS

A. Sessler, J. Tennyson, A. Piwinski, E. Keil and B. Zotter are acknowledged for useful comments and discussions.

REFERENCES

1. J. Augustin, Orsay, 36-69 (1969).
2. A. Piwinski, IEEE Trans Nucl. Sci. **NS-24**, 1408 (1977).
3. M. Bassetti, *Longitudinal Energy Changes in a Bunch*, LNF-T-105 (1978).
4. M. Bassetti, *Come Migliorare gli Attuali Limiti di Luminosita'*, LNF-ARES-18 (1989).
5. K. Hirata, H. Moshammer, F. Ruggiero and M. Bassetti, *Synchro-Beam Interaction*, CERN SL-AP/90-02 (1990).
6. S. Kheifets, DESY, Petra Note 119 (1976).
7. B.W. Montague, CERN ISR-GS 36 (1975).
8. K. Hirata and F. Ruggiero, *Treatment of Radiation for Multiparticle Tracking in Electron Storage Rings*, CERN LEP-TH/89-43 (1989), presented at the XIV Int. Conf. on High Energy Accelerators, Tsukuba, 1989.
9. E. Keil and R. Talman, Part. Accel. **14**, 109 (1982).
10. S. Krishnagopal and S. Siemann, *Bunch Length Effects in the Beam-Beam Interaction*, CLNS 89/936 (1989).
11. M. Bassetti and G. Erskine, CERN ISR TH/80-06 (1980).
12. A. J. Dragt, in *Physics of High Energy Accelerators*, proceedings of the Summer School on High Energy Particle Accelerators, Fermilab, 1981, edited by R.A. Carrigan, I.R. Huson, and M. Month (AIP Conf. Proc. No. 87) (AIP, New York, 1982), p. 147.
13. A. Piwinski, *Synchro-Betatron Resonances*, Proc. 11th Int. Conf. on High Energy Accelerators, CERN, 1980, p. 638.

DETERMINATION OF THE HIGH-FREQUENCY BEHAVIOR OF THE IMPEDANCE FROM LOW-FREQUENCY DATA

S. A. Heifets
Continuous Electron Beam Accelerator Facility,
12000 Jefferson Avenue, Newport News, Va. 23606

ABSTRACT

The finite frequency sum rule is given defining the high-frequency roll-off of the impedance from the low-frequency modal analysis.

Impedance is an important characteristic of an accelerator describing the interaction of a particle with the beam environment. For example, the longitudinal impedance considered henceforth is directly related to the energy loss of a particle given by an integral of the impedance over frequency. For structures with axial symmetry it will suffice to find the longitudinal impedance because the transverse impedance (describing the transverse kick) and the longitudinal impedances are related.[1]

For bunches long compared to the beam pipe radius the impedance is dominated by the low frequency modes which can be found numerically. For very short bunches, however, the number of modes giving substantial contribution to the energy loss is very large and that makes the modal analysis impractical. At the same time the impedance rolls-off slowly, as $Z(\omega) \propto \omega^{-1/2}$ for $\omega \to \infty$. That makes the high frequency tail contribution dominant. In calculations the high frequency tail is usually parameterized and added to the contribution of the low frequency modes. In this paper we show that this procedure can be done rigorously provided the high frequency behaviour of the impedance is known. The method not only eliminates the dependence of the result on the number of modes taken into account but also defines the parameter of the high-frequency behavior of the impedance of a given structure from the low-frequency data for the structure. The method is based on the finite energy sum rule well known in particle physics[2], and called here the finite frequency sum rule.

Impedance $Z(\omega)$ is a Fourier transform of the wake potential $W(t)$,

$$W(t) = \int \frac{d\omega}{2\pi} Z(\omega) e^{-i\omega t}. \tag{1}$$

Because $W(t)$ is real,

$$Z^*(\omega) = Z(-\omega) \tag{2}$$

real and imaginary parts of the impedance are even and odd functions of ω correspondingly:

$$Z'(\omega) = Z'(-\omega), \quad Z''(\omega) = -Z''(-\omega). \tag{3}$$

It follows from causality that $W(t) = 0$ for $t < 0$. Hence, all singularities of $Z(\omega)$ are in the lower half plane of complex ω.

It has been found for a number of models[3] that in the high frequency limit $Z'(\omega)$ may be presented as series over $\omega^{-1/2}$ with the leading term

$$Z'(\omega) \propto \frac{\alpha}{\sqrt{\omega}}, \quad \omega \to \infty \tag{4}$$

The arguments based on diffraction theory[4] allow the expectance of such behavior in a very general case.

These properties are enough to yield the usual dispersion relations:

$$Z'(\omega) = \frac{1}{\pi} \int_{-\infty}^{\infty} \frac{d\nu\, Z''(\nu)}{\nu - \omega}, \tag{5}$$

$$Z''(\omega) = -\frac{1}{\pi} \int_{-\infty}^{\infty} \frac{d\nu\, Z'(\nu)}{\nu - \omega}. \tag{6}$$

The integrals here are understood as a principal value.

In compliance with Eq. (2) we can write the expansion in the high frequency limit as

$$Z(\omega) = \frac{\alpha}{\sqrt{\omega}} \exp^{i\pi/4} + \frac{i\beta}{\omega} + o\left(\frac{\omega_0}{\omega^{3/2}}\right) \tag{7}$$

with real parameters $\alpha = \alpha^*$, $\beta = \beta^*$. The cut is implied along the real axis ω.

Consider now the function

$$\tilde{Z}(\omega) = Z(\omega) - \frac{\alpha}{\sqrt{\omega}} \exp^{i\pi/4} - \frac{i\beta}{\omega} \tag{8}$$

It satisfies the relation

$$\tilde{Z}(-\omega) = \tilde{Z}^*(\omega) \tag{9}$$

and is analytic in the upper half of the complex plane ω. Thus the dispersion relation for $\tilde{Z}(\omega)$ have the same structure as that for $Z(\omega)$ except the additional pole at $\omega = 0$:

$$\tilde{Z}'(\omega) = \frac{2}{\pi} \int_0^{\infty} \frac{\nu\, d\nu\, \tilde{Z}''(\nu)}{\nu^2 - \omega^2}, \tag{10}$$

$$\tilde{Z}''(\omega) = -\frac{\beta}{\omega} - \frac{2\omega}{\pi} \int_0^{\infty} \frac{d\nu\, \tilde{Z}'(\nu)}{\nu^2 - \omega^2}. \tag{11}$$

Asymptotically, at $\omega \to \infty$,

$$\tilde{Z}'' \to \frac{1}{\omega^{3/2}}$$

Therefore, Eq. (11) yields the superconvergence relation

$$\int_0^{\infty} d\omega\, \tilde{Z}'(\omega) = \pi\beta/2. \tag{12}$$

Otherwise, $Z''(\omega)$ would roll off as ω^{-1}.

For sufficiently large $\omega > \omega_0$,

$$Z'(\omega) = \frac{\alpha}{\sqrt{\omega}} \left[1 + o\left(\frac{\omega_0}{\omega}\right)\right]$$

For $\omega_c > \omega_0$ Eq. (12) gives the finite sum rule

$$\int_0^{\omega_c} d\omega \left[Z'(\omega) - \frac{\alpha}{\sqrt{\omega}}\right] = \frac{\pi\beta}{2}\left[1 + o\left(\frac{\omega_0}{\omega_c}\right)\right], \tag{13}$$

or

$$\alpha = \frac{1}{\sqrt{2\omega_c}} \int_0^{\omega_c} \left[d\omega\, Z'(\omega) - \frac{\pi\beta}{2} \right] \left[1 + o\left(\frac{\omega_0}{\omega_c}\right) \right]. \tag{14}$$

The finite frequency sum rule Eq. (14) has to be saturated by the low frequency modes. The low frequency impedance of the modes with frequencies ω_n, quality factors $Q_n = \omega_n/2\gamma_n$, and loss factors ξ_n is

$$Z(\omega) = i \sum_n \xi_n \left(\frac{1}{\omega - \omega_n + i\gamma_n} + \frac{1}{\omega + \omega_n + i\gamma_n} \right). \tag{15}$$

The loss factor is related to the shunt impedance of the structure:

$$\xi_n = \frac{L\omega_n}{2} \left(\frac{R}{Q}\right)_n$$

where L is the length of the structure, and (R/Q) is defined as in the code URMEL[5] (in units Ohm/m). Thus, Eq. (14) yields the parameter α:

$$\alpha = \frac{\pi}{\sqrt{2\omega_c}} \left[\sum_{\omega_n < \omega_c} \xi_n - \frac{\beta}{2} \right]. \tag{16}$$

The impedance is given then by the parameters of the modes with frequencies $\omega_n < \omega_c$:

$$Z(\omega) = i\theta(\omega_c - \omega) \sum_{\omega_n < \omega_c} \xi_n \left(\frac{1}{\omega - \omega_n + i\gamma_n} + \frac{1}{\omega + \omega_n + i\gamma_n} \right)$$
$$+ \theta(\omega - \omega_c) \frac{\pi(1+i)}{2\sqrt{\omega\omega_c}} \left[\sum_{\omega_n < \omega_c} \xi_n - \frac{\beta}{2} \right]. \tag{17}$$

The sum Eq. (17) and the parameter α are independent of the choice of ω_c provided ω_c is large enough. Practically that can be checked calculating the sum in Eq. (16) as function of ω_c and parameterizing the result in the form

$$\Sigma(\omega_c) = \sum_{\omega_n < \omega_c} \xi_n = \frac{\beta}{2} + \frac{2\alpha}{\pi} \sqrt{\omega_c} + o\left(\frac{\omega_0}{\omega_c}\right). \tag{18}$$

The theorem has been checked for two cases: a pill box cavity with attached pipes and for the CEBAF cavities.

The narrow band impedance of a pill box cavity with pipes has been discussed in the paper[6]. For frequencies close to the cut-off frequency $k_{c.off} a \simeq 1$ the impedance may be described as given by a sum of modes with the field pattern similar to that in a closed pill-box cavity but with the widths due to the openings into the beam pipes. The link to the beam pipes gives also coupling of the modes which may be taken into account by perturbations. The explicit form of the impedance obtained neglecting the mode coupling is given by Eqs. (10), (11), and Eq. (12) of the reference[6]. This approximation works usually very well except relatively rare cases of close (degenerated) modes when the coupling of the degenerated modes may give a noticeable correction to the impedance. The impedance of the reference [6] can be represented in the form Eq. (17) with loss factors

$$\xi = \frac{16}{g} \frac{J_0^2(\nu p)}{\nu^2 J_1^2(\nu)} \sin^2\left(\frac{\omega_{n,\nu} g}{2c}\right) \tag{19}$$

for all odd modes with frequencies close to the resonance frequencies

$$\frac{\omega_{n,\nu}}{c} = \sqrt{\left(\frac{\nu}{b}\right)^2 + \left(\frac{2\pi n}{g}\right)^2},$$

or

$$\xi = \frac{16}{g} \frac{J_0^2(\nu p)}{\nu^2 J_1^2(\nu)} \cos^2\left(\frac{\omega_{n,\nu} g}{2c}\right) \tag{20}$$

for even modes with the resonance frequencies

$$\frac{\omega_{n,\nu}}{c} = \sqrt{\left(\frac{\nu}{b}\right)^2 + \left(\frac{2\pi}{g}\right)^2 (n+1/2)^2}.$$

Here g is the cavity length, a and b are radii of the pipe and the cavity, ν-s are roots of the Bessel function $J_0(\nu) = 0$, n is an integer. The shunt impedances $R/Q = 2\xi_n/\omega_n$ of the first 10 modes calculated from Eqs. (19), (20) are compared in the following table with the shunt impedances given by URMEL for a pill-box cavity with parameters $a = 10$ cm, and $g = b = 60$ cm. The metallic boundary conditions were used to close pipes at $z = 120$ cm for URMEL calculations.

f URMEL	R/Q URMEL	f (MHz)	R/Q (Ohm)
535.010	4.132	535.385	4.101
667.022	13.529	665.561	15.464
691.841	4.329	774.053	1.320
855.383	2.871	851.021	3.120
940.020	0.043	869.190	7.376
1014.757	0.589	1018.177	0.547
1018.212	3.969		
1066.043	0.043	1063.239	0.085
1089.907	4.154	1092.254	3.797

Parameter α Eq. (16) can be obtained from Eqs. (19), (20) calculating the sum

$$\sigma(ka) = \frac{a}{\sqrt{ka}} \sum_{k<k_n,\nu} \xi_{n,\nu} \tag{21}$$

where $k = \omega/c$. The function $\sigma(ka)$ is depicted in Fig. 1 as the function of dimensionless parameter ka for a cavity with parameters $a/b = 0.2, g/a = 10.0$. The asymptotic value σ_{as} should be compared with

$$\sigma_{as} = \left(\frac{2}{\pi}\right)^{3/2} \sqrt{\frac{g}{a}} \tag{22}$$

predicted by the impedance[3] of a pill box cavity with pipes in the high frequency limit:

$$Z(\omega) = \frac{\alpha_{PB}}{\sqrt{\omega}}, \quad \alpha_{PB} = \frac{2}{a}\sqrt{\frac{g}{\pi c}}. \tag{23}$$

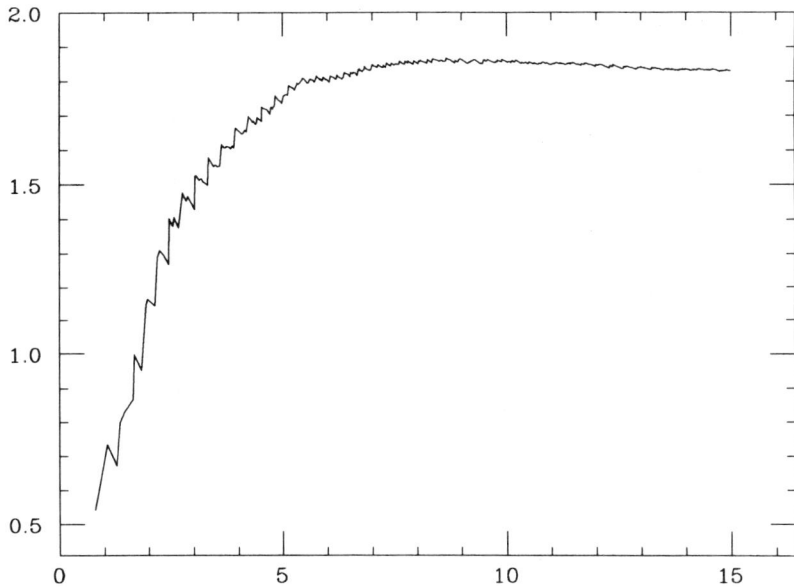

Fig. 1. The sum of the loss factors with $k < k_{\max}$ over $\sqrt{(k_{\max})}$ vs. $ka_{\max}$ for a cavity with pipes, $a/b = 0.2$, $g/a = 10.0$.

The asymptotic is reached at frequencies $2 - 3$ times of the cut-off frequency $k_{c.off}a = 2.405$ and is in a reasonable agreement with Eq. (21) taking into account all approximations we used.

The sum

$$\sum_{\omega_n < \omega_c} \frac{\omega_n}{2\pi} \left(\frac{R}{Q}\right)_n \tag{24}$$

has been calculated as a function of ω_c for CEBAF cavities ($L = 0.5$ m, the fundamental frequency 1.5 GHz, the cut-off frequency 3.2 GHz). The loss factors of the modes up to 6 GHz have been calculated with URMEL[7] and compared with experimentally measured loss factors. In calculations the beam pipes were closed at distances large enough to make the loss factors of the modes above cut off frequency independent on the length of the pipes. The sum behaves in accordance with Eq. (16) as $\Lambda\sqrt{\omega_c}$ with $\Lambda = 2.67 \, 10^6$ Ohm.MHz/m/cm$^{1/2}$. The ω_c dependence of the sum is depicted in Fig. 2 which shows that Eq. (16) is saturated quite well by the modes below the doubled cut-off frequency.

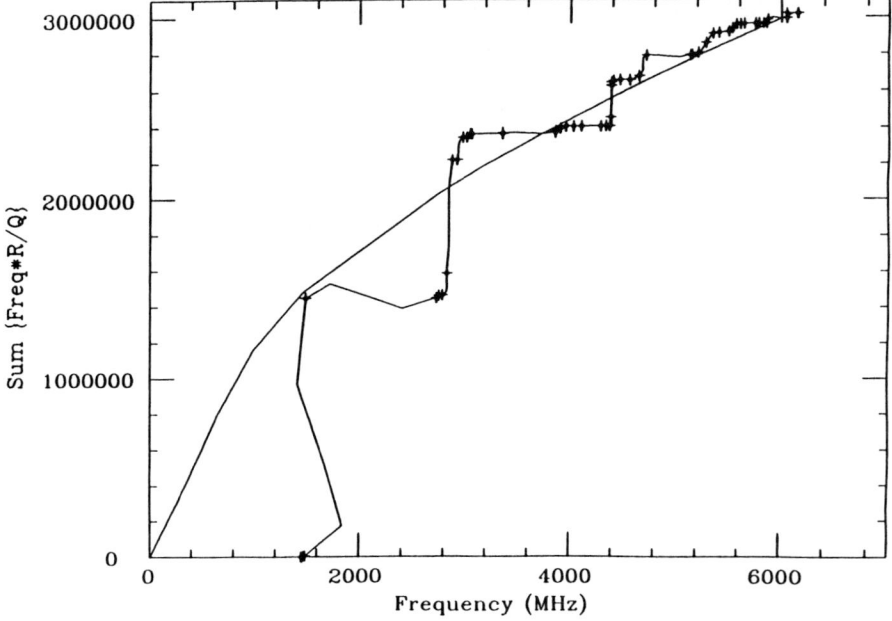

Fig. 2. Dependence of the sum Eq. (24) on the maximum frequency ω_c.

ACKNOWLEDGEMENTS

The author wishes to thank V. Novikov for his comments on the paper and to B. Yunn for the data on CEBAF cavities.

REFERENCES

1. K. L. Bane, P. B. Wilson, T. Weiland, SLAC-PUB-3528 (1984).
2. R. Dolen, D. Horn, C. Schmid, "Finite-Energy Sum Rules and Their Application to $\pi - N$ Charge Exchange", Phys. Rev., V. 166, Number 5 (1968).
3. G. D. Dome, IEEE Trans. Nucl. Sci., Vol. NS-32, No. 5 (1985), p. 2531.
   J. D. Lawson, RHEL/M 144 (1963).
   S. A. Heifets, S. Kheifets, Particle Accelerators, Vol. 23 (1989).
4. K. Bane and M. Sands, SLAC-PUB-4441 (1987).
   R. B. Palmer, SLAC-PUB-4433 (1987).
   S. Heifets, Phys. Rev. D, 40, 9 (1989).
5. T. Weiland, Nucl. Instrum. Methods, Vol. 216, 329 (1983).
6. S. A. Heifets Particle Accelerators, Vol. 23 (1989).
7. B. Yunn, CEBAF, unpublished.

CURVATURE EFFECTS TO BEAM DYNAMICS—APPLIED TO THE ASYMMETRIC B FACTORY

King-Yuen Ng

Fermi National Accelerator Laboratory,[*] *Batavia, IL 60510*

ABSTRACT

The resonances in the toroidal beam pipes of the asymmetric storage rings are computed. The change in tunes due to pipe's curvature is studied. These effects are found to be negligibly small. The issue of free-space radiation is discussed.

I. INTRODUCTION

The curvature of the beam pipe can lead to residual longitudinal and transverse forces on the particle beam even if the velocity of the particles approaches the velocity of light. In Section II, we compute the resonances that exist in the toroidal beam pipes of the asymmetric storage rings. In Section III, the modifications to the betatron tunes are studied. We find that the effects of these residual forces are negligibly small. Finally in Section IV, free-space radiation is reviewed and its relation to the radiation inside a closed vacuum chamber is discussed.

II. RESONANT IMPEDANCE

In a circular accelerator or storage ring, the beam pipe has the topology of a toroid. If the beam travels with velocity βc at a radius R, the electromagnetic wave traveling with the beam will have a phase velocity $r\beta c/R$ at a radius r. When this phase velocity exceeds c, the electromagnetic wave will be able to propagate (in analogy to a *straight* waveguide). Because the toroidal beam pipe is closed, these are eigenwaves with discrete frequencies. These waves will act back on the beam and the beam sees a resonant impedance. As described above, the condition for this to happen is[1]

$$\frac{R_+\beta}{R} > 1 , \qquad (2.1)$$

---

[*]Operated by the Universities Research Association, Inc., under contract with the U.S. Department of Energy.

where R_+ is the radius of the outer edge of the beam pipe. In other words, $R_+ = R + b$, where b is the radial distance from the beam to the outer wall of the pipe. The asymmetric B factory consists of storage rings for electrons and positrons of a few GeV. As a result, Criterion (2.1) is always satisfied.

These toroidal resonances have been studied extensively in the literature.[1,2] Here,[3] the energies of the two rings are 3.1 GeV and 9.0 GeV with corresponding ring circumferences 733.3 m and 2200.0 m. The cross sections of the beam pipes are taken to be rectangular with full height $h = 3$ cm and half-width $b = 6$ cm. The walls of the pipes are assumed to be stainless steel with a conductivity of $\sigma = 1.37 \times 10^6$ (ohm-m)$^{-1}$. The frequencies and impedances $Z_{i,j}^{TE,TM}$ of the first few resonances are listed in Table I and Table II. Here, (i,j) specifies the ordering of the modes, respectively in the radial and vertical directions. Unlike the usual definitions of TE and TM, they are defined here with respect to the vertical axis of the toroid.

Mode	Frequency GHz	Harmonics	$Z_{ij}^{TE,TM}/n$ ohms	Q
TE_{10}	182	4.46E+5	6.24E−3	3.20E+4
TM_{10}	228	5.58E+5	8.61E−3	1.67E+4
TE_{20}	287	6.57E+5	9.71E−2	9.57E+4
TM_{20}	314	7.67E+5	2.00E−2	1.95E+4
TE_{30}	359	8.78E+5	1.90E−1	1.35E+5
TM_{30}	408	9.97E+5	2.17E−2	2.23E+4

Table I: The first six lowest frequency modes of the 3.1 GeV ring.

In our model, the cutoff frequencies (harmonics) of the two rings are, respectively, $f_c = 2.50$ GeV ($n_c = 6120$) and $f_c = 2.50$ GeV ($n_c = 18300$). We see that the resonances occur at very high frequencies but the impedances per harmonic can be appreciable. The impedance per harmonic appears to increase for higher modes as depicted in Tables I and II. In fact, it will fall off very fast after some modes. For example, Z_{ij}/n reaches a maximum of 0.213 ohms (TE_{40}) for the low-energy ring and 0.0311 ohms (TE_{40}) for the high-energy ring. These resonances appear to be different from those impedances arising from discontinuities of the vacuum chamber, because beam particles at different radii R see different sets of toroidal resonances. Therefore, the impedance of the ij-th mode $Z_{ij}(R)$ is a

Mode	Frequency GHz	Harmonics	$Z_{ij}^{\text{TE,TM}}/n$ ohms	Q
TE_{10}	316	2.31E+6	9.13E−4	4.22E+4
TM_{10}	395	2.90E+6	1.26E−3	1.19E+4
TE_{20}	465	3.41E+6	1.42E−2	1.26E+5
TM_{20}	543	3.99E+6	2.93E−3	2.57E+4
TE_{30}	622	4.56E+6	2.79E−2	1.78E+5
TM_{30}	706	5.18E+6	3.17E−3	2.93E+4

Table II: The first six lowest frequency modes of the 9.0 GeV ring.

function of R, the radius of curvature of the particle orbit. However, it has been shown[4] that if $Z_{ij}(R)$ varies slowly across the beam, $\langle Z_{ij}/n \rangle$ is exactly the longitudinal Z/n that drives the self-bunching (or microwave) instability of the beam. Therefore, random currents (Schottky noise) which exist at arbitrary frequencies on a bunched beam can in principle generate internal bunch instabilities. Here, for the TE_{10} mode of the low-energy ring, Z_{10}^{TE} varies by 9.7% across a beam width of 1 mm.

Because of the intense synchrotron radiation from the electrons or positrons, intensive pumping often requires a rather wide width of the beam pipe. For example, the beam orbit of the 9.0 GeV ring can be $b = 41$ cm from the outer wall of the beam pipe. The lowest resonance (TE_{10}) then occurs at 108 GeV only, much lower than the corresponding 316 GeV for $b = 6$ cm. However, the impedance per harmonic becomes much smaller, $Z/n = 1.16 \times 10^{-22}$ ohms. As a result, theses resonances play no role in the beam dynamics at all and can be safely neglected. Table III shows the frequency and Z/n of the lowest resonance (TE_{10}) as a function of the radial distance b from the beam orbit to the outer wall of the vacuum chamber.

III. CENTRIFUGAL SPACE-CHARGE FORCE

In a *straight* beam pipe, a particle beam also induces transverse space-charge force on a beam particle. This force is of the order γ^{-2} due to the near cancellation of the electric and magnetic fields. However, in curved geometry, this cancellation

b	Frequency	Harmonic	Z_{10}^{TE}/n	Q
cm	GeV		ohms	
6	316	2.32E+6	9.14E−04	4.22E+4
12	210	1.54E+6	1.30E−06	4.50E+4
18	168	1.23E+6	9.11E−10	4.66E+4
24	143	1.05E+6	4.75E−13	4.77E+4
30	127	9.33E+5	2.10E−16	4.84E+4
36	115	8.47E+5	8.37E−20	4.89E+4
42	106	7.80E+5	3.10E−23	4.92E+4

Table III: Mode TE_{10} for the 9 GeV ring as a function of b, distance from beam orbit to outer wall of beam pipe.

is incomplete, leaving behind[5,6]

$$\left|\vec{E} + \vec{v} \times \vec{B}\right|_{sc} \sim \frac{\lambda}{4\pi\epsilon_0 R}, \qquad (3.1)$$

even when the particle velocity $\vec{v}$ equal c. In the above, R is the radius of the beam orbit and λ is the line charge density of the beam. This residual force has been termed "centrifugal space-charge force" (CSCF). Lee[7] pointed out that there is another transverse force of equal magnitude in the curved beam pipe. This second force is a result of oscillations of the particle's kinetic energy in the present of the beam's electric potential, as the particle undergoes betatron oscillations. Although these two transverse forces will cancel each other considerably so that excitation high-order resonances may no longer be a concern, nevertheless, they can still affect the betatron tunes and chromaticities.

If we denote by δr the radial deviation of a particle from the ideal beam orbit, the equation governing the radial motion can be linearized to

$$\delta r'' = \left[-\frac{1}{r} - \frac{e}{\gamma m c}\left(\frac{E_r}{cr} + \frac{1}{c}\frac{\partial E_r}{\partial r} + \frac{\partial B_z}{\partial r}\right)\right]_{r=R} \delta r, \qquad (3.2)$$

where m is the mass of the particle, 'prime' is the derivative taken along the beam orbit, and we have let $v \to c$. The radius of the orbit R is given by

$$\gamma m c = eR\left(\frac{E_r}{c} + B_z\right) \qquad (3.3)$$

In Eq. (3.2), the radial electric field E_r arises from space charge while the vertical magnetic flux density B_z contains a space-charge part and an external part. The term E_r/cr is the space-charge force due to kinetic-energy oscillation while the rest is CSCF. Following Lee, we separate out the space-charge (sc) parts of the fields and define at $r = R$

$$F = -\left(\frac{E_r}{c} + B_z\right)_{sc}\bigg|_{r=R} , \tag{3.4}$$

$$\frac{\partial G}{\partial R} = -\left(\frac{E_r}{cR} + \frac{1}{c}\frac{\partial E_r}{\partial r} + \frac{\partial B_z}{\partial r}\right)_{sc}\bigg|_{r=R} \tag{3.5}$$

Then, Eq. (3.2) can be rewritten as

$$\delta r'' = \left[-\frac{1}{R} - \frac{1}{BR(1-F/B)}\left(-\frac{\partial G}{\partial R} + \frac{\partial B}{\partial R}\right)\right]\delta r , \tag{3.6}$$

where B is the vertical guide field of the dipoles and quadrupoles. Lee showed that

$$F = \mathcal{O}\left(\frac{Z_0\lambda}{4\pi R}\right) , \tag{3.7}$$

and is positive, while

$$\frac{\partial G}{\partial R} = \mathcal{O}\left(\frac{F}{R}\right) , \tag{3.8}$$

with $Z_0 = 376.7$ ohms and is also positive. The asymmetric rings carry average beam currents of 3 amp with bunches containing $N = 1.589 \times 10^{11}$ particles each. In the higher- (lower-) energy ring, there are 864 (288) bunches of rms length $\sigma_\ell = 1.0$ cm (1.4 cm). Expressing in terms of the electron classical radius $r_e = 2.818 \times 10^{-15}$ m, we obtain

$$\frac{F}{B} \sim \frac{r_e N}{4\sigma_\ell \gamma} = \begin{cases} 6.36 \times 10^{-7} & \text{higher-energy ring} \\ 1.32 \times 10^{-6} & \text{lower-energy ring ,} \end{cases} \tag{3.9}$$

showing that the effects of curvature should be very small. In computing the modification to the betatron tunes $\Delta\nu$, $\partial G/\partial R$ can be neglected at the quadrupoles where $\partial B/\partial R$ is large, but must be retained elsewhere. We get

$$\Delta\nu \sim \mathcal{O}\left[-\frac{\nu}{2}\left(\frac{F}{B}\right)\right] + \mathcal{O}\left[-\frac{1}{4\pi}\int_0^{2\pi R}\frac{\beta(s)}{BR}\left(\frac{\partial G}{\partial R}\right)ds\right]$$

$$\sim \mathcal{O}\left[-\frac{\nu}{2}\left(\frac{F}{B}\right)\right] + \mathcal{O}\left[-\frac{\bar{\beta}}{2B}\left(\frac{\partial G}{\partial R}\right)\right]$$

$$\sim \mathcal{O}\left[-\frac{\nu}{2}\left(\frac{F}{B}\right)\right] + \mathcal{O}\left[-\frac{\pi}{\nu}\left(\frac{F}{B}\right)\right] , \tag{3.10}$$

where $\beta(s)$ is the horizontal beta-function. We see that the modifications to the tunes are extremely small. The same applies to the vertical tunes.

IV. FREE-SPACE RADIATION

In free space without any beam pipe, the power spectrum radiated by a particle carrying charge e and traveling along a curve with radius of curvature ρ, as derived by Schwinger,[8] is (in mks units)

$$P(\omega) = \left(\frac{Z_0 e^2 c}{\rho}\right) \left(\frac{3^{\frac{1}{6}}\Gamma(\frac{2}{3})}{4\pi^2}\right) \left(\frac{\omega}{\omega_0}\right)^{\frac{1}{3}} \left\{1 - \frac{1}{2}\Gamma(\frac{2}{3})\left(\frac{\omega}{\omega_0}\right)^{\frac{1}{3}} + \cdots\right\} \quad (4.1)$$

for $\omega \ll \omega_{fs}$, and drops exponentially as

$$P(\omega) = \left(\frac{Z_0 e^2 c \gamma^4}{4\pi\rho}\right) \left(\frac{2}{\pi}\right)^{\frac{1}{2}} \left(\frac{\omega_0}{\omega_{fs}}\right) \left(\frac{\omega}{\omega_{fs}}\right)^{\frac{1}{2}} e^{-4\omega/3\omega_{fs}} \left\{1 + \frac{55}{96}\frac{\omega_{fs}}{\omega} + \cdots\right\} \quad (4.2)$$

when $\omega \gg \omega_{fs}$. In the above, the angular frequency is defined as $\omega_0 = \beta c/\rho$ and the critical angular frequency is $\omega_{fs} = 2\gamma^3 \omega_0$. Note that ω_{fs} is only an order of magnitude; it has also been defined as $\frac{3}{2}\gamma^3 \omega_0$ by some authors. The power radiated into each harmonic $n = \bar{R}\omega/\beta c$ is $P_n = \omega_0 P(\omega)$, where $\bar{R}$ is the average radius of the particle orbit. The impedance at the n-th harmonic seen by beam particle is given by $Z_n = 2P_n/I_n^2$, where $I_n = e\beta c/\pi\bar{R}$ is the n-th harmonic Fourier current amplitude of the δ-function charge under consideration. Including the reactive part[9] but neglecting the higher-order terms, we obtain for $n \ll n_{fs}$,

$$\frac{Z_n}{n} = Z_0 n^{-\frac{2}{3}} \left(\frac{\bar{R}}{\beta\rho}\right) \left[\frac{3^{\frac{1}{6}}\Gamma(\frac{2}{3})}{\sqrt{3}}\right] \left(\frac{\sqrt{3}}{2} - i\frac{1}{2}\right). \quad (4.3)$$

The squared-bracketed term gives 0.93889. Since $\bar{R}$ is usually larger than ρ, therefore very closely

$$\left|\frac{Z_n}{n}\right| \sim Z_0 n^{-2/3} \quad (4.4)$$

with $Z_0 = 376.7$ ohms.

There is a fundamental difference in the quality of the free-space radiation and the radiation inside a beam pipe. In free space, the radiation spectrum is continuous, whereas inside a beam pipe it consists of discrete resonances in order to satisfy the boundary condition at the pipe's walls, as was demonstrated in Section II. In a storage ring, the beam is shielded by a beam pipe so that radiation into the infinite free space is not possible below cutoff frequency of the beam pipe. Above cutoff, however, electromagnetic waves can propagate inside the vacuum chamber. When the wavelength of the radiation is a few times less than the radius of the beam pipe, it appears that the presence of the pipe's walls is irrelevant. This implies that the coupling impedance should be given roughly by the free-space radiation value.

The resonances inside a toroidal beam pipe start with the TE_{10} mode at a harmonic n_s given approximately for large h by[9]

$$\frac{R_+\beta}{R} = 1 + 0.80862 n_s^{-\frac{2}{3}}, \qquad (4.5)$$

where $R = \rho$ is the radius of the beam orbit and $b = R_+ - R$ its distance to the outer wall of the vacuum chamber. Note that criterion (2.1) has been included in Eq. (4.5). We obtain

$$n_s \sim \frac{1}{\sqrt{2}}\left(\frac{b}{R} - \frac{1}{2\gamma^2}\right)^{-\frac{3}{2}}, \qquad (4.6)$$

where we have approximated 0.7271 by $1/\sqrt{2}$. It was also shown in Eq. (4.28) of Ref. 1 that the shunt impedance over Q of the resonances falls off exponentially

$$\left(\frac{Z}{Q}\right)_{ij}^{TE,TM} \sim e^{-4n/3n_{fs}} \qquad (4.7)$$

according to the free-space cutoff harmonic $n_{fs} = 2\gamma^3$ in the same way as the free-space radiation in Eq. (4.2). From Eq. (4.6), we see that the toroidal resonances can begin at a large range of values. When $\frac{b}{R} \gg \frac{1}{2\gamma^2}$,

$$n_s \sim \frac{1}{\sqrt{2}}\left(\frac{R}{b}\right)^{\frac{3}{2}} \ll \frac{1}{\sqrt{2}}\left(2\gamma^2\right)^{\frac{3}{2}} \sim 2\gamma^3 = n_{fs}. \qquad (4.8)$$

Thus the toroidal resonances exist from $n_s \sim \frac{1}{\sqrt{2}}\left(\frac{R}{b}\right)^{3/2}$ to n_{fs} where they roll off exponentially. When $\frac{b}{R} - \frac{1}{2\gamma^2} = \frac{1}{2\gamma^2}$ or $\frac{1}{\sqrt{2}}\left(\frac{R}{b}\right)^{3/2} = 2^{-3/2} n_{fs}$, the start of the resonances moves to $n_s = n_{fs}$. Thus Z/Q of the resonances drop off rapidly. When $\frac{b}{R} - \frac{1}{2\gamma^2} \ll \frac{1}{2\gamma^2}$ or $\frac{1}{\sqrt{2}}\left(\frac{R}{b}\right)^{3/2} \sim n_{fs}$, the start of the resonances is very much above n_{fs} and moves to infinity eventually, implying that these resonances are of negligibly small values. Finally when $\frac{b}{R} < \frac{1}{2\gamma^2}$ or $\frac{1}{\sqrt{2}}\left(\frac{R}{b}\right)^{3/2} > n_{fs}$, there is no solution to n_s in Eq. (4.5) implying that the resonances do not exist at all.

Coming back to the situation where toroidal resonances are possible. From Table III, we know that the frequency will be at least 106 GHz or the wavelength at most ~ 3 mm. Thus the pumping ports may have openings bigger than the wavelengths of these resonances. The beam will therefore see a rather diffused vacuum-chamber. The result is that the resonances will be heavily de-Qued and overlap each other. The spectrum will become more and more continuous. If the port openings are made still larger, the radiation will pass through those openings as if there were no beam pipe at all. The radiation therefore resembles free-space radiation. We may therefore conjecture that if the Z/n of these resonances were averaged over the range of harmonics in Eq. (4.8), the average would be given by the free-space Z_n/n of Eq. (4.3).

A very wide beam pipe of full height h with pump-port openings at the outer wall is very similar to two infinite parallel plates separated by h. The peak value of the resistive component of the coupling impedance is found to be[11]

$$\mathcal{R}e\left(\frac{Z_n}{n}\right) \sim 300 \left(\frac{h/2}{R}\right) \text{ ohms },\qquad(4.9)$$

at roughly the harmonics

$$n = \left(\frac{R}{h/2}\right)^{\frac{3}{2}} \qquad (4.10)$$

when the synchrotron radiation is fully unshielded. Note that Eqs. (4.9) and (4.10) are compatible to Eq. (4.4). For the low- (high-) energy ring, this amounts to 0.039 ohm (0.013 ohm) and is the peak impedance loss per harmonic due to radiation.

REFERENCES

1. K.Y. Ng, Particle Accelerators **25**, 153 (1990).

2. R.L. Warnock and P. Morton, Particle Accelerators **25**, 113 (1990).

3. *Feasibility Study for an Asymmetric B Factory Based on PEP*, LBL PUB-5244 (SLAC-352 or CALT-68-1589), 1989.

4. K.Y. Ng, R. Ruth, and R.L. Warnock, unpublished.

5. R. Talman, Phys. Rev. Lett. **56**, 1429 (1986).

6. G.A. Decker, Cornell University Thesis, *The Centrifugal Space Charge Force in Circular Accelerators*, 1986.

7. E.P. Lee, Particle Accelerators **25**, 241 (1990).

8. J. Schwinger, Phys. Rev. **75**, 1912 (1949).

9. A.G. Bonch-Osmolovsky, JINR Report P9-6318, Dubna, 1972.

10. A. Faltens and L.J. Laslett, Proc. of the 19975 ISABELLE Summer Study, Vol. II, p.486, Brookhaven National Laboratory, 1975.

11. E.P. Lee *et al*, Lawrence Berkeley Laboratory Report LBL-15116 (1982).

TRAPPED ION EFFECTS IN BFI

R. Cappi
CERN, 1211 Geneva 23, Switzerland

ABSTRACT

The aim of this note is to estimate the effects, if any, of trapped ions in the e-rings of a B-Factory (in the ISR ring) machine, actually under study.

Figures found in this note will probably be obsolete, at the time of its publication, as the machine and beam parameters are still, at the moment, in a "transitory state".

However formulae and qualitative reasoning, given here as a reminder, should still be valid (we hope).

ION TRAPPING (A COLLECTION OF DEFINITION AND FORMULAE)

Ions will be trapped by the e- beam potential well if their atomic weight is larger than the critical mass A_c given by[1,2].

$$A_c = \frac{N_b \, r_p \, \pi \, R}{K_b \, \sigma_x \, \sigma_y \, (1 + \sigma_y / \sigma_x)} \quad (1)$$

where

N_b is the number of particles in the bunch
R is the machine radius
r_p is the classical proton radius ($= 1.53 * 10^{-18}$ m)
K_b is the number of bunches
σ_x, σ_y are the rms horizontal and vertical beam dimensions

Values of A_c related to different machine options are shown in table 1). From these values it appears that CO ions (A = 28) are the best candidates for trapping.

The ions induce tune shift and spread given by

$$\Delta Q_y = \frac{r_e <\beta_y> N_i}{2\pi \, \gamma \, \sigma_x \, \sigma_y \, (1 + \sigma_y / \sigma_x)} \quad (2)$$

where

r_e is the classical electron radius ($= 2.82 * 10^{-15}$ m)
$<\beta_y>$ is the average beta function
γ is the usual relativistic factor
N_i is the total number of trapped ions

The time τ_i required to create N_i ions is

$$\tau_i = \frac{N_i \, \tau_m}{N_b \, K_b} = \eta \, \tau_m \quad (3)$$

where

η is the neutralisation factor

and τ_m is the ionisation time, i.e. the time required by an electron to create one ion and is given by

$$\tau_m = (d_m \sigma_m c)^{-1} \qquad (4)$$

where

σ_m is the ionisation cross section ($\cong 1.77 * 10^{-22}$ m^2 for CO)
d_m is the molecular density $= 3.3 * 10^{22}$ P_m[Torr]
P_m is the ion partial pressure $\approx 1/3 \, 10^{-8}$ Torr

Values of neutralisation factor η producing tune shifts of $\cong 0.05$ as well as the corresponding τ_i values are shown in table 1).

CURES

1. CLEARING ELECTRODES

A possible way to get rid of the ions, at least partially, is to install as many clearing electrodes as possible around the ring with voltage given by

$$V_{ce} > \frac{d \, N_b \, K_b \, e}{(2\pi)^2 \, R \, \varepsilon_0 \, \sigma_y \, \sqrt{2}} \qquad (5)$$

where

d is the electrode spacing
e is the electron charge
ε_0 is the vacuum permittivity ($= 8.85 * 10^{-12}$ m^{-3} kg^{-1} s^2 C^2)

Approximate value of V_{ce} are shown in table 2). More precise computations of V_{ce} should take into account local beam dimensions, depening on the actual beam emittances and beta functions, both unknown at the moment.

2. MISSING BUNCHES

Another method to avoid ion trapping consists in modulating the bunch intensity such that the ion motion will result unstable[3,4,5]. Such a modulation is usually realised by creating a gap in the bunch distribution along the ring.

The ion motion is characterised by the transfer matrix M given by

$$M = \{M_0 M_b\}^{k_b \cdot k_m} \begin{vmatrix} \cosh(g\tau_h) & \sinh(g\tau_h)/g \\ g \sinh(g\tau_h) & \cosh(g\tau_h) \end{vmatrix} \qquad (6)$$

where

$$M_0 = \begin{vmatrix} \cosh(g\tau) & \sinh(g\tau)/g \\ g\sinh(g\tau) & \cosh(g\tau) \end{vmatrix} \qquad (7)$$

is the transfer matrix of the drift space between consecutive electron bunches, where the ions feel only the space charge force of the other ions

$$g^2 = \frac{N_b\, K_b\, r_p\, c^2\, \eta}{\pi\, R\, A\, \sigma_x\, \sigma_y\, (1+\sigma_y/\sigma_x)} = \eta\, \omega_i^2 \qquad (8)$$

where

ω_i is the rotation frequency of the ions in the electron beam potential
A is the ion mass
c is the speed of light

$$M_b = \begin{vmatrix} 1 & 0 \\ -a & 1 \end{vmatrix} \qquad (9)$$

is the transfer matrix of an electron bunch crossing and producing a kick

$$a = \frac{2\, N_b\, r_p\, c}{A\, \sigma_x\, \sigma_y\, (1+\sigma_y/\sigma_x)} \qquad (10)$$

k_m is the number of missing bunches
and

$$\tau_h = \frac{k_m\, 2\pi\, R}{k_b\, c} \qquad (11)$$

is the length of the gap.
The motion is stable (i.e. the ions are trapped) if the trace (Tr) of M is

$$-2 < \mathrm{Tr}\, M < 2 \qquad 12)$$

A plot of Tr M versus N_b for the e- ring at $L=10^{34}$ cm^{-2} s^{-1} with 16 missing buches out of 160 is shown on fig 1).
Only very narrow intervals of Nb allow $|T_r\, M| < 2$ (these intervals are even smaller due to the stop bands creation by non uniform bunch intensity[6]).

In table1) the approximate number of missing bunches required for ion clearing are given for each machine options.

CONCLUSIONS

Ion trapping (CO) is possible and very probable in all the proposed options.

Due to the high intensities and poor vacuum conditions, strong tune shifts (coherent and incoherent[7]) producing beam blow-up will reduce strongly the luminosity.

Installing clearing electrodes with voltages of many kV, providing very low coupling impedances, tolerating high temperatures etc. appears as a difficult task. RF-shaking[8] also seems impractical. Due to the very small values of τ_i, an almost continuous shaking of the beam is required, providing beam oscillation amplitudes of the order of the beam dimensions[7].

The missing bunch method, where only about 10% of the total intensity is "lost", looks promising. Investigations and more detailed experiments should be done to confirm its validity.

REFERENCES

1. Y. Baconnier and G. Brianti, CERN/SPS/80-2 (DI), March, 1980.
2) Y. Baconnier, CERN 85-19, November 1985.
3) M. Barton, Nucl. Instr. and Methods A243 (1986) 278-280.
4) S. Sakanaka, KEK Preprint 86-17, June 1986.
5) A. Poncet, CERN /PS 88-14, April 1988.
6) H. Schonauer, IEEE Tr. on N.S. NS-20, 866 (1973).
7) R. Cappi and J.P. Riunaud, to be published in EPAC 1990.
8) Y. Orlov, CERN PS/AR/Note 88-20.

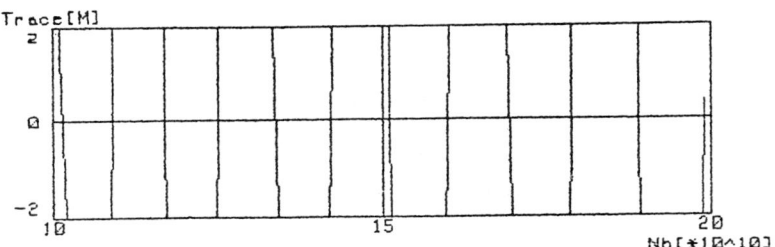

Fig 1) A plot of Tr M versus Nb in the region of nominal intensity, for the 8GeV e-beam in the 10^{34} option.
16 bunches out of 160 are missing. Tr M lies most of the time outside the +- 2 region implying ion instability.

L [cm^{-2} s^{-1}]	10^{33}	4 * 10^{33}	10^{34}
E[GeV]	8	5.3	8
A$_c$	3.5	22.7	3.2
V$_{ce}$ [kV] >	9	10	30
η [10^{-3}] (ΔQ=0.05) τ$_i$ [μs]	5 980	3 640	1.4 280
f$_i$ [MHz]	2.8	2.8	5.3
~k$_m$ / k$_b$	8 / 80	3 / 32	16 / 160

Table 1: 1st and 2nd rows: machine options (e$^-$ rings only).
    3rd row: critical mass (all ions with larger mass will be trapped).
    4th row: approx. minimum clearing voltage.
    5th row: required neutralisation to get a ΔQ = 0.05, while τ$_i$ is the time required to produce such a neutralisation
    6th row: oscillation frequency of the ions
    7th row: required number of missing bunches k$_m$ with respect to the total number of bunches k$_b$ to get ion clearing

Symmetrization of the Beam-Beam Interaction in an Asymmetric Collider

Yong Ho Chin

Exploratory Studies Group
Accelerator & Fusion Research Division
Lawrence Berkeley Laboratory, Berkeley, CA 94720

Introduction

The attainable luminosity in an asymmetric storage-ring collider will be determined to a large extent by the physics of the beam-beam interaction. Nothing is known experimentally about the beam-beam tune shift limit under asymmetric energy conditions. The situation is complicated, since two beams with unequal energies naturally tend to behave differently. Indeed, what is often observed in computer simulations is that one beam blows up badly while the other beam suffers practically no blowup. This is a serious problem, since the significant blowup in the weaker beam imposes an unnaturally low beam-beam tune shift limit on the stronger beam.

Probably the best cure is to bring the beam-beam interaction into the "strong-strong" regime where the two beams blow up in a similar manner, reducing the beam-beam force on both beams simultaneously. In this way, putting the two beams on an equal footing as far as transverse dynamics is concerned, we might expect to reach the same maximum beam-beam tune shift limit set by nature in equal-energy colliders. A possible set of conditions to achieve such a circumstance is generally referred to as the "energy transparency conditions" [1, 2, 3]. The idea of the energy transparency conditions results from two facts:

- We know about the actual behavior of the beam-beam effect only under symmetric conditions – the beam-beam tune shift limit, ξ, in equal-energy electron-positron colliders.

- The beam-beam interaction in the strong-strong regime is not well understood in a quantitative sense at present. The only systematic tool to understand it is provided by computer simulations. However, there is no simulation program to date that can consistently explain the experimental data, even from various symmetric machines, in a quantitative sense.

Therefore, by adopting the energy transparency conditions, one can hope to design an asymmetric collider in a "rational" way, without relying in detail on any particular theory or simulation code. Several B-Factory designs have adopted variants of the concept of energy transparency as a design guideline [1, 4, 5].

Energy Transparency Conditions

Two possible sets of energy transparency conditions have been proposed [1, 5]. The author has proposed the following set of four conditions [1] (the superscripts label the electron (-) and positron (+) beam):

- Same nominal linear beam-beam tune shift parameters:

$$\xi_{0x}^- = \xi_{0x}^+, \quad \xi_{0y}^- = \xi_{0y}^+ \tag{1}$$

- Same nominal cross sectional areas at the IP:

$$\sigma_{0x}^- = \sigma_{0x}^+, \quad \sigma_{0y}^- = \sigma_{0y}^+ \tag{2}$$

- Same radiation damping decrements:

$$\delta^- = \delta^+ \tag{3}$$

where the damping decrement, δ, is defined as the product of the absolute radiation damping rate and the time interval between collisions.

- Same betatron phase modulations due to synchrotron motion:

$$\left(\frac{\sigma_s Q_s}{\beta_x^*}\right)^- = \left(\frac{\sigma_s Q_s}{\beta_x^*}\right)^+, \quad \left(\frac{\sigma_s Q_s}{\beta_y^*}\right)^- = \left(\frac{\sigma_s Q_s}{\beta_y^*}\right)^+ \tag{4}$$

where σ_s is the rms bunch length, Q_s is the synchrotron tune, and β^* is the beta function at the IP.

The validity of these criteria is demonstrated in references 1-3 by applying a modified version of Yokoya's beam-beam simulation program to the APIARY-I lattice and showing that the two unequal energy beams maintain symmetric behavior.

The first condition equalizes the beam-beam kicks in the two rings; any remaining difference in beam dynamics must then come from the difference of beam parameters elsewhere in the rings. The second condition is necessary for complete overlap of the two beams at the IP. The fourth condition guarantees the same strength of synchro-betatron resonances, which are supposed to be a source of beam blowup. Radiation damping is an important effect that suppresses external perturbations of beams [6]. There are many experimental [7] and computer simulation results [6] that indicate the damping decrement dependence of the luminosity in symmetric colliders. In the simulations, the effect is not simply that the larger the damping rate, the larger the beam-beam limit will be. If one starts with two identical rings, and increases the damping decrement of one beam, keeping that of the other beam constant, the beam with larger damping decrement shrinks, while the beam with smaller damping decrement blows up. The luminosity will start to drop when the asymmetry of the damping decrements exceeds a certain value. Figure 1 shows an example of the luminosity L and the dynamic emittance (after blowup) as a function of the asymmetry of the damping decrement of two beams when the damping decrement of Beam+ is changed while that of Beam- is kept fixed. The main parameters used are shown in Table 1 below. In order to isolate the effect of the damping decrement, only this parameter is different in the two rings.

Table 1. Main parameters of the sample asymmetric collider used.

Parameters	*Beam+*	*Beam-*
Energy, E (GeV)	8	8
Circumference, C (m)	2200	2200
Nominal emittance, $\epsilon_{0x} = \epsilon_{0y}$ (nm·rad)	68.133	68.133
Bunch length, σ_s (cm)	1.0	1.0
Beta function at IP $\beta_x^* = \beta_y^*$ (m)	3.0	3.0
Damping decrement, δ	—	1.643×10^{-4}
Bunch current, I_b (mA)	5.188	5.188
Synchrotron tune, Q_s	0.089	0.089
Nominal beam-beam tune shift, $\xi_{0x} = \xi_{0y}$	0.05	0.05
Betatron fractional tunes, $Q_x = Q_y$	0.72	0.72

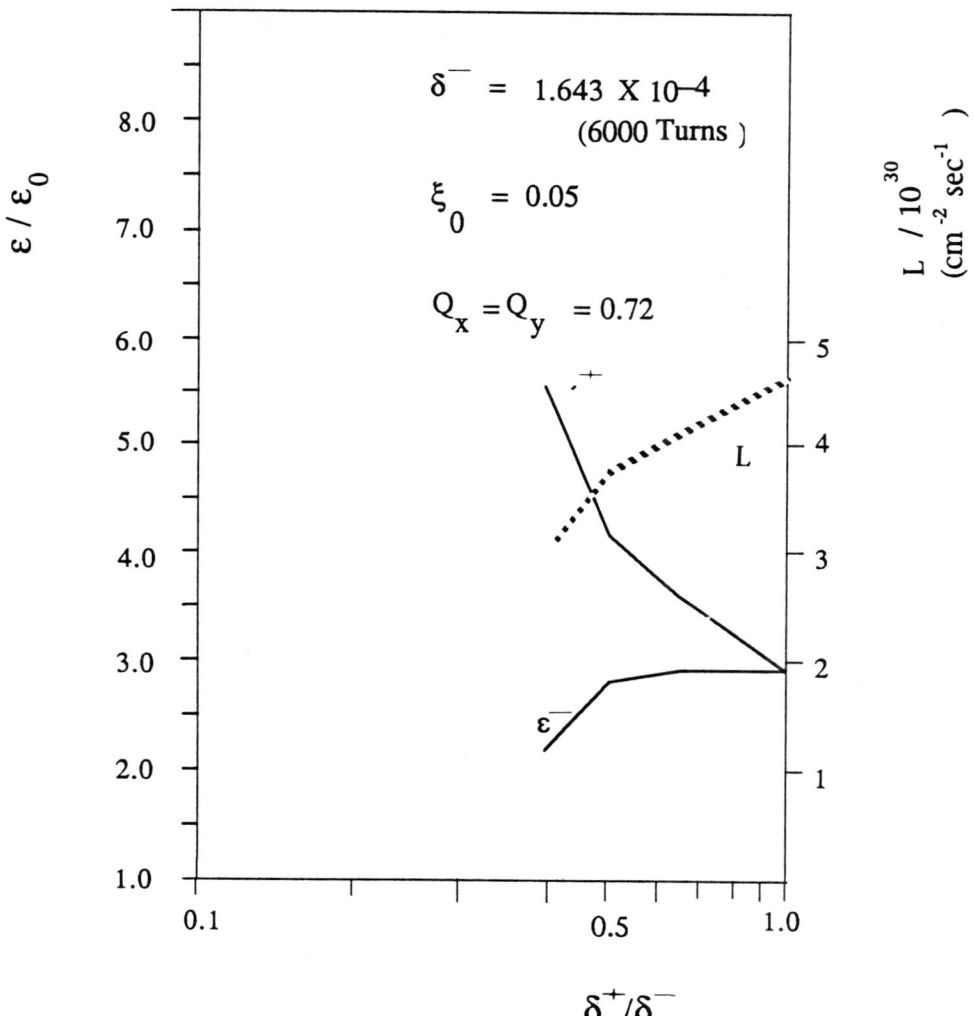

Figure 1: The luminosity L and the dynamic emittance ϵ as a function of the asymmetry of the damping decrement. Here, ϵ_0 is the nominal emittance, and ξ_0 is the nominal beam-beam parameter in the absence of beam blowup.

It can clearly be seen from Fig.1 that the unequal damping decrement causes asymmetric behavior of the beam sizes. We know that there is another report that shows a weaker damping decrement dependence of the luminosity [8]. In that calculation, however, other parameters are also changed when the damping decrement is changed in order to maximize the luminosity. Moreover, an unrealistically short bunch ($\sim 100~\mu$ m) is used, making the betatron phase modulation due to synchrotron motion, the source of beam blowup, extremely small. Therefore, the effect of the damping decrement alone is obscured, and a fair comparison with the present results is difficult.

Siemann and Krishnagopal have proposed a stricter set of energy transparency conditions than discussed here. They require [5]: the same products of the number of particles N and the Lorentz factor, the same beta functions at the IP, the same emittances, the same bunch length, the same synchrotron tunes, and the same fractional parts of the betatron tunes. We haven't yet studied these conditions in detail.

Discussion

At present, when there are no existing asymmetric colliders, it is not known how strictly such symmetrization conditions must be satisfied, or how much they can be relaxed in real machines. We need further work on this problem of the minimum set of requirements for the practical design of rings.

Another question is whether one could relax such strong constraints by compensating for one asymmetry with another. The answer is not straightforward. Any such compensation scheme would require a credible theory and a computer simulation program that can quantitatively predict how much asymmetry in one parameter is needed to compensate for an asymmetry in another parameter. At present, neither of these is available. Moreover, there is some evidence that the stability of such a delicately compensated beam-beam mode would be unpredictable [9, 10].

Let us examine a possible compensation. The idea is the following: when one beam blows up more than the other beam, one tries to equalize the beam sizes of the two beams by reducing the nominal emittance of the blown-up beam. Table 2 shows the parameters of the sample lattice used in a simulation test of this idea. The main asymmetric parameter in the two rings is the beta function. The emittances and the bunch currents are chosen so that the rms beam sizes at the IP and the nominal beam-beam parameters are equal in the two rings.

Table 2. Parameters of the sample lattice used in the simulation test of a possible compensation scheme.

Parameters	Beam+	Beam-
Energy, E (GeV)	8	3.5
Circumference, C (m)	2200	2200
Nominal emittance, $\epsilon_{0x} = \epsilon_{0y}$ (nm·rad)	68.133	136.3
Bunch length, σ_s (cm)	1.0	1.0
Beta function at IP $\beta_x^* = \beta_y^*$ (m)	3.0	1.5
Natural beam sizes, $\sigma_x = \sigma_y$ (μm)	45.21	45.21
Damping decrement, δ	$.657 \times 10^{-3}$	$.657 \times 10^{-3}$
Bunch current, I_b (mA)	4.54	5.188
Synchrotron tune, Q_s	0.089	0.089
Nominal beam-beam tune shift, $\xi_{0x} = \xi_{0y}$	0.05	0.05
Betatron fractional tunes, $Q_x = Q_y$	0.70	0.70

The simulation result for the above configuration shows that the beam size of Beam- increases to 59.97 μm (33% blowup) while that of Beam+ increases only to 48.02 μm (6% blowup). We then reduced the nominal emittance, ϵ^-, of Beam- and ran additional simulations. The results for various values of ϵ^- are summarized in Table 3, where σ_0 and σ are the nominal and the dynamic beam sizes, respectively.

Table 3. Simulation results for five different emittances of Beam-.

Emittance, ϵ^- (nm·rad)	σ_0^+	σ_0^-	σ^+	σ^-
81.76	45.21	35.02	48.61	57.73
95.39	45.21	38.42	49.12	59.65
109.0	45.21	40.44	48.6	61.1
122.6	45.21	42.89	48.75	56.15
136.3	45.21	45.21	48.02	59.97

It can clearly be seen from Table 3 that the dynamic beam sizes of the two beams depend very weakly on the nominal Beam- emittance. This is because the dynamics of the beam-beam interaction tends to make up the difference between the equilibrium and the nominal beam sizes no matter what the nominal emittance is. The dynamic beam size in the beam-beam limit is a result of the beam-beam interaction determined by all the other parameters. It is not a free parameter that may be controlled by changing its nominal value. This result agrees with observations at PEP, where the dynamic vertical beam size is seen to remain nearly constant when the x-y coupling of the beams is changed to reduce the nominal vertical beam size (in an attempt to improve the luminosity). For this reason, the suggested compensation scheme using the beam size or the emittance as a free parameter does not appear to work in this parameter regime. Of course, we have not yet studied the plausibility of other possible compensation schemes, involving more or different parameters, so no statement can be made about the efficacy of compensation schemes in general.

There is another worthwhile point to be mentioned here. Figures 2(a) and (b) show the time evolution of beam sizes of Beam+ and Beam-, respectively, for the case of $\epsilon^- = 81.76$ nm·rad in Table 3, up to 24000 turns corresponding to 16 damping times. In this particular case of large asymmetry of the nominal beam sizes, 16 damping times was necessary to reach the true equilibrium. When the asymmetry is smaller, for example as in the third case in Table 3, only several damping times are needed to reach equilibrium. In the plot, Beam+ and Beam- are denoted as bunch # 1 and bunch # 2, respectively. The points x, y and o represent the horizontal, vertical, and longitudinal beam sizes, respectively, in units of their nominal rms sizes. Each point represents the average over 400 turns. There is some beam blowup, on the order of 10%, in the first 400 turns in both beams, so that the plotted points start from around 1.1. At the beginning, Beam- blows up while Beam+ suffers almost no blowup. They appear to be in a steady state. However, after about 8000 turns Beam- shrinks to nearly its original size while Beam+ blows up; this is the true equilibrium state. If the simulation were halted at 8000 turns, the results would be misleading. This example clearly indicates the need to be very careful in conducting computer simulations and interpreting their results.

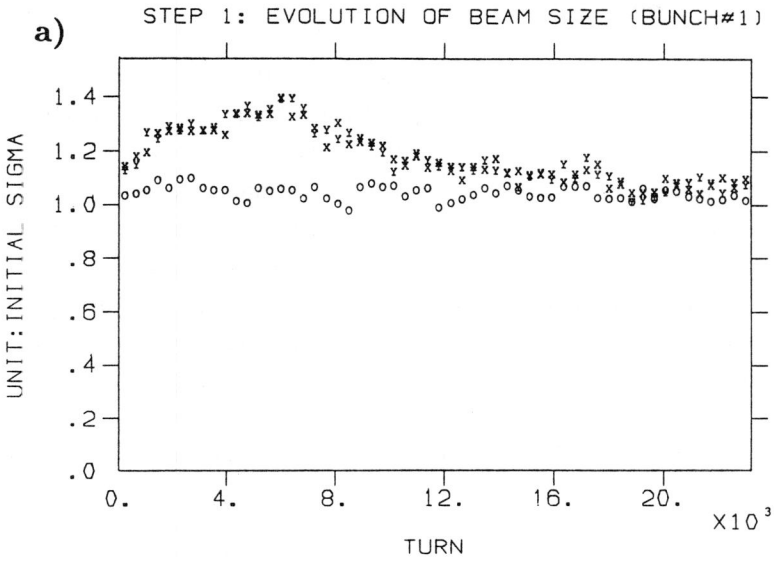

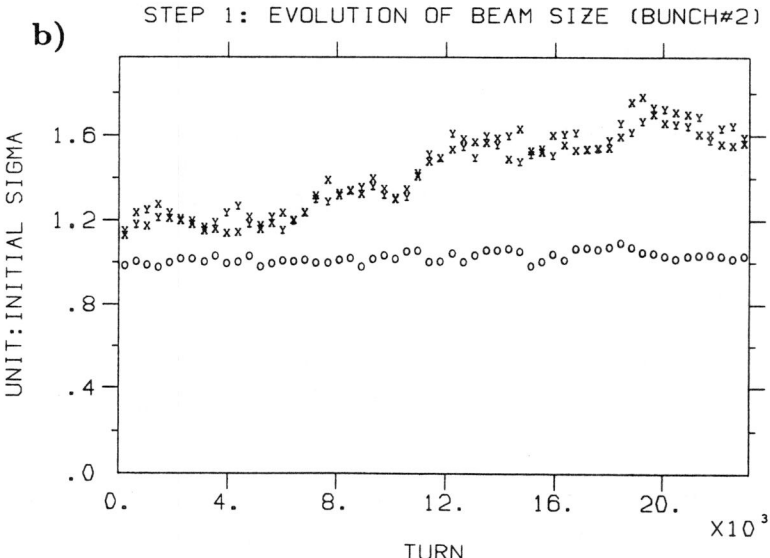

Figure 2: The time evolution of beam sizes of (a) Beam+ and (b) Beam-, respectively, for $\epsilon^- = 81.76$ nm·rad, up to 24000 turns. The points x, y and o represent the horizontal, vertical, and longitudinal beam sizes, respectively, in units of their nominal rms sizes.

Conclusions

We have studied the idea of symmetrizing both the lattice and the beams of an asymmetric collider, and have discussed why this regime should be within the parametric reach of the design in order to credibly ensure its performance. We have also examined the effectiveness of a simple compensation method using the emittance as a free parameter and shown that it does not work in all cases. At present, when there are no existing asymmetric colliders, it seems prudent to design an asymmetric collider so as to be similar to a symmetric one (without relying on a particular theory of the asymmetric beam-beam interaction that has not passed tests of fidelity). Nevertheless, one must allow for the maximum possible flexibility and freedom in adjusting those parameters that affect luminosity. Such parameter flexibility will be essential in tuning the collider to the highest luminosity.

Acknowledgement

The author would like to thank M. Zisman for helpful discussions and careful proofreading of the manuscript.

This work was supported by the Director, Office of Energy Research, Office of High Energy and Nuclear Physics, High Energy Physics Division, of the U.S. Department of Energy under Contract No. DE-AC03-76SF00098.

References

[1] Y. H. Chin, Lawrence Berkeley Laboratory Report LBL-27665, August, 1989. Presented at *the XIV Int. Conf. on High Energy Accelerators*, Tsukuba, Japan, August, 1989 (to be published).

[2] *Feasibility Study for an Asymmetric B Factory Based on PEP*, Lawrence Berkeley Laboratory Report LBL PUB-5244, October, 1989.

[3] *Investigation of an Asymmetric B-Factory in the PEP Tunnel*, Lawrence Berkeley Laboratory Report LBL PUB-5263, March, 1990.

[4] K. Oide, private communication.

[5] S. Krishnagopal and R. Siemann, CESR Report, CLNS 89/96, 1989.

[6] S. Myers, in *Nonlinear Dynamics Aspects of Particle Accelerators, Lecture Notes in Physics* **247**, (Springer-Verlag, Berlin, 1985), p.176.

[7] J. Gareyte, in *Proc. of the Third Advanced ICFA Beam Dynamics Workshop on Beam-Beam Effects in Circular Colliders*, edited by I. Koop and G. Tumaikin (INP, Novosibirsk, May, 1989), pp.135-139.

[8] J. L. Tennyson, these proceedings.

[9] M. Furman, in *Proc. of the Third Advanced ICFA Beam Dynamics Workshop on Beam-Beam Effects in Circular Colliders*, edited by I. Koop and G. Tumaikin (INP, Novosibirsk, May, 1989), pp.52-57.

[10] P. J. Channell, these proceedings.

Beam-Beam simulation with Non-Gaussian beams

E. Kikutani

National Laboratory for High Energy Physics

1. Introduction

In the field of the simulation of the beam-beam force phenomena, S. Meyers' method[1] has been one of the standard methods. With this technique, the particle distribution in a bunch is assumed to be a gaussian for transverse (horizontal and vertical) and longitudinal directions. If one wants to calculate a beam-beam force with more 'realistic' manner, he or she must devise a new technique.

At KEK, K. Yokoya and K. Oide[2] developed a new technique to calculate the beam-beam force. By this technique potential is calculated without assuming the gaussian distribution. In this article we introduce the technique and show an example of the method.

2. Calculation method

Generally, change of the direction of particle motion is given by the expression,

$$\Delta(\frac{dx}{ds}) = \frac{1}{\gamma mc^2}(-\frac{\partial V(x,y)}{\partial x}), \qquad (1)$$

where γmc^2 is the nergy of the particle and $V(x,y)$ is the two-dimensional potential, which is the solution to the poisson equation,

$$(\frac{\partial^2}{\partial x^2} + \frac{\partial^2}{\partial y^2})V(x,y) = \frac{1}{\epsilon_0}\rho(x,y), \qquad (2)$$

where $\rho(x,y)$ is the charge density distribution. To solve the equation, we use the Green function, which satisfies the equation,

$$(\frac{\partial^2}{\partial x^2} + \frac{\partial^2}{\partial y^2})G(x-x\prime, y-y\prime) = \delta(x-x\prime, y-y\prime), \qquad (3)$$

where δ is Dirac's delta function. The solution of (3) is well known and is given as

$$G(x,y) = \frac{1}{2\pi}\frac{1}{2}\log(x^2+y^2). \tag{4}$$

After $G(x,y)$ is obtained, $V(x,y)$ will be calculated as

$$V(x,y) = \frac{1}{\epsilon_0}\int\int G(x-x',y-y')\rho(x',y')dx'dy'. \tag{5}$$

Instead of perfoming the integration (5) actually, we calculate $V(x,y)$ in Fourier-transformed manner, i.e.

$$\tilde{V}(k_x,k_y) = \tilde{G}(k_x,k_y)\cdot\tilde{\rho}(k_x,k_y), \tag{6}$$

where the tilde indicates the Fourier transform of each function. If we inversely Fourier-transform $\tilde{V}(k_x,k_y)$, we get $V(x,y)$ at last and then $\Delta(dx/ds)$ by using expression(1).

3. Calculation technique

In actual calculation with computers, the plane where the collision occurs is divided into many rectangular sections. The ratio of the section height to the its width is selected to be the same value as the aspect ratio of the beam.

In this meshed plane, the Green function (4), which is calculated prior to the tracking loop, is replaced with the following averaged one,

$$G_{mesh}(i,j) = \frac{1}{2\pi}\frac{1}{2}\int\int_{(i,j)-th\ mesh}\log(x^2+y^2)dxdy. \tag{7}$$

With this method, the bad effects which caused by the small number of super-particles will be avoided. Then the Fourier transform of the potentail, $\tilde{V}(k_x,k_y)$ is obtained by a discrete version of expression (6), i.e.

$$\tilde{V}_{mesh}(i,j) = \tilde{G}_{mesh}(i,j)\cdot\tilde{\rho}_{mesh}(i,j). \tag{8}$$

Finally we can obtain $\Delta(\frac{dx}{ds})$, by making difference, $\frac{V(i+1,j)-V(i,j)}{\Delta x}$ at each point after inverse Fourier transformation.(Δx is the mesh size.)

When we use Fourier transform methods in computers, the integration is not performed from $-\infty$ to $+\infty$. This means that functions have finite periods corresponding to integral regions. In our case, the obtained potential with the periodic charge density will introduce an unrealistic periodic force. To avoid this effect we prepare a plane which is two times (in length) larger than a plane where the particles are distributed. As shown in Fig. 1, the particles distributed in only 1/4 of the prepared plane. Then the force is actually not affected by the neighboring peak of the potential.

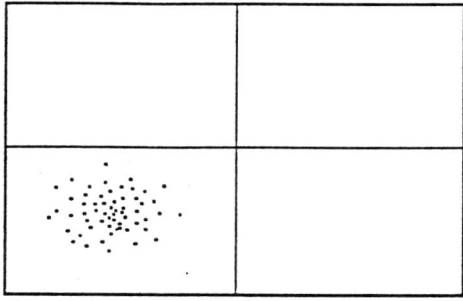

Fig. 1

4. Example

Motivation

As an example, we show preliminary results of a simulation on the coherent beam-beam tune shift in the TRISTAN Accumulation Ring. Recently we wrote a paper[2] on the coherent tune shift observed in the TRISTAN Accumulation Ring and its theoretical interpretation. Now we make a simulation to check how the tune shift behave as a function of the beam current.

Simulation condition

To make the simulation we used the machine parameters of TRISTAN Accumulation Ring. The main parameters are shown below.

circumference	377.63m
beam energy	5 GeV
harmonic number	640
emittance H/V	$1.73 \times 10^{-7}m / 3.47 \times 10^{-9}m$
β_x^*/β_y^*	1.2m/0.03m
ν_x/ν_y	$5.449 \times 2 / 5.148 \times 2$
# of int. points	2
# of bunches	1×1

In the simulation of this kind, the mesh size is of quite critical. In our case mesh size is chosen to be 0.4 × RMS initial beam size for both horizontal and vertical direction. The number of super-particles is also a quite significant parameter. Figure 2 shows the number of particles in each 32 × 32 mesh out of 1000 super-particles in total.

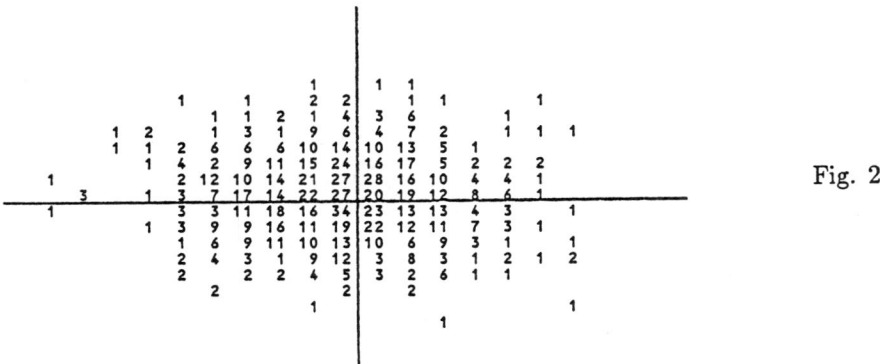

Fig. 2

Program

The program has two parts, a praparation part and a tracking loop part. In the preparation part, the Green function is calculated based on the expression (7)

438   Beam-Beam Simulation

for each section and initial particle distributions are given.

The tracking loop part has 4 components, beam-beam traversal, arc traversal, RF acceleration and a coherent kick. The structure is similar to that of S. Meyers'[1] except that there is the coherent kick part.

Results

Figure 3 shows the horizontal coherent tune shift without assumig the gaussian distribution. The tune shifts obtained by simulation ($\delta\nu$) are plotted with a symbol "." and two reference lines are shown in the figure. The lower line corresponds to the $\delta\nu = \xi$ and the upper line corresponds to $\delta\nu = 1.33\xi$, where ξ is socalled beam-beam parameter(space charge parameter). In the previous work[2], we showed that the experimental data indicates $\delta\nu = 1.33\xi$ and that theoretical calculations support the experimental data. In this simulation, as shown in Fig. 3, $\delta\nu = 1.33\xi$ is reproduced.

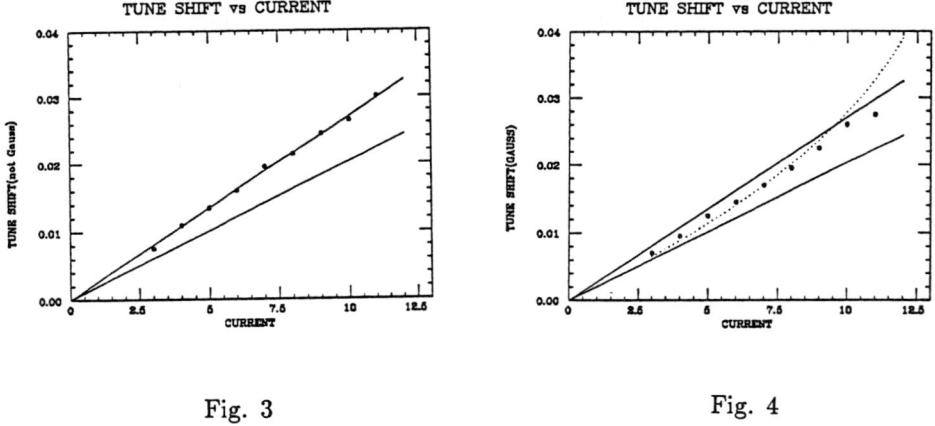

Fig. 3   Fig. 4

For the forces assuming the gaussian distribution, result is shown in Fig. 4. In this figure, there is one more guide line (doted line), which indicates the expected tune shift in case it is related to ξ through the expression,

$$\cos 2\pi(\nu + \delta\nu) = \cos 2\pi\nu - 2\pi\xi \sin 2\pi\nu, \qquad (9)$$

where ν is unperturbed or 0-mode tune. As shown in the figure, the obtained tune shift does not fit to the $\delta\nu = 1.33\xi$ line and there is an indication that the obtained tune shift is attracted by the guide line mentioned above.

For the vertical direction, on the other hand, the simulation result does not reproduce the theoretical prediction as shown in Fig. 5 (without assuming gaussian) and Fig. 6 (with gaussian assumption). In the previous paper[2], the ratio of $\delta\nu$ to ξ is 1.285(upper reference line), but simulation result prefers the ratio $\simeq 1$ (the lower reference line). We must study this subject further, for the experimental data suggests that the ratio is larger than 1.

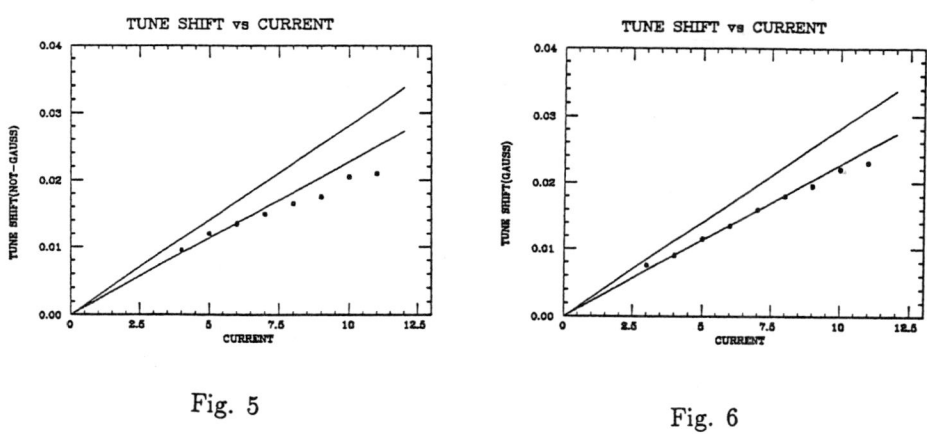

Fig. 5   Fig. 6

CPU time

One of the interesting things in those simulations is the CPU time consumed. We do not distinguish the CPU time for the calculation of collision from other parts. Then we show only gross CPU time by HITAC M680. For a program assuming the gaussian distribution CPU time is $2 \sim 3$ minuts for 1000 super-particles and 1100 turns. On the other hand, the CPU time for the new technique is $4 \sim 5$ minutes. The new technique is not so tremendously wasting CPU time.

5. Summary

A new method of calculating the beam-beam force is given and the method does not assume that the particle distribution is gaussian. With the method, the coherent beam-beam tune shift as a function of beam current is reproduced for the horizontal case.

References

1. S. Meyers, Nucl. Instr. Meth. 211 (1983) 263.
2. K. Yokoya and K. Oide, private communication.
3. K. Yokoya *et al.*, KEK Preprint 89-14 and K. Yokoya *et al.*, Proceedings of the 14-th International Conference of High Energy Accelerators 1989, Tsukuba Japan, to be published.

The Long-Range Beam-Beam Force:
Effects on the B factory*

Kohji HIRATA

KEK, National Laboratory for High Energy Physics
Oho, Tsukuba, Ibaraki 305, Japan

25 June 1990

ABSTRACT

For the KEK B factory, the necessary horizontal separations between beams at the peripheral collision points are given by means of the rigid Gaussian model. For the present design parameters, (1) in the head-on collision scheme, bunches can be filled in every three buckets and 2mm separation is enough at the point 1.8 meter distant from the interaction point, (2) all buckets can be filled if the crossing has an angle of 3mrad (half angle). The limitation comes from a reduction of the luminosity and the excitation of the closed orbit at the arc and (1+2) The combination of (1) and (2) with crossing angle of 2mrad and 0.6T separation magnets (all buckets being filled) provides enough luminosity and enough space for the final focus system.

1. Introduction 2
2. The Head-On Scheme 4
3. The Crossing Scheme 5
4. Mixed Scheme 7
5. Discussions and Conclusion 8
   5.1 Dynamical Closed Orbit
   5.2 Beam-Beam Modes
   5.3 Crab Crossing
   3.4 Conclusion

* Published as an internal B-Factory Report of KEK, June 25, 1990.

1. Introduction

The beam-beam force is the Coulomb force and is roughly proportional to r^{-1}, outside the bunch, where r is the distance from the center of the bunch. This is a long range force. The focusing force, on the contrary, is proportional to r^{-2} so that its effect becomes weak for r larger than the r.m.s. beam sizes. This long range force — dipole force — plays an important role in the high luminosity two ring colliders, where the bunch spacing is small so that the beams can feel the beam-beam kick not only at the interaction point (IP) but also around it: we call the latter "peripheral collision points (PCP)".

The luminosity is expressed as

$$L = \frac{f_{rev}}{2\pi} \sum_{col} \frac{N_+^i N_-^i}{\Sigma_x \Sigma_y} \exp\left[-\frac{(\bar{x}_+^i - \bar{x}_-^i)^2}{2\Sigma_x^2} - \frac{(\bar{y}_+^i - \bar{y}_-^i)^2}{2\Sigma_y^2}\right], \quad (1)$$

where x and y refer to the horizontal and vertical coordinates, $\bar{x}$ and $\bar{y}$ are the averages

$$\bar{z} = <z>, \quad (2)$$

over the bunch and Σ's are the effective beam sizes

$$\Sigma_z \equiv \sqrt{(\sigma_z^+)^2 + (\sigma_z^-)^2}. \quad (3)$$

Here the sum extends over all the pairs of e^+ and e^- bunches which collide with each other at the IP. From Eq. (1), the reduction of the luminosity comes from separation between bunches $\bar{x}_+^i - \bar{x}_-^i$ and $\bar{y}_+^i - \bar{y}_-^i$ at the IP, and/or enhancement of the beam sizes Σ_x and Σ_y. We consider mainly the first aspect here.

The main effect of the peripheral collisions is that it can make a closed orbits at the IP, which can be different between beams and even between bunches: two bunches do not collide head-on so that the luminosity is reduced. These closed orbits depend on the beam-beam force at the PCP's as well as at the IP. The force itself, on the contrary, depends on the closed orbits.

When two bunches collide, each bunch feels a kick[1],

$$\delta \bar{y}' + i\delta \bar{x}' = -\frac{N_* r_e}{\gamma} f_G(\bar{x} - \bar{x}_* + D_x, \bar{y} - \bar{y}_* + D_y; \Sigma_x, \Sigma_y), \quad (4)$$

where the function f_G is the Bassetti-Erskine function[2]:

$$f_G(x, y; \sigma_x, \sigma_y) = \sqrt{\frac{2\pi}{\sigma_x^2 - \sigma_y^2}} \\ \times \left\{ w\left(\frac{x + iy}{\sqrt{2(\sigma_x^2 - \sigma_y^2)}}\right) - \exp\left(-\frac{x^2}{2\sigma_x^2} - \frac{y^2}{2\sigma_y^2}\right) w\left(\frac{\frac{\sigma_y}{\sigma_x}x + i\frac{\sigma_x}{\sigma_y}y}{\sqrt{2(\sigma_x^2 - \sigma_y^2)}}\right) \right\}. \quad (5)$$

Here D_z is the separation between nominal closed orbits and w is the complex error function.

The criteria for the "enough separation" are

1. The luminosity efficiency

$$R = \frac{L}{L_0} \tag{6}$$

should be considerably close to the unity (98%, say). Here L_0 is the nominal luminosity defined as L without any COD at the IP, without any change in the beam size, nor without any error in the bunch current.

2. The closed orbit at the arc is also produced by the beam-beam interaction. We define

$$A_z = \sqrt{\frac{z^2 + (\alpha_z z + \beta_z z')^2}{\beta_z \epsilon_z}}, \tag{7}$$

at the arc: A_z is constant there. The amount of the distortion A should be reasonably small, 10% of the beam sizes, say. (This can be corrected in principle by the usual correction magnets. This, however, is current dependent and the re-correction should be done according to the change of the current. Thus it is better to avoid a large COD of the beam-beam origin.)

Another important issue is the beam size effect: beam blow up due to the peripheral collisions. This cannot be studied by the rigid Gaussian model. This effect is characterized by the beam-beam parameter at the PCP:

$$\xi_{peripheral} = F\xi_{IP}, \tag{8}$$

where F is a function of the separations and beam sizes at the PCP and can be obtained in terms of the first derivative of f_G there. It seems almost clear that the effect is due to the nonlinear nature of the beam-beam force. In the head-on collision, the force field changes a lot within the beam distribution. With enough horizontal separation, however, the horizontal and vertical beam-beam focusing force is no longer nonlinear: it does not change considerably within the range of the beam size. Thus, the peripheral tune shift parameter contributes to the tune shift but does not contribute to the enhancement of the beam sizes directly, apart from the dynamic betas and dynamic emittances[3].

In order to evaluate the luminosity at the steady-state of the system, a FORTRAN code BCD (Beam-beam Closed orbit Distortion) was made. This is a tracking code based on the rigid Gaussian model. The beam-beam force is evaluated in terms of Eq. (5).

2. The Head-On Scheme

In our B factory design[4], the distance between the neighbouring buckets is 0.6 meter. Other relevant parameters are

β_x	β_y	1.0	1.0×10^{-2}	(meter)
ϵ_x	ϵ_y	1.8×10^{-8}	1.8×10^{-10}	(meter radian)
$E(HER)$	$E(LER)$	8	3.5	(GeV)
ξ_x	ξ_y	0.05	0.05	

In order to maximize the luminosity, we had better fill as many buckets as possible. The interaction region is considered as consisting with two bending magnet in both side of the IP. (from $s = 0.5m$ to $1.5m$). Thanks to the difference in energy between two beams, we can separate beams by them. We assume that the beam-beam force can be shielded for $|s| > 1.5m$.

We denote the distance between two successive bunches in each beam S_B. When all buckets are filled, $S_B = 0.6m$. The cases of $S_B = 0.6m$ and $S_B = 1.2m$ are hopeless in this scheme. Let us consider the case of $S_B = 1.8m$. The peripheral collision takes place at $s = \pm 0.9m$ where the horizontal separation D_x is 3.84 mm if the separation magnet is 1T. Here we have

$$(\Sigma_x, \Sigma_y) = (0.255\text{mm}, 0.171\text{mm}), \tag{9}$$

so that the separation is large enough comparerd to Σ_x.

In Fig. 1, the efficiency R is plotted as a function of D_x at the PCP. The big error bar for some data comes from the fact that the separation does not fall into an equilibrium and performs a multi-period behavior: The error bars do not decrease even with more number of turns. It is observed that the reduction comes mainly from the fact that some bunches collide in head-on whereas other bunches have a large separation at the IP. That is, the system is not in equilibrium and a certain portion of the bunches collide in head-on. From the luminosity point of view, $D_x = 1mm$ appears to be enough.

Let us consider the second criterion: We evaluate the amount of the closed orbit of each bunches at the arc (A_x, A_y) in Table 1. The estimates are based on the assumption that the two separation magnets have the opposite polarity. Even when they are identical the results does not change considerably.

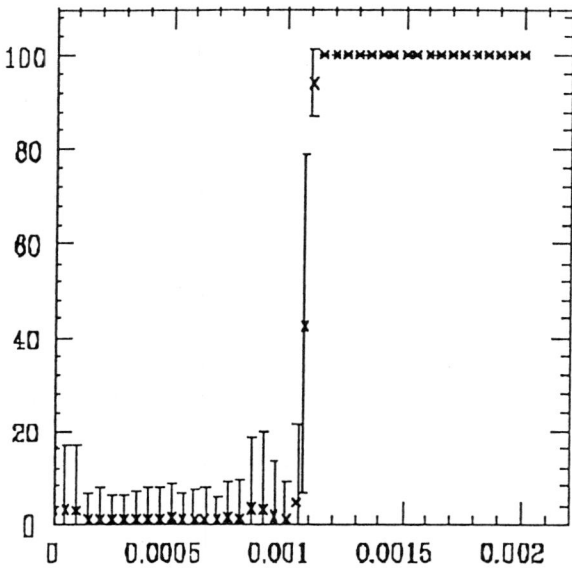

Fig. 1. The luminosity efficiency $R(\%)$ as a function of the horizontal separation $D_x(m)$. Parameters are: $\nu_x^\pm = 0.2$, $\nu_y^\pm = 0.15$.

D_x (mm)	R (%)	F_x	F_y	A_x	A_y
0.5	1.06±6.35	-1.1×10^{-1}	15.0	0.01	22.8 ± 6.9
1.	98.8	-7.4×10^{-2}	3.34	0.11	2.25
1.5	100.0	-3.1×10^{-2}	1.38	0.08	4.1×10^{-9}
2.	100.0	-1.7×10^{-2}	0.76	0.06	1.4×10^{-9}

Table 1. The closed orbits at the arcs. The focusing factors are also shown for the reference.

3. The Crossing Scheme

Let us consider a possibility of filling all the buckets. Since the bunch distance S_B (now 0.6m) is too small, the head-on scheme does not work. Two beams should collide with an angle. The problem is what is the necessary and sufficient angle. As is well known, the crossing angle leads the synchro-beta resonance[5]. If the crossing angle is so large that the instability is unignorable, we need a special scheme[6].

The separations D_x and D_y at each collision are denoted by

$$D_x(s) = D_x(0) + (\Phi + D'_x(0))s \text{ and } D_y(s) = D_y(0) + D'_y(0)s, \tag{10}$$

where s is the distance from the IP, and Φ is the full crossing angle. The $D_x(0)$ and $D'_x(0)$ give the errors in the crossing. When $D_x(0)$, $D'_x(0)$, $D_y(0)$ and $D'_y(0)$ are zero, we have the results shown in Table 2.

N_{CP}	0	1	2	3	4	5	6	9
1 mrad	100	19.7±35	46.7±39.0	22.1±33.5	14.0 ± 28.0			
2 mrad	100	100	100	44.9±32.3	44.6±36.6	16.2±29.8		
3 mrad	100	100	100	100	100	100	100	99.8

Table 2. Luminosity efficiency (%) with perfect crossing for some values of Φ and numbers of the PCP (in one side), N_{CP}. For reference, Σ's at each collision point are shown.

From these results, it can be said that $\Phi = 3 mrad$ is enough from luminosity point of view. From closed orbit point of view, however, we have

$$A_x = 0.24, \quad A_y = 1. \times 10^{-4} \text{ for } N_{CP} = 5. \tag{11}$$

The A_x is a little too large. It is reduced to be 0.12 when $\Phi = 6 mrad$.

Let us consider the case where there are some errors in crossing optics. When $D'_x(0)$ and $D'_z(0)$ are 10% of the $\Sigma_x(0)$ and $\Sigma_x(0)/\beta_x(0)$, respectively, the results are not so much different with the previous (no error) case. See Table 3.

	$\Phi = 3$ mrad		6 mrad	
N_{CP}	5	6	5	6
R	98.5	98.6	98.2	98.3
A_x	0.24	0.28	0.12	0.14
A_y	0.07	0.07	0.043	0.046

Table 3. Luminosity efficiency with 10% spacial and momentum miscrossing.

From above discussions, it can be said that $\Phi = 6 mrad$ is sufficient.

4. Mixed Scheme

When Φ is small enough, the crab cavity is not necessary but the final focus system becomes difficult because of too small separation at the focusing magnets. One possible way is that

1. we make a small crossing angle at the IP
2. we further separate beams by the separation magnet

In Fig.2, the interaction region is illustrated.

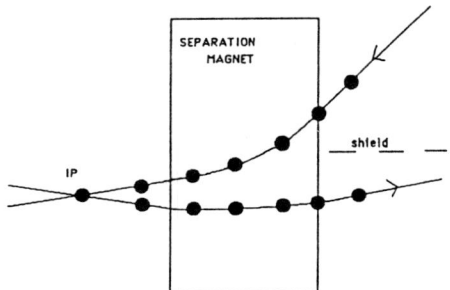

Fig. 2. The interaction region with the mixed scheme.

In Table 4, we show R, A_x and A_y for the several choices of Φ and the strength of the separation magnets. For $\Phi = 4mrad$ and strength is $0.8T$, for example, the horizontal separation at $s = 1.5m$ is $25.2mm$, which seems to be large enough to put final focus magnets.

Even with 10% spacial and momentumm errors and with 10% population error, the result does not change considerably. For $\Phi = 4mrad$ and strength is $0.8T$, we have

$$R = 98.2 \pm 14, \quad A_x = 0.056, \quad A_y = 0.027.$$

These are still acceptable.

For a reference, some parameters are shown in Table 5 for the case of $\Phi = 4mrad$ and strength is $0.8T$.

In the actual operation, tune spread (or the tune distribution) is also important, as well as the coherent tune shift. If the tune spread is too large, some particles can be trapped by nonlinear resonance. It is impossible to study this effect completely in the present method. It, however, seems to be self-evident to assume that the tune spread lies within the nominal tune and the most shifted coherent tune. In Fig. 3, the spectrum is shown for the vertical betatron oscillation in the case without any errors in trajectory and population. The upper figure corresponds to the case of weaker beam-beam effect. In the lower figure, the highest mode is at around 0.25. Thus, under the present assumption, the tune spread extends within the region from 0.15 to 0.25. This amount seems to be within the allowable range. Also see Chap.5.2.

	0T	0.2T	0.4T	0.6T	0.8T
0 mrad	0.27 ± 13.6 0 47 ± 24	0.46 ± 17.4 0.2 ± 0.1 26 ± 13			
1 mrad		52 ± 47 0.11 ± 0.03 19 ± 4.1			
2 mrad		99 ± 0.2 0.26 ± 0.054 1.93 ± 0,94	100 0.22 0	100 0.2 0	100 0.19 0
4 mrad				100 0.12 0	100 0.11 0

Table 4. Luminosity efficiency R (%) and closed orbit amplitudes A_x and A_y.

PCP	s(m)	D_x(mm)	Σ_x(mm)	Σ_y(mm)	F_x	F_y
0	0.	0	0.190	1.90×10^{-3}	1	1
1	0.3	1.2	0.198	5.70×10^{-2}	-3.0×10^{-2}	0.247
2	0.6	2.6	0.221	0.114	-7.5×10^{-3}	0.198
3	0.9	6.7	0.255	0.171	-1.5×10^{-3}	0.0663
4	1.2	14.2	0.296	0.228	-4.4×10^{-4}	0.0259
5	1.5	25.2	0.342	0.285	-1.9×10^{-4}	0.0130

Table 5. Some parameters in the case of $\Phi = 4mrad$ and strength is $0.8T$. F_x and F are evaluated at the nominal closed orbits.

5. Discussions and Conclusion

5.1. Dynamical Closed Orbit

When the beam-beam force is strong, the closed orbit does not necessarily exist.

1. The case with head-on collision but without peripheral collision was considered in Ref. 7: we expect the spontaneous separation of beams (period one or two or a limit cycle) when the ξ exceeds a certain value.

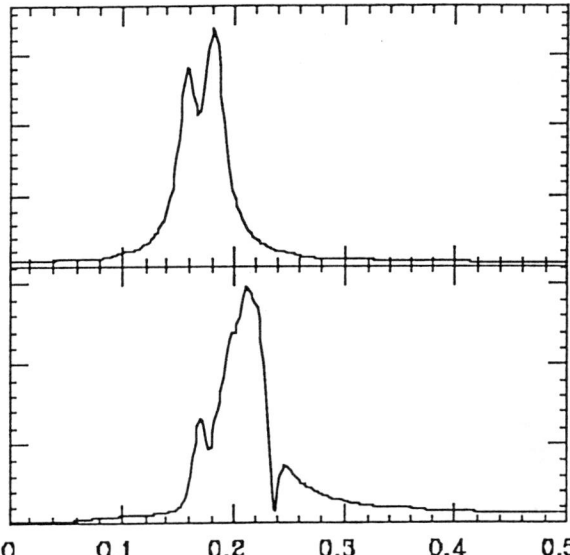

Fig. 3. The power spectrum for the case of $\Phi = 4mrad$ and strength $0.8T$ without errors. The case $\xi = 0.02$, (upper) and 0.05 (lower): $\nu_y = 0.15$.

2. The case with initial off-set between beams (but without the peripheral collision) was studied a little in Ref. 8. The period one fixed point appears always when ξ is small enough. The equilibrium off-set is smaller or larger than the initial value depending on the tune. When ξ exceeds a certain value, period one and period two fixed points can coexist. That depends on the initial positions of the beams. For larger value of ξ, more variety appear.

The present case is more complicated. Whenever the R is reduced considerably, it is observeed that the different bunch falls into different trajectories. The average luminosity reflects a portion of trajectories which collide almost head on at the IP. The difference between cases with $\Phi = 2$, $N_{CP} = 3$ and $\Phi = 2$, $N_{CP} = 4$ (See Table 2) illustrates one of the most typical instability mechanism: See Fig. 4. There, the vertical separation $\bar{y}_+ - \bar{y}_-$ is shown for each PCP. In both cases, almost half of the bunches contribute to the luminosity. The average and variance of R are almost identical between these two cases. The lower graph is more chaotic than the upper. They differ in A. We have

	$N_{CP} = 3$	$N_{CP} = 4$
A_x	0.26 ± 0.11	0.38 ± 0.19
A_y	4.38 ± 2.02	4.89 ± 2.10

This shows also the importance of the second criterion.

450   The Long-Range Beam-Beam Force

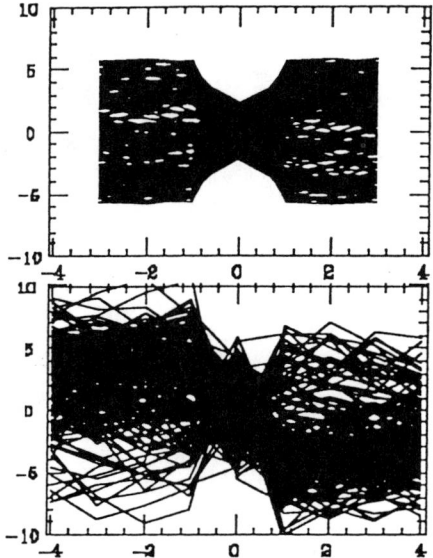

Fig. 4. The steady state of the vertical separations at each collision points. The separation is normalized by Σ_y. For $\Phi = 2mrad$ and for $N_{CP} = 3$ (upper) and 4 (lower).

5.2. BEAM-BEAM MODES

If the collision occurs only at the IP, an e^+ bunche couples with only one particular e^- bunch. In this case, they form two beam-beam modes[1], σ and π. Each mode has N_B (number of bunches in one beam) fold degeneracy. Through the peripheral collision, all bunches couple with each other. The number of the beam-beam family[10] is one. Thus, some nontrivial beam-beam modes can possibly exist. This can be studied by a linear analysis[9]: an eigenvalue analysis around the closed orbits. This, however, is possible only when there is an equilibrium trajectory for each bunches. For more general cases, the tracking such as BCD would be the only possible way.* The mode structure can roughly be seen by performing spectrum analysis (see Fig. 3) and tune survey. In Fig. 5, ν_y^+ is changed while ν_y^- and $\nu_x^\pm$ being fixed. Some remarkable points are

1. by the spectrum analysis, only two modes are seen when ξ is small. When ξ becomes large, the degeneracy is broken and some intermediate modes appear. There, however, does not seem to exist such a strange mode as that found in the case of the asymmetric collider with different circumferances[12]: the fundamental mode structure does not seem to depend on the number of bunches. Thus, we conclude that there is no drastic

---

★ Furman studied similar problem in the context of the SSC[11]. He studied the focusing effect only and did not consider the closed orbit effect.

change of the mode structure when the PCP's turned on and the modes appear just as a mixing between two fundamental modes.

2. the lowest mode is also shifted from the nominal tune (see Fig. 3). We should be careful with this fact when operating the B factory.

3. in the tune survey,the sum resonance at $\nu_y^+ + \nu_y^- =$ integer is dominant. This is the usual π mode resonance. Big error bars there are due to the fact that the π mode falls into a limit cycle there.

4. half integer resonances are not remarkable. In addition, these resonances occur somewhat below the half integral values of the tune. At just below the integral tunes, the beams are stable. This reminds us the case studied by Chao and Keil[13]: similar structure exists in the case of more than one IP and with broken superperiodicity.

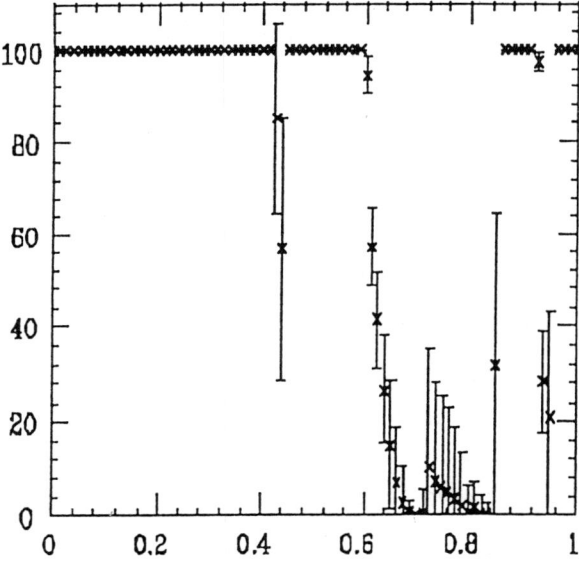

Fig. 5.   The luminosity efficiency R as a function of the ν_y^+. Parameters are: $\nu_x^\pm = 0.2$, $\nu_y^- = 0.15$.

5.3. Crab Crossing

With the mixed version discussed in Chap. 4, we can separate beams enough without loss of the luminosity and without large excitation of the closed orbits at the arcs. The important question is that whether we need the crab crossing then. According to a multiparticle tracking[14], 2mrad crossing is small enough and does not excite a remarkable synchro-beta resonance. (One should not, however, believe any multiparticle tracking too much: there are several factors omitted, which can possibly be important).

5.4. Conclusion

1. In the head-on collision scheme with the separation magnets, bunches can be filled in every three buckets and 2mm separation is enough at the peripheral collision point, 1.8 meter distant from the interaction point. This corrersponds to 0.5T magnet strength.
2. All buckets can be filled if the beams collide with a crossing angle of 3mrad (half angle). The final focus system, however, is dirfficult to design.
3. The combination of the crossing angle of 2mrad and 0.6T separation magnets (all buckets being filled) provides enough luminosity and enough space for the final focus system.

The limitation comes from the trajectory distortion due to the beam-beam interaction: the reduction of the luminosity and the excitation of the closed orbit at the arc.

Acknowledgements: K. Oide is acknowledged for useful discussion.

REFERENCES

1. K. Hirata, Nucl. Instrum. Methods Phys. Res. **A269**,7 (1988).
2. M. Bassetti and G. Erskine, CERN/ISR-TH/80-06.
3. K. Hirata and F. Ruggiero, Particle Accel. Vol.28, 137(1990).
4. Y. Funakoshi, et al, *Asymmetric B-Factory Project at KEK*, KEK Preprint 90-30 (1990)
5. A. Piwinski, IEEE Trans. on Nucl. Sci. NS-24, 1408 (1977).
6. K. Oide and K. Yokoya, Phys.Rev.**D40** 315 (1989).
7. K. Hirata and E. Keil, *Barycentre Motion of beams due to Beam-Beam Interaction in Asymmetric Ring Colliders*, CERN/LEP-TH/89-76(1989). To appear in Nucl. Instrum. Methods Phys. Res.
8. K. Hirata, *Barycenter Motion of Beams under the Beam-Beam Interaction*, TRISTAN Construction Note, TN-880010 (1988).
9. K. Hirata and E. Keil, *A Program for Computing Beam-Beam Modes*, CERN/LEP-TH/89-57(1989).
10. K. Hirata and E. Keil, to be published.

11. M. Furman, *Results of Coherent Dipole Beam-Beam Interaction Studies for SSC Lattices*, The Superconducting Super Collider Report, SSC-62 (1986).
12. K. Hirata and E. Keil, Phys.Lett. **B232**,413 (1989).
13. A. Chao and E. Keil, *Coherent Beam-Beam Effect*, CERN-ISR-TH/79-31(1979).
14. H. Koiso and K. Oide, private communication, to be written down soon (1990).

INCOHERENT BEAM-BEAM EFFECT - THE RELATIONSHIP BETWEEN TUNE-SHIFT, BUNCH LENGTH AND DYNAMIC APERTURE

C.D. Johnson and L. Wood
CERN, Geneva Switzerland

ABSTRACT

Simulation studies of the influence of long bunches on the beam-beam effect in particle colliders suggest that, despite the risk from synchro-betatron resonances, the attainable luminosity may be greater that that obtained for short bunches.

INTRODUCTION

According to the thin-lens analysis of beam-beam effect in particle colliders, the bunch length, σ_z, the beta value at the interaction point, β^*, the linear tune-shift parameter, ξ, and the disruption parameter, D are related as follows:

$$\frac{\sigma_z}{\beta^*} = \frac{D_{x,y}}{4\pi\xi_{x,y}}$$

Note that the bunch intensity and transverse dimensions are contained in the parameter D:

$$D_{x,y} = \frac{2r_e N \sigma_z}{\gamma \sigma_{x,y}(\sigma_x + \sigma_y)}$$

where r_e is the classical electron radius, N the number of particles per bunch, σ_x, σ_y are the transverse beam dimensions (Gaussian distribution) and γ is the Lorentz factor.

Disruption is essentially the extent to which one bunch is focused or defocused within the length of the other bunch, with or without the mutual pinch. For a design tuneshift of, say, 0.05, it may be required that the ratio σ_z/β^* should be less than 0.5 so that D does not exceed 0.3, to avoid excessive phase-space dilution of the colliding beams. In this note we demonstrate that beam blow-up is not so simply related to longitudinal bunch length and that longer bunches could be beneficial.

THICK-LENS SIMULATIONS

To investigate the bunch crossing in the range of σ_z/β^* from 0.05 to 1.0 we have used a program developed by one of us (L.Wood) for the CERN Compact Linear Collider (CLIC) study. This program simulates the beam crossing and includes the mutual pinch - the so-called strong/strong effect. To break down the overall effect, the mutual pinch can be turned off to simulate the weak/strong case, and the natural beta variation over the bunch length can be set to a constant value for the target bunch, leaving just the the thick-lens effect . Both bunches are sampled longitudinally during the crossing, however it is a single-pass event and so synchro-betatron effects are not simulated. This has been done using a single-particle (weak/strong) tracking program that follows the particle in transverse and longitudinal phase space for typically 1000 revolutions. The particle is tracked with a variable longitudinal step size through the target bunch and a linear lattice transform is used between beam-beam crossings.

PARAMETERS

In this study we have computed the values of the linear tuneshift parameter that corresponded to a 10% increase in the beam divergence. This is equivalent to a value of $\xi = 0.04$ for short bunches - not far from the design limit for many e^+e^- colliders. From an arbitrarily selected transverse bunch size and a chosen σ_z/β^*, the number of particles (and hence the tuneshift) is increased until the computed value of X'/X'_0 is equal to 1.1. Note that the values of X' that are quoted are rms values. Both round and flat beams (aspect ratio 20) have been studied. Both bunches contain the same number of particles, although this only matters in the strong/strong cases.

PHENOMENOLOGY

In the weak/strong case and for a fixed number of particles, as the length of the target bunch is increased the focusing effect decreases simply because the incoming particles are pulled into a weakening field as they are drawn towards the axis of the target bunch (thick lens effect). Compared to the thin lens X'/X'_0 is reduced, but X/X_0 is no longer unity. The variation of beta function over the length of the bunch further reduces the computed X'/X'_0 (and of course the luminosity). Both of these effects are compensated by the mutual pinch of the strong/strong interaction, but in our range of parameters the compensation is only partial and the same pattern of reduced X'/X'_0 with increasing bunch length is observed, although the pinch restores most of the luminosity lost due to beta variation in the weak/strong case.

ILLUSTRATIONS

Figures 1 to 4 illustrate the effects described in the previous section. The thick lens effect and the beta variation contribute approximately equally to the increased tolerance to the beam-beam effect as the ratio σ_z/β^* increases from 0.1 to 1.0. The mutual pinch reduces this tolerance but, as the strong/strong simulation of Fig. 4 shows, for $\sigma_z/\beta^* = 1.0$ the thin-lens tuneshift of 0.04 can be increased to 0.08. Figures 5 and 6 show the equivalent cases for flat bunches. We find essentially identical results for round and flat bunches. The weak/strong and strong/strong simulations are indistinguishable for $\sigma_z/\beta^* \leq 0.2$. Longer bunches display differences between the two simulation techniques, these differences increasing with bunch length.

The tune shift (round bunch) is given by

$$\xi = r_e N/4\pi\varepsilon$$

This term contains no dependence on the transverse dimension or on energy. Weak/strong and strong/strong simulations returned the expected result that the behaviour of X'/X'_0 was independent of beam energy and beam widths.

Figure 7 illustrates the variation of X'/X'_0 with tuneshift for several values of σ_z/β^*. For short bunches the results can be superimposed. We find roughly an exponential increase in X'/X'_0 for short bunches ($\sigma_z/\beta^* \leq 0.2$) at ξ limits of $0.2 \leq \xi \leq$

0.05. At larger values of z the increase is linear. If $X'/X'_0 = 1.1$ represents the tuneshift limit then this exponential behaviour near $\xi = 0.04$ can explain why even the addition of extra damping, e.g. via wigglers, has little effect on the tuneshift limit.

In Fig. 8 for the same range of tuneshift and bunch length we see that because of the mutual pinch the ratio of the luminosities calculated with and without pinch increases with bunch length and with tuneshift. This can be enough to offset the luminosity loss due to beta variation along the bunch.

SYNCHRO-BETATRON COUPLING

Of course, all of the advantages of long bunches may be wiped out by synchro-betatron effects. Figures 9 and 10 for the weak/strong and the strong/strong cases show the extent of variation of beam-beam effect (still expressed as X'/X'_0) over the length of the bunch for a single crossing. Since the weak and strong cases are not vastly different (mostly the variation over bunch length comes from the geometry of the interaction), we have applied a weak/strong particle tracking program to study the evolution over 1000 crossings in a linear machine with no transverse coupling.

Single particles of selected synchrotron amplitude are tracked through the target bunch repeatedly for previously chosen good values of the synchrotron and betatron tunes (i.e. away from low-order betatron and synchro-betatron resonances). The beam-beam effect is characterised by the increase in the transverse amplitude, A, of particles with starting amplitudes, A_0, equal to $2\sigma_x$. The target bunch is represented by a six-dimensional Gaussian distribution and in Fig. 11 the ratio A/A_0 is plotted against synchrotron amplitude (in units of σ_z) for short and long bunches ($\sigma_z/\beta^* = 0.1$ and 1.0, respectively). To amplify the observed effects the example given is for a tuneshift parameter ξ equal to 0.1. Values of D are 0.13 and 1.3. At zero synchrotron amplitude particles passing through the long target bunch are blown up much less than by the short bunch. The difference becomes less marked at larger amplitudes, but the long bunch requires a lower dynamic aperture than the short bunch over the entire range of synchrotron amplitudes.

SUMMARY

The synchro-betatron study has just begun and we are examining ways of obtaining similar results for the strong/strong interaction. If, as we would expect, the long bunch is at least no worse than the short bunch (for the same overall luminosity), then it offers the advantage of economy of rf power. There could also be a net advantage in that larger values of the tuneshift parameter (i.e. higher luminosity) would be allowed. The future study must include a lattice transform containing non-linear elements.

FIGURE CAPTIONS

1. Tuneshift parameter ξ versus bunch length corresponding to $X'/X'_0) = 1.1$ weak/strong, beta off (see text) round beams.

2. Tuneshift parameter ξ versus bunch length corresponding to $X'/X'_0 = 1.1$ weak/strong, beta on (see text) round beams.

3. As Fig. 1, but for strong/strong case.

4. As Fig. 2, but for strong/strong case.

5. As Fig. 2, with flat beams (aspect ratio 1:20).

6. As Fig. 2, but strong/strong case with flat beams (aspect ratio 1:20).

7. The ratio X'/X'_0 versus tuneshift parameter ξ for various bunch lengths σ_z/β^* from 0.1 to 0.9, strong/strong, round beams.

8. The computed luminosity divided by the theoretical luminosity without pinch versus the tuneshift, corresponding to X'/X'_0 of 1.1 for various bunch lengths as in Fig. 7, strong/strong, round beams.

9. Variation in X'/X'_0 along the bunch versus position along the bunch (in units of σ_z) for various bunch lengths, weak/strong, round beams.

10. As Fig. 9, but strong/strong.

11. Dynamic aperture versus synchrotron amplitude (in units of σ_z) for test particles in a weak/strong simulation after 1000 crossings with a tuneshift parameter $\xi = 0.1$, linear lattice, no transverse coupling, fractional part of transverse tune q = 0.295, synchrotron tune Qs = 0.0651. Upper curve: short bunch, $\sigma_z/\beta^* = 0.1$, D = 0.13; lower curve: long bunch, $\sigma_z/\beta^* = 1.0$, D = 1.3.

Incoherent Beam-Beam Effect

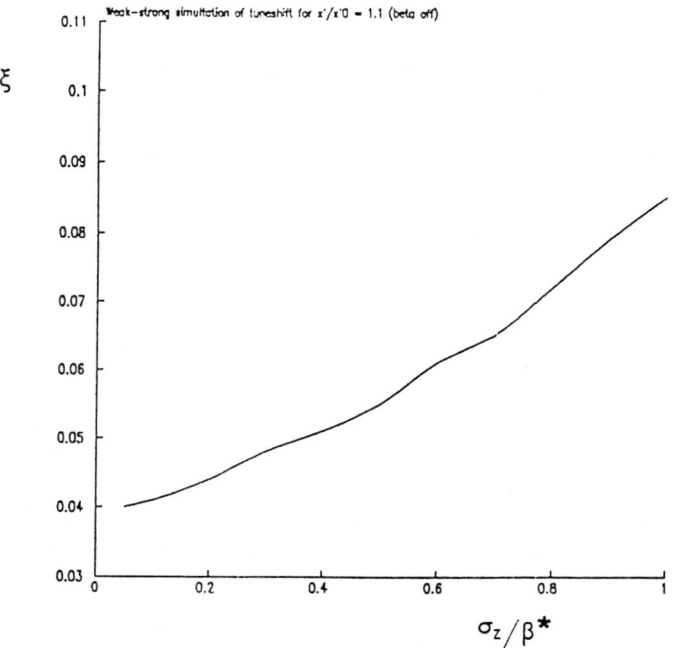

Figure 1

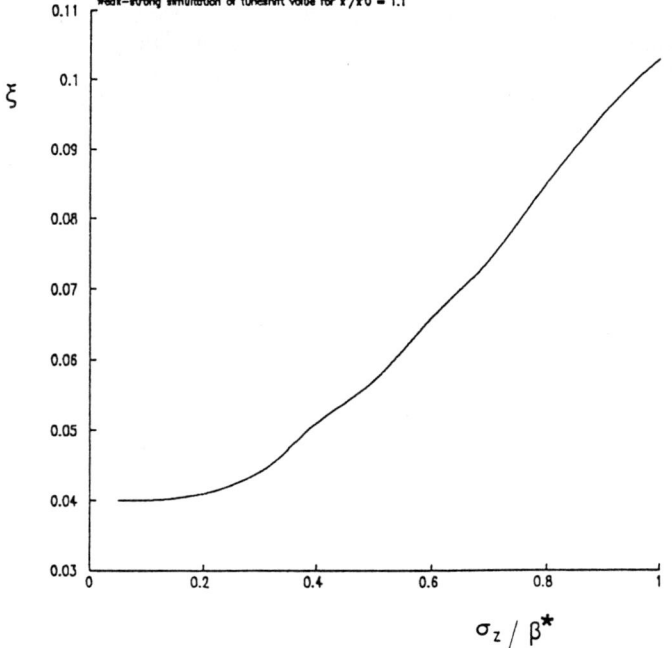

Figure 2

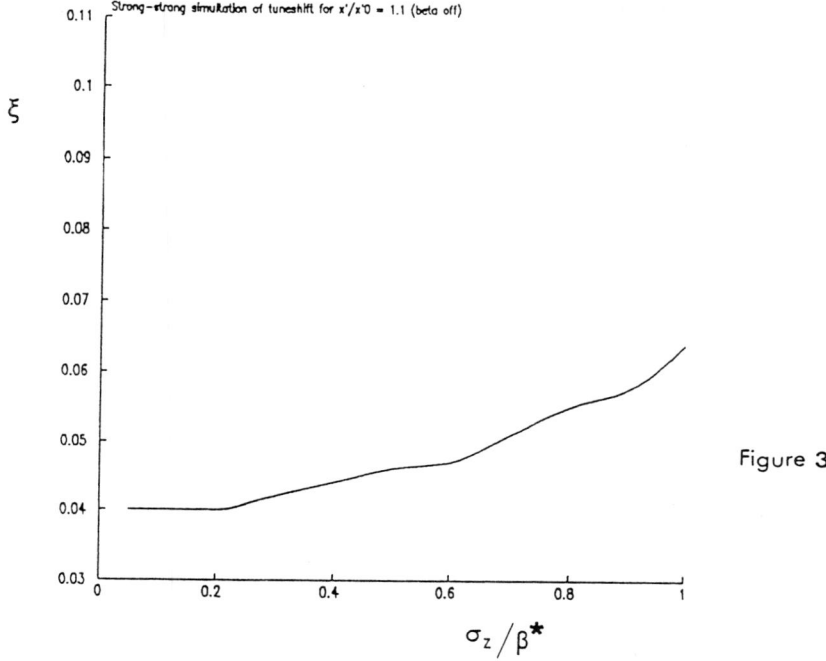

Figure 3

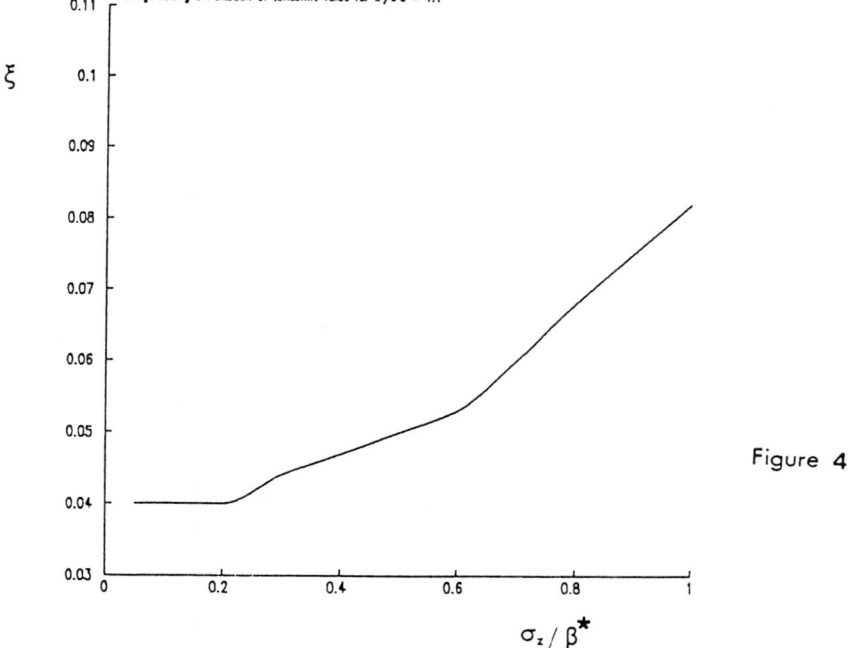

Figure 4

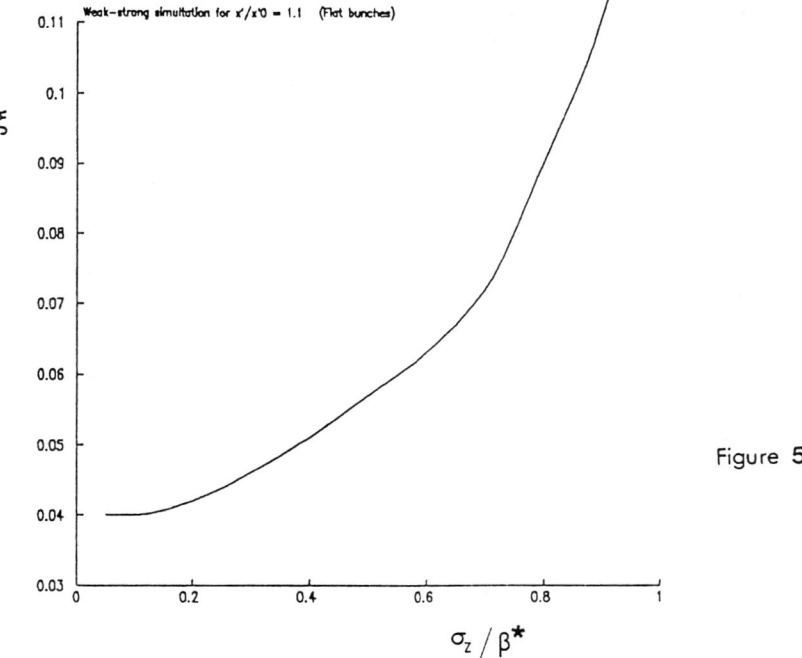

Figure 5

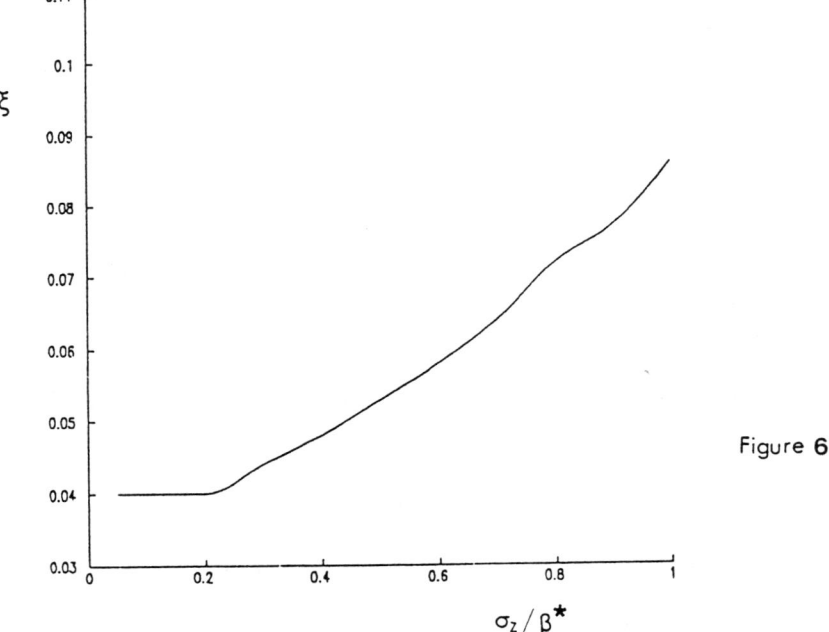

Figure 6

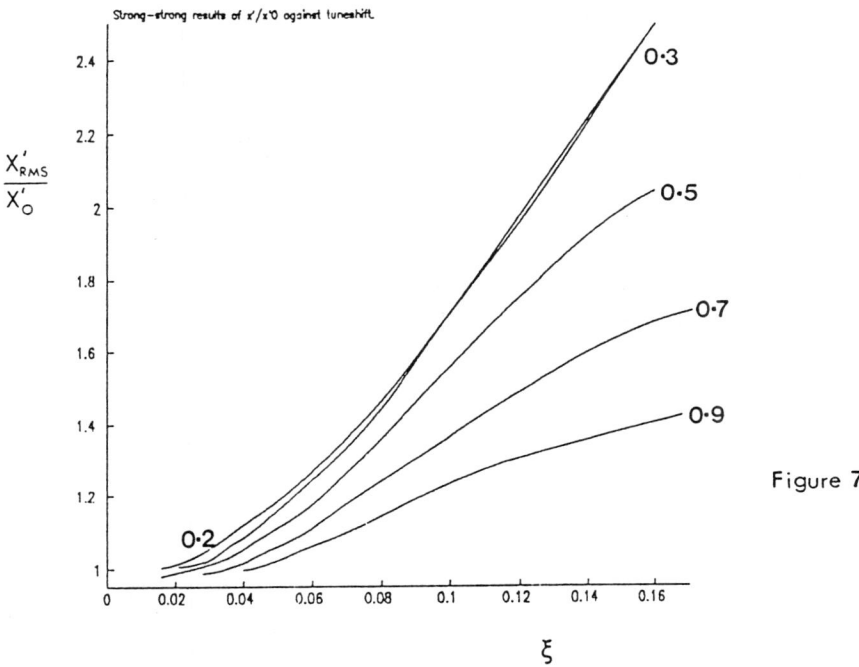

Figure 7

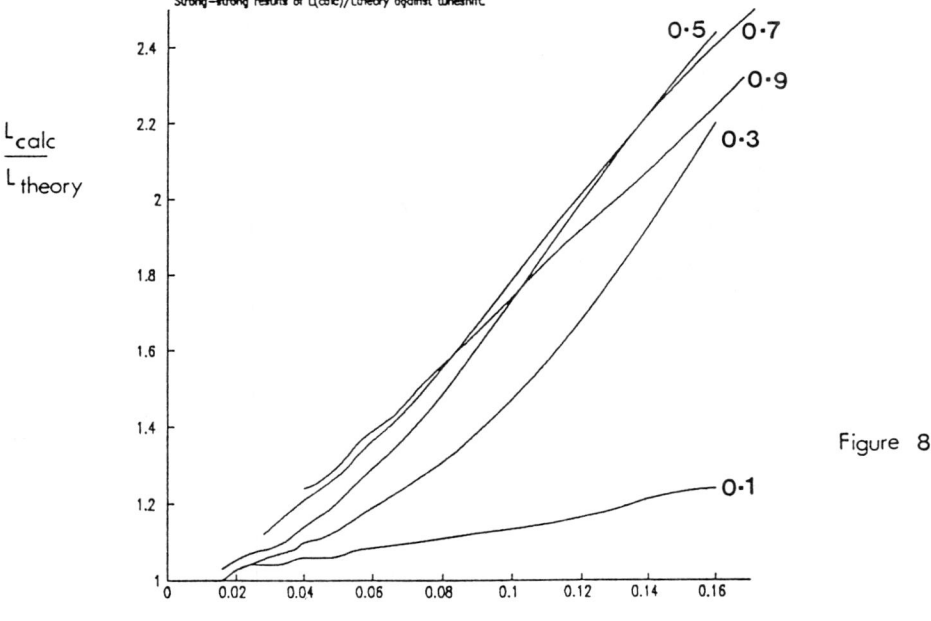

Figure 8

462 Incoherent Beam-Beam Effect

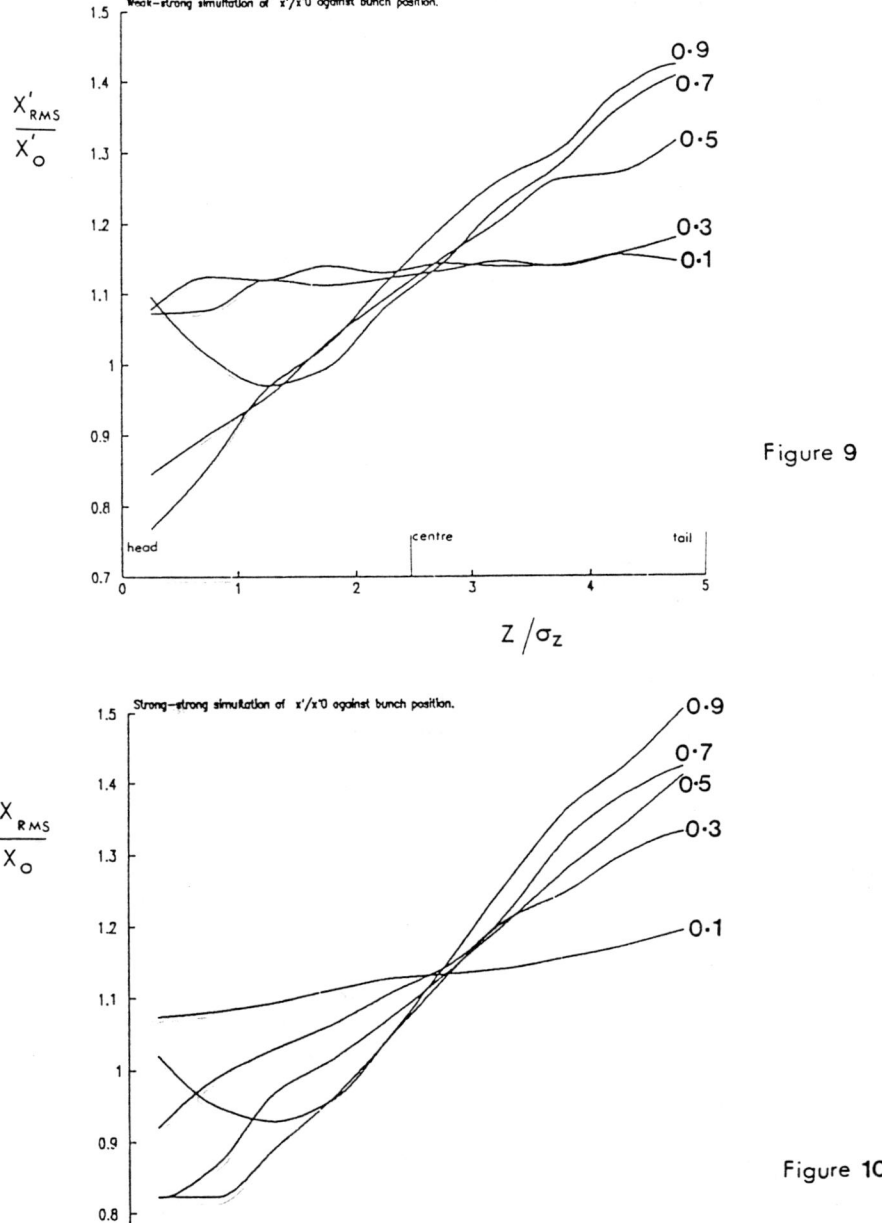

Figure 9

Figure 10

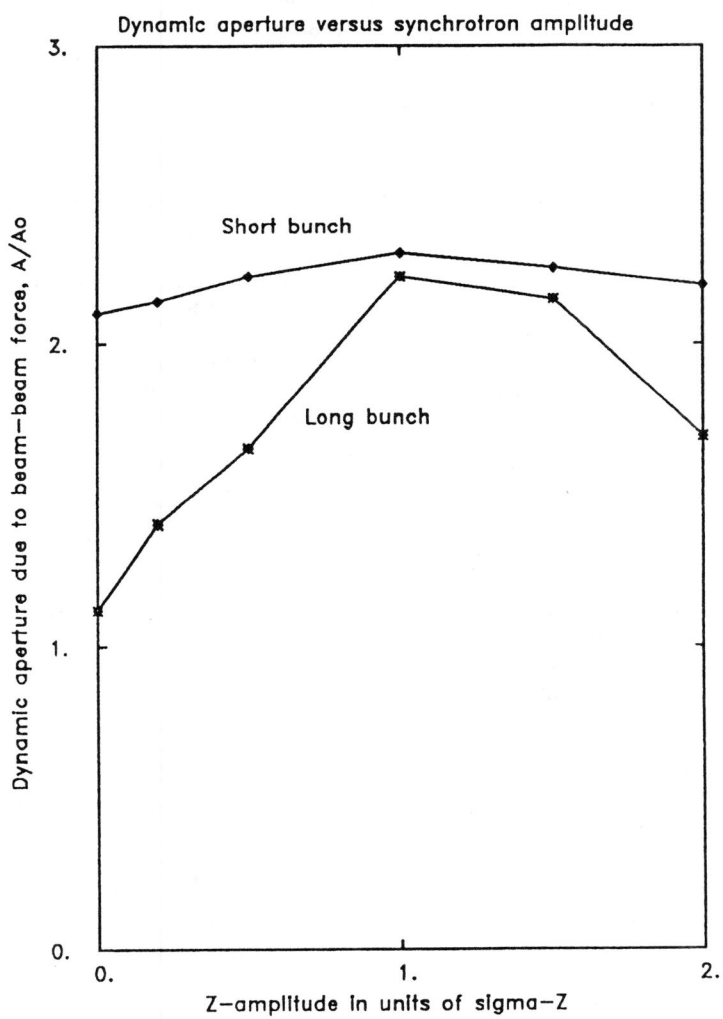

FIGURE 11:Variation of dynamic aperture with synchrotron amplitude

The Beam-Beam Interaction as a Discrete Lie-Poisson Dynamical System

Paul J. Channell
Los Alamos National Laboratory
Los Alamos, NM 87545

1. Introduction

The nonlinear beam-beam interaction in colliding beam storage rings is thought to be one of the fundamental limitations on the achievable luminosity of these machines. As a result it has been studied extensively, see references [4],[5],[6] and [7] and references contained therein for recent work. Present machines and machines under construction were designed using a conservative beam-beam limit so that it is not a major problem for these machines. However, designs are now being considered for a B-factory to make precision measurements of rare phenomena such as CP violation; these machines require luminosities several orders of magnitude higher than existing machines of the same energy. For these machines achieving the luminosity required in a storage ring design is very difficult, if not impossible, using conservative values of the beam-beam limit. Thus it is important to understand how far the beam-beam limit can be pushed and what strategies are optimal for achieving the highest luminosities.

Much of the analytic and numerical work that has been done on the beam-beam limit has used the so-called weak-strong model in which one beam is of low intensity and thus does not affect the strong beam significantly; the trajectories of particles in the weak beam under the influence of the nonlinear forces due to the strong beam (whose distribution can be assumed known) are studied and stochastic instability of these trajectories is taken as evidence that the beam-beam limit has been reached. Of course, to achieve the maximum luminosity both beams will usually be taken to be at the beam-beam limit so that the weak-strong model is not applicable and a strong-strong model is appropriate. There are numerous reasons to suspect that the weak-strong beam-beam limit might not be applicable to the strong-strong case. It is clearly possible for collective instabilities to occur [2] that will set a lower beam-beam limit than in the weak-strong case. Even if collective instabilities do not occur or can be avoided it is possible that the beam distributions that evolve in the strong-strong case have nonlinearities or time dependence that are sufficiently different from the weak-strong case that the qualitative behavior of the trajectories is entirely different. It is possible that the full self-consistent fields lead to trajectories that are in fact integrable when the trajectories in the non-self-consistent fields are stochastic; this possibility can be rephrased as the conjecture that in the strong-strong limit beams 'heal' themselves. On the other hand, one might imagine that self-consistent evolution would lead to persistent oscillations with an additional time dependence that could make the particle trajectories considerably more stochastic than in the weak-strong case. It is not known which of these possibilities occurs, or whether both might occur in different circumstances for the beam-beam interaction.

My purpose in this paper is to provide additional tools to study the dynamics of the beam-beam interaction. First, I will show that at least one good model of the beam-beam interaction fits into the mathematical category of discrete Lie-Poisson dynamical systems, a category that, in the continuous time case, has recently attracted a great deal of attention because it contains such important systems as the Vlasov-Poisson equations, the Vlasov-Maxwell equations, the equations of an ideal gas, the equations of ideal MHD, elastodynamics, and even a classical version of QCD; see [11] and [12] for a summary of the general framework and the way many of these systems fit into the framework. Particularly intriguing and possibly very important is the application of the Lie-Poisson structure to find numerous nonlinear stability results for particular solutions of these systems [9]. If an analogue of this so-called Energy-Casimir method can be found for the beam-beam system it might point the way to beam distributions with higher beam-beam limits than are usually assumed.

After I show that the beam-beam interaction is a discrete Lie-Poisson dynamical system of infinite dimensionality, I develop a basis for the Lie algebra that has very nice algebraic features; in particular, it has a straightforward Levi decomposition with a finite dimensional semisimple part and a solvable part with an infinite sequence of nested ideals. The practical implication of this algebraic structure is that the full infinite dimensional Lie algebra and the dynamical system can be truncated systematically to a sequence of finite dimensional Lie algebras and associated Lie-Poisson dynamical systems that more closely approximate the full system as the dimensionality increases. (Note that this last assertion is a formal statement, I have not studied the delicate functional-analytic questions that always plague systems in infinite dimensions.)

The ability to find finite dimensional Lie-Poisson systems that approximate the beam-beam interaction allows us to write practical simulation codes for the beam-beam interaction that exactly preserve the Lie-Poisson structure (to round-off). I will give some preliminary results from such a code at the end. Clearly, many studies remain to be done with this code and with extensions to determine the dependence of the beam-beam interaction on parameters.

2. The Physical Model of the Beam-Beam Interaction

In this paper I will ignore the dissipative and noise effects associated with radiation. For proton beams this is an excellent approximation. For electron and positron beams the approximation implies that I can examine only 'fast' phenomena, i.e. phenomena that take place in a time shorter than a radiation damping time which is typically a few thousand revolutions. Dissipation and noise destroy the underlying Hamiltonian nature of the particle motion and it cannot be expected that the Lie-Poisson structure will survive, except possibly in some perturbative sense.

I will also assume that the beams are highly relativistic and are short enough that I can ignore changes in the distribution functions during one beam crossing; in other words I assume that the beams are 'stiff' enough and short enough that during one crossing no significant changes in the distributions can occur, changes occur only as the result of multiple beam crossings. If this assumption fails then the beam-beam interaction is so strong that the beams probably won't survive anyway; in any case this approximation is excellent for all the designs being considered for storage rings.

I will also ignore effects due to synchrotron oscillations and assume that the beams collide head-on so that I can describe the beams using only two space dimensions instead of three. Most of the constructions I describe can be extended to three space dimensions, but the restriction to time scales short compared to a radiation damping time would make any three dimensional effects from the model have questionable validity since the synchrotron oscillation period is typically several to tens of revolutions and thus only moderately short compared to a radiation damping time. Of course, a three dimensional model would apply very well to a proton collider.

With these approximations we can now compute for a single particle in one bunch the effect of a single passage through the other bunch, say, for example, the effect on a particle of beam I of its passage through beam II. This interaction serves as the basis for our description of the collective behavior of both beams. It is convenient to transform to the rest frame of the beam (beam II) that the single particle is passing through. In this frame of reference the fields of beam II are electrostatic. Denoting the particle position in this frame by $\vec{q}$, the electric field of beam II is

$$\vec{E}(\vec{q}) = e_2 \int \frac{(\vec{q}-\vec{q}\,'')n_2(\vec{q}\,'')}{|\vec{q}-\vec{q}\,''|^3} d\vec{q}\,'',$$

where e_2 and n_2 are the charge and density of the particles in beam II.

It is a good approximation to compute the transverse momentum imparted to the particle in this field using the Born approximation, with the result that the change in momentum, $\Delta \vec{p}_\perp$, is given by

$$\Delta \vec{p}_\perp \simeq \frac{e_1}{c} \int_{-\infty}^{\infty} \vec{E}_\perp(\vec{q}_\perp, z) dz,$$

where e_1 is the charge of particles in beam one, c is the speed of light, $\perp$ indicates the perpendicular components only, and where I have assumed that the velocity is approximately the speed of light.

Because I am ignoring dynamical effects along the direction of bunch travel it is reasonable to assume that the beam densities factor into a part that depends only on transverse variables and a part that depends only on the longitudinal variable; to be specific let us take

$$n_i(\vec{q}) = \frac{1}{\sqrt{\pi}\sigma_i} \bar{n}_i(\vec{q}_\perp) e^{-z^2/\sigma_i^2} \quad \text{for} \quad i=1,2,$$

where σ_i is related to the lengths of the beams. It is now easy to do the integration in z and to see that the momentum change is given by

$$\Delta \vec{p}_\perp = \frac{2e_1 e_2}{c} \vec{\nabla}_\perp \int \ln(\frac{|\vec{q}_\perp - \vec{q}\,'_\perp|}{r_0}) \bar{n}_2(\vec{q}\,'_\perp) d\vec{q}\,'_\perp,$$

where r_0 is an arbitrary constant with dimensions of length to make the argument of the logarithm dimensionless and $\vec{\nabla}_\perp$ denotes the gradient with respect to tranverse variables only. The transverse positions and transverse momenta are invariant under relativistic transformations so that the transverse momentum change of a particle in the laboratory frame after passing through the opposing beam is also given by the above expression. To find the transverse momentum change of a particle of beam II after one passage through beam I one can simply replace $\bar{n}_2$ by $\bar{n}_1$ in the above expression. During the beam intersection we have assumed that the transverse position does not change significantly; we can thus write the transformation for particles in each beam as

$$\vec{q}_\perp(\text{after}) = \vec{q}_\perp(\text{before}),$$

$$\vec{p}_\perp(\text{after}) = \vec{p}_\perp(\text{before}) - \vec{\nabla}_\perp \phi(q_\perp),$$

with $\phi(q_\perp)$ defined in the obvious way for each beam. Note the important fact that because of the gradient structure of the momentum change this transformation is exactly symplectic.

The transformation of the particles around the remainder of the machine depends on the detailed design of the machine. In general it is almost a linear transformation with small nonlinear terms; in the remainder of this paper I will take it to be given by a linear transfer matrix alone, i.e.

$$\begin{pmatrix} \vec{q}_\perp \\ \vec{p}_\perp \end{pmatrix}(\text{after}) = \mathcal{M} \begin{pmatrix} \vec{q}_\perp \\ \vec{p}_\perp \end{pmatrix}(\text{before}),$$

where $\mathcal{M}$ is the usual transfer matrix. All of the derivations in the next section remain unchanged if this transformation is replaced by a nonlinear transformation, possibly obtained from a code such as TRANSPORT or MARYLIE, so long as the transformation is exactly symplectic.

3. The Lie-Poisson Structure of the Beam-Beam Interaction

In this section I will show that the transverse distribution functions of the two beams with the particle dynamics outlined in the previous section form a discrete Lie-Poisson dynamical system. Briefly, continuous Lie-Poisson dynamical systems are Hamiltonian systems in which the symplectic structure has been hidden. Namely, there is a bracket on functions of the dynamical variables that satisfies

$$\{F, G\} = -\{G, F\} \quad \text{(antisymmetry)},$$

and

$$\{F, \{G, H\}\} + \{H, \{F, G\}\} + \{G, \{H, F\}\} = 0 \quad \text{(the Jacobi identity.)}$$

Given a Hamiltonian, H, and the bracket, the equation of motion for any function of the dynamical variables is

$$\dot{F} = \{F, H\}.$$

468   The Beam-Beam Interaction

If we supplement the above requirements by the condition that the bracket be nondegenerate then Jost [10] has shown that the bracket is just the usual Poisson bracket of functions on a symplectic space and the dynamics is just the usual Hamiltonian dynamics. It is the fact that the bracket can be degenerate that leads to the Energy-Casimir method mentioned previously. For discrete Lie-Poisson systems the equation of motion doesn't apply and is replaced by the requirement that the discrete transformations preserve the bracket.

For the beam-beam interaction the dynamical variables are the transverse beam distribution functions, $f^{(1)}(\vec{q}_\perp^{(1)}, \vec{p}_\perp^{(1)}) d\vec{q}_\perp^{(1)} d\vec{p}_\perp^{(1)}$ and $f^{(2)}(\vec{q}_\perp^{(2)}, \vec{p}_\perp^{(2)}) d\vec{q}_\perp^{(2)} d\vec{p}_\perp^{(2)}$. The Lie-Poisson bracket for functionals of these variables (called the beam-beam bracket) is just the algebraic direct sum of the brackets for each beam found for the Vlasov-Poisson equations [11]; namely, the bracket of two functions F and G evaluated at a pair of distributions, $(f^{(1)} d\vec{q}_\perp^{(1)} d\vec{p}_\perp^{(1)}, f^{(2)} d\vec{q}_\perp^{(2)} d\vec{p}_\perp^{(2)})$, is given by

$$\{F,G\}(f^{(1)} d\vec{q}_\perp^{(1)} d\vec{p}_\perp^{(1)}, f^{(2)} d\vec{q}_\perp^{(2)} d\vec{p}_\perp^{(2)}) =$$
$$\int \{\frac{\delta F}{\delta f^{(1)}}, \frac{\delta G}{\delta f^{(1)}}\}^{(1)} f^{(1)} f^{(2)} d\vec{q}_\perp^{(1)} d\vec{p}_\perp^{(1)} d\vec{q}_\perp^{(2)} d\vec{p}_\perp^{(2)}$$
$$+ \int \{\frac{\delta F}{\delta f^{(2)}}, \frac{\delta G}{\delta f^{(2)}}\}^{(2)} f^{(1)} f^{(2)} d\vec{q}_\perp^{(1)} d\vec{p}_\perp^{(1)} d\vec{q}_\perp^{(2)} d\vec{p}_\perp^{(2)},$$

where $\delta F/\delta f^{(i)}$ and $\delta G/\delta f^{(i)}$ are the functional derivatives with respect to the respective distribution functions and where $\{,\}^{(i)}$ is the usual symplectic Poisson bracket with respect to the variables $(\vec{q}_\perp^{(i)}, \vec{p}_\perp^{(i)})$. The antisymmetry of this bracket is obvious. It is also clear that the Jacobi identity is satisfied because it is satisfied by the conventional brackets.

Note that if the functionals F and G are sums of separate functionals of $f^{(1)}$ and $f^{(2)}$, for example if they are both linear functionals, then the bracket simplifies to

$$\{F,G\}(f^{(1)} d\vec{q}_\perp^{(1)} d\vec{p}_\perp^{(1)}, f^{(2)} d\vec{q}_\perp^{(2)} d\vec{p}_\perp^{(2)}) =$$
$$\int \{\frac{\delta F}{\delta f^{(1)}}, \frac{\delta G}{\delta f^{(1)}}\}^{(1)} f^{(1)} d\vec{q}_\perp^{(1)} d\vec{p}_\perp^{(1)}$$
$$+ \int \{\frac{\delta F}{\delta f^{(2)}}, \frac{\delta G}{\delta f^{(2)}}\}^{(2)} f^{(2)} d\vec{q}_\perp^{(2)} d\vec{p}_\perp^{(2)}.$$

The space of $(\vec{q}_\perp^{(1)}, \vec{p}_\perp^{(1)}, \vec{q}_\perp^{(2)}, \vec{p}_\perp^{(2)})$ has the obvious symplectic structure induced from its components. We then have the following

Lemma A symplectic transformation leaves the beam-beam bracket invariant.

Proof: This follows straightforwardly from the fact that a symplectic transformation leaves the conventional brackets inside the integrands invariant and that the volume element is invariant under symplectic transformations.

This lemma implies that the single particle transformation of the previous section induces a Lie-Poisson transformation of the distribution functions.

4. A Basis for the Lie Algebra and Finite Dimensional Truncations

The Lie-Poisson structure for the beam-beam interaction that I detailed in the previous section is interesting in its own right and may provide the setting for some variant of the Energy-Casimir method. However, if one wants to numerically simulate the beam-beam interaction, one has to find some way to truncate from infinite dimensions to finite dimensions. One would like this truncation to preserve the Lie-Poisson structure and dynamics. In this section I will show that there is a natural basis for the Lie algebra of functionals that has a natural truncation to a sequence of finite dimensional Lie algebras and Lie-Poisson dynamical maps that more closely approximate the full system as the dimensionality increases.

It is possible and in most physical situations a good approximation [1] to work with a subalgebra of the full algebra given by the linear moment functionals on the distributions; i.e. we can consider the dynamical variables to be the quantities

$$<q^{(i)m}p^{(i)n}>^{(i)} \equiv \frac{1}{N^{(i)}} \int q^{(i)m}p^{(i)n} f^{(i)}(q^{(i)},p^{(i)},t)dqdp,$$

with $i = 1, 2$, where $N^{(i)}$ is the total number of particles and where $q^{(i)}$ and $p^{(i)}$ can have multiple components and m and n are then multiindices. The lowest moments have simple physical interpretations; for example $<q>$ is the average position of the system, $<p>$ is the average momentum of the system, $<q^2>$ measures, indirectly, the average size of the system, $<p^2>$ is related to the average 'temperature' of the system, etc.

The moments form an infinite dimensional subalgebra of the full algebra with a Lie-Poisson bracket inherited from the full algebra [8]. For example, in one space dimension, the bracket of two moments is easily computed to be

$$\{<q^{(i)m}p^{(i)n}>^{(i)}, <q^{(i)r}p^{(i)s}>^{(i)}\}^{(i)} = (ms - nr) <q^{(i)m+r-1}p^{(i)n+s-1}>^{(i)}.$$

The bracket of a moment of beam I with a moment of beam II is zero and vice versa. In two space dimensions the bracket has two terms that are similar and in three space dimension there are three terms.

In this paper I will assume that both beams are centered so that the transverse first moments of both beams are zero. This assumption is not necessary and the misaligned case can be treated [1], but for simplicity I will not include it here. With this assumption, the algebra of moments can be decomposed into a finite dimensional semisimple subalgebra, the subalgebra $\mathcal{M}_2$ of second moments of both beams, and an infinite dimensional nilpotent subalgebra, the algebra $\mathcal{R}_3$ of third and higher moments, which further decomposes into a nested sequence of ideals, $\mathcal{R}_n$, of moments of order n and higher; i.e.

$$\mathcal{M} = \mathcal{R}_3 \oplus_s \mathcal{M}_2,$$

where $\oplus_s$ denotes the algebraic semidirect sum, and

$$\mathcal{R}_3 \supset \mathcal{R}_4 \supset \mathcal{R}_5 \supset \cdots,$$

is a nested sequence of ideals. If one sets all the elements of an ideal to zero ('mods out by the ideal') then the remaining algebra is again a Lie algebra. Thus, the above nested sequence of ideals induces an increasing sequence of finite dimensional Lie-Poisson dynamical systems that more closely approximate the full system as the dimensionality increases.

5. Expression for the Collective Interaction

At this stage we have, in principle, completely specified the model. To investigate the dynamics we

1) Find the densities, $\bar{n}_i$ as functions of the moments,

2) Put the single particle transformations given in terms of these densities into the definitions of the moments,

3) Compute the induced moment transformations, and then

4) Set to zero all moments higher than some given order; with higher orders giving greater accuracy, but requiring more work.

The difficulty lies with step number 1; in principle the complete set of moments determines the distribution function (again, ignoring singular phenomena), but in practice a finite (i. e. truncated) set of moments is consistent with more than one distribution. We are thus free to arbitrarily choose a model of the density that produces the correct set of moments through the order we are keeping and that agrees with our notion of a 'reasonable' distribution. Note, by the way, that only a model of the spatial distribution is required; the momentum moments don't enter into the collective model.

One possible (nonunique) model is to write

$$\bar{n}_i(q,t) = \sum_i a_i(t) e^{-\mu(q-q_i)^2},$$

where the a_i are time-dependent Gaussian amplitudes and the q_i are arbitrary fixed points chosen to give a good representation of the beam profiles. The a_i are linearly related to the moments and once this relationship is specified (i. e. the q_i and μ are chosen), the density is then given in terms of the moments and this relationship can be inserted into the single particle transformations to compute the moment advance.

To be more specific, let us consider our two dimensional model and keep second through fourth moments of the distribution functions. This means that we are including the linear and nonlinear behavior of all the transverse quadrupole, sextupole, and octupole 'modes' and ignoring higher order motions. There are then 65 moments that describe each beam for a total of 130 dynamical quantities. The relation of the interaction terms to the moments can be computed by choosing density models with 17 Gaussians for each beam; there are twelve spatial moments through fourth order, fixing the total number of particles requires one more Gaussian, fixing the centers of the beams requires two Gaussians, and requiring that the collective deflection on axis be zero requires the final two Gaussians.

6. Results

The procedure of the preceeding section was implemented for a representation of the beams as second through fourth moments in two dimensions, resulting in 130 dynamical variables. Note that fixing the beams to be exactly centered eliminates any coherent dipole modes. If dipole modes might be significant, it would be necessary to rewrite the code including off-axis effects. The machine lattice was, for simplicity, given by the linear transfer matrix. In addition to tracking the collective behavior for each beam, I tracked two spectator particles, one in each beam. The spectator particles did not affect the collective behavior of the beams, they simply evolved under the forces of the linear lattice map and the time-dependent collective beam-beam forces. The initial conditions for the spectator particles were that their positions and momenta were taken to be at the rms values for each beam.

The lattice used was taken from reference 3, pages 3-2 and 3-21; it is summarized in table 1.

	Low-energy ring	High-energy ring
Energy, E [GeV]	3.1	9.0
Emittances [nm-rad]	123	41
Bunch radius at IP [cm]	0.005	0.005
Beta functions at IP		
β_x [cm]	2.0	6.0
β_y [cm]	2.0	6.0
Lattice tunes		
ν_x	16.905	21.34
ν_y	15.71	18.205

Table 1

The results are shown in figures 1-10. In each figure two of the four phase space variables are shown for the spectator particle in each beam, the top two plots in each figure. Also, two of the sixty five collective variables are shown for each beam, the bottom two plots in each figure. In all the figures the plots on the left are for beam I, the low-energy beam, and the plots on the right are for beam II, the high-energy beam. The initial values of the moments were identical in all cases and were computed by generating Gaussian beams that were approximately matched to the lattice and then computing its moments.

In figure 1 the beam intensities used were small; note that the APIARY design at full intensity has approximately $1.6 * 10^{11}$ particles per bunch in each ring so that this case has about 160 times fewer particles than in the design. As a result, the single particle phase spaces show uncoupled linear motion. The collective plots would have been points if the beams were perfectly matched; the results show that there was a slight oscillation due to mismatch, but the oscillation is regular and completely stable. The simulation was truncated at 1616 turns because nothing new seemed to be happening.

Figures 2-6 are runs in which both beams had equal numbers of particles. In figure 2 the intensity is high enough that a coherent quadrupole has been excited, though the growth rate is slow enough that damping effects would probably eliminate it. Note that the single particle trajectories show no evidence of stochasticity on this time scale. In figure 3 the intensity is higher still, with the result that the quadrupole mode growth rate is fast enough (less than 1000 turns) that radiation damping would have little effect. Note that the single particle trajectories now show some evidence of either stochasticity or coupling between the degrees of freedom. In figure 4 the intensity is again higher resulting in an even faster quadrupole mode growth rate and more particle stochasticity. Note that the collective variables are now beginning to show evidence of stochastic wander in addition to growth, indicating mode nonlinearity by coupling to other modes. In figure 5 the intensity is about the design intensity for APIARY, and the growth rate is higher, the mode nonlinearity is pronounced, and the particle trajectory is stochastic with indications of instability. In figure 6 the intensity exceeds the APIARY design value. The growth rate is not quite as large as in figure 5, but the mode nonlinearity is greater and the particle trajectory in the high-energy beam has been driven unstable.

In figures 7-10 the effects of different numbers of particles in each ring was explored. In figures 7 and 8 the low-energy beam was less intense, and in figures 9 and 10 the high-energy beam was less intense. Comparing these with figure 4, we see that in all these cases the growth rate is about the same as in the equal beam case, but that the less intense beam is now the most unstable, with the higher intensity beam being somewhat more stable than in the equal beams case.

7. Summary

It is clear that a more extensive and systematic study of the various parameter regimes of the beam-beam interaction is necessary. Nonetheless, it is already evident that strong-strong simulations of the beam-beam effect that contain higher order modes reveal new physics and different limits than weak-strong simulations. In the strong-strong regime, the limit can arise from instability of the collective variables or from instability of the single particle variables. In some cases the strong-strong regime can exhibit greater stability than the weak-strong regime. Other more complicated possibilities probably occur for other parameter regimes.

There are obviously many extensions that can be made to the model I have used. The dimensionality could be increased in order to investigate the effects of synchro-betatron resonances and nonzero crossing angle. Different models of the collective behavior could be investigated to find out the sensitivity of the results to this model. Finally, even higher moments could be included to improve the accuracy and possibly reveal the effects of higher modes on collective stability.

No attempt has been made to derive analytic consequences from the fact that the beam-beam interaction is a Lie-Poisson system, though the possible application of the Energy-Casimir method is particularly intriguing and should be pursued. In any case, it seems that the reformulation of the beam-beam interaction as a Lie-Poisson dynamical system is a useful and possibly important tool for investigating the beam-beam limit.

References

1. Channell, P. J., and Scovel, J. C., *Integrators for Lie-Poisson Dynamical Systems*, Los Alamos National Laboratory internal report, 1989.

2. Chao, A. W., and Ruth, R. D., *Coherent Beam-Beam Instability in Colliding-Beam Storage Rings*, Particle Accelerators **16** (4), 1985, p. 201.

3. Chattopadhyay, S. and Zisman, M., *Feasibility Study for an Asymmetric B Factory Based on PEP*, LBL PUB-5244, SLAC-352, CALT-68-1589, 1989.

4. Chin, Y. H., *Renormalization Theory of Beam-Beam Interaction in Electron-Positron Colliders*, Lawrence Berkeley Laboratory internal report LBL-27478, 1989.

5. Furman, M. A., *A Symplectic Coherent Beam-Beam Model*, Lawrence Berkeley Laboratory internal report LBL-27351, 1989.

6. Hirata, K., *Beyond Gaussian Approximation for Beam-Beam Interaction -an attempt-*, CERN internal report CERN/LEP-TH/88-56, 1988.

7. Hirata, K., *Stratonovich Expansion and Beam-Beam Interaction*, CERN internal report CERN/LEP-TH/89-14, 1989.

8. Holm,D., *Lie-Poisson Formulation of the Dynamics of Moments*, preprint, 1988.

9. Holm, D.D., Marsden, J.E., Ratiu, T., Weinstein, A., *Nonlinear Stability of Fluid and Plasma Equilibria*, Physics Reports, (1985).

10. Jost, R., *Poisson Brackets*, Rev. of Mod. Physics, 36, p. 572 (1964).

11. Marsden, J.E., Weinstein, A., Ratiu, T., Schmid, R., Spencer, R.G., *Hamiltonian Systems With Symmetry, Coadjoint Orbits, and Plasma Physics*, Proc. IUTAM-ISIMM Symposium on Modern Developments in Analytical Mechanics, Atti Accad. Sci. Torino, Suppl., Vol. 117, 289-340 (1983).

12. Schmid, R., *Infinite Dimensional Hamiltonian Systems*, Bibliopolis, Napoli, 1987.

474 The Beam-Beam Interaction

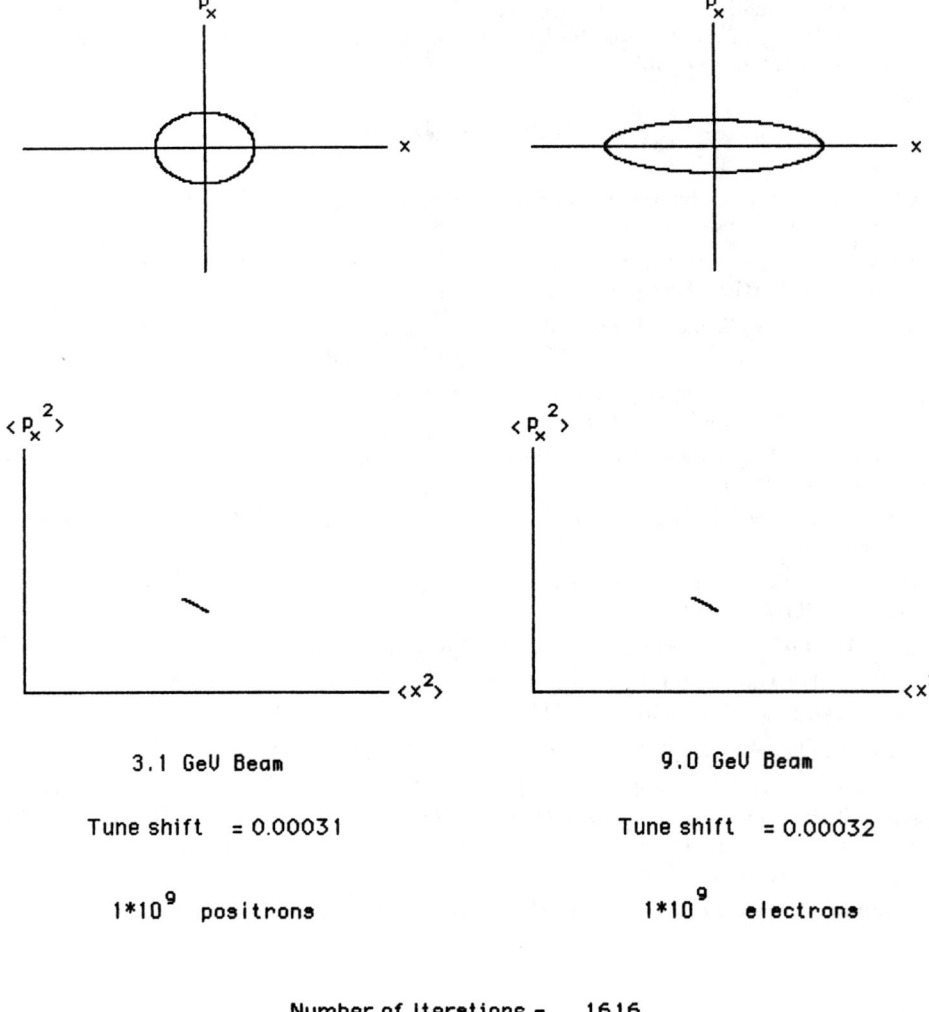

Figure 1

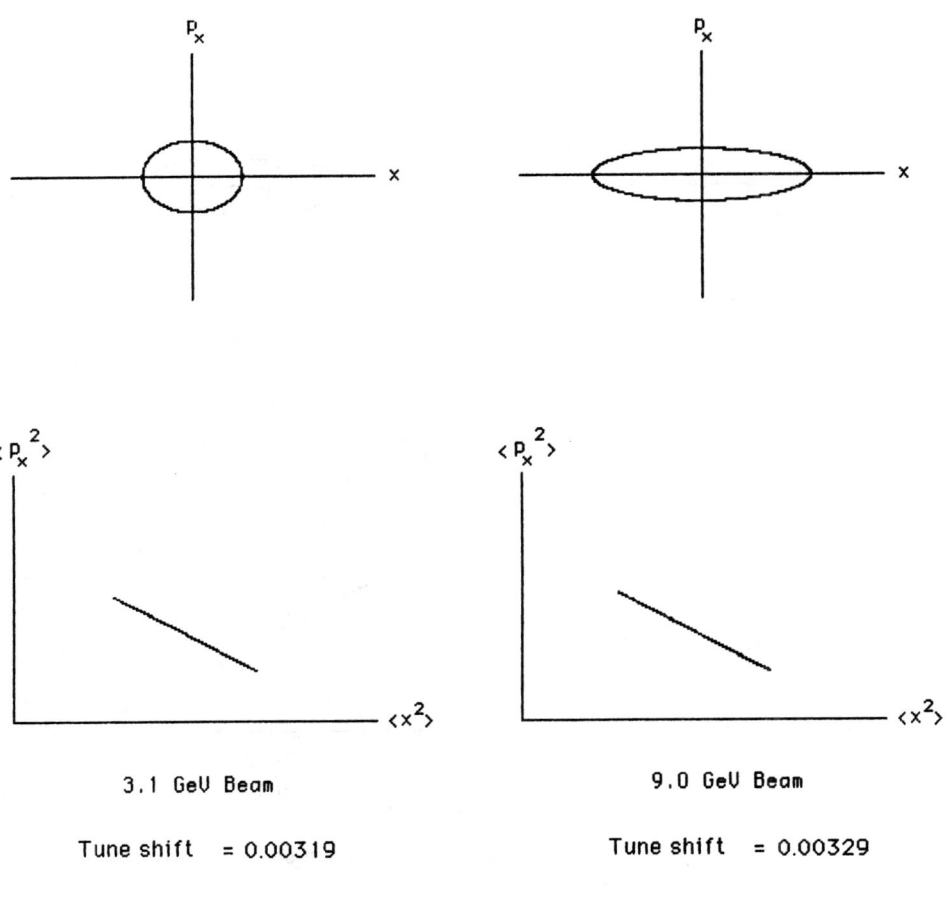

Figure 2

476 The Beam-Beam Interaction

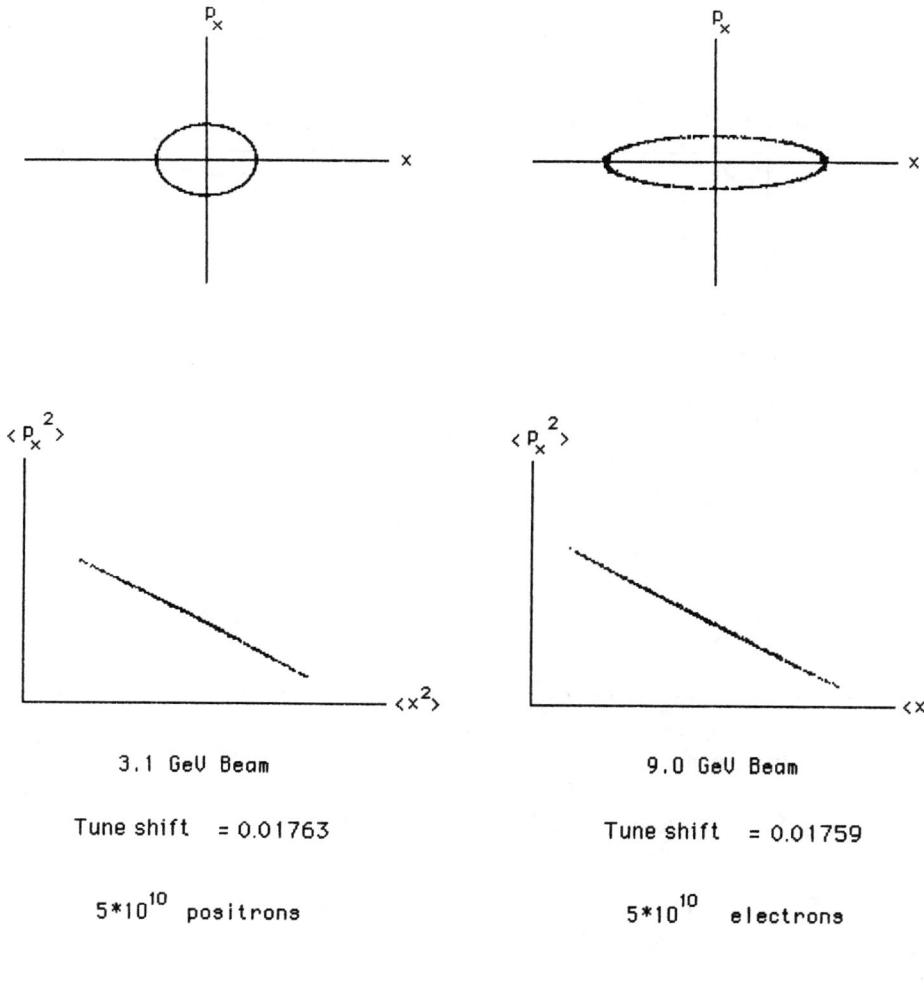

Figure 3

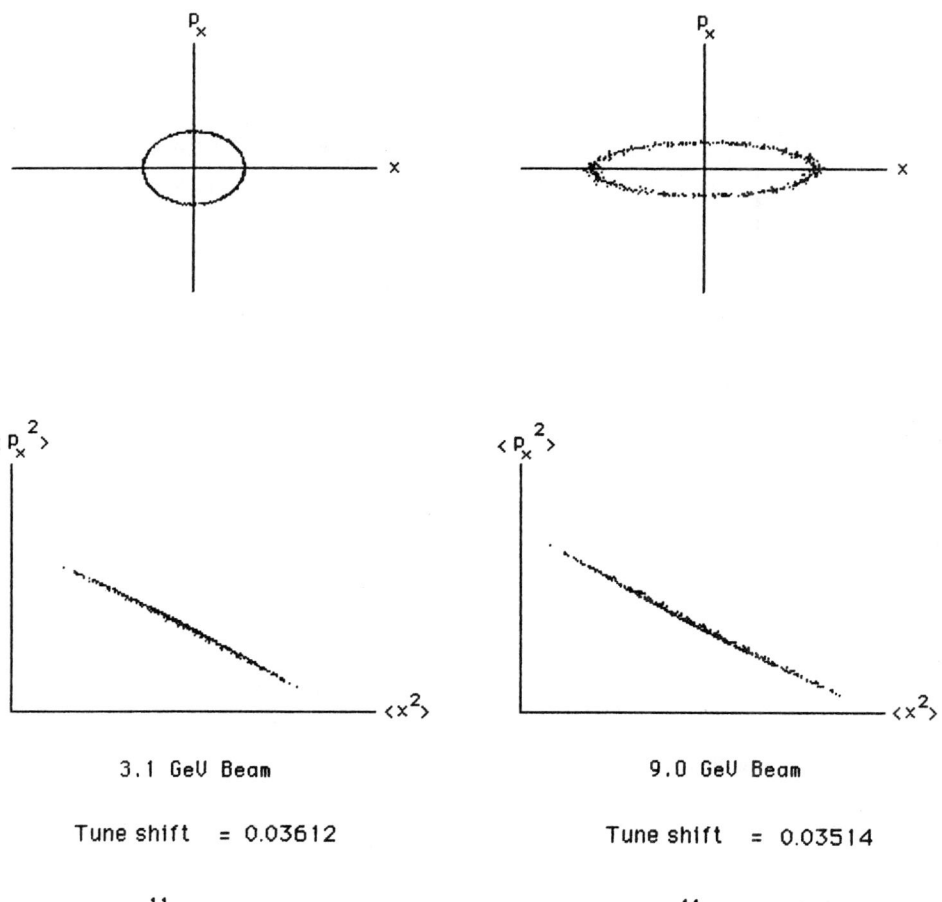

Figure 4

478 The Beam-Beam Interaction

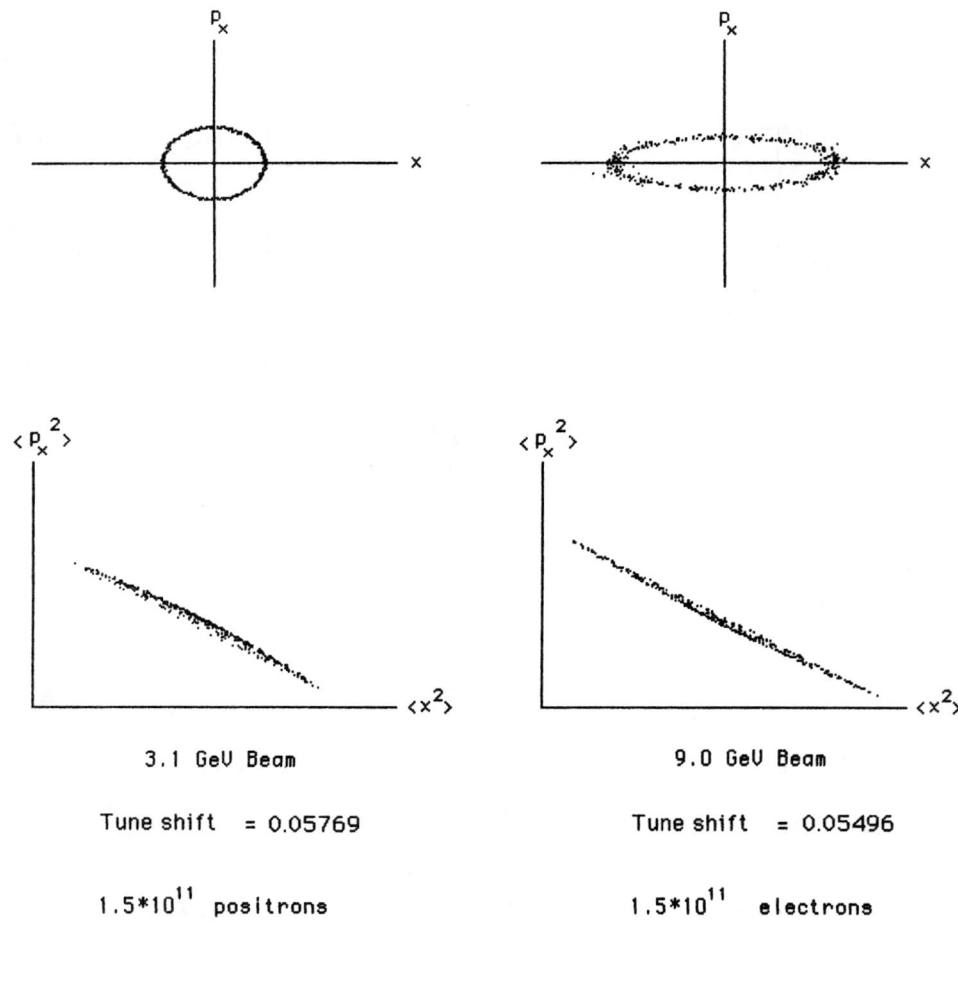

Figure 5

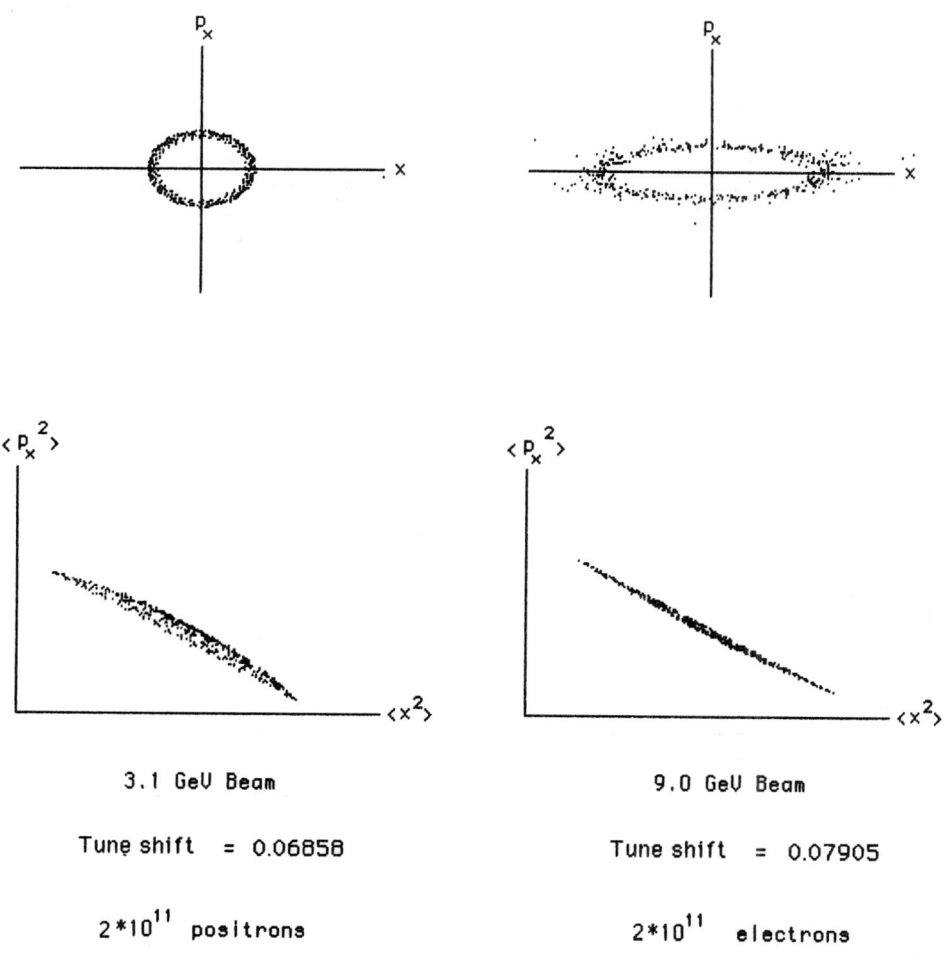

Figure 6

480 The Beam-Beam Interaction

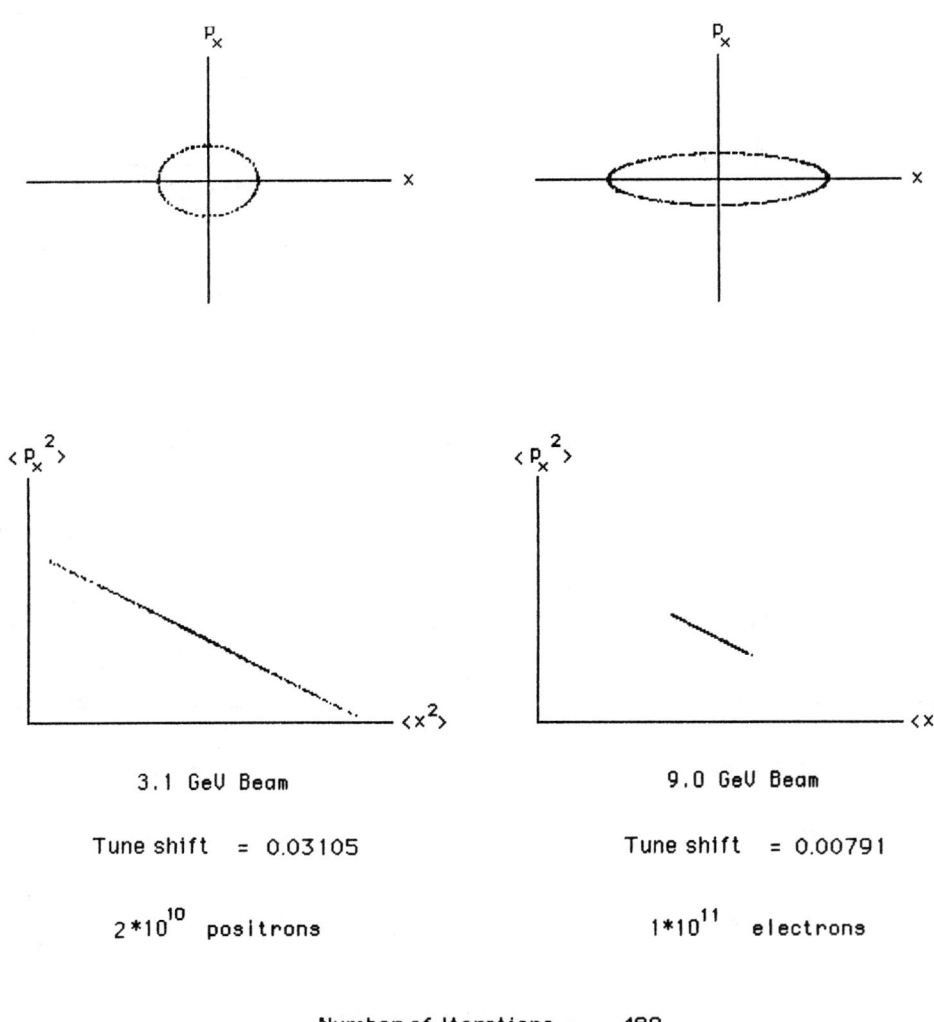

3.1 GeV Beam

Tune shift = 0.03105

$2*10^{10}$ positrons

9.0 GeV Beam

Tune shift = 0.00791

$1*10^{11}$ electrons

Number of Iterations = 489

Figure 7

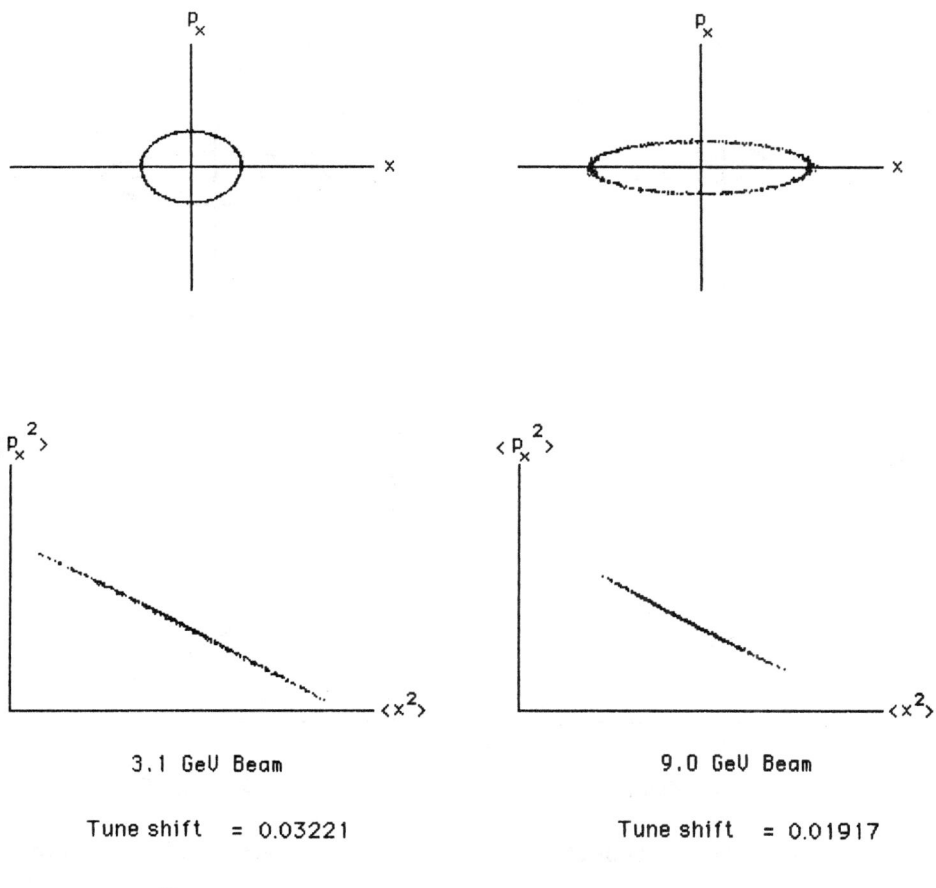

Figure 8

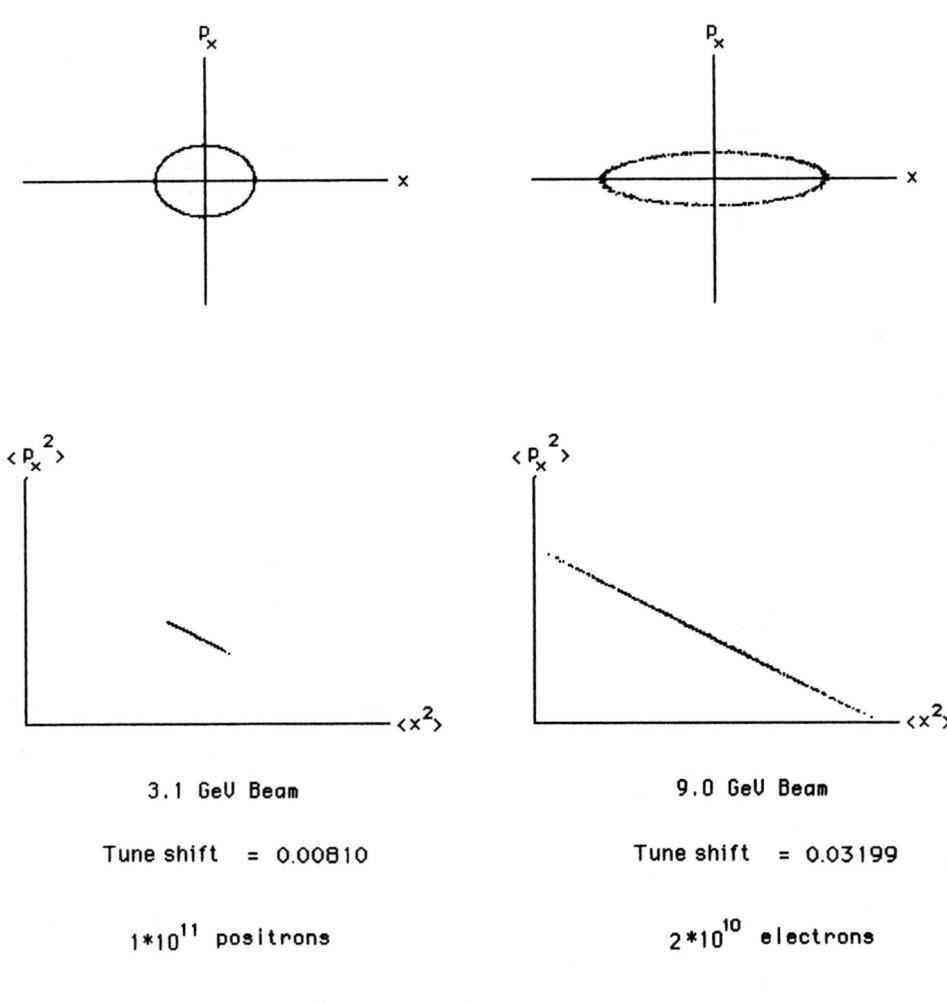

Figure 9

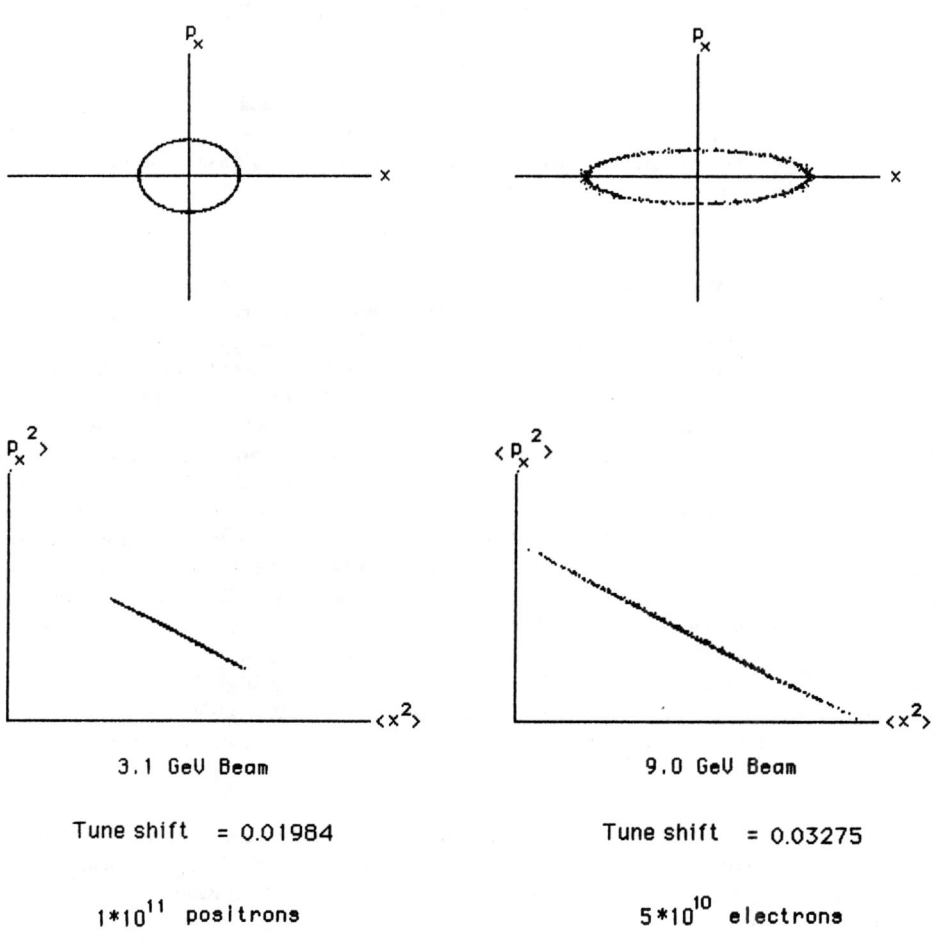

Figure 10

COMMENTS ON A LINAC BASED BEAUTY FACTORY

S. A. Heifets, G. A. Krafft, C. McDowell, M. Fripp
Continuous Electron Beam Accelerator Facility,
12000 Jefferson Avenue, Newport News, Va. 23606

ABSTRACT

A consistent set of parameters is given for a B-factory based on collisions of an electron beam from a SRF linac with the positron beam in a storage ring. An optimized lattice, an impedance estimate, a study of beam stability, and a discussion of collisions with large disruption parameters are included.

INTRODUCTION

Several recent workshops on B-factories (Cal Tech, Blois) clearly indicate that this problem has caught the interest of the high energy and accelerator communities. Preliminary estimates of the parameters of a SC linac based B-factory were presented at the Williamsburg conference in 1988[1]. The B-factory design is based on the asymmetric collisions of an electron beam from the SC linac with the positron beam accelerated and stored in an independent storage ring (SR)[2]. Here we give the results of a more detailed (but, of course, still incomplete) study of the feasibility and possible parameters of such a B-factory. As a model we use parameters of the CEBAF SC linac. The main goal is to achieve the luminosity $L \geq 10^{34}$ cm^{-2} sec^{-1} at a total energy equal to $m(\Upsilon) = 10.58$ GeV. We will show that these criteria, together with the conditions of stability and some practical limitations (on the emittance, the SR impedance, the bunch spacing), basically define the whole set of B-factory parameters. The main problem in such an approach is the large disruption parameter of the electron bunches. We discuss and model such collisions to demonstrate that it is possible to preserve stability of the positrons under these conditions. The problem of obtaining high luminosity is very challenging, and all parameters of the machine marginally achievable. First we discuss the various constraints, and then choose a consistent set of machine parameters.

LINAC

The main advantages of a superconducting linac are the relatively low emittance of the electron beam compared to a storage ring, and the relatively low wall losses compared to a room temperature linac. Since a large aperture is allowed, the transverse impedance is reduced.

Having asymmetric collisions gives preferable kinematics for analysis of the CP violation in B decays and allows a low energy electron beam. We choose the energy of the electron beam to be 3.5 GeV. Lower electron energy increases the energy of the SR, the power consumption in the SR, and the emittance of the positron beam.

The main limitation on the electron current is given by the single bunch transverse beam break-up (BBU) instability. The multibunch instability gives a less severe constraint. The beam breakup code TDBBU[3] has been used for numeric studies of the BBU instability. Fig. 1 shows the emittance degradation obtained for parabolic bunches with the density profile

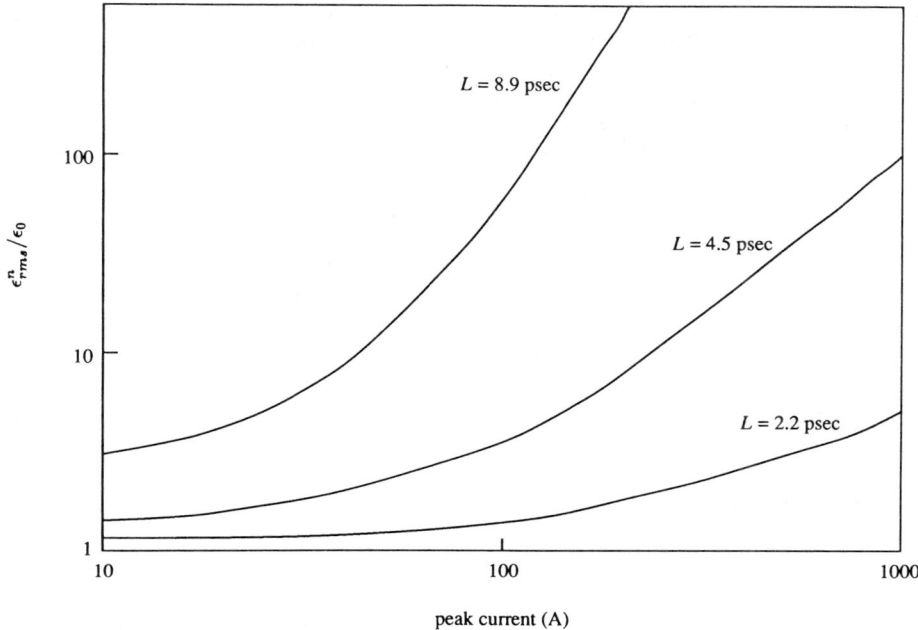

Fig. 1. Emittance degradation in the linac.

$$\rho(z) = \frac{3N}{2l}\left[1 - \left(\frac{2z}{l}\right)^2\right], \quad |z| < L/2, \quad l = 4.47\ \sigma$$

with different full length l as a function of the peak current I_p. In the simulation the slope w of the transverse wake field is equal to $w = 30.0$ V/pC/cm² per cavity, and the number of cavities is 1600. The emittance of the $l_e = 2.2$ psec bunch doubles with the peak current exceeding 571 A, or

$$w_{total} * I_p = 2.74\ 10^7\ \text{A V/pC/cm}^2. \tag{1}$$

This corresponds to $N_e = 5.26\ 10^9$ particles per bunch. For the same bunch length the emittance increases 4.8 times at $I_p = 1000$ A. Hence, the luminosity which is proportional to the ratio I_p/ϵ_e decreases if I_p increases above the doubling threshold.

The doubling threshold rapidly decreases with the bunch length. For example, the normalized emittance doubles for a 4 psec full length bunch if $I_p * w_{total} > 5 \cdot 10^5$ A V/pC/cm². Therefore, we choose a 2.2 psec bunch length of the electron beam for the B-factory.

The effective CEBAF impedance gives[4] the slope $w = 94.22$ kV/pC/cm²/pass or 290 V/pC/cm²/cavity. Scaling the result of Eq. (1) to that slope, we get the doubling threshold $N_e = 5.44\ 10^8$ for the final energy 4 GeV.

The estimate shows that the main degradation of the emittance occurs on the first pass. Hence, neglecting the weak dependence on energy we assume

$$N_e = 0.544\ 10^9, \quad I_{av} = 1.6\ \text{mA}. \tag{2}$$

The dependence on the energy may be studied analytically by solving the equation of motion[5]. For the first pass at CEBAF the betatron wave length is kept constant: $k_0 = 2\pi/\lambda_\beta, \lambda_\beta = 56.4$ m. For the next passes $k^2(s) = k_0^2 \, (\gamma_1(s)/\gamma(s))$ where the index 1 refers to the first pass and s gives the location along the orbit. For small amplitude betatron oscillations the δ- functional transverse wakefield per unit length for the short bunches is defined by the slope W'':

$$W(z, x, s) \simeq W''(s) \, x \, z$$

where z is the distance from the head of the bunch to the tail. The equation of motion takes the form:

$$\frac{d}{ds}(\gamma x') + k_0^2 \gamma_1(s) \, x = N_B r_0 W'' \int_z^\infty dz' \rho(z')(z'-z) x(z', s). \tag{3}$$

(The actual dependence on the distance between particles z, is, probably, $\sqrt{z}$. In this case W'' should be considered as the "effective" slope giving the correct wake at the end of a bunch). Here ρ is the normalized charge density.

The solution is obtained by iterations:

$$x = \frac{A}{\gamma(s)} \left\{ \cos\left(\int_0^s k \, ds\right) + \lambda \sin\left(\int_0^s k \, ds\right) + \ldots \right\}$$

where

$$k^2(s) = k_0^2 \frac{\gamma_1(s)}{\gamma(s)} + \left(\frac{\gamma'}{2\gamma(s)}\right)^2.$$

The second term, proportional to the acceleration rate ($\gamma' = 5$ m^{-1}), is small even for the first pass and can be neglected.

The first iteration could have been computed from the WKB approximation. Averaging the fast oscillating terms gives the enhancement factor λ at the tail of the bunch with the length l_B as

$$\lambda = \frac{\Lambda}{\sqrt{k(s)}} \int_0^s \frac{ds' \, W''(s')}{\gamma(s')\sqrt{k(s')}}, \qquad \Lambda = \frac{N_B r_0 l_B}{2}. \tag{4}$$

Estimating λ for the first pass and introducing the total slope w we obtain

$$\lambda = \frac{\Lambda}{k_0 \gamma_{1f}} \left[w_{linac} \ln \frac{\gamma_{2f}}{\gamma_{1i}} + w_{arc}\left(1 + \frac{\gamma_{1f}}{\gamma_{2f}}\right) \right] \tag{5}$$

where indexes 1, 2 refer to the first and to the second linacs of the first pass, and i, f refer to the beginning and the end of the linacs. According to Eq. (4) the dependence on the final energy is weak: the main contribution may be expected from the first pass.

The total slope of the wake is the sum over all impedance generating elements of different types

$$w = \int ds W'' = \sum W_i n_i \, .$$

It has been estimated[4] as

$$w_{linac} = 8.45 \, 10^3 \text{ cm}^{-3}$$

for a linac, and as
$$w_{arc} = 4.87 \ 10^4 \ \text{cm}^{-3}$$
for an arc.

The threshold of instability corresponds to $\lambda \simeq 1$, giving $N_B = 0.2 \ 10^9$ in agreement with the numeric simulations.

One might attempt to increase the peak current by BNS phasing the linac[6] to minimize the emittance degradation*. If the head of a bunch has the phase ϕ_0 behind the crest of the RF voltage in the cavities, then the energy of a particle depends on the distance z of the particle from the head of the bunch:

$$\gamma(s,z) = \gamma(s,0)\left[1 - \frac{2\pi z}{\lambda_{RF}}\tan\phi_0\right].$$

A dependence of $\gamma_1 k_0^2$ in Eq. (3) on z is generated which can compensate the part of the wake linearly dependent on z in the right hand side of the equation if

$$\tan\phi_0 = \frac{N_B r_0 W'' \lambda_{RF}}{4\pi\gamma(0,s)k_0^2}.$$

With this choice of the phase ϕ_0 the motion of following particles in the bunch is the same as the motion of the particle at the head of the bunch. However, for $N_B = 10^9$, $\lambda_{RF} = 20$ cm, $w_{linac} = 16.9 \ 10^3 \ \text{cm}^{-3}$, $\gamma = 10^3$, and for the length of the linac $L = 240$ m, the phase $\phi_0 = 84.2°$ would drastically reduce the acceleration rate making BNS phasing unacceptable.

The power in the beam in the linac is

$$P = 0.16\left(\frac{N_e}{10^9}\right)\left(\frac{f}{\text{MHz}}\right)\left(\frac{E}{\text{GeV}}\right)\left[\frac{MW}{\eta}\right].$$

For a SC linac the efficiency $\eta \simeq 1$. This gives $P = 5.65$ MW for the repetition rate $f = 20$ MHz and $E_e = 3.25$ GeV.

Power deposition in the cold section of the linac may be a limiting factor. For example, the loss factor estimated for CEBAF cavities[4] is $k_l = 9.15$ V/pC for 2.2 psec bunches. Most of the loss comes from the high frequency modes (3.2 V/pC from the modes with frequency $f < 6.5$ GHz). The losses in the fundamental power coupler and the HOMs increase this factor to 15 V/pC per cavity. The power deposition per cavity with N_e given by Eq. (5), and $f = 20$ MHz is

$$P = e^2 N_b^2 k_l f = 2.0 \text{ W/cavity}.$$

For comparison, the CEBAF cryounit can handle 5 W/cavity. The cryogenic load could be reduced by increasing the bunch length,

$$k_l \sim 1/\sqrt{\sigma_l}.$$

However, that would reduce luminosity.

---

* In the simulations[5] the phasing minimized the energy spread.

STORAGE RING

At the $\Upsilon(4S)$ resonance ($M = 10.58$ GeV, $\Gamma(\Upsilon) = 24$ MeV) the energy of the positron beam E_p is related to E_e:

$$E_e E_p = 27.984 \text{ (GeV)}^2, \qquad \hat{\gamma}_e \hat{\gamma}_p = 107.169. \tag{6}$$

Here and later we use for convenience $\hat{\gamma}$ defined as $\gamma = \hat{\gamma}\, 10^3$.

The optics of the SR has to be chosen to minimize the transverse emittance, allowing storage of large average current. The single mode longitudinal instability limits the average current in the SR:

$$\frac{I_{av}}{I_A} \leq \sqrt{\frac{\pi}{2}} \frac{Z_0}{s_B} \frac{\alpha \gamma_p \sigma_\delta^2 \sigma_{p,z}}{\left(\frac{Z}{n}\right)_{eff}}$$

where $Z_0 = 377$ Ohm, $I_A = ec/r_0 = 1.7\, 10^4$ A, s_B is the bunch spacing, and $\left(\frac{Z}{n}\right)_{eff}$ is the effective impedance.

We can rewrite this as

$$\frac{I_{av}}{A} \leq 0.86\, 10^{12}\, \frac{\alpha \sigma_\delta^2 \sigma_{p,z}}{s_B \hat{\gamma}_e \left(\frac{Z}{n}\right)_{eff}} \text{ Ohm.} \tag{7}$$

As is clear from Eq. (7), I_{av} increases with the energy spread σ_δ, compaction factor α, and the bunch length $\sigma_{p,z}$. For example, $I_{av} = 0.5$ A can be obtained with $s_B = 15$ m, $\hat{\gamma}_e = 6$, $\sigma_{p,z} = 1$ mm, and $\left(\frac{Z}{n}\right)_{eff} = 0.5$ Ohm if $\alpha \sigma_\delta^2 = 2.6\, 10^{-8}$. However, large $\alpha \sigma_\delta^2$ requires high RF voltage, i.e. a large number of RF cavities. That increases the contribution of the cavities to the effective impedance. If this contribution becomes dominant I_{av} does not increase with $\alpha \sigma_\delta^2$ any further. We assume that the rest of the effective impedance (the contribution of the beam pipe discontinuities) could be as low as

$$\left(\frac{Z}{n}\right)^{eff}_{lattice} = 0.25 \text{ Ohm} \tag{8}$$

and use that for the comparison with the total impedance of the RF cavities.

We discuss first the choice of the lattice for the SR, and then give the estimate of the impedance of the cavities.

LATTICE

The equilibrium horizontal emittance ϵ_x, energy spread σ_δ, and the compaction factor α are given[7] for decoupled x and y motion in terms of the horizontal dispersion function D_x and the β_x function:

$$\epsilon_x = \frac{\lambda_0}{J_x} \frac{\langle K^3 f(s) \rangle}{\langle K^2 \rangle}$$

$$\sigma_\delta^2 = \frac{\lambda_0}{J_z} \frac{\langle K^3 \rangle}{\langle K^2 \rangle}$$

$$\alpha = \langle K D_x \rangle.$$

Here $K = 1/\rho$ is the curvature in the bend magnets,

$$\lambda_0 = \frac{55\sqrt{3}}{48}\Lambda_c\gamma^2 = 3.06\ 10^{-6}\ (E/\text{GeV})^2,$$

J_x, J_z are the partition functions:

$$J_x = \frac{1}{\langle K^2 \rangle}\langle K^2[1-(1-2n)KD_x]\rangle, \quad J_z + J_x = 3,$$

and

$$f(s) = \frac{1}{\beta_x(s)}\left[D_x^2 + \left(\beta_x D_x' - \frac{1}{2}\beta_x' D\right)^2\right].$$

For a periodic structure with the cell length L and the phase advance per cell $\mu_c = 2\pi\nu$,

$$D_x(s) = \frac{\sqrt{\beta_x(s)}}{2\sin(\pi\nu)}e^{i(\mu(s)+\pi\nu)}\int_s^{s+L}ds' B(s') + c.c$$

where

$$B(s) = K(s)\sqrt{\beta_x(s)}e^{-i\mu(s)}.$$

The function $f(s)$ takes the form

$$f(s) = \frac{1}{4\sin^2(\pi\nu)}\left|\int_s^{s+L}ds' B(s')\right|^2.$$

The function $B(s) \neq 0$ only in the bend magnets. For short dipoles, and taking into account that $\beta_x \propto L$ we obtain the well known scaling:

$$\alpha \propto \theta_c^2, \quad \sigma_\delta^2 \propto \frac{\lambda_0 \theta_c}{J_z L}, \quad \epsilon_x \propto \frac{\lambda_0 \theta_c^3}{4 J_x}.$$

Here we used the bend angle of the cell $\theta_c = KL$.

Notice, that the emittance and $\alpha\sigma_\delta^2$ are related:

$$\epsilon_x = \left(\frac{J_z}{J_x}\right)\alpha\sigma_\delta^2 L \Phi(\nu)$$

where $\Phi(\nu) \simeq 1$ depends on the details of the cell structure.

The B-factory lattice has to be designed to minimize the emittance with relatively large $\alpha\sigma_\delta^2$. Emittance depends basically on the bend angle. Small emittance requires short bends. Given emittance, $\alpha\sigma_\delta^2$ can be increased reducing the cell length L. Additional reduction of the emittance with simultaneous increase of $\alpha\sigma_\delta^2$ can be obtained by reducing J_z. This requires combined function dipoles focusing in the horizontal plane.

We choose a short cell with combined function dipoles focusing in the x-plane and a defocusing quad. We do not use more sophisticated lattices because they require longer cells. To get enough flexibility, the dipole is split into two halves with a thin focusing quadrupole between them. That makes the cell similar to the usual FODO structure. We give as an example the results for the two model cells:

$$L = 5 \text{ m}, \ 2L_b = 4.0, \ L_{q,D} = 0.40 \text{ m}, \ L_{q,F} = 0.20 \text{ m}, \ \rho = 45.0 \text{ m}$$

$$K1_{q,D} = 2.0 \text{ m}^{-2}, \ K1_{q,F} = -1.8 \text{ m}^{-2}, \ K1_b = -0.12 \text{ m}^{-2}.$$

This gives at the energy 7.94 GeV

$$\nu_x = 0.283, \ \nu_y = 0.255, \ \beta_x = 9.8 \text{ m}, \ \beta_y = 1.4 \text{ m}, \ D_x = 0.255 \text{ m}, \ J_x = 2.65, \ J_z = 0.344,$$

$$\alpha = 0.273 \ 10^{-2}, \ \sigma_\delta = 0.244 \ 10^{-2}, \ \epsilon_x = 6.18 \ 10^{-9} \text{ m}, \ \Delta E = 0.110 \text{ MeV/cell},$$

$$\tau_x = 0.9 \text{ msec}, \ \tau_y = 2.4 \text{ msec}, \ \tau_E = 6.9 \text{ msec}.$$

The lattice is flexible: using $\rho = 43$ m, and $K1_b = -0.06$ m^{-2} we obtained $\alpha = 0.208 \ 10^{-2}$, $\sigma_\delta = 0.124 \ 10^{-2}$, $\epsilon_x = 7.45 \ 10^{-9}$ m at $E = 8.02$ GeV.

IMPEDANCE

Let us now estimate the impedance of the RF cavities. We start with the impedance of a single cavity. A high frequency RF system minimizes the RF voltage needed for short bunches. To minimize impedance and the losses we choose for our studies SC CEBAF 1.5 GHz cavities. In the modal analysis the impedance is

$$Z(\omega) = i \sum_n \chi_n (\omega - \omega_n + i\gamma_n)^{-1}$$

where the loss factors

$$\chi_n = \frac{L\omega_n}{2} \left(\frac{r}{Q}\right)_n$$

depend on the length of the structure L and the (r/Q) for each mode as they are given by URMEL (in units Ohm/m). Hence, the loss is

$$Re \int^\omega d\omega \ Z(\omega) = \frac{\pi L}{2} \sum_{\omega_n < \omega} \omega_n \left(\frac{r}{Q}\right)_n.$$

The sum has been calculated for CEBAF cavities ($L = 0.5$ m) for the modes given by URMEL[8] with frequencies up to 6 GHz. The sum behaves as $\sqrt{k}$, $k = \omega/c$:

$$\sum_{\omega_n < \omega} \frac{\omega_n}{2\pi} \left(\frac{r}{Q}\right)_n = \Lambda \sqrt{k}$$

where $\Lambda = 2.67 \ 10^6$ (Ohm/m) MHz $\sqrt{\text{cm}}$, as in Fig. 2. This means that the average $Re \ Z(\omega)$ is inversely proportional to $\sqrt{k}$:

$$Re \ Z(k) = A/\sqrt{k}, \qquad A = \frac{\pi^2 L \Lambda}{2c} = 220 \text{ Ohm}/\sqrt{\text{cm}}.$$

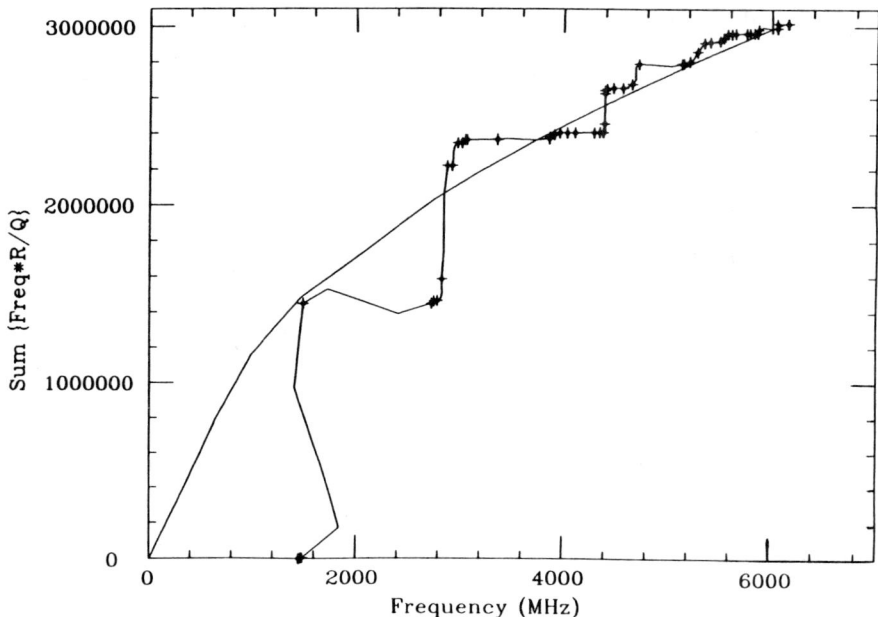

Fig. 2. The sum $\sum_{f<f_{max}} f_n(r/Q)_n$ for CEBAF cavities vs. frequency f_{max}.

Let us compare this result with the estimated[4] total loss $k_l = 17.5$ V/pC/cavity for a $l_b = 2.2$ psec bunch. For a Gaussian bunch with rms length σ

$$k_l = \int \frac{d\omega}{2\pi} Z_l(k) e^{-k^2\sigma^2}.$$

Assuming $Z(k) = A/\sqrt{k}$ we obtain

$$k_l = 0.577 \frac{A\,c}{\sqrt{\sigma}}.$$

This gives 17.5 V/pC for $\sigma = 2.2$ psec $*c/4 = 165\ \mu$m if $A = 130$ Ohm/$\sqrt{\mathrm{cm}}$, agreeing within a factor of 2 with the previous estimate.

For the following discussion we assume

$$Z(k) = A/\sqrt{k}, \quad A = 200\ \mathrm{Ohm}/\sqrt{\mathrm{cm}}. \qquad (9)$$

The effective impedance is dominated by the lowest collective mode

$$\left(\frac{Z(n)}{n}\right)_{eff} = \frac{\int (Z/n) k^2 \sigma^2 e^{-k^2\sigma^2} dk}{\int e^{-k^2\sigma^2} dk}.$$

For the model Eq. (9) this gives

$$(Z/n)_{eff} = 1.5\,10^2\,\frac{\sigma^{3/2}}{R}\ \mathrm{Ohm}/\sqrt{\mathrm{cm}}/\mathrm{cavity}.$$

A Linac Based Beauty Factory

The number of RF cavities in a storage mode is given by the losses per turn ΔE:

$$\Delta E = eV \cos \phi_s,$$

and the bunch length. The rms bunch length is related to the rms energy spread:

$$\sigma_z = \frac{\alpha R}{\nu_s} \sigma_\delta$$

where ν_s is the synchrotron tune

$$\nu_s = \sqrt{\frac{qeV\alpha \sin \phi_s}{2\pi E_p}}, \quad q = 2\pi R/\lambda_{RF}.$$

Assuming that $\Delta E \ll eV$ we obtain:

$$eV = \frac{\alpha \sigma_\delta^2 R \lambda_{RF}}{\sigma_z^2} E_p$$

Therefore, n_{cav} CEBAF-type cavities are required with the acceleration rate $E' = 2.5$ MeV/cavity with

$$n_{cav} = \frac{\alpha \sigma_\delta^2 \gamma_p R \lambda_{RF}}{5\sigma_z^2}. \tag{10}$$

Notice, that the $\alpha \sigma_\delta^2 R$ depends on the bending angle of the dipoles but not on the cell length and is fixed for a given transverse emittance.

The total effective impedance of the cavities is

$$(Z/n)_{cav,\,eff} = 1.03\,10^7 \frac{\alpha \sigma_\delta^2 \lambda_{RF}}{\hat{\gamma}_e \sqrt{\sigma_z/\text{mm}}} \frac{\text{Ohm}}{\text{cm}}. \tag{11}$$

The maximum current given by Eq. (7) and Eq. (11) is

$$\left(\frac{I_{av}}{0.5\,\text{A}}\right) = 167.0\,\kappa \frac{(\sigma_z/\text{mm})^{3/2}}{(s_B/\text{m})(\lambda_{RF}/\text{cm})} \tag{12}$$

where

$$\left(\frac{1}{\kappa}\right) = 1 + \frac{(\frac{Z}{n})^{eff}_{latt}}{(\frac{Z}{n})^{eff}_{cav}}. \tag{13}$$

The parameter $\kappa \simeq 1$ for large $\alpha \sigma_\delta^2$, and is equal to 0.5 if $(\frac{Z}{n})^{eff}_{cav} = (\frac{Z}{n})^{eff}_{latt}$.
The number of cavities is

$$n_{cav} = \frac{(Z/n)^{eff}_{latt}}{\text{Ohm}} \left(\frac{\kappa}{1-\kappa}\right) \frac{(R/\text{m})}{(\sigma_z/\text{mm})^{3/2}}.$$

For $(Z/n)^{eff}_{latt} = 0.25$ Ohm this gives

$$n_{cav} = 5.2 \left(\frac{\kappa}{1-\kappa}\right) \frac{(R/\text{m})}{(\sigma_z/\text{mm})^{3/2}}. \tag{14}$$

COLLISIONS

BEAM DISRUPTION

One of the main phenomenon affecting the choice of the machine parameters is the disruption of the bunches during a collision. In the scheme under consideration the electron bunch is dumped after the collision. The disruption of the electron beam affects the kinematics of the collisions, creates a problem with the background in the detectors, and makes handling of the electron beam after collisions more difficult. However, the main problem is that the disrupted electron bunch affects the dynamics of the positrons generating orbit distortion and a tune shift. Different positrons see different cross-sections of the electron bunch. At all times during the collision a given positron sees the same cross section area for the electrons depending on the distance of the positron from the head of the positron bunch. Therefore, the adverse effects can not be compensated in the SR. The kink instability affects collisions with a non-zero offset.

The problem has been considered many times since the Hollebeek paper[9] both analytically and numerically, but mostly for symmetric collisions. The situation for the asymmetric collisions is very different from that in storage rings and, in a sense, is simpler. Qualitative consideration of the dynamics of the collision is analytically allowed using a simple model although the importance of numeric study of the problem can not be overestimated. It should be noted, however, that such simulations have to include the dynamics of the distorted beam in the storage ring rather than the beam dynamics at collision only.

If the transverse rms bunch sizes at the IP are matched $\sigma_{ex} = \sigma_{px}$, $\sigma_{ey} = \sigma_{py}$, then the luminosity

$$L = \frac{N_e N_p f}{4\pi \sigma_{px} \sigma_{py}} . \tag{15}$$

On being expressed in terms of N_e and the electron disruption parameter $D_{e,y}$, where

$$D_{e,y} = \frac{2 r_0 N_p \sigma_{pz}}{\gamma_e \sigma_{p,x} \sigma_{p,y}} \tag{16}$$

Eq. (15) takes the form:

$$L = \frac{N_e f \gamma_e D_{e,y}}{8\pi r_0 \sigma_{pz}} . \tag{17}$$

The repetition rate f is limited by the multibunch instabilities, power limitations, and the rise time of the kickers. Assuming $N_e = 0.5 \, 10^9$, $f \leq 20$ MHz, and $\gamma_e = 6.0 \, 10^3$, we obtain

$$L = \frac{10^{34}}{117.3} \frac{D_{e,y}}{(\sigma_{pz}/\text{mm})} \, \text{cm}^{-2}\text{sec}^{-1}. \tag{18}$$

Hence, the desired luminosity $L = 10^{34}$ cm^{-2}sec^{-1} may be achieved with CEBAF-type linac and for the bunch length $\sigma_{p,z}$ of the order of 1 mm, only if $D_{e,y} \simeq 120$. This is at least two orders of magnitude larger than that usual in storage rings. The feasibility of the collisions with such large disruptions has to be studied very carefully.

At the same time, the disruption parameter is small for positrons because the number of electrons per bunch is about two orders of magnitude lower than that for positrons. In this case, in the first approximation we may neglect the disruption of the positron beam. For simplicity we additionally assume flat beams. In this case the electric field of a bunch only has a vertical component E_y and is defined by the density ρ (normalized to one) of the encountering bunch:

$$\frac{\partial E_y}{\partial y} = 4\pi e N_p \rho(x, y, s, t). \tag{19}$$

Let us consider first a collision of an electron with a rigid positron bunch. We assume that the distribution of particles in a positron bunch

$$\rho(z, x, y) = \rho(z)\rho(x)\rho(y)$$

is Gaussian in the transverse directions

$$\rho(x) = \frac{e^{-x^2/2\sigma_x^2}}{\sqrt{2\pi\sigma_x^2}}.$$

For simplicity we assume the Gaussian longitudinal distribution:

$$\rho_z(z) = \frac{1}{\sqrt{2\pi\sigma_{pz}^2}} e^{-z^2/2\sigma_{pz}^2}. \tag{20}$$

The equation of motion in the y-plane for an electron is

$$m\gamma_e \frac{d^2 y}{dt^2} = -2e E_y(s, y, t)|_{x=0, s=ct+z_e}.$$

Here z is the distance of the electron from the center of the electron bunch ($z > 0$ for the head of the bunch), and the factor 2 takes into account the magnetic field of the positron bunch. Eq. (19) and Eq. (20) give

$$\frac{d^2}{ds^2}\left(\frac{y}{\sigma_{py}}\right) + \frac{2D_e}{\sigma_{pz}}\rho_z(2s + z_e) \int_0^{y/\sigma_{py}} dy\, e^{-y^2/2} = 0 \tag{21}$$

where D_e is defined in Eq. (16). More accurate calculations only replace σ_{px} in Eq. (16) with $\sigma_{px} + \sigma_{py}$.

The solution of Eq. (21) gives the trajectory of an electron in the form

$$y_e(s, z_e) = f\left(y_0, y_0', \left(\frac{s + z_e/2}{\sigma_{pz}}\right)\right) \tag{22}$$

where y_0 and y_0' are the initial conditions for the trajectory at $s \to -\infty$.

For small y, Eq. (21) is the equation of plasma oscillations:

$$y'' + k^2(s + z_e/2)y = 0,$$

where

$$k^2(z) = \frac{2D_e}{\sigma_{pz}} \rho_z(2z). \tag{23}$$

The total number of oscillations during the collision is

$$n_{osc} = \int_{-\infty}^{\infty} \frac{kds}{2\pi} = 0.252\sqrt{D_e}. \quad (24)$$

(The coefficient is $(2\pi)^{-3/4}$). At the center of the positron bunch, the amplitude of the oscillations adiabatically decreases as $1/\sqrt{k(s)}$ due to increasing density. The positron bunch may be considered as a "transport line". The beta function of the line $1/k(s)$ is on average

$$\beta_{eff} = \left(\frac{1}{2\pi}\right)\left(\frac{2\sigma_{pz}}{n_{osc}}\right) = 1.26\frac{\sigma_{pz}}{\sqrt{D_e}}. \quad (25)$$

The frequency of plasma oscillations rapidly decreases with increasing amplitude $y_e(-\infty)$ as is clear from the expansion of the second term in Eq. (21) over y. For very large $y \gg \sigma_{pz}$ Eq. (21) may be simplified to

$$\frac{d^2}{ds^2}\left(\frac{y}{\sigma_{p,y}}\right) + \frac{D_e\sqrt{2\pi}}{\sigma_{pz}}\rho(2s+z_e) = 0. \quad (26)$$

Oscillations disappear if an electron trajectory crosses the line of collisions $y = 0$ at $s > \sigma_{pz}$ behind the positron bunch. It is easy to find from Eq. (26) that this happens for the trajectory with the initial condition $y(-\infty) = y_0$, $y'(-\infty) = 0$ for

$$y_0 \simeq D_e \sigma_{p,y}. \quad (27)$$

For the trajectories with $y(-\infty) < y_0$ there is at least one crossing point.

An example of trajectories found by numeric integration of Eq. (21) is shown in Fig. 3 for zero initial emittance and disruption parameter $D_e = 120$. The number of oscillations and nodes of the function f in Eq. (22) agrees with the estimate Eq. (24). The frequency of the nonlinear oscillations ought to decrease for large amplitudes. Hence, decoherence of the oscillations is expected, as depicted in Figures 4 and 5. The initial conditions are chosen in such a way that free particles are focused to match the transverse size of the positron bunch at the IP having the beam divergence $\beta_y^* = \sigma_{p,z}$. Decoherence, as well as the nodes in the distribution of the electrons, is explicit.

The nodes may drastically affect the dynamics of the positrons. The equation of motion for a positron is

$$m\gamma_p\frac{d^2y}{dt^2} = -2eE_y(s,x,y,t)|_{x=0,y=y_p(t),s=-ct-z_p} \quad (28)$$

where z_p is the distance of the positron from the center of the bunch ($z > 0$ in the head of the bunch). The field here is given by the electron density ρ_e as in Eq. (19). The density at the moment t is the transform of the initial distribution ρ_0 in time:

$$\rho_e(x=0) = \int \frac{dy_0'dy_0\,dz_0}{\sqrt{2\pi\sigma_{ex}^2}}\rho_0(y_0,y_0',z_0)\delta(y-y_e(y_0,y_0',z_0,t))\delta(s-s_e(y_0,y_0',z_0,t)).$$

496 A Linac Based Beauty Factory

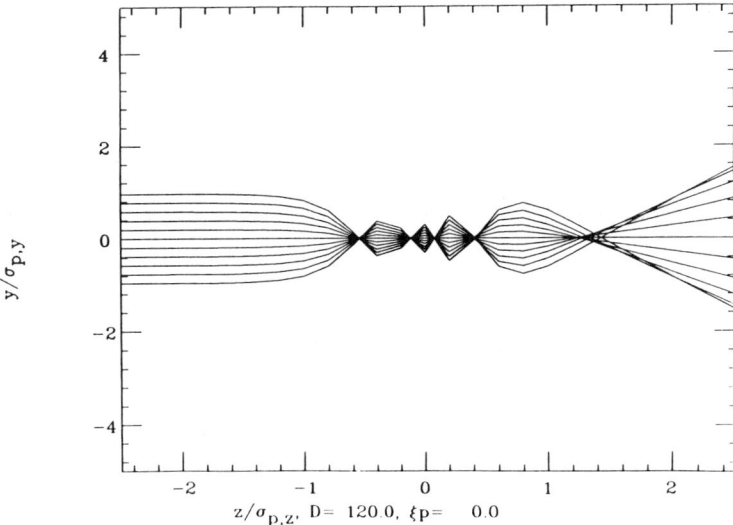

Fig. 3. Pinch of a beam with zero emittance: electron trajectories along the IR.

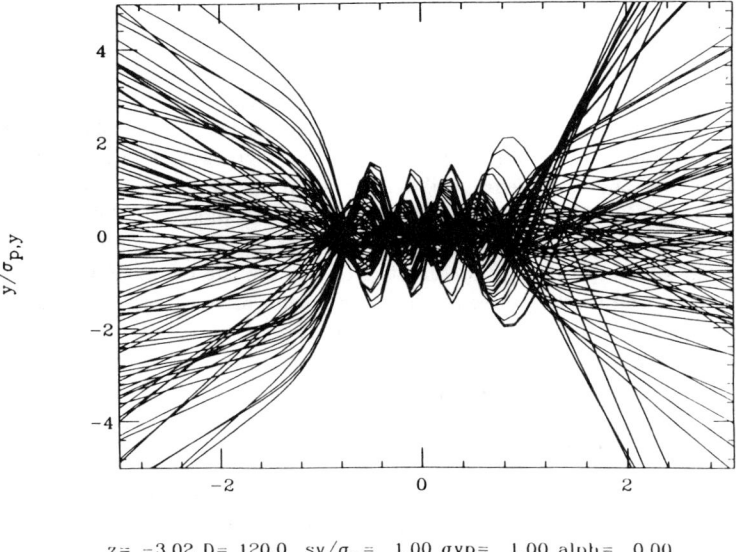

Fig. 4. Disruption of the electron beam with nonzero emittance. Decoherence is the result of the dependence of the plasma frequency on the amplitude of the oscillations for the Gaussian bunch. The choice of the initial conditions is explained in the text.

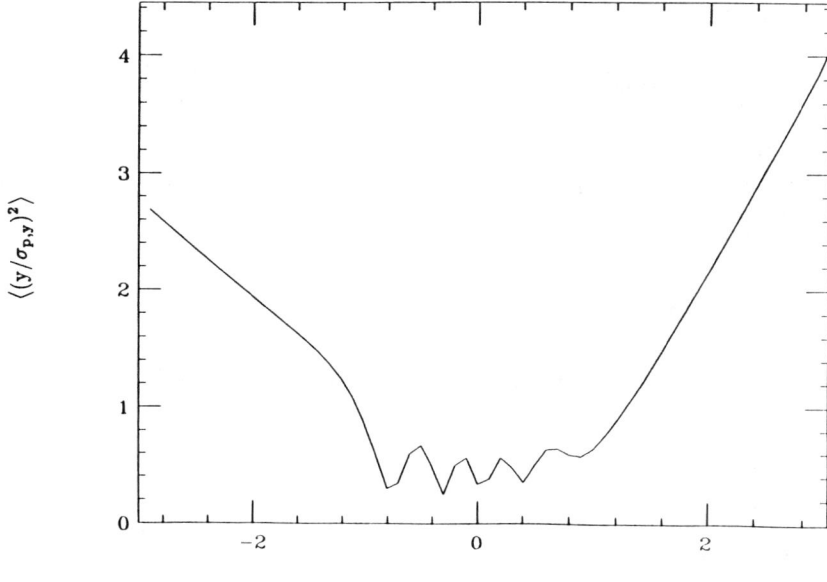

Fig. 5. Transverse rms beam size along the IR for an unmatched beam of Fig. 4.

Here $s_e(y_0, z_0, t) = z_0 + ct$, and

$$y_e(y_0, y_0', z_0, t) = f\left(y_0, y_0', \left(\frac{ct + z_0/2}{\sigma_{pz}}\right)\right)$$

are the electron trajectory, see Eq. (22). For small N_e we can neglect the transverse motion of a positron during the collision: $y_p(t) = y_p$. Considering for simplicity $y_p > 0$ we obtain after integration over z_0 and y:

$$eE_y(x = 0, s = -vt - z_p, y = y_p) =$$

$$\frac{4\pi e^2 N_e}{\sqrt{2\pi\sigma_{ex}^2}} \int dy_0' dy_0 \; \rho_0(y_0, y_0', z_0 = -2ct - z_p) \Theta\left[y_p > f\left(y_0, y_0', \left(\frac{-z_p}{2\sigma_{pz}}\right)\right) > 0\right]. \quad (29)$$

Here $\Theta[..] = 1$ if the condition in the brackets is fulfilled, and $\Theta[..] = 0$ otherwise. Let us assume the factorized distribution $\rho_0(y_0, y_0', z_0) = \rho_z(z_0)\rho_\perp(y_0, y_0')$ and Gaussian distribution in y_0. Eq. (11) takes the form:

$$\frac{d^2 y_p}{ds^2} + \frac{2D_p}{\sigma_{ez}} \rho_z(2s + z_p) \int dy_0 dy_0' \rho(y_0') e^{-y_0^2/2\sigma_{ey}^2} \Theta\left[y_p > f\left(y_0, y_0', \left(\frac{-z_p}{2\sigma_{pz}}\right)\right) > 0\right] = 0. \quad (30)$$

Where D_p is the disruption parameter for positrons:

$$D_p = \frac{2r_0 N_e \sigma_{ez}}{\sigma_{ex}\sigma_{ey}\gamma_p}. \quad (31)$$

For small D_p the collision does not change y_p giving a positron the kick

$$\Delta y_p' = -\frac{D_p}{\sigma_{ez}} \int dy_0 dy_0' \rho(y_0') e^{-y_0^2/2\sigma_{ey}^2} \Theta \left[y_p > f\left(y_0, y_0', \left(\frac{-z_p}{2\sigma_{pz}}\right)\right) > 0 \right]. \quad (32)$$

The kick depends on the position of the positron in the bunch z_p and the number of the electron trajectories with offsets $y_0 < y_p$ at a given z_p. Let $(y_0)_{max} = \phi(-z_p/2\sigma_{pz})$ be the envelope of the electron trajectories at the given z_p. Then, an estimate of the integral Eq. (32) gives

$$\Delta y_p' = -\frac{D_p}{\sigma_{e,z}} \, minim \left\{ \frac{y_p}{\phi(-z_p/2\sigma_{pz})}, \sigma_{ey}\sqrt{\frac{\pi}{2}} \right\}. \quad (33)$$

The envelope $\phi = 1$ for small D_e. Hence, in this case Eq. (33) gives

$$\Delta y_p' = -\frac{D_p}{\sigma_{e,z}} y_p \quad (34)$$

corresponding to a thin lens with focal length $\sigma_{e,z}/D_p$. This gives a tune shift $\Delta \nu = \xi$, where the beam-beam parameter ξ is related to D_p:

$$\xi_{py} = \frac{D_p}{4\pi} \frac{\beta_{py}^*}{\sigma_{e,z}}. \quad (35)$$

The maximum value of $\Delta y_p'$ given by the second term in Eq. (33) is less than the natural divergence of the trajectories in the beam at the IP $y_{nat}' = \sigma_{p,y}/\beta_{p,y}^*$ if

$$\xi_{py} < \frac{1}{(2\pi)^{3/2}} \frac{\sigma_{py}}{\sigma_{ey}} = 0.06 \frac{\sigma_{py}}{\sigma_{ey}}. \quad (36)$$

For $\sigma_{p,y} = \sigma_{p,x}$ the usual constraint $\xi_{p,y} < 0.06$ is obtained.

For large D_e, λ is large near the nodes of the function ϕ. Positrons located in the bunch in the vicinity of the nodes are unstable. For large electron disruption parameters D_e the number of nodes increases, but it is enough to have a single node to lose positrons. The synchrotron motion shuffles positrons along the bunch and pumps new positrons to the nodes where they get lost. Instability means diffusion to larger transverse amplitudes. The rate of the diffusion is affected by the synchrotron motion and is stabilized by the synchrotron radiation, but these effects are too weak to change the dangerous situation.

Transverse instability of positrons around the nodes generates a periodic perturbation of the longitudinal density in the positron bunch with the wave length $2\pi/k$:

$$\rho(s) = \rho_0[1 + \Delta \cos(ks)]. \quad (37)$$

Because k^2 in the equation of motion for electrons with small amplitudes Eq. (23) depends on $\rho(2s)$, the linearized equation of motion takes the form typical for the parametric resonance:

$$y'' + k^2[1 + \Delta \cos(2ks)]y = 0. \quad (38)$$

Thus, the well known "kink instability" may result. When the electron beam is dumped out after a collision the kink instability should not cause a problem.

For collisions with large D_e, synchrotron radiation is a serious problem. The radius of curvature R for a trajectory $y(s)$ is

$$\frac{1}{R} = -y'' = \frac{2D_e\sigma_{p,y}}{\sigma_{p,z}}\rho(2s)\int_0^{y/\sigma_y} dt\, e^{-t^2/2}. \tag{39}$$

Minimum R corresponds to $y \geq \sigma_{p,y}$. The maximum energy loss during the collision

$$\Delta E = \frac{1}{3}r_0\gamma_e^4 \int ds(1/R)^2 \text{ MeV} \tag{40}$$

is proportional to

$$\int ds(1/R)^2_{max} \simeq \left(\frac{D_e\sigma_{p,y}}{\sigma_{p,z}}\right)^2 \frac{\sqrt{\pi}}{2\sigma_{p,z}}. \tag{41}$$

The variation in the invariant mass

$$\frac{\Delta M}{M} \simeq \frac{r_0}{6}\gamma_e^3 \left(\frac{D_{ey}\sigma_{py}}{\sigma_{pz}}\right)^2 \frac{\sqrt{\pi}}{2\sigma_{pz}} \tag{42}$$

is less than $0.5\, 10^{-3}$ if

$$\frac{D_{ey}\sigma_{py}}{\sigma_{pz}} < 6.63\, 10^{-2}\sqrt{\sigma_{p,z}/\text{mm}} \tag{43}$$

giving a limitation on $\sigma_{p,y}$.

The size of the effect of synchrotron radiation on the decoherence of the electron oscillations has to be studied separately.

Variation of the transverse momentum is not so significant. If an electron has momentum $\vec{p} = -\vec{p}_0 + \vec{\Delta}p$ where $\Delta p = p_0\theta$ then

$$\frac{\Delta M}{M} \simeq \frac{\theta^2}{8}.$$

Estimating θ as

$$\theta \simeq 2\sigma_{py}n_{osc}/\sigma_{pz}$$

we found that $\Delta M/M \simeq 10^{-3}$ if

$$\frac{\sigma_{p,y}}{\sigma_{p,z}}\sqrt{D_{ey}} < 0.18. \tag{44}$$

BEAM MATCHING*

The feasibility of collisions with large electron disruption $D_e \gg 1$ might be questioned due to the instability of positrons at the nodes of the electron distribution. However, there are at least three reasons to expect that narrow waists in the electron

---

* Prof. N. Kroll mentioned to one of us (S. H.) that he suggested the idea of matching at the workshop at Blois, June 1989.

distribution may be avoided: nonlinearity of the oscillations, dependence of the phase of the oscillations on the location of an electron in the electron bunch, and synchrotron radiation.

The effect of the pinching of the electron beam on the stability of positrons can be minimized by the proper choice of the initial conditions for the electron beam entering the interaction region (IR). The simplest way to find the best initial conditions would be to track an electron bunch, matched to the positron bunch at the IP, back to the beginning of the IR. However, for $D_e \gg 1$ this method does not work due to the instability of the electron trajectories. We use the following approach. Let us specify the ellipse of the electron beam

$$\left(\frac{y}{\sqrt{2\epsilon/k}}\right)^2 + \left(\frac{y' + \alpha y}{\sqrt{2\epsilon k}}\right)^2 = 1 \qquad (45)$$

where $\alpha = k'/2k$, and

$$k^2(s) = \frac{D_e}{\sigma_{p,z}^2}\sqrt{\frac{2}{\pi}} e^{-2(z/\sigma_z)^2}. \qquad (46)$$

If two bunches are matched at the IP: $\sigma_{e,y}^2 = \langle y^2 \rangle = \sigma_{p,y}^2$ where $\alpha = 0$ then the emittance and $\sigma'_{p,y}$ are defined:

$$\epsilon = k(0)\sigma_{p,y}^2, \quad \sigma_{y'} = k(0)\sigma_y. \qquad (47)$$

For large D_e, close to the center of the positron bunch, electrons oscillate rapidly. As long as

$$\left|\frac{d}{ds}\frac{1}{k(s)}\right| \ll 1 \qquad (48)$$

emittance is preserved and the ellipse changes adiabatically.

Eq. (48) is valid for

$$k < k_{min} \simeq \frac{2}{\sigma_{p,z}}, \quad |z| < z_{min} = \sigma_{pz} \ln\left[\frac{k(0)}{2k_{min}}\right]. \qquad (49)$$

The ellipse on the phase plane at $z = z_{min}$ is given by Eq. (45) with $k = k_{min}$. For larger $|z|$ the positron density decreases, and oscillations degenerate into free motion:

$$y = y_{min} + (z - z_{min})y'_{min}, \quad y' = y'_{min}.$$

Therefore the ellipse at the first quad of the IR is defined, i.e. at $z = -L$:

$$\left(\frac{y + ly'}{\sqrt{2\epsilon/k_{min}}}\right)^2 + \left(\frac{y' + \alpha_{min}(y + ly')}{\sqrt{2\epsilon k_{min}}}\right)^2 = 1 \qquad (50)$$

where $l = L - z_{min}$.

In the simulations the initial conditions at $z = -L$ for Eq. (21) have been generated by

$$y = \sigma_\xi(1 + \alpha_{min}l)\xi - \sigma_\eta l\eta$$
$$y' = -\sigma_\xi \alpha_{min}\xi + \sigma_\eta \eta$$

where ξ and η are random numbers uniformly distributed in the interval from -1 to 1, and

$$\sigma_\xi = \sigma_{py}\sqrt{\frac{2k(0)}{k_{min}}}, \quad \sigma_\eta = \sigma_{p,y}\sqrt{2k(0)k_{min}}, \quad \text{and} \quad \alpha_{min} = \alpha(k_{min}).$$

The trajectories given by Eq. (21) and these initial conditions are shown in Fig. 6 for $D = 120$. The pinches have almost disappeared, and the distribution of electrons within the interaction area $|z| < \sigma_{p,z}$, $|y| < \sigma_{p,y}$ is practically uniform. The result is robust: small variations of k_{min} do not change this result significantly. Fig. 7 shows that the longitudinal variation of the transverse rms size σ_y of the electron beam is small for $k_{min}\sigma_{pz} \simeq 2$. The absolute value of $y(-L)$ and $y'(-L)$ are moderate:

$$y(-L) \simeq \sigma_{p,y}\left(\frac{L}{\sigma_{pz}}\right)D^{1/4}, \quad y'(-L) \simeq \left(\frac{\sigma_{p,y}}{\sigma_{p,z}}\right)D^{1/4}. \tag{51}$$

Practically, $y(-L)$ is of the order of a mm. Figures 8 and 9 are the same as Figures 6 and 7 but for $D = 360$.

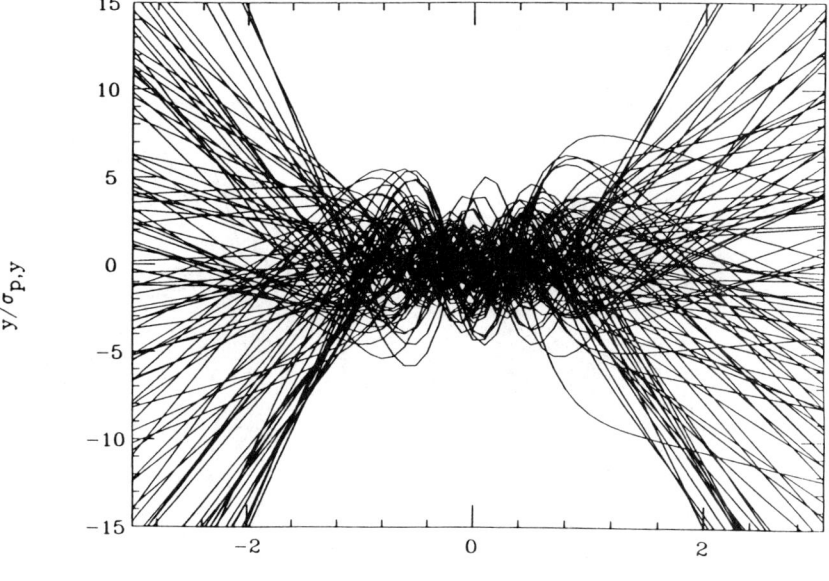

Fig. 6. Trajectories for the matched beam. $D_e = 120.0$. The electron distribution within $|z| < 2\sigma_z$, $|y| < \sigma_{py}$ is rather uniform.

502 A Linac Based Beauty Factory

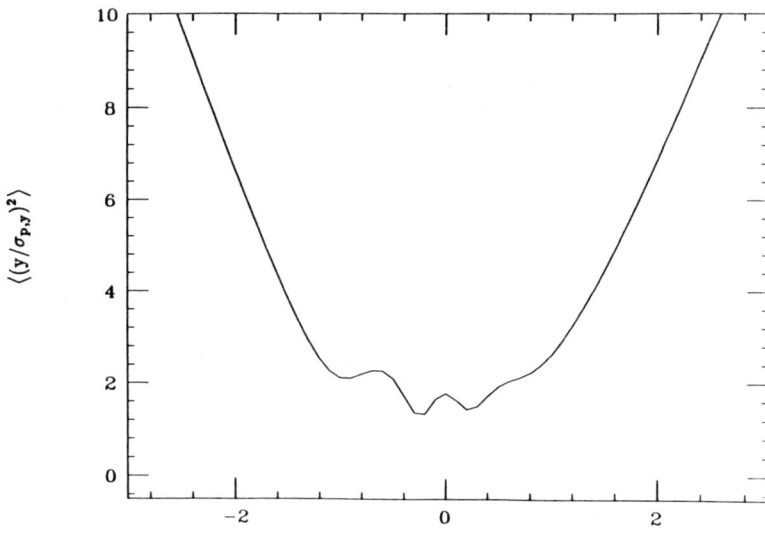

z= -3.02 D= 120.0 sy/σ_p= 3.13 σyp= 6.26 alph= 1.26

Fig. 7. Variation of the rms $\sigma_{e,y}$ for the matched electron beam along the IR; trajectories are shown in Fig. 6. $D_e = 120.0$.

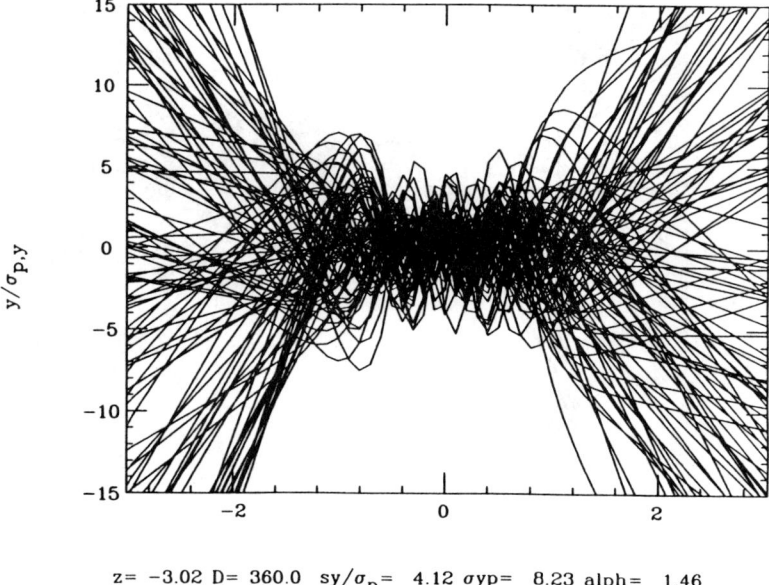

z= -3.02 D= 360.0 sy/σ_p= 4.12 σyp= 8.23 alph= 1.46

Fig. 8. The same as in Fig. 6 for $D_e = 360.0$.

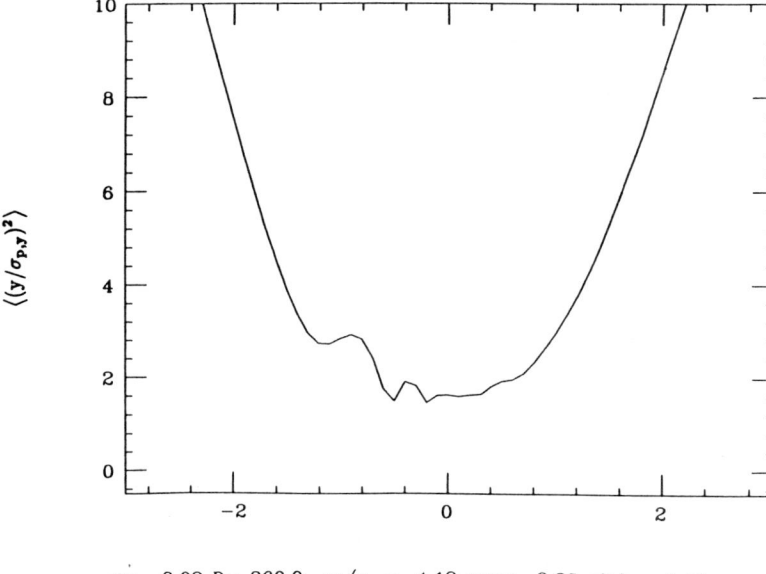

z= -3.02 D= 360.0 sy/σ_p= 4.12 σyp= 8.23 alph= 1.46

Fig. 9. The same as in Fig. 7 for $D_e = 360.0$

Therefore, assume that collisions with large disruption parameters are feasible. We will use the dimensionless bunch length and the SR emittance

$$\varsigma = \sigma_{p,z}/\text{mm}, \quad \varepsilon = \epsilon_{p,x}/\text{nm} \tag{52}$$

and consider electron bunches with $N_e = 0.5 \ 10^9$ per bunch. Then the remaining parameters may be chosen in the following way. To get $L = 10^{34}$ cm^2/sec the disruption parameter has to be (see Eq. (18))

$$D_{e,y} = 108.65 \ \varsigma. \tag{53}$$

Eq. (43) then gives the relation between σ_{pz} and $\sigma_{p,y}$:

$$\sigma_{p,y} = 0.61\sqrt{\varsigma} \ \mu\text{m}. \tag{54}$$

Introducing the aspect ratio of the beam sizes in the SR $R_p^2 = \epsilon_{px}/\epsilon_{p,y}$ we obtain

$$\beta_{py}^* = R_p^2 \beta_{px}^* \left(\frac{\sigma_{py}}{\sigma_{px}}\right)^2. \tag{55}$$

To avoid degrading the luminosity, the divergence of the positron beam near the IP has to be limited:

$$\beta_{p,x}^* \geq \sigma_{pz}. \tag{56}$$

Bunch rotation could be used to change the bunch length prior to a collision. That, however, would increase the energy spread of the positrons σ_δ and decrease the average luminosity

$$L \propto \Gamma(\Upsilon)/\sigma_\delta$$

if the resonance width $\Gamma(\Upsilon) > \sigma_\delta$. Therefore this option is not considered further.
If we choose the equal sign in Eq. (56):

$$\beta^*_{p,x} = \varsigma \text{ mm}, \tag{57}$$

then

$$\sigma_{px} = \sqrt{\varepsilon\varsigma} \ \mu\text{m}, \tag{58}$$

and

$$\beta^*_{py} = \frac{R_p^2}{\varepsilon}\left(\frac{\sigma_{p,y}}{\mu\text{m}}\right)^2 \text{ mm} = 0.3 R_p^2 \frac{\varsigma}{\varepsilon}. \tag{59}$$

The beam-beam parameter ξ in the vertical plane

$$\xi_y = \frac{N_e r_0 \beta^*_{p,y}}{2\pi\gamma_p\sigma^*_{e,y}(\sigma^*_{e,y}+\sigma^*_{e,x})}, \tag{60}$$

as was discussed before, has to be limited: $\xi_y \leq \xi_{max} = 0.06$. This gives

$$\frac{\sigma_{py}\sigma_{px}}{\beta^*_{py}} \geq 2.26 \ 10^{-4} \mu\text{m},$$

or

$$\varepsilon \leq 0.266 \ R_p^{4/3}, \tag{61}$$

or $\varepsilon = 5.75$ for typical $R_p = 10$.

The beam-beam parameter ξ_x in the horizontal plane is small anyway $\xi_x \ll \xi_y$.
Eq. (53) and the definition Eq. (16) give the number of positrons per bunch:

$$N_p = 1.256 \ 10^{11}\left(\frac{\sigma_{p,x}\sigma_{p,y}}{(\mu\text{m})^2}\right), \tag{62}$$

and the average current in the SR:

$$I_{av} = 0.40\left(\frac{\sigma_{p,x}\sigma_{p,y}}{(\mu\text{m})^2}\right) \text{ A}. \tag{63}$$

The condition of stability Eq. (12) for $\lambda_{RF} = 20$ cm and $s_B = 15$ m takes the form:

$$\left(\frac{\sigma_{p,x}\sigma_{p,y}}{(\mu\text{m})^2}\right) \leq 0.69\kappa\varsigma^{3/2}$$

or

$$\varepsilon \leq 0.69\kappa R_p\sqrt{\varsigma}. \tag{64}$$

The number of RF cavities, Eq. (14) gives another relation between κ and ς. Choosing $n_{cav} = 50$ and the circumference of the SR $C = 450$ m, we obtain from Eq. (14)

$$\varsigma^{3/2} = 7.446\left(\frac{\kappa}{1-\kappa}\right). \tag{65}$$

Eq. (61), (64) and Eq. (65) define all parameters as function of R_p. For $R_p = 10$ we obtain
$$\varepsilon = 5.75, \quad \kappa = 0.454, \quad \varsigma = 3.33.$$
This defines
$$\epsilon_{p,x} = 5.75 \text{ nm}, \quad \sigma_{px} = 4.37 \text{ } \mu\text{m}, \quad \sigma_{py} = 1.11 \text{ } \mu\text{m}$$
$$\beta_{px}^* = 3.33 \text{ mm}, \quad \beta_{py}^* = 21.55 \text{ mm}.$$
The disruption parameter is $D_{e,y} = 361.8$. Eq. (13), (11) and $\kappa = 0.454$ specify the lattice parameter in the SR:
$$\alpha \sigma_\delta^2 = 1.2 \text{ } 10^{-8}.$$

The energy spread has to be comparable with the width of the Υ resonance. Choosing $\sigma_\delta = 2.45 \text{ } 10^{-3}$ we obtain the compaction parameter
$$\alpha = 2.0 \text{ } 10^{-3}.$$

Parameters of the storage ring ϵ_{px}, α, and σ_δ are close to the sample lattice given before. The ring requires 70.68 cells of that type, with the total length 353.43 m. This is consistent with the circumference of the ring being 450 m, the parameter which has been used for the estimate. The radiated power is 7.77 MeV/turn/particle. The emittance of the electron beam is approximately

$$\epsilon_{e,y} = k(0)\sigma_{p,y}^2 = \sigma_{py}^2 \frac{\sqrt{D_{e,y}}}{\sigma_{pz}} \left(\frac{2}{\pi}\right)^{1/4} \tag{66}$$

or $\epsilon_{e,y} = 5.66$ nm. The disruption parameter in the horizontal plane is σ_{py}/σ_{px} times smaller than D_{ey}. Consequently
$$D_{ex} = 92.0, \quad \text{and} \quad \epsilon_{e,x} = 49.1 \text{ nm}.$$

PARAMETERS

A final set of parameters is as follows:
Linac:
$$E_e = 3.5 \text{ GeV}$$
$$N_e = 0.544 \text{ } 10^9$$
$$I_{\text{av}} = 1.6 \text{ mA}$$
$$f_{RF} = 1.5 \text{ GHz}$$
$$l_z^e = 2.2 \text{ psec}$$
$$\epsilon_{e,y} = 5.66 \text{ } nm; \epsilon_{e,x} = 49.1 \text{ nm}$$
$$P = 5.65 \text{ MW}$$
$$\text{power loss} = 2.0 \text{ W/cav}$$
$$\text{repetition rate} = 20 \text{ MHz}$$

SR:

$$E = 8.0 \text{ GeV}$$
$$I_{aver} = 1.94 \text{ A}$$
$$N_B = 6.1 \ 10^{11}$$
$$n_B = 30$$
$$S_B = 15 \text{ m}$$
$$P = 15.1 \text{ MW}$$
$$n_{cav} = 50$$
$$E' = 2.5 \text{ MeV/cavity}$$
$$C \simeq 450 \text{ m}$$
$$R_p = 10$$
$$\epsilon_{p,x} = 5.75 \text{ nm}, \quad \epsilon_{p,y} = 0.057 \text{ nm}$$
$$\alpha = 2.0 \ 10^{-3}$$
$$\sigma_\delta = 2.45 \ 10^{-3}$$
$$\sigma_z^p = 3.33 \text{ mm}$$
$$(Z/n)_{tot} = 0.5 \text{ Ohm}$$
$$\rho_{bend} = 45 \text{ m}$$
$$\tau_x = 0.9 \text{ msec}$$
$$\tau_y = 2.4 \text{ msec}$$
$$\tau_\delta = 6.9 \text{ msec}$$
$$L_{cell} = 5 \text{ m}$$
$$\beta_x = 9.8 \text{ m}, \beta_y = 1.4 \text{ m}, D_x = 0.255 \text{ m}$$

Collisions:

$$L = 10^{34} \text{ cm}^{-2}\text{sec}^{-1}$$
$$D_{ey} = 361.8, \quad D_{pe} = 6 \ 10^{-3}$$
$$\xi = 0.06$$
$$\text{rep. rate } 20 \text{ MHz}$$
$$\beta_{e,y}^* = \beta_{p,x}^* = 3.33 \text{ mm}$$
$$\beta_{p,y}^* = 21.55 \text{ mm}$$
$$\sigma_{e,x}^D = \sigma_{p,x}^* = 4.37 \ \mu\text{m}$$
$$\sigma_{e,y}^D = \sigma_{p,y}^* = 1.11 \ \mu\text{m}$$
$$\text{beam size at 1 m from IP } y = 1.45 \text{ mm}$$

DISCUSSION

A beauty factory based on a SRF linac with CEBAF parameters can be designed only with very large disruption parameters. We have presented arguments and results of model simulations which indicate that such a design may be feasible. Confirmation that the kink instability is suppressed would be a significant argument in favor of the asymmetric collisions. Much more elaborate simulations are needed before a sound statement may be made but the situation does not look hopeless. With this implication we give as an example a consistent set of parameters for the B-factory with asymmetric kinematics of collisions and luminosity 10^{34} cm^2/sec.

A lattice with low emittance, large compaction factor and large energy spread is necessary. Test lattices have been designed with parameters close to those specified. A new tunnel will be necessary for the SR, but the parameters of the CEBAF-type linac are satisfactory. The existing CEBAF cavities can be used; only additional power has to be added.*

There are many problems to study: design of the lattice, injector, and the interaction region, the minimum value of the impedance possible in the SR, multibunch stability in the SR, dynamic aperture and acceptance of the SR, the enhancement factor of asymmetric collisions, etc. Nevertheless, a linac-SR scheme promises very high luminosity and seems to be feasible.

ACKNOWLEDGEMENTS

We appreciate J. R. Rees' comments on the effect of collisions with large disruption parameters on the SR stability, and valuable discussions with J. Bisognano, J. Kewisch, D. Douglas, and R. Rossmanith.

REFERENCES

1. J. J. Bisognano, J. Boyce, D. Douglas, S. Heifets, J. Kewisch, G. Krafft, R. Rossmanith, "A Beauty Factory Using an SRF Linac and a Storage Ring", 1989 Linear Accelerator Conference, Williamsburg (1988).
2. P. Grosse-Wiesmann, "Colliding a Linear Electron Beam with a Storage Ring Beam", SLAC-PUB-4545 (February 1988).
3. G. A. Krafft, J. J. Bisognano, "On Using a Superconducting Linac to Drive a Short Wavelength FEL", CEBAF-PR-89-019, Proceedings of the Particle Accelerator Conference, Chicago (1989).
4. J. J. Bisognano, S. A. Heifets, B. C. Yunn, "Impedances of CEBAF Superconducting Accelerator", CEBAF-TN-0059 (1988).
5. A. W. Chao, B. Richter, C. G. Yao, Nucl. Instr. and Methods, 178, 1–8 (1980).
6. V. E. Balakin, A. V. Novokhatsky, Proc. of the 12th Int. Conf. on High Energy Acceler., Fermilab (1983).
   K. Bane, Wakefield Effects in a Linear Collider, SLAC-PUB-4169 (1986).
7. M. Sands, The Physics of Electron Storage Rings, SLAC-121, UC-28 (ACC) (1970).
8. B. C. Yunn, Calculations of the CEBAF Modes with the Code URMEL (unpublished).
9. R. Hollebeek, "The Linear Collider Beam-Beam Problem".

---

* J. Bisognano noticed that the resetting of the external Q might be necessary as well to decrease the reflected power.

Superconducting Cavities for a B-factory
-Interim Progress Report

H. Padamsee, W. Hartung, J. Kirchgesner, D. Moffat,
D. Rubin, D. Saraniti, Y. Samed, J. Sears and Q. S Shu

Laboratory of Nuclear Studies
Cornell University

Introduction

Recently there has been significant worldwide attention focussed on a B-factory, with a luminosity roughly 100 times higher than CESR's world record of 10^{32} per cm^2 persec^2. For such a future high luminosity B-factory, the approach actively under consideration relies on bunches that are factor of two shorter than used in CESR[1], eg, $\sigma_l \sim 0.85$ cm. To achieve the short bunches will require a factor of 4 higher voltage than used for CESR. If two rings are used, as seems likely to optimize the conditions to push for the highest possible luminosity, the total voltage demand for a B-factory would be 56 MV as compared to the 7 MV presently supplied by the copper RF system for CESR.

Superconducting cavities can economically provide this high voltage. The total RF power (including higher order mode losses) consumed by a copper cavity system similar to the CESR RF would be 23 Mwatts. By comparison, the total RF power plus refrigerator power to maintain a superconducting system at 4.2 K would be 1.4 Mwatts, more than an order of magnitude lower. Beam power is not included in either case.

Among other factors, the maximum current in storage rings is limited by disruptive beam cavity interactions. A key feature essential to achieving high luminosity is to lower the impedance presented to the beam by the cavity and other components. Use of high voltage superconducting cavities minimizes the overall accelerating structure length, and thereby the accompanying cavity impedance. Moreover, superconducting cavities can have larger beam holes and geometries more suitable for higher beam currents7 than copper cavities. A comparison of CESR and superconducting cells indicates that the total beam induced HOM loss factor for CESR cells is 3 times higher.

Development of a superconducting RF system for a B-factory poses many challenges. For short bunches, it is desirable to operate at the highest possible gradient to minimize the overall cavity impedance. Today, 5-cell superconducting structures built by industry for the storage rings TRISTAN, HERA and LEP and the Cornell/CEBAF structures are capable of 5 - 12 MeV/m in acceptance tests. In beam tests show lower values. Relying on continuing technology advancements forthcoming over the next several years, we may (optimistically) expect to operate a B-factory SRF system at gradients as high as 10 MeV/m. Efforts to understand and suppress field emission will remain of high importance to ensure success in this regard.

A major development challenge lies in advancing the capabilty of sc cavities to carry beam currents of several amps necessary for a high luminosity storage ring. The maximum current handled by sc cavities to date is ~ 30 mA[2]. New fundamental and HOM couplers with substantially enhanced power capability will be needed to safely couple in the required beam power (plus HOM power) of nearly 1.5 Mwatt/meter, and to safely couple out the beam induced HOM power at the level of 100 - 200 Kwatts/meter. In addition, the higher modes will need to damped much more heavily. Prelimnary estimates call for Qext values of ~ 100 for the longitudinal modes. These are heavy demands over the capability of components presently used in sc cavities, which can handle 100 Kwatts/input coupler, 10 - 100 watts/output coupler and provide much lighter damping than needed, for eg. Qext = severalx1000, for the dominant longitudinal mode.

To improve chances for success, an attractive approach is to distribute the input and output powers over a large number of couplers by using single cell cavities with individual input and output couplers. One possible parameter list for a B-factroy calls for a total of 20 cells in two rings, each cell providing 3 MV. In spite of the strong coupling requirements of this new application, we must try to continue to place the couplers on the beam pipe, outside the cells, so as not to perturb the ideal fields within the cells which have special anti-multipacting shapes evolved over years of development. This choice also avoids the inevitable surface magnetic field enhancement associated with coupler holes that could lead to premature thermal breakdown. Because of the high power handling capability needed, waveguide couplers need to be considered even at this low frequency, over the more popular coaxial versions used so far for 350 and 500 MHz storage ring cavities.

Choosing the Best Cell shape

Choosing the best cell shape for a B-factory cavity is trade-off between several considerations:

(a) A high fundamental mode R/Q is desired, to hold down the refrigerator heat load.
(b) R/Q for the HOMs should be kept as low as possible to reduce the beam induced HOM power and avoid longitudinal and transverse multi-bunch instabilities.
(c) On the beam pipe, the number of holes and the size of the holes for the fundamental and higher mode couplers should be kept small so that the accompanying increase in the R/Q of the cells can be kept low. This is important to avoid both single and multibunch instabilities. To achieve (c), it is desirable to increase the fields outside the cell in the fundamental and HOMs.
(d) Ep/Eacc and Hp/Eacc should be kept small to avoid field emission and thermal breakdown.

With these considerations in mind, we searched for the best cell shape by varying the following features of the cell shape:

(a) Beam pipe diameter from 18 cm to 28 cm (step size = 2 cm)
(b) Iris radius from 1 cm to 5 cm (step size = 1cm)
(c) Cell length from 24 cm to 30 cm (step size = 1 cm)
(d) Cell diameter to keep the fundamental frequency at 500 MHz for each set of choices between a - c.

The wall slope has been maintained constant at 75°.

In total, 150 parameter sets were studied, but the actual number of URMEL T runs was near 500 to arrive at the correct cell diameter for 500 MHz for the fundamental frequency.

URMEL T was used to calculate R/Q for the fundamental (TM010 mode) and one HOM mode which has the highest R/Q (the TM011 mode). Using the field tables from the URMEL calculations, we also calculated the effective Qext for these two modes from a hypothetical coupler on the beam pipe. The loading of such a coupler is simulated with a resistive band.

$Q_{ext} = \omega U/P_d$,

Where Pd is the power dissipated in a 13 cm wide resistive band located on the beam pipe just past the end of the iris rounding.

$Pd = 1/2 \, Rs \int H^2 da$

Rs is a scaling factor chosen (arbitrarily) to give $Qext = 2\times10^4$ in the fundamental mode for the first set of cell shape parameters (Beam tube radius = 18 cm, iris radius = 1cm and cell length = 30 cm). The same Rs is chosen for the TM011 mode, as well as for all the other parameter sets.

In the past, we have used this technique to predict how the Qext changes from one higher order mode to the next, within individual pass-bands for a multi-cell cavity, as well as across different pass-bands. In addition we have used it to estimate the effect on Qext's as the number of cells is increased. The results from such simulations have been checked with measurements on machined Aluminum 5 cell and 10 cell cavities for two dangerous longitudinal and two dangerous deflecting mode pass-bands. Results from typical comparisons are shown in Fig 1. We see that the calculations are remarkably predictive, although there are occasional deviations which is not too surprising, considering the simplicity of the physical model used to simulate the action of a coupler. In particular, field perturbations from the coupling hole are not taken into account.

The cell parameter phase space can conveniently be divided in two broad categories. In the first, the dominant HOM is non-propogating and will have to be damped with coupler(s) near the cell as for sc cavities for the present generation of storage rings. However, for a B-factory cavity, the corresponding HOM coupler system would have to extract several 10's of Kwatts of power through the liquid He vessel, heat shields and vacuum vessel, out to a room temperature load. In the second category the TM011 mode (as well as all other monopoles) propogate down the beam tube. This does not emiminate the couplers but opens the attractive possibility of damping the modes by placing high power HOM coupling device(s) completely outside the cryostat, greatly simplifying the power extraction problem. In addition, as discussed below, once the dominant longitudinal mode starts to propogate, the R/Q drops by more than a factor of 2. Both considerations indicate a preference for a cell shape from the propogating class.

Non-Propogating HOMs

The effect of increasing the beam tube diameter on the R/Q and Qext of the TM010 (fundamental accelerating) mode and the TM011 (most dangerous longitudindal HOM) is shown in Fig 2a and 2b. At a fixed iris radius of 1 cm and fixed cell length of 30 cm, the R/Q of the fundamental decreases by a factor of 1.9 on increasing the beam tube diameter from 18 cm to 28 cm. However, the HOM R/Q decreases much faster, by a factor of 4. Similarly, the Qext decreases much faster for the HOM mode (by x16) than for the fundamental (by x10). In both cases the lower Qext implies a smaller coupling hole perturbation on the R/Q of the cell. In view of the seriousness of the HOM power dissipation and HOM induced multi-bunch instabilities for a B-factory, these calculations clearly emphasize the desriability of using larger beam tubes.

The effect of increasing the iris radius is shown in Fig. 3. for a fixed beam tube diameter of 24 cm and fixed cell length of 30 cm. Note that whereas the fundamental R/Q is essentially unaffected, the R/Q of the HOM is reduced by a factor of 3.8 in increasing the iris radius from 1 cm to 5 cm. Also the Qext of the HOM decreases by a factor of 3. However for the fundamental, the Qext increases by a factor of 2. On the whole it appears desirable to choose larger iris radii.

The effect on the fundamental R/Q and Qext of shortening the cell length from $\lambda/2$ (= 30 cm) to 24 cm is not substantial as the other parameters. A shorter cell length is preferred for decreasing the R/Q and Qext of the HOM. Ultimately, as the beam tube radius increases, the TM011 mode (and all other monopoles) become propogating for the shorter cell lengths studied.

Propogating HOMs

Once the TM011 HOM mode under study becomes propogating, it is difficult to single it out from other monopoles with similar field patterns. Fig. 4 compares the distribution of HOM monopole R/Q values for the propogating and non-propogating cases, both with the same beam tube diameter of 24 cm, but with different cell length and iris radius. Note the drop from 11 Ω/cell for the TM011 mode to below 4 Ω/cell typical of HOM monopoles in a propogating example.

Among the various propogating cases studied, we show in Fig. 5a, the R/Q and Qext of the fundamental for a fixed cell length of 24 cm. as a function of iris radius for various beam tube diameters. Our desire to obtain the lowest possible fundamental mode Qext eliminates beam tube diameters < 24 cm. The 10% drop in R/Q for every increment in beam tube diameter influences the selection of 24 cm. Also the drop in R/Q with increasing iris radius pushes for a small iris diameter. However 1 cm is considered too small to be practical from deep drawing and welding considerations.

A parallel set of curves for the cell length of 26 cm in Fig. 5b shows the same trends with beam tube diameter and iris radius, but both the values of R/Q and Qext are less favorable than for the cell length of 24 cm, which is therefore the one finally selected.

Fundamental Mode Properties of the B-factory Cell Shape.

The above considerations have led us to select the following as the B-factory sc cavity cell shape : Beam tube diameter = 24 cm, iris radius = 2 cm and cell length = 24 cm. Fig. 6 compares the half cell shapes for the nc CESR cavity, a sc cavity shape suitable for the current generation of storage rings, and the B-factory cell shape. Note that the loss factor for HOMs at 1 cm bunch length in the new shape is a factor of 3 lower than for the CESR shape. Also note that the highest R/Q for any monopole is 4 Ohms/cell as compared to 55 Ohms/cell for the CESR shape or 28 Ohms/cell for the LEP shape. Fig. 7 shows the corresponding half-cells.

f= 500 MHz
R/Q = 89 Ohms/cell, 342 Ohms/meter
E_p/E_{acc} = 2.5
H_p/E_{acc} = 55.8 Oe/MV/m
Esurface for 3 MV/cell = 25 MV/m
Hsurface for 3 MV/cell = 558 Oersted

Fundamental Coupler Studies

Power Handling Requirement
For a B-factory based on sc cavities, the input power is the sum of the synchrotron radiated power and the higher order mode power. The power dissipated in the cavity walls is negligible. For the parameters of a B-factory assumed in ARW[1], the synchrotron radiation is power is 3.32 Mwatts/beam. (1.1 MeV/particle per turn as in CESR, 2.91 A/beam). For HOM losses, BCI calculations are used. In the new one cell sc cavity, the calculated G value is 21.5 m^{-1} at a bunch length of 0.85 cm. Here G is loss factor related to k (Volts/coulomb), G= $4\pi\varepsilon_0 k$. The loss factor for the fundamental mode alone is

$$G \text{ (fund.)} = 4\pi\varepsilon \, \omega/4 \, R/Q \exp -(\omega\sigma/c)^2$$

$$= 7.8 \text{ m}^{-1}, \text{ for } R/Q = 89 \, \Omega/\text{cell}.$$

Then $G(\text{HOM}) = 21.5 - 7.8 = 13.7 \text{ m}^{-1}$

HOM power is given by:

$$P \text{ (HOM)} = k \, (Ne)^2 \, n_b \, f_{rev}.$$
$$= 47 \text{ Kwatts/cell}$$

where k is the loss factor, N the number of particles per bunch = 8.3×10^{11}, n_b the number of bunches = 56, frev the revolution frequency = 3.9×10^5 Hz.

With 10 cells per ring, the total input power per cell then becomes :

332 Kwatts (fund.) + 47 Kwatts (HOM) = 379 Kwatts ~ 400 kwatts/cell.

In the present generation storage rings, the input power requirements of maximum 100 kwatts per coupler are handled with coaxial input couplers from room temperature to the coupler opening, where the center conductor serves as an antenna. The center conductor is cooled either by a separate cold He gas circuit, or by water in one case (KEK)[3]. Because of the substantially higher powers involved, we have chosen to use a waveguide to bring power from the klystron into the liquid He vessel. An additional advantage of waveguide over coax is the simplicity that only the outer wall needs cooling, and may be accomplished by cold gas heat exchangers similar to that used for the beam tubes. The overall size and static heat load can be reduced using a half-height guide. The detail design of the coupling between the cavity and the waveguide is still under development. (see below)

One of the most demanding needs is the development of a window to handle 1/2 Mwatt of RF power (travelling wave). In the absence of beam, the window will have to tolerate 100 kwatts of reflected power, and the full 400 Kwatts of reflected power for a short time interval in case of beam loss. At present we are engaged in discussions with several klystron manufacturers to seek their participation in developing and testing a planar high power RF window that may be used at a waveguide flange. Such a window also opens the possibility of a double window (with an evacuated waveguide section vacuum in between) for added safety factor.

Fundamental Mode Qext

The required Qext is calculated from:

$$Q_{ext} = V^2/(R/Q\ P_{beam}) = 2.7 \times 10^5$$

using $P_b = 4 \times 10^5$ watts, $R/Q = 89$ Ω/cell and $V = 3$ MV/cell

Careful attention must be given to minimize the additional impedance associated with the hole for the power coupler. Prelimnary calculations using the BCI code indicate this to be a serious consideration. For a bunch length of 0.85 cm, the G value increased from 23 m^{-1} to 37 m^{-1} upon introduction of a 75 mm slot opening on a trial 22 cm diam beam pipe outside the cell. For the cell shape used, Gfund was 15 m^{-1}, so G(HOM) increased from 8 to 22, implying that the HOM power increases from 40 to 110 Kwatts/cell. However, the restriction of cylindrical symmetry characteristic of the BCI code made it necessary to examine a much larger disturbance than presented by a real coupler hole. To make a realistic evaluation, we need to use 3D code MAFIA. This is already installed on the CONVEX mini-supercomputer in the Materials Science Center. Initial calculations have started.

A bench modelling effort is underway to determine the minimum hole size necessary to provide the needed input external Q with a waveguide type coupler. This effort was started before the best cell shape and beam pipe diameter were fixed. Existing S-band model cavities were used. When scaled to 500 MHz, the model has an equivalent beam pipe = 20 cm, iris radius = 3.8 cm, and cell length = 30 cm.

For slot coupling between the cell and the input waveguide, we find that a slot 75 mm wide x 50 mm long (parallel to beam axis) between the waveguide end-wall and the beam pipe is large enough for $Q_{ext} = 3 \times 10^5$, provided proper matching elements are located in the waveguide near the coupling hole. This opening is already smaller than the diameter of hole (100 mm) for the antenna type input couplers in working versions of 500 MHz superconducting structure for TRISTAN storage ring, for which a weaker Qext of ~ 10^6 is adequate[3]. Model measurements so far, refer to a beam pipe diameter of 20 cm. Calculations discussed in the section on choosing the best cell shape show that the increased beam pipe to 24 cm should provide increased coupling by a factor of 6, i.e. an even smaller hole size should be possible.

An antenna type coupler between cell and waveguide is also under study. An 85 mm diameter hole with a 50 mm diameter antenna provides the needed Qext of 3×10^5. Again, we expect to be able to reduce the coupler opening as we increase the beam pipe diameter of 24 cm.

Higher Order Mode Couplers.

Longitudinal Modes

For the cell shape chosen all monopoles propogate out the cell down the beam pipe. It is still necessary to provide mechanisms to damp the modes to low Q values needed to avoid resonant power deposition and instabilities. As discussed before the advantage here is that the damping and power extraction may now be provided outside the cryostat.

The distribution of R/Q values of many modes up to 2 GHz for the cell shape with 24 cm beam pipe and taper to vacuum chamber dimensions (11 cm) is shown in Fig. 8. The maximum R/Q reached is 4 Ω/cell. For multibunch stability, Q_L for a mode with the maximum R/Q needs to be lowered to 150[4].

The power deposited by the beam in a HOM monopole is given by :

$$P_n = q_b^2 \tfrac{1}{T_b} B(\omega_n,\sigma) \tfrac{\omega_n}{4} (R/Q)_n F(\tau_n,\delta_n)$$

- P_n = power transmitted from the beam to the cavity in the mode n.
- ω_n = angular resonant frequency of the mode n.
- q_b = charge per bunch. = 1.32×10^{-7} coulombs
- T_b = time between two bunches. = 4.6×10^{-8} secs
- $(R/Q)_n$: R/Q for the mode n. = 4Ω/cell
- $B(\omega_n,\sigma)$: Bunch form factor = $e^{\omega_n^2 \sigma^2/c^2}$ for a Gaussian bunch.
- σ = r.m.s. length of the bunch. = 8.5 mm
- $c = 3 \cdot 10^8$ m/s.
- $F(\tau_n,\delta_n)$ is a function which takes into account the cumulative effect and the decrease of the induced voltage between two bunches

$$F(\tau_n,\delta_n) = (1 - e^{-2\tau_n})/(1 - 2e^{-\tau_n}\cos\delta_n + e^{-2\tau_n})$$

with

$\tau_n = T_b/T_{fn}$

T_{fn} (filling time of the cavity) = $\tfrac{2Q_L}{\omega_n}$

$O_l = \tfrac{Q_0 Q_{ext}}{Q_0 + Q_{ext}}$, Q_0 and Q_{ext} being respectively the unloaded and external quality factors

$\delta_n = (\omega_n - \omega_g) T_b$ ω_g = generator angular frequency

The highest losses correspond to the resonant condition:

$\delta_n = 0$ and $F(\tau_n,0) = (1 + e^{-\tau_n})/(1 - e^{-\tau_n})$

$Q_0 = 10^9$

Fig. 9 shows the variation of the power deposited with Qext results for a mode with R/Q = 4 Ω/cell. At Qext = 150, the power deposited in the worst case of overlap of the HOM frequency with a harmonic of the bunch revolution frequency is 4 Kwatts. For a mode with QL < 50, the resonant power deposition reaches its asymtotic value of 2 kwatts.

Using the field table output from URMEL T we calculated for various modes the Qext obtained if a 15 cm section of the beam pipe were replaced with an absorbing material with effective RF resistivity 10^4 times higher than Cu. By this method it is theoretically possible to lower the Qext of the sc cell to between 45 and 75 for all the HOMs. A technical material (Ferrite-50) has been found which has the desired loss properties. At this stage, we are still investigating whether the accompanying losses from the image currents of the beam or the impedance presented to the beam by such a section will be tolerable.

Transverse Modes

For the new shape, all the monopoles and most of the deflecting modes are propogating in the beam tube. However two of the low frequency deflecting modes (TE111 and TM110) still remain non-propogating. R^t/Q (Ohms/cell) for all the deflecting modes up to 2500 MHz is given in Fig.10 . As mentioned, the two lowest frequency modes do not propogate. Using field tables from URMEL T output we have determined that, like the monopoles, Qext for all the propogating deflecting modes will also be damped to < 100 by a 15 cm section of Ferrite-50 beam pipe.

The properties of the nonpropogating modes as calculated by URMEL T are given in Table 1. The beam induced power for a 1 mm displacement off beam axis will be roughly 5 Kwatt if the modes are damped to Qext of ~ 10^4.

Table 1: Properties of non-propogating deflecting higher modes.

Mode	Freq (MHz)	R/Q for 3.5 cm off axis (Ohms)	R^t Ohms/cell
TE111	644	1	2.8
TM110	680	4.4	25

These two deflecting modes will need to be coupled near the cell and through the cryostat. In keeping with the need to minimize the number of coupling holes, a single HOM coupler for these modes is desirable. However with only one coupler, the two polarizations of each deflecting mode will be fixed by the coupler itself, leaving one polarization inadequately damped. This difficulty can be overcome by deliberate breaking of the cell's symmetry by minor perturbations to the cell geometry. To break the cell symmetry, the azimuthal boundary of the cavity is made up of 6 chords joined by circular arcs which meet the chords tangentially. Two chords fix the orientation of the dipole modes and four chords are used for the quadrupole modes. The width of each chord is reduced linearly with the wall radius, vanishing at the iris curvature. The cavity remains cylindrically symmetric at the beam hole. Thus the modes are oriented at an angle $\pi/2m$ about the coupler axis (Fig. 11).

The polarization approach has already been extensively tested in connection with the development program for a sc structure suitable for a future TeV linear collider[5]. Here cell polarization is also under considerations to reduce the complexity and cost associated with multiple couplers per cavity unit. Bench top RF measurements were conducted on a 5-cell S-band copper cavity with polarized cells and equipped with a single waveguide coupler on the beam pipe. Parallel measurements were conducted on a similar un-polarized cavity also with only one coupler. Both TE111 and TM110 deflecting modes were studied (20 modes). As expected, in the un-polarised cavity one polarization of each mode had a poor Qext (10^6 or higher). On the other hand, for the polarized cavity all, but two, of the 20 dipole modes had Qext between 10^4 and 10^5. The exceptions had Qext of ~ 5×10^5. Detail examination showed that for these exceptions the coupler perturbation overwhelmed the chord perturbations introduced in the cells. To avoid this polarization pulling effect we have tentatively increased the cell polarizations by a factor of 2. In any case, we expect the pulling to be significantly less for the smaller hole size contemplated for the B-factory coupler.

An important question regarding polarized cells is whether the shape distortions introduced will regenerate any multipacting problems. Several 1-cell and 3-cell S-band polarized Nb cavities were built and cold tested a few times. No multipacting at all was encountered up to surface electric fields of 900 Oe, which is well beyond the field necessary for 1st order one surface multipcating to exist at 3 GHz(~ 700 Oe). This settles the key question of the suitability of polarized cells for superconducting cavities.

If the fundamental coupler could be used to damp the transverse modes that do not propogate out the beam pipe, and couple out any power deposited in these modes, it would reduce the number of couplers and the additional associated impedance. In this respect, we have found the antenna coupler superior to the slot coupler. With Qext = 3×10^5 for the fundamental, we measured Qext = 1.5×10^4 and 3.6×10^4 for the TE111 and the TM110 deflecting modes. By contrast, in several slot shapes we were unable to obtain Qext < 3 - 4 $\times 10^5$ for either deflecting mode.

A Conceptual Design of an Superconducting Cavity for a B-factory.

Based on the above studies, Fig. 12 summarizes the present state of the conceptual design. Much work clearly remains to flesh out the concept in the areas of high power window development, fundamental, deflecting mode couplers, beam pipe universal coupler, etc. These studies are continuing with the goal of testing a sc cell in CESR.

References

[1] J. Alexander, D. Rubin and J. Welch, CON 89-1, Cornell University Internal Report.
[2] D. Proch, Proc. of the 3rd Workshop on RF Superconductivity, Argonne, Illinois, 1984, Ed. K. Shepard.
[3] Y. Kojima, ibid refs. 2
[4] "Prospects of a CESR B-Factory Upgrade", CESR/CLEO staff, CLNS 89/962, Cornell University Internal Report.
[5] J. Kirchgessner, et al., Proc. of the 1989 Particle Acclerator Conference, Chicago, Illinois, 1989 and CLNS 89/896.

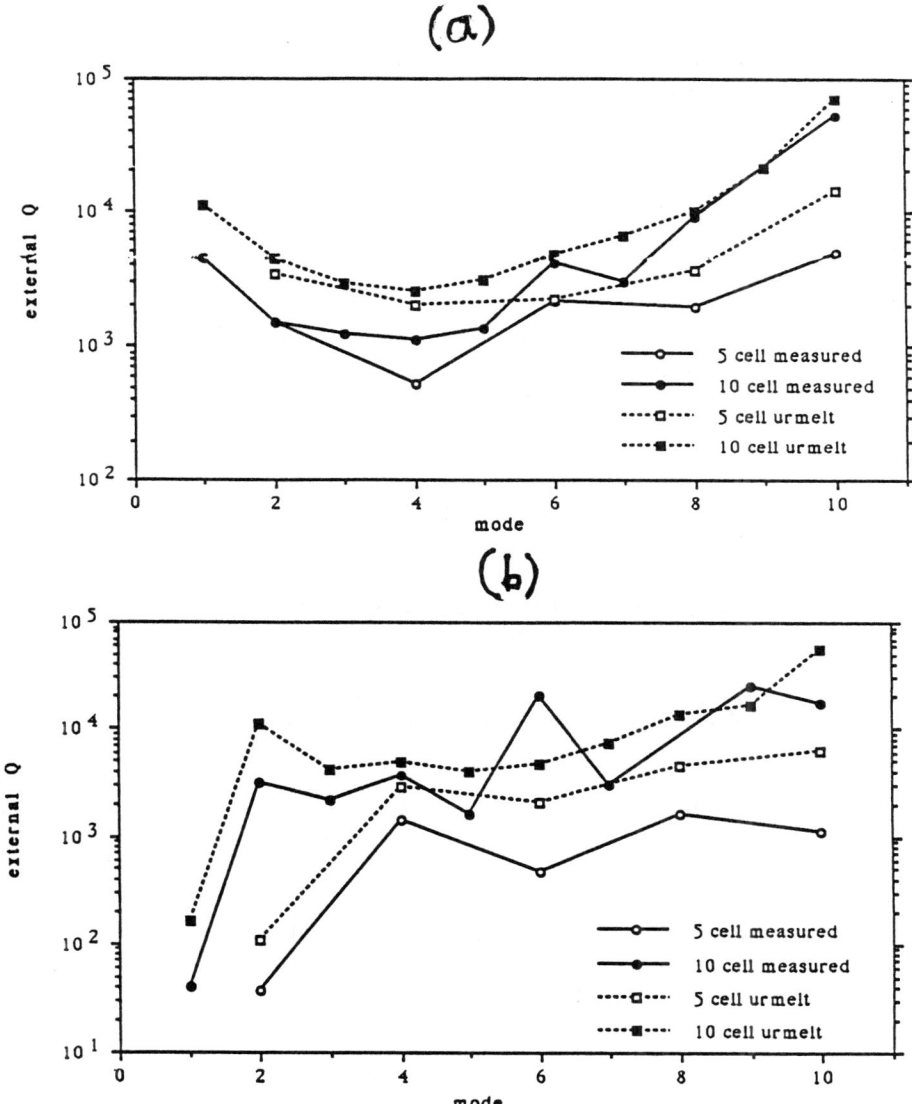

Fig. 1 A comparison between calculated and measured Qext values.
  (a) For a particular end cell shape, Qext for the TM011 mode is calculated from the field output of URMELT as discussed in the text for a 5-cell and for a 10-cell cavity, and compared with measured Qext for the same mode using machined Aluminum model 5 and 10 cell S-band structures with a single coupler located on the beam tube past the end cell.
  (b) Same as (a) but for a different end-cell shape.

Superconducting Cavities for a B-Factory

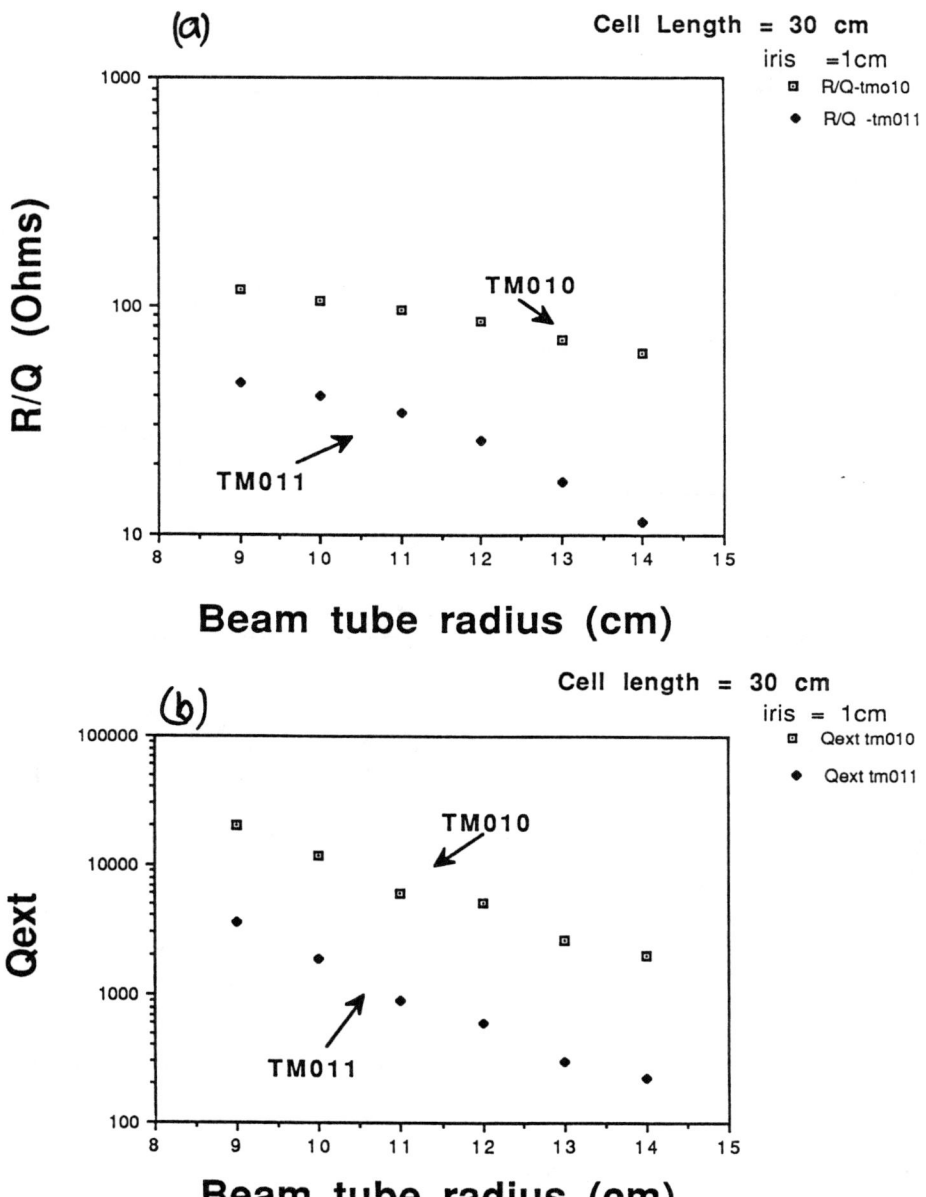

Fig.2a Calculated R/Q for the fundamental (TM010) mode and the most dangerous longitudinal higher mode (TM011) vs. beam tube diameter.

2b Calculated Qext for the same two modes vs. beam tube diameter.

In (a) and (b) the cell length is kept fixed at 30 cm and the iris radius is kept fixed at 1 cm.

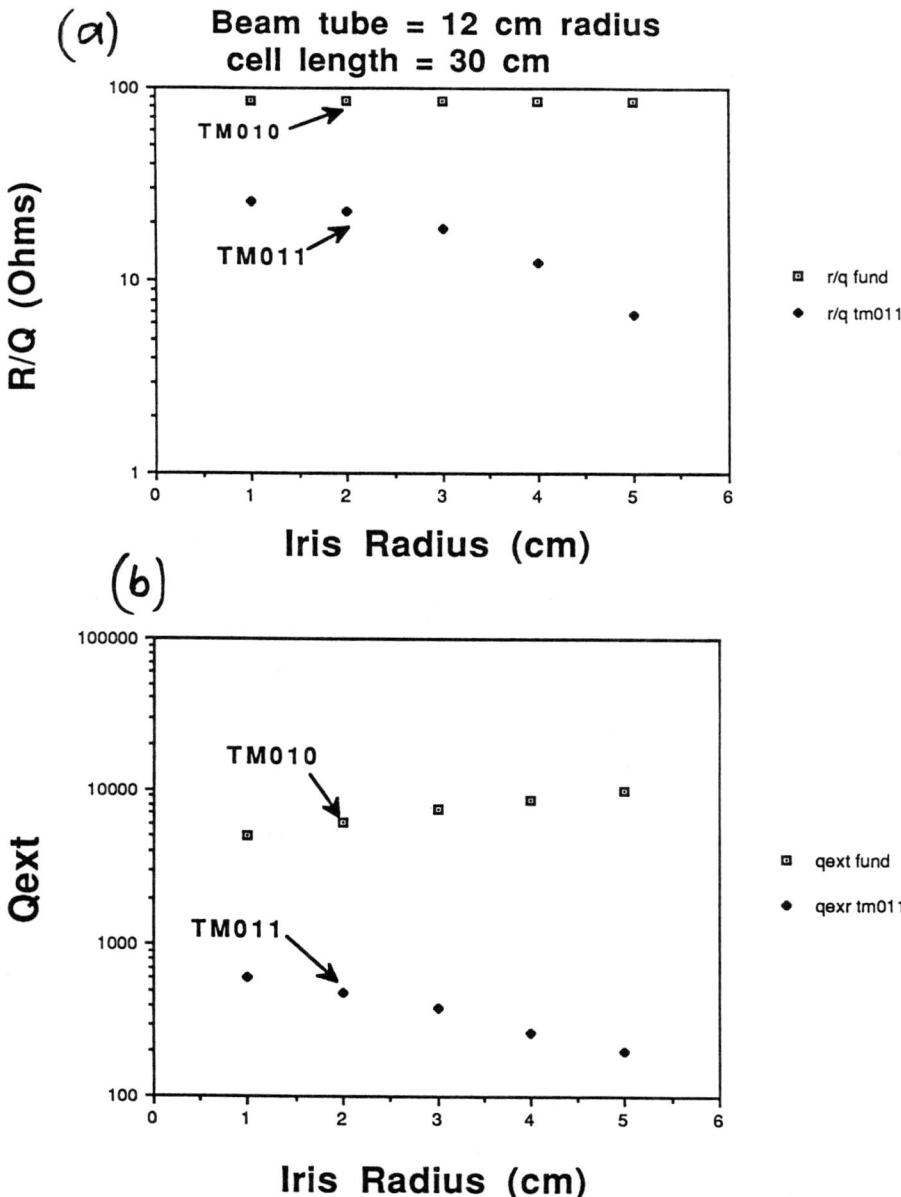

Fig.3a Calculated R/Q for the fundamental (TM010) mode and the most dangerous longitudinal higher mode (TM011) vs. iris radius.

3b Calculated Qext for the same two modes vs. iris radius.

In (a) and (b) the cell length is kept fixed at 30 cm and the beam tube diameter is kept fixed at 24 cm.

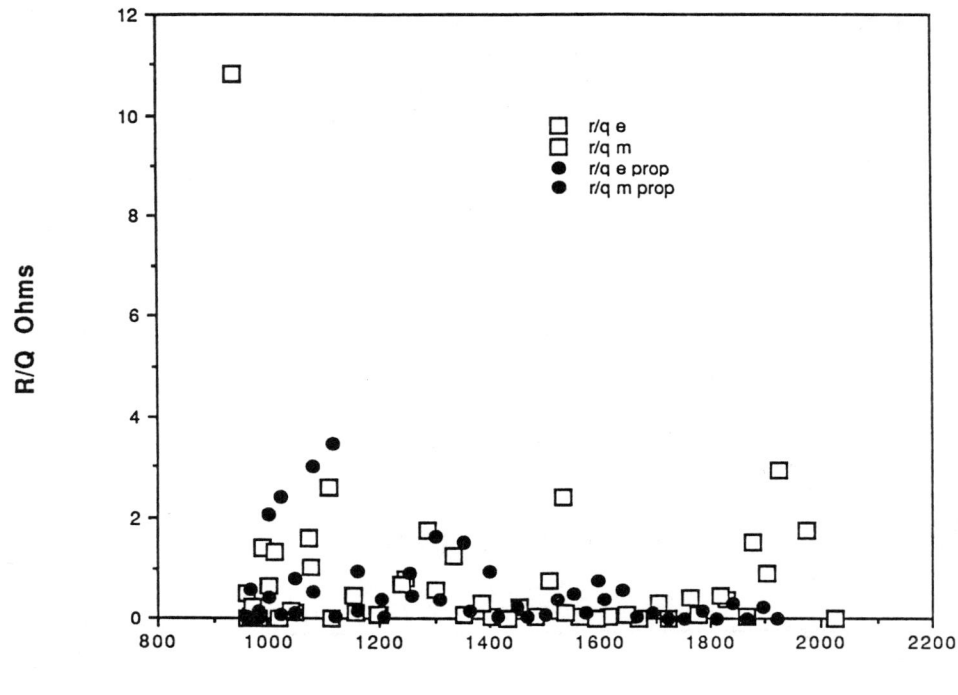

Fig. 4 A comparison of R/Q of the higher modes between two different cell shapes, one for which the TM011 mode is propogating (closed circles) and one for which it is non-propogating (open squares).

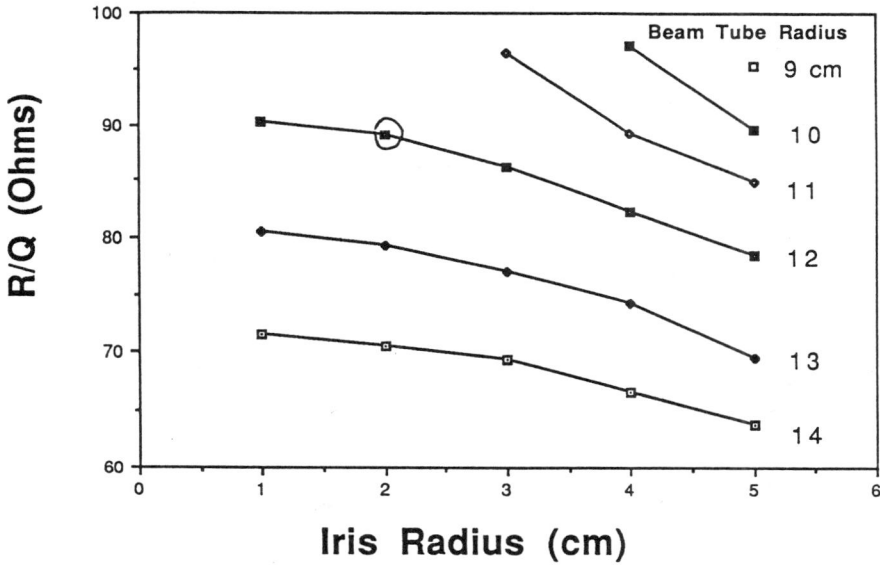

Propogating TM011

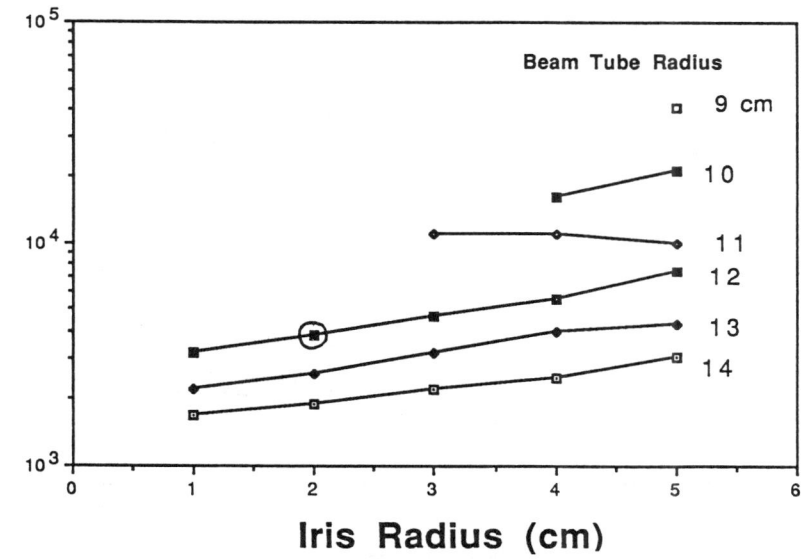

Fig.5a R/Q and Qext for the fundamental mode as a function of iris radius for various beam tube diameters, the cell length is kept fixed at 24 cm. In each case the cell shape parameters are such that the TM011 mode is propogating in the beam tube. The final shape selected is circled.

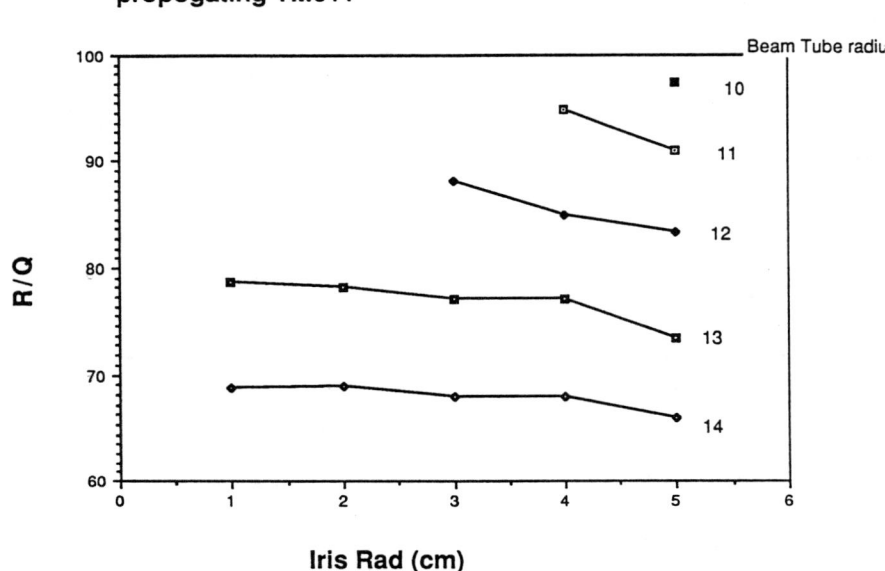

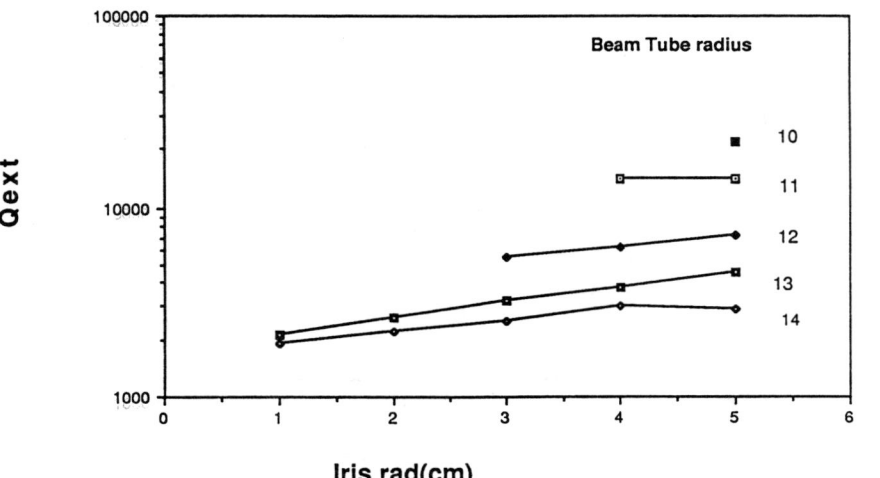

Fig. 5b Same as Fig. 5a, except for a fixed cell length of 26 cm.

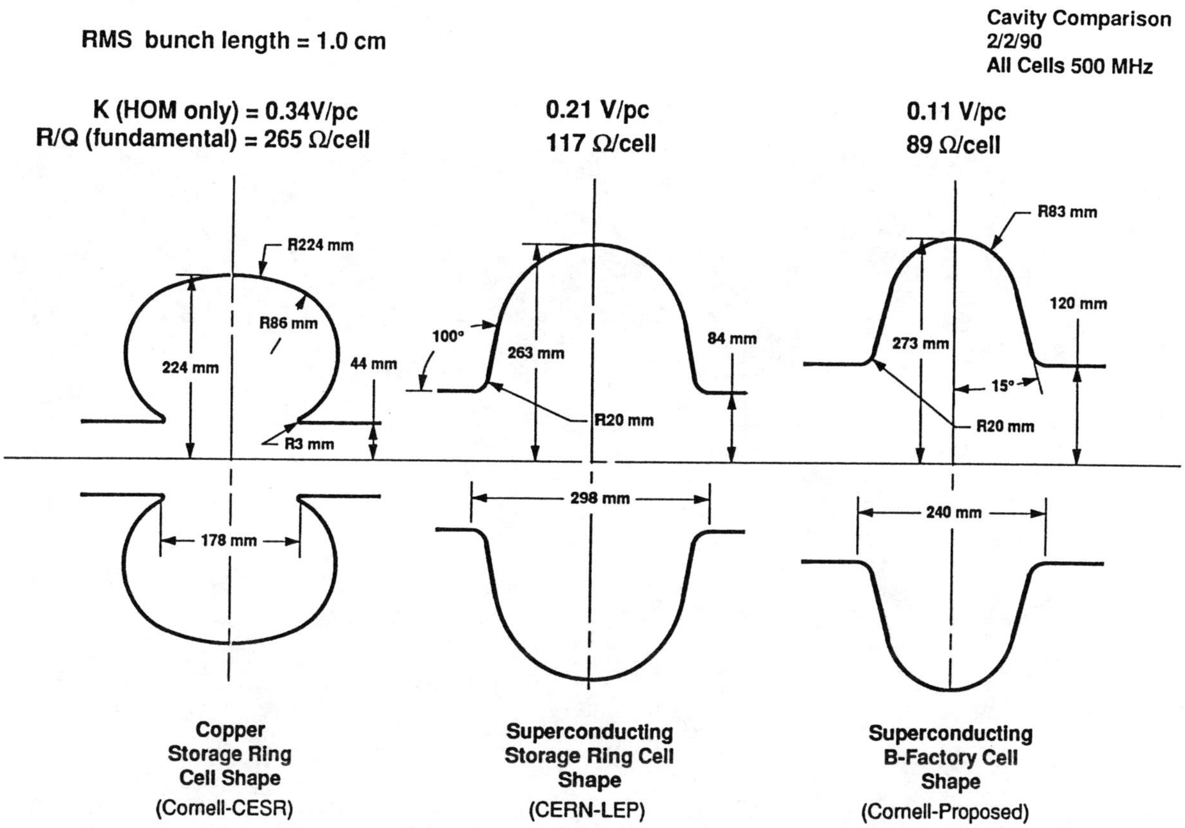

Fig. 6 A comparisons of cell shapes. Left: the CESR cell shape for copper cavities; middle: typical cell shape for a sc cavity used in the present generation storage rings; right: the B-factory cell shape.

Fig. 7 A comparision of half cells. Left: the CESR half-cell for copper cavities; middle: typical half-cell for a sc cavity used in the present generation storage rings; right: the B-factory half-cell.

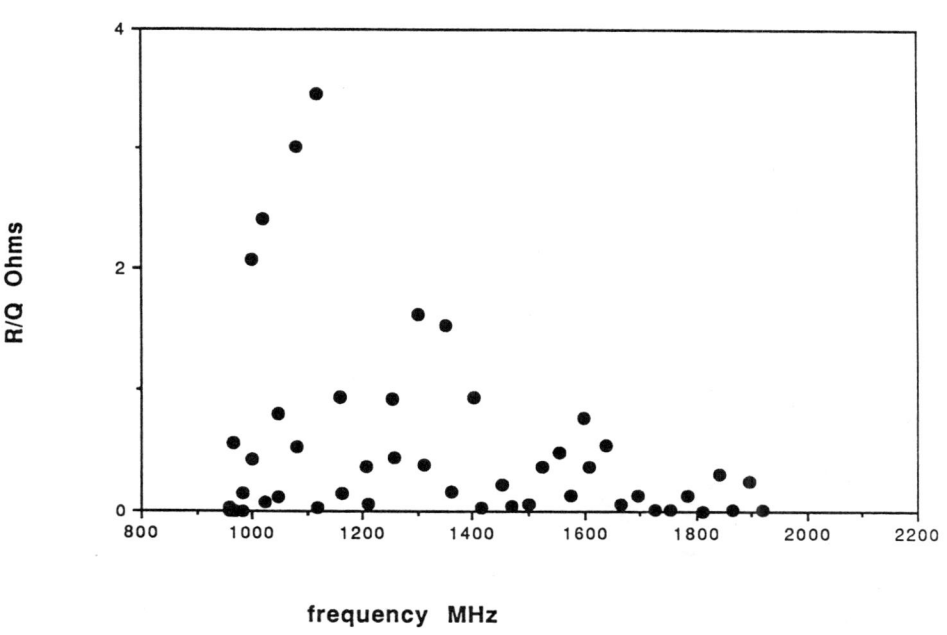

Fig. 8 R/Q for all longitudinal higher order modes for the B-factory cell up to 2 GHz.

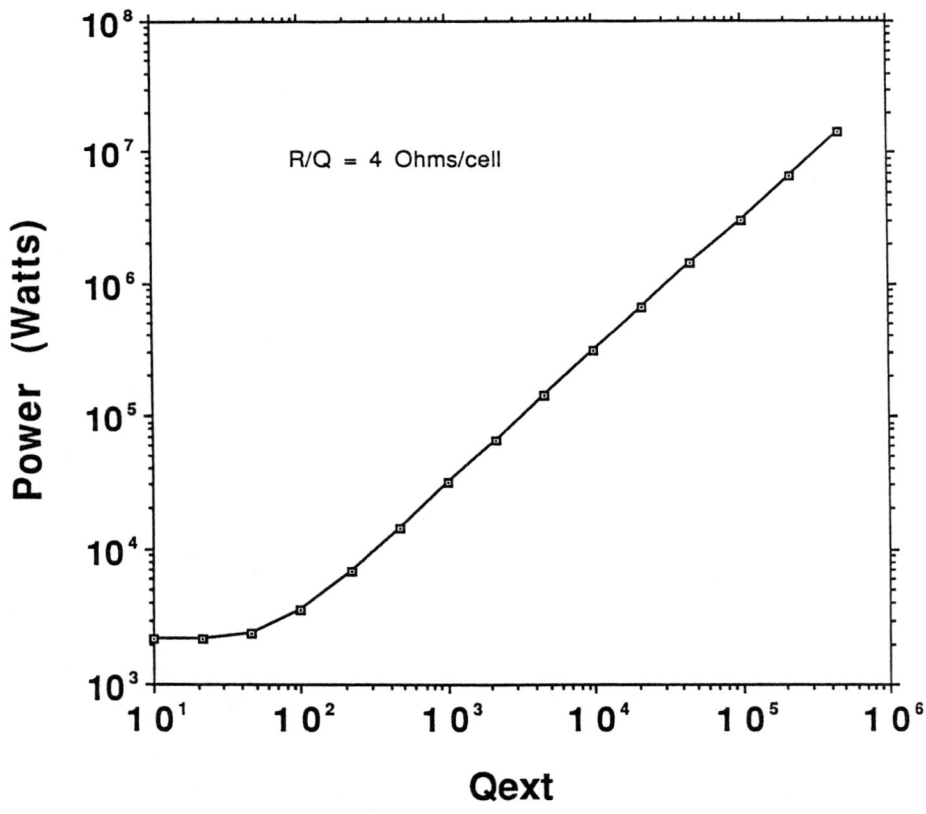

Fig. 9 Beam induced higher mode power vs Qext for resonant (worst case) excitation of a longitudinal HOM with R/Q = 4 Ω/cell.

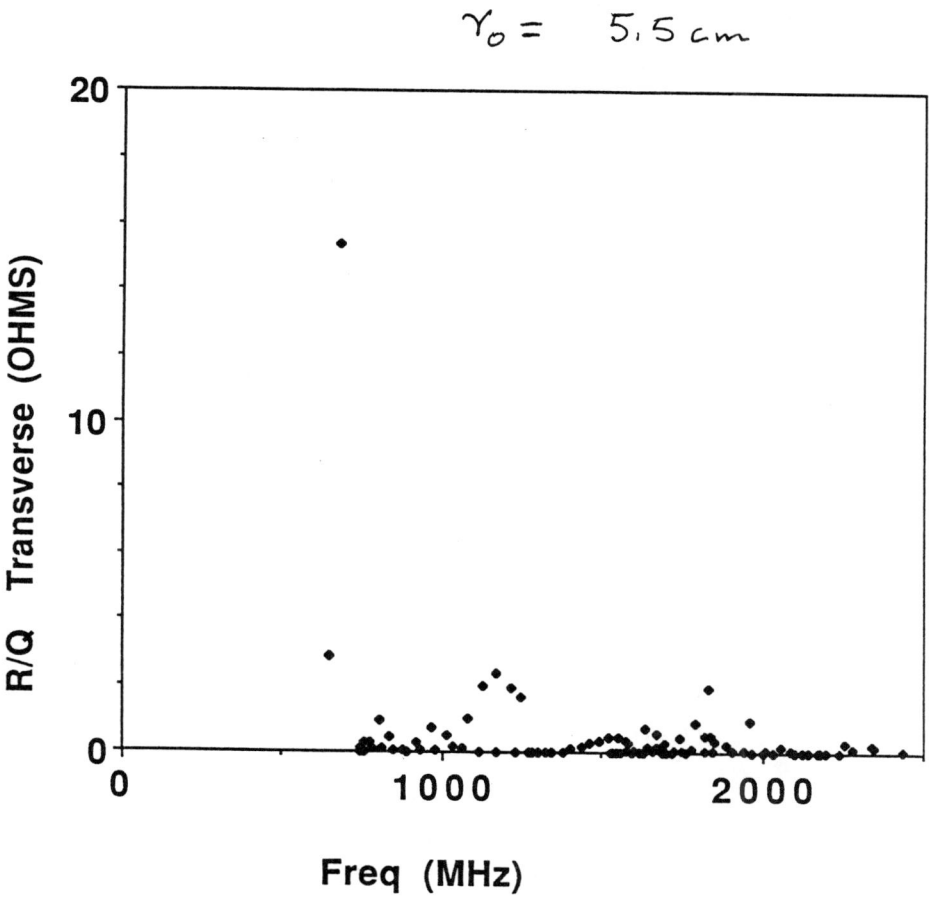

Fig. 10 Calculated values of R^t/Q from URMELT for all modes up to 2500 MHz

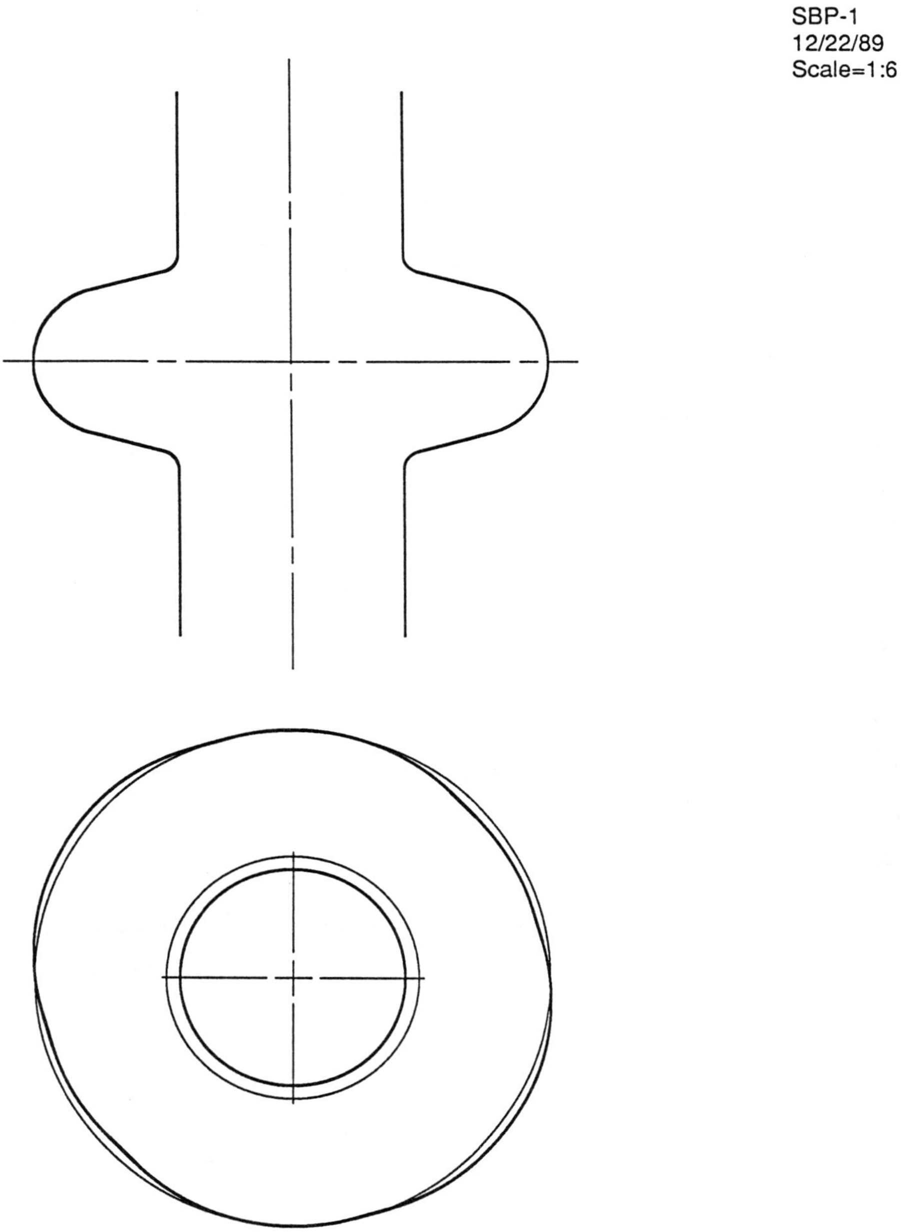

SBP-1
12/22/89
Scale=1:6

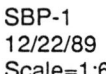

Fig. 11 Polarized cell shape

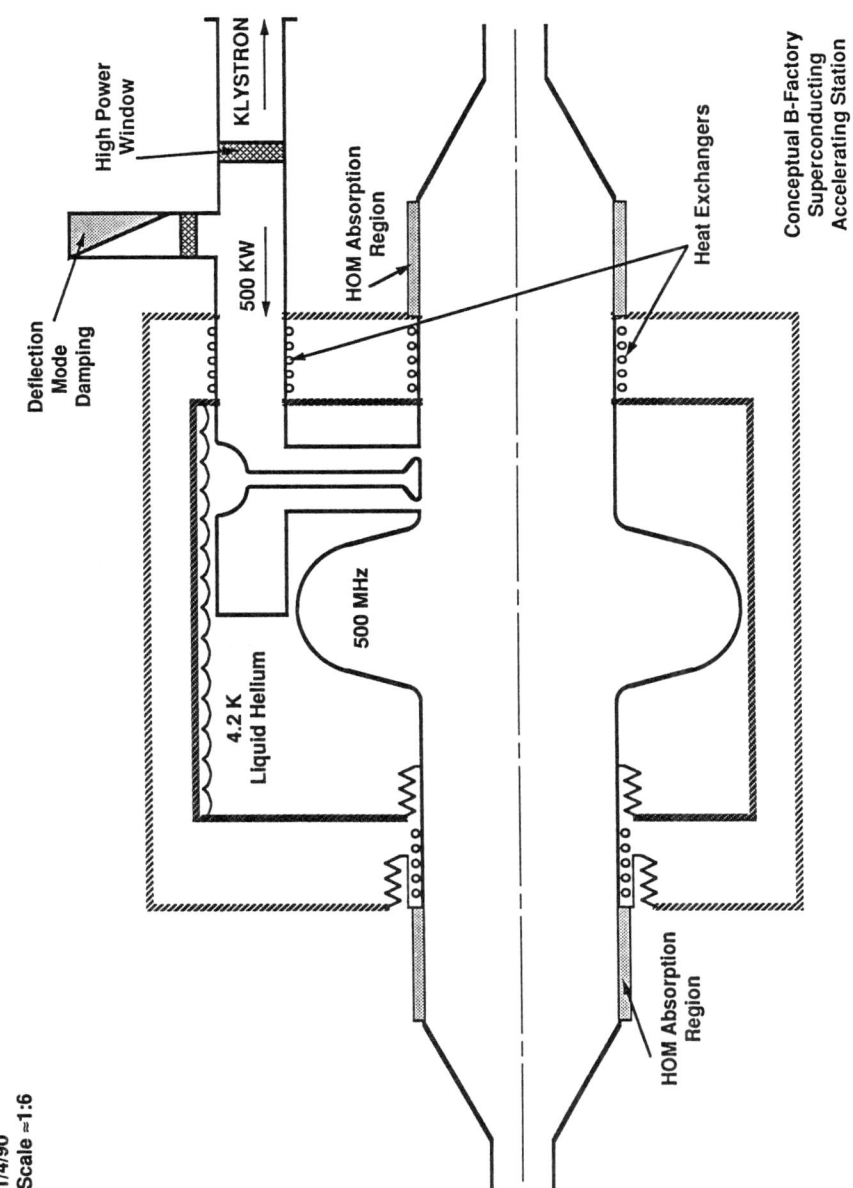

Fig.12 Current status of the conceptual design for a sc cavity for a B-factory.

Part IV

B-FACTORY PROJECTS

B-Meson Factory in the CERN-ISR Tunnel

L. Rivkin for CERN/PSI Collaboration[1]
Paul Scherrer Institute, Villigen, Switzerland

1 Introduction

The present feasibility study of the B-meson factory in the ISR tunnel is predated by several events:

- Studies of a B-meson factory plans at the Paul Scherrer Institute resulted in publication of a "Proposal for an Electron Positron Collider for Heavy Flavour Particle Physics and Synchrotron Radiation"[2] in **July 1988**.

- In **September 1989** the Swiss authorities decided not to fund the project at PSI due to different priorities in the government research policies.

- Immediately following that decision, in **October 1989**, the directors of CERN and PSI decided to start a joint effort to study the feasibility and the costs of a similar collider project using the ISR tunnel and the CERN electron positron injector complex.

The results of this feasibility study have been published in **March 1990** in a report titled "Feasibility Study for a B-meson Factory in the CERN ISR Tunnel"[3], which contains both the physics motivations and the machine design. The major points from the latter part are presented in this paper. The detailed list of references is given in the report.

1.1 Specifications

The approach adopted has been to study an asymmetric (3.5 GeV on 8 GeV) collider which can start with a luminosity of 10^{33} cm^{-2}s^{-1}, but choosing major components such that it would have the potential to go to 10^{34} cm^{-2}s^{-1}. The possibility of conversion to a symmetric machine is introduced in the design because of the lack of experimental data about the physics of the beam-beam interaction in asymmetric machines. Reaching the luminosity of $6 \cdot 10^{33}$ cm^{-2}s^{-1} in a symmetric mode of operation has been studied as well.

2 General Layout and Injectors

The ISR was installed in a circular tunnel of 15 m width and 300 m diameter as shown in fig. 2.1.

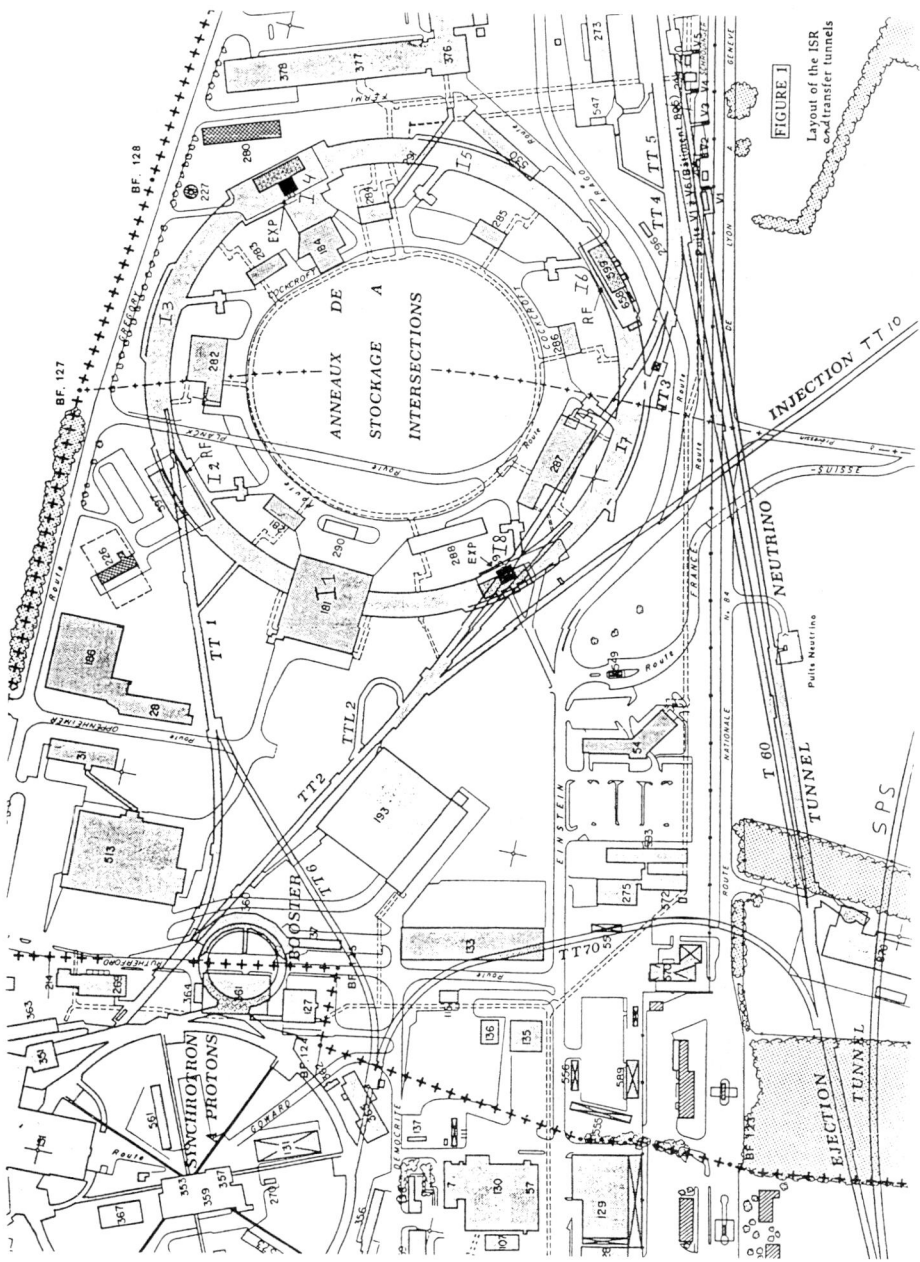

Figure 2.1: Layout of the ISR and transfer tunnels.

It is proposed to use the tunnel TT2 to inject into the B-factory 3.5 GeV positrons accelerated in the PS. The injection of 8 GeV electrons will require successive acceleration to 3.5 GeV in the PS and 8 GeV in the SPS using the LEP electron channels. The injection of the electrons will then be done from the SPS using as a transfer line TT10, TT2, the PS ring (used as a transfer line), then TT6 and TT1. No new civil engineering work is required.

For the symmetric mode of operation the acceleration of both beams to 5.3 GeV can be accomplished in the PS. The general layout of the injection transfers is shown in Fig. 2.2.

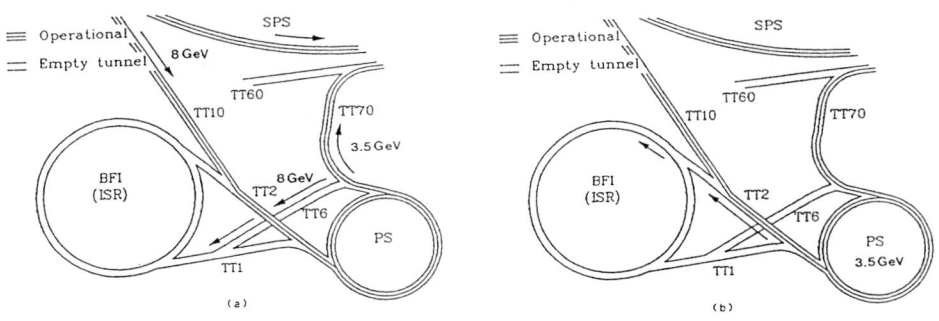

Figure 2.2: Transfer of electrons (a) and transfer of positrons (b).

It appears possible, thanks to the flexibility of the CERN injector complex, to use the LEP injector with practically no interference with the CERN programme.

The ISR beams collided in 8 interaction areas numbered I1 to I8 (Fig. 2.3). It is proposed to install two interaction areas for the beauty factory in I4 and I8. The RF straight sections would be placed in I2 and I6. Four short straight sections in I3,I7 and I1,I5 would be used for injection and for the installation of wigglers respectively. The new ring installed in the ISR tunnel is also shown in Fig. 2.3. No modification of the ISR tunnel is required.

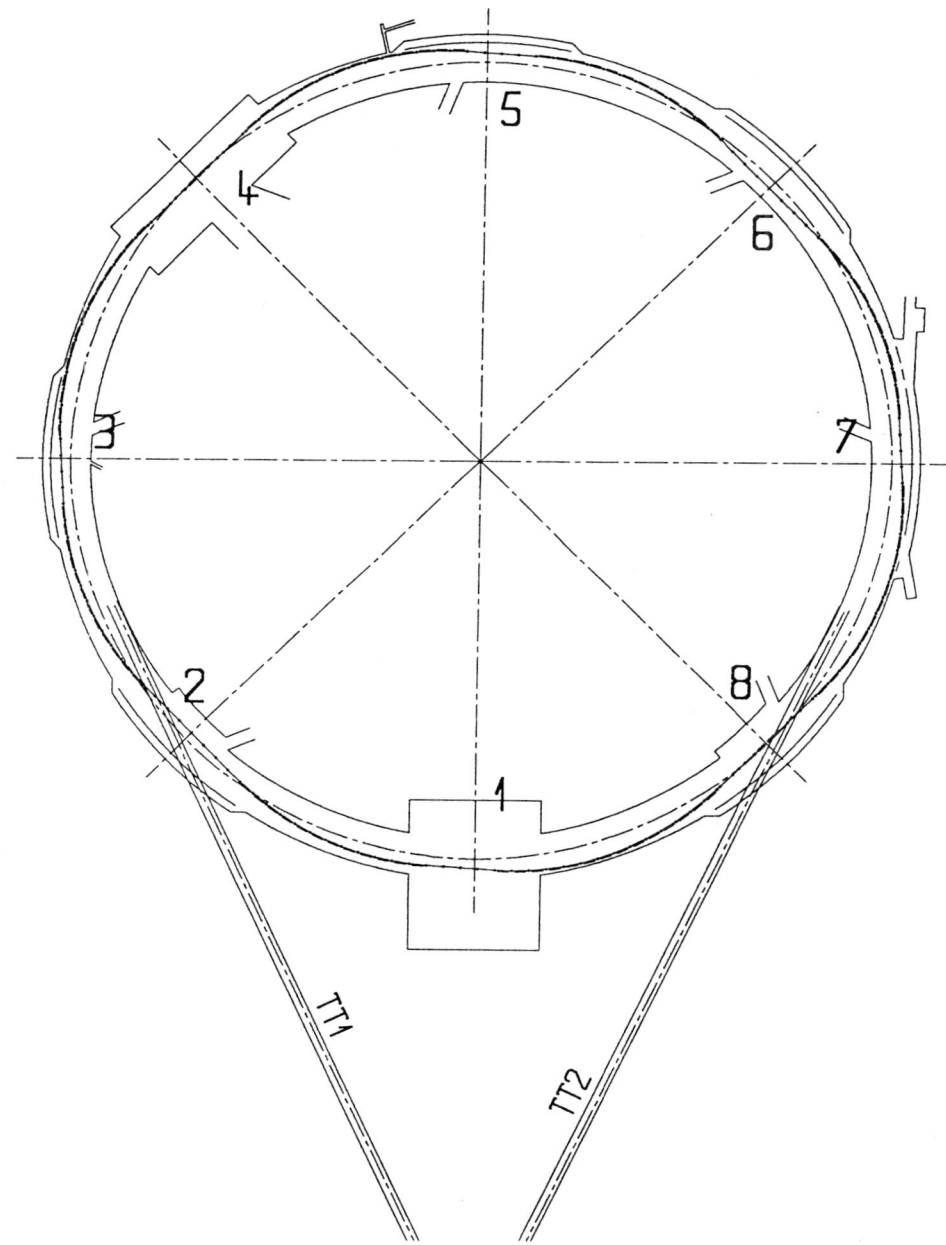

Figure 2.3: Layout of the B-factory in the ISR tunnel and the old interaction regions labelled 1-8.

3 Choice of Parameters

The proposed sets of parameters for the double–ring colliders at the center-of-mass energy of 5.3 GeV reflect several basic choices that have been made in view of the present state of experience in and prejudices about the existing single ring colliders.

The biggest uncertainty in the design of asymmetric machines is their performance limitations coming from the beam-beam interaction.

Already in the existing symmetric single-ring colliders the currents of the two beams are equalized in order to avoid blowing up the transverse dimensions of the weak beam by the strong beam due to the beam-beam interaction, with a corresponding loss in luminosity. Often an additional requirement of equal vertical and horizontal tune shift parameters is imposed in an attempt to optimize the luminosity (so–called "magic coupling").

In order to match the beam-beam interaction of two unequal energy beams, several attempts have been made to put down a set of constraints on the parameters of the two beams [4, 5]. In the case of this study we have adopted a subset of these rules[6]. We require:

a) complete overlap of the beams at the interaction point, i.e. equal transverse beam sizes, equal bunch lengths, equal beta functions and equal emittances of the two beams

b) equal tune shift parameters in both planes and for both beams

c) an additional constraint on the bunch length, coming from the requirement that the **disruption**

$$D = 4\pi\xi\frac{\sigma_s}{\beta_V^*} < 0.25 - 0.35,$$

Under these assumptions, and assuming flat beams, the luminosity of both asymmetric and symmetric modes of operation of the collider can be written as:

$$\mathcal{L} = \frac{1}{2\,e\,r_e}\frac{\xi\,I_i\,\gamma_i}{\beta_V^*}\,,\ i=1,2 \tag{3.1}$$

$$\mathcal{L} = \frac{\pi}{r_e^2}\gamma_1\cdot\gamma_2\,\xi^2\,\frac{\varepsilon_o\,f_b}{\beta_V^*}, \tag{3.2}$$

where e and r_e are the charge and classical radius of the electron, I_i the average current in ring i, ε_o the equilibrium emittance in the absence of coupling, β_V^* the vertical beta function at the interaction point, γ_i the relativistic factor $\gamma_i = E_i/m_e c^2$ of beam i and f_b is the bunch frequency. The last equation is independent of the degree of the asymmetry, since $\gamma_1\cdot\gamma_2$ is a constant for the operation of the machine at the $\Upsilon(4S)$ resonance.

Critical issues in the asymmetric mode
Since the product of the beam current and the beam energy is equal for the two rings, the current in the low energy ring of an asymmetric machine will be higher than in a symmetric machine by the ratio of the energy (50% higher for the 3.5 GeV ring as compared to the 5.3 GeV ring of the symmetric machine). This places stricter requirements on the impedance of the low energy ring to keep the beam instabilities under control.

Furthermore, the peak synchrotron radiation power loss in the high energy ring will be 3.5 times higher than in the symmetric case (it varies as E^3 at constant luminosity), placing severe constraints on the design of the vacuum system in that ring.

Critical Issues in the symmetric mode
Beam separation in the symmetric machine with head–on collisions is done either by electrostatic or RF deflectors, resulting in a much smaller number of bunches per ring.

Fewer bunches in the machine means higher current per bunch and requires a higher emittance for a given value of the tune shift parameter, or a lower value of the beta function at the interaction point, compared to the asymmetric option. In the "low" luminosity range ($\mathcal{L} \approx 10^{33}$ cm^{-2}s^{-1}) the required emittance is still compatible with a standard magnet size, but for higher luminosity configurations of a symmetric machine ($\mathcal{L} \approx 10^{34}$ cm^{-2}s^{-1}) the magnet cost becomes important. Another serious drawback of a small number of bunches will come from the higher order mode losses in the RF cavities and in the rest of the vacuum chamber. The situation will be improved, if the development of RF magnetic separators allows a higher number of bunches.

3.1 Parameter lists

The parameter list of the reference machine is given in table 3.1, and possible parameter list of the "ultimate" asymmetric machine with $\mathcal{L}=10^{34}$ cm^{-2}s^{-1} and the "ultimate" symmetric machine with $\mathcal{L}=6\cdot 10^{33}$ cm^{-2}s^{-1} is presented in Table 3.2.

The Ultimate Machines
The design of the interaction areas restricts the maximum number of bunches in the ultimate machines to 320 for the asymmetric and 48 for the symmetric machine.

For the ultimate machines we considered, that beam-beam tune shifts of $\xi=0.05$ could be achieved after a long period of machine improvement. This is a reasonable estimate for the symmetric machine, but assumes that no new phenomena limit this value in asymmetric machines.

Table 3.1: Parameter list of the reference machine with a luminosity of 10^{33} cm^{-2}s^{-1}

		Ring 1	Ring 2
Particles		e^+	e^-
Energy E	[GeV]	3.5	8.0
Circumference L	[m]	963.430	963.430
Bending radius ρ	[m]	65	65
Number of bunches n_b		80	80
Harmonic number		1600	1600
RF frequency	[MHz]	497.9	497.9
Momentum compaction factor α		0.0086	0.0086
Horizontal tune Q_H		14.3	12.3
Vertical tune Q_V		16.4	13.4
Aspect ratio σ_V/σ_H		0.03	0.03
Vertical tune shift ξ_V		0.03	0.03
Horizontal tune shift ξ_H		0.03	0.03
Vertical beta at interaction point β_V^*	[m]	0.03	0.03
Horizontal beta at interaction point β_H^*	[m]	1.0	1.0
Vertical emittance $\varepsilon_V = \sigma_V^2/\beta_V^*$	[10^{-6}m]	0.009	0.009
Horizontal emittance $\varepsilon_H = \sigma_H^2/\beta_H^*$	[10^{-6}m]	0.30	0.30
Disruption parameter $\mathcal{D}$		0.25	0.25
Bunch length σ_s	[m]	0.02	0.02
Energy spread σ_ε/E	[10^{-3}]	0.52	0.84
Longitudinal damping time	[ms]	37	4.6
Total current I	[A]	1.28	0.56
Current per bunch I_b	[A]	0.016	0.007
Particles per bunch	[10^{11}]	3.21	1.40
Radiation loss per turn U_o	[MeV]	0.3	5.6
Synchrotron radiation power	[MW]	0.39	3.1
Radiation loss per meter in bending magnets	[kW/m]	0.64	7.6
Peak RF voltage V_{RF}	[MV]	2.0	13
Total RF power P_{RF}	[MW]	0.70	4.4
Number of 1 MW klystrons		1	5
Number of cavities		4	20
Beam power loss P_{beam}	[MW]	0.6	3.2
Dissipated power per cavity P_{diss}	[kW]	34	60
Higher order mode losses per cavity P_{HOM}	[kW]	8.2	1.6
HOM power in vacuum chamber P_{VAC}	[kW]	140	27
Total input power per cavity	[kW]	175	220
Accelerating field	[MV/m]	1.6	2.2

Table 3.2: The ultimate machines

Type Luminosity	[cm^{-2}s^{-1}]	Asymmetric 10^{34}		Symmetric 6·10^{33}
		Ring 1	Ring 2	
Particles		e^+	e^-	
Energy	[GeV]	3.5	8	5.3
Circumference L	[m]	963.43	963.43	963.43
Bending radius ρ	[m]	65	65	65
Number of bunches n_b		320	320	48
Harmonic number		1600	1600	1600
RF frequency	[MHz]	497.9	497.9	497.9
Momentum compaction factor α		0.0086	0.005	0.017
Horizontal tune Q_H		14.3	18.3	8.3
Vertical tune Q_V		16.4	16.4	14.3
Aspect ratio σ_V/σ_H		0.03	0.03	0.03
Vertical tune shift ξ_V		0.05	0.05	0.05
Horizontal tune shift ξ_H		0.05	0.05	0.05
Vertical beta at interaction point β_V^*	[m]	0.01	0.01	0.01
Horizontal beta at interaction point β_H^*	[m]	0.33	0.33	0.33
Vertical emittance $\varepsilon_V = \sigma_V^2/\beta_V^*$	[10^{-6} m]	0.003	0.003	0.011
Horizontal emittance $\varepsilon_H = \sigma_H^2/\beta_H^*$	[10^{-6} m]	0.09	0.09	0.36
Disruption parameter $\mathcal{D}$		0.30	0.30	0.30
Bunch length σ_s	[m]	0.0048	0.0048	0.0048
Energy spread σ_ε/E	[10^{-3}]	0.41	0.85	0.56
Longitudinal damping time	[ms]	50	4.6	16
Total current	[A]	2.56	1.12	1.0
Current per bunch I_b	[mA]	8	3.5	21
Particles per bunch	[10^{11}]	1.6	0.70	4.2
Radiation loss per turn	[MeV]	0.22	5.6	1.1
Synchrotron radiation power	[MW]	0.56	6.24	1.1
Peak radiation loss in bending magnets	[kW/m]	1.28	15.3	2.7
Peak RF voltage V_{RF}	[MV]	20.5	120	115
Total RF power P_{RF}	[MW]	1.4	6.8	3.4
Number of 1 MWatt klystrons		2	8	5
Number of cavities		8	64	40
Beam power loss P_{beam}	[MW]	1.4	6.8	3.4
Dissipated power per cavity P_{diss}	[kW]	s.c.	s.c.	s.c.
HOM losses per cavity P_{HOM}	[kW]	34	6.5	36
HOM power in vacuum chamber P_{VAC}	[kW]	590	110	850
Total input power per cavity	[kW]	180	110	85
Accelerating field	[MV/m]	8.5	6.3	9.5

The minimum value of β_V^* of the ultimate machines can be deduced from the lattice and low beta study. We have selected somewhat arbitrarily 1 cm for both the symmetric and asymmetric machines. A more precise number will result from the detailed study of a low beta insertion and from lattice optimization.

The vacuum chamber and magnet aperture is fixed using equation (1.2) which determines the emittance. Because the number of bunches is much smaller, it is the symmetric machine which determines the magnet gap with an emittance four times larger than the asymmetric machine (0.4 mm mrad against 0.1 mm mrad) even though the luminosity is reduced from 10^{34} to $6 \cdot 10^{33}$ cm^{-2}s^{-1}.

From equation (1.1) the maximum currents and the resulting synchrotron radiation power are deduced. This determines the choice of copper for the vacuum chamber material. The bunch length σ_s for the ultimate machines has been fixed so that the disruption, somewhat arbitrarily, $\mathcal{D}_V=0.3$. This determines the RF system, and as a consequence the length of the straight sections required to install the RF cavities.

The Reference Machine

A parameter list for a 10^{33} cm^{-2}s^{-1} asymmetric machine is produced, based on technology which exists, or can easily be developed. The essential features allowing an evolution towards the high luminosity asymmetric machine or conversion to the symmetric machine are introduced in the design. A preliminary design of the reference machine has been made in order to estimate the cost.

The straight section lengths, vacuum chamber aperture and therefore magnet gap of the reference machine have been specified by the ultimate machines.

To define the parameters of the reference machine a safer choice of beam-beam tune shift $\xi=0.03$ and disruption $\mathcal{D}_V=0.25$ is made. The number of bunches can be reduced to 80 since the vacuum chamber is large enough to accept an emittance of 0.3 mm mrad and this eases the design of the feedback system.

4 Lattice

Geometrical constraints on the lattice coming from the existing shape and dimensions of the ISR tunnel result in the following modifications to the original lattice proposed for the PSI project :

- The straight sections of the original racetrack design were too long to fit in the circular ISR tunnel. They have been shortened to 50 m, still compatible with the separation schemes for the symmetric option. Magnetic separation employing a tilt of the detector solenoid is used in the asymmetric cases, as described below.

- The dispersion suppressors using the missing magnet scheme were replaced with ones that use only quadrupoles, making it possible to increase the bending radius of the machine to 65 m.

- The asymmetry of 3.5 on 8 GeV requires more straight section space with zero dispersion for the installation of RF cavities.
- The existing injection lines from the PS require additional straight sections for injection into the ring at specific locations in the tunnel.

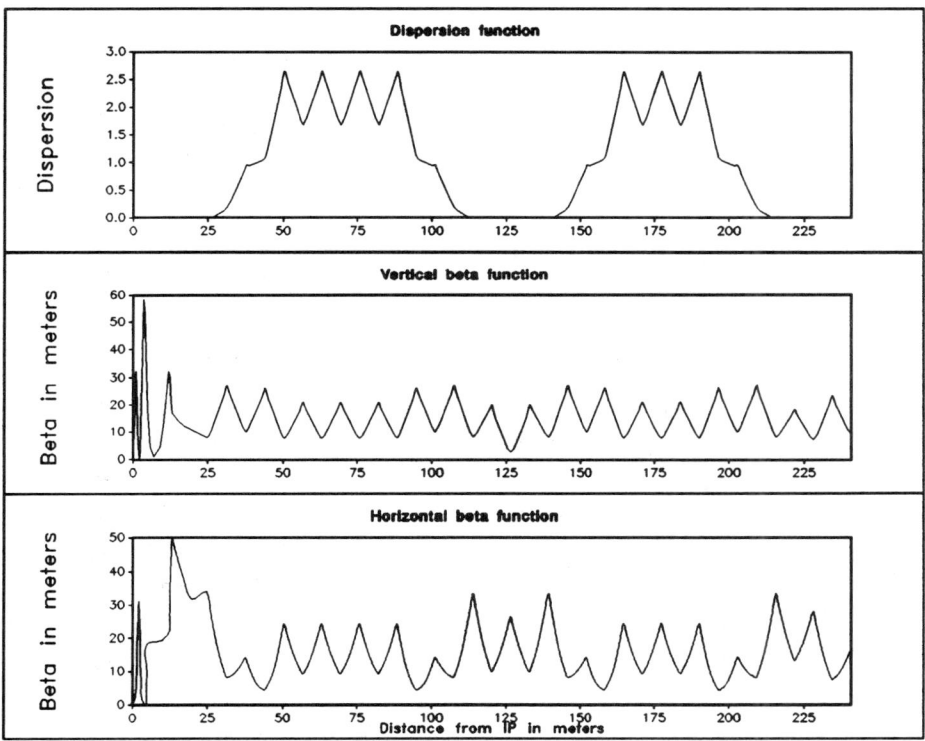

Figure 4.1: Dispersion, vertical and horizontal beta functions of one quadrant.

The lattice provides 100 m of straight sections for the RF, as well as 100 m of space for injection and other special inserts. The optics of the new dispersion suppressors provides a wide variety of lattice functions, both with zero and with large dispersion for the emittance control with wigglers. A classical FODO cell structure is used in the arcs, the cell length being 12.7 m. The phase advance per cell is variable over a wide range of more than a factor of two, providing variation in emittance of at least a factor of ten and a factor of more than five in the momentum compaction factor.

The lattice functions for the reference machine are shown in fig. 4.1 for a phase advance per cell of 50 degrees in the horizontal and 60 degrees in the vertical planes. The corresponding values of the lattice parameters are summarized in tables 3.1 and 4.1.

Table 4.1: Parameters of the wiggler magnets and impact on the lattice

		$\mathcal{L}=10^{33}\,\mathrm{cm}^{-2}\mathrm{s}^{-1}$		$\mathcal{L}=10^{34}\,\mathrm{cm}^{-2}\mathrm{s}^{-1}$		$\mathcal{L}=6\cdot 10^{33}\,\mathrm{cm}^{-2}\mathrm{s}^{-1}$
		Low Energy	High Energy	Low energy	High energy	Symmetric[2]
Wigglers off						
Natural emittance $\varepsilon_H = \sigma_H^2/\beta_H^*$	$[10^{-6}\,\mathrm{m}]$	0.06	0.3	0.06	0.09	0.36
Phase advance per cell		50°	50°	50°	75°	36°
Momentum compaction α		0.0086	0.0086	0.0086	0.0051	0.017
Radiation loss per turn U_o	[MeV]	0.2	5.4	0.2	5.4	1.1
Energy spread σ_ϵ/E	$[10^{-3}]$	0.37	0.84	0.37	0.84	0.56
Wigglers on						
Emittance with wigglers $\varepsilon_H = \sigma_H^2/\beta_H^*$	$[10^{-6}\,\mathrm{m}]$	0.3	0.3	0.09	0.09	
$\mathcal{H}$ in the Wiggler[1]	[m]	0.7	0	0.5	0	
Wiggler field strength	[T]	0.9	0	0.5	0	
Wiggler length	[m]	4.2	0	4.2	0	
Radiation loss per turn U_o	[MeV]	0.3	5.4	0.2	5.4	
Energy spread σ_ϵ/E	$[10^{-3}]$	0.52	0.84	0.41	0.84	

Note 1: $\mathcal{H}$ is defined as $\mathcal{H} = \gamma D^2 + 2\alpha DD' + \beta D'^2$
Note 2: The emittance in the symmetric machine is adjusted with the phase advance in the arcs

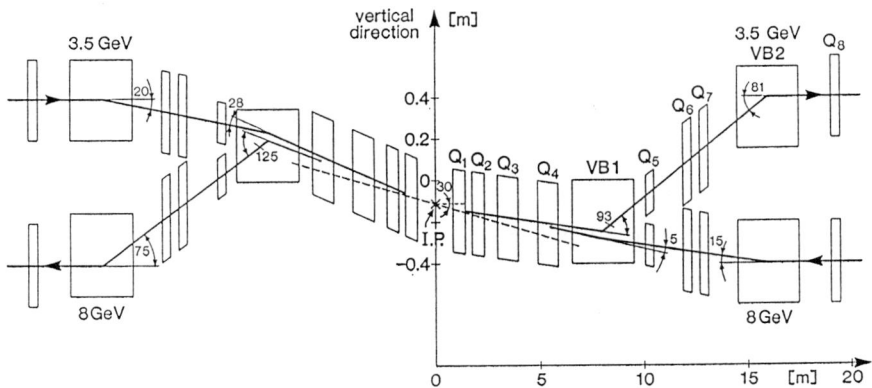

Figure 5.1: Magnetic separation of the 3.5 and 8 GeV beams. VB1 is a magnetic septum. (angles in mrad)

The emittance in the low energy ring is adjusted using wiggler magnets. The magnets, totaling 4.2 m in length and providing maximum magnetic field of 1.5 T, affect not only the emittance but also the energy spread and the length of the bunch as well as the radiation loss per turn and the damping times. The present solution for wiggler insertion optics minimizes their impact on the RF system requirements. The main parameters affected by wigglers are summarized in table 4.1.

Among the outstanding issues are the matching of the vertical dispersion, compensation of coupling and extensive tracking studies.

5 Interaction Region

The general layout of the interaction region is shown in fig. 5.1. The concept described by K.Wille [7] to focus simultaneously beams of unequal energy has been extended to apply to beams of 3.5 and 8 GeV. The first four quadrupoles are shared by the two beams, and are centered on the 3.5 GeV beam. The rest of the quadrupoles in the straight sections are used for matching into the dispersion suppressors. In order to minimize the peak values of the beta functions in the final focus quadrupoles and their chromaticity contribution, the first quadrupole is placed very close to the interaction point. A free space of ±0.7 m is available between the flanges of the cryostats encasing the superconducting quadrupoles Q1 and Q2.

The optics solution with the values of 0.03 m and 1 m for β_V^* and β_H^* respectively in the reference machine has been used in the backgrounds and masking calculations. Preliminary optics solutions with 0.01 m for β_V^* required in the ultimate machine have been worked out. The superconducting quadrupoles Q1 and Q2 are helium bath cooled, ironfree magnets of similar design to that of the

Table 5.1: Parameters of the superconducting mini-beta quadrupoles

nominal gradient	50	T/m
magnetic length	0.6	m
warm bore radius	60	mm
coil inside radius	80	mm
cryostat outside radius	200	mm
nominal current	1000	A

LEP low-β quadrupoles. The major characteristics of these magnets are given in table 5.1.

5.1 Beams separation

It is the advantage of asymmetric machines that the beam separation may be achieved by magnetostatic fields. It is proposed in this study that the required vertical separation between beams of $5 \cdot \sigma_H$ for a distance between bunch crossing points of 3 m is produced by the horizontal field component resulting from a 5° rotation of the detector solenoid[8]. This provides for operation with up to 160 bunches per ring. The separation between beams is already $2 \cdot \sigma_H$ at 1.5 m from the interaction point and with optimization it should be possible, by increasing the rotation of the detector to less than the maximum acceptable value of about 10°, to obtain sufficient bunch separation for 320 bunch operation in the ultimate asymmetric machine. A simplified model of the tilted solenoid fields has been used; more detailed picture, including the radial fields and the proper superposition of the quadrupole and the solenoidal fields effects are presently under study.

A 3 m long magnetic septum VB1 centered at 8 m from the interaction point serves to kick the beams to their nominal separation of 0.8 m. In order to minimize the photon background from the septa, these are arranged to give only 5 (28) mrad kicks to the incoming 8 (3.5) GeV beams, which leads to the 30 mrad tilt of the beam axes (and the experiment) at the interaction point as shown in fig. 5.1. In order to have equal path length in the interaction region, the interaction point is located some 10 cm below the mid-plane of the two rings. Dedicated vertical bending magnets VB2 are required to bring the trajectories parallel.

In the symmetric mode of operation the separation would be done with the RF dipole separators, placed after the superconducting doublet; the normal quadrupoles are rearranged for this optics. For 48 bunch operation each such device requires an input power of 300 kW.

6 The RF System

The RF system has been designed [9] for the reference machine, taking into account the ultimate machines parameters in the choice of the RF frequency and in the amount of space reserved for the RF cavities.

The final choice of the RF frequency f_{RF} will need a more detailed study. In particular double frequency accelerating systems (a lower frequency part providing for the power loss compensation and a higher harmonic part providing for the longitudinal focusing) could be advantageous in the ultimate machines. In this study, in order to take advantage of existing technologies, we have restricted the choice to the two standard frequencies (350 MHz and 500 MHz), compatible with the length of the injected bunches. The RF-voltage V_{RF} is mainly determined by the longitudinal focusing needed to get short bunches:

$$V_{RF} \propto \frac{\alpha}{f_{RF}\sigma_s^2} \quad (6.1)$$

where α is the momentum compaction factor. This favors the 500 MHz system which requires about 30% lower voltage (and therefore less straight section space) than the 350 MHz system.

The power requirementsare determined by synchrotron radiation power losses, HOM losses in the cavities and the vacuum chamber and dissipation in the cavities. Cavities contribution to the HOM losses have been calculated for the selected cavity shape; the vacuum chamber contribution was estimated using the results of measurements in CESR storage ring at Cornell University[10]. The results are presented in tables 6.1 and 7.1.

The Reference Machine

The RF requirements of the reference machine are compatible with the use of a normal conducting system. Monocell type cavities have been selected for easier attenuation of parasitic resonances and coupling of high power into each cell. A shape characterized by a smooth accelerating gap region and a large pipe diameter is proposed instead of the conventional "nose cone shape". It allows to achieve higher accelerating fields, to reduce the parasitic transverse impedances and to lower the propagation cutoff frequency at the expense of a relatively low increase in total power consumption. Passive HOM dampers installed on each cavity and a complementary active system will ensure the beam stability against coupled bunch instabilities. A model cavity designed according to these guidelines is being tested at PSI. It is represented in fig. 6.1, and its computed characteristics are listed in table 6.1; the parameters of the RF system are found in table 3.1.

Six sets of 1 MW klystrons powering four cavities each (one set in the low energy ring, five sets in the high energy ring) lead to a performance comparable to presently operating systems (see table 6.2).

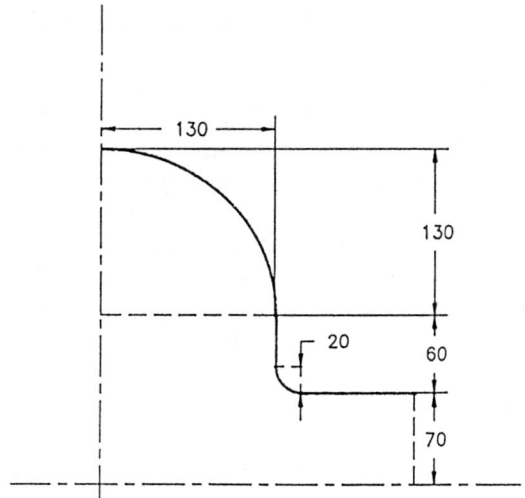

Figure 6.1: Shape and dimensions of a model cavity being tested at PSI.

Table 6.1: Computed characteristics of the model RF cavity

Frequency	500	MHz
R/Q	73	electr. Ω
Q_0	50 000	
Shunt impedance R_s	3.65	MΩ
E_{max}/E_{acc}	2.1	
HOM loss factor ($\sigma_s = 2$ cm)	0.125	V/pC
transverse cut-off frequency	1.3	GHz
longitudinal cut-off frequency	1.6	GHz

Table 6.2: Required performance of the normal conducting ($\mathcal{L}=10^{33}$ cm^{-2}s^{-1}) and superconducting cavities

		n.c.	s.c.
Accelerating voltage V_{acc}	[MV]	0.75	3.0
Accelerating gradient	[MV/m]	2.5	10.0
Power dissipation in the walls	[kW]	80	-
Input power	[kW]	250	200
HOM power	[kW]	10	40

Ultimate machines
The RF parameters presented in Table 3.2 are based on the use of a superconducting system (500 MHz, monocell cavities); a maximum accelerating field of 10 MV/m is assumed (see table 6.2).

The high energy ring of the asymmetric machine requires a very high voltage (120 MV). The use of a superconducting system is imposed by space constraints. The voltage can be provided by 64 superconducting cavities (the maximum number that can be installed in the 16 half cell straight sections reserved for the RF) for a total RF power of about 7 MW. The HOM power to be extracted per cavity is about 7 kW.

The low energy ring requires less voltage (20 MV, 8 superconducting cavities) but the current is higher, leading to 5 times more HOM power per cavity. A normal conducting system could fit within the available straight sections. It is probably easier than with a superconducting system to provide the damping of the higher order modes and to extract 35 kW of power per cavity. However, it requires 40 cavities, leading to large expected impedance values. We therefore propose a superconducting system.

In the symmetric high luminosity machine the required RF voltage is 115 MV. This voltage can be supplied with 40 superconducting cavities, the total RF power being 3.3 MW. The HOM power deposited per cavity again reaches 35 kW.

Critical issues
The development of the HOM damping devices with characteristics well exceeding the presently operating systems is necessary. The question of operating the superconducting cavities in the extremely harsh environment due to high beam currents and high synchrotron radiation power has not been addressed.

6.1 Coherent Synchrotron Radiation Emitted by Short Bunches

The synchrotron radiation at wavelengths greater than the bunch length is emitted coherently. In the extreme case $\lambda \gg \sigma_s$, all the N_b electrons in a bunch can be considered as a single particle with charge $N_b \cdot e$. The power emitted per electron is N_b times larger in the coherent case than in the incoherent one.

This effect could increase the radiated power in many machines if it were not partially suppressed by the presence of the conducting vacuum chamber. The ratio of the coherent radiation power to the incoherent one is approximately [11, 12]

$$\frac{P_{coh}}{P_{incoh}} \sim 0.5 \left(\frac{a}{\sigma_s}\right)^2 \left(\frac{\rho}{\sigma_s}\right)^{2/3} \frac{N_b}{\gamma^4},$$

where a is the half height of the vacuum chamber and ρ is the bending radius. Since the incoherent power itself is proportional to γ^4 the above expression indicates that the coherent power is independent of the energy, as expected for the low frequency part of the synchrotron radiation spectrum. For some of the high luminosity parameters, the coherent radiation could lead to a sizable increase of the RF power necessary in the beauty factory. The RF parameters given in table 3.2 do not include the additional power requirements due to this effect. However, the above expression is rather approximate and the coherent power can be reduced relatively fast by increasing the bunch length.

7 Beam Instabilities

To avoid the lengthening of the bunch with its attendant loss in luminosity, strict limits must be placed on the longitudinal broad band impedance of the machine. The limiting values of the impedance coming from the turbulent bunch lengthening threshold are computed [13] using the Chao-Gareyte correction factor for short bunches [14]. The results, usually more critical for the low energy ring, are given in table 7.1 together with the estimates of the impedance we hope to be able to achieve.

With these values of longitudinal impedance the potential well bunch lengthening has been estimated to be small.

The transverse broad band impedance requirements for transverse mode coupling stability are less severe than the corresponding longitudinal requirements.

Ions will be trapped in the e^--beams unless specific measures are taken [15]. The installation of an efficient system of clearing electrodes appears to be very difficult. It is proposed to clear the ions by the technique of missing bunches. In addition we propose to store the electrons in the high energy machine in which the more rigid beam is less sensitive to ion induced effects.

Multibunch instabilities are mostly driven by parasitic resonances in the RF cavities. The parasitic modes in the 500 MHz cavities, equipped with passive

Table 7.1: Loss factors and longitudinal broad band impedances

		10^{33}cm^{-2}s^{-1}	$6 \cdot 10^{33}$cm^{-2}s^{-1}	10^{34}cm^{-2}s^{-1}
		Asymmetric Low Energy	Symmetric	Asymmetric Low Energy
Current I	[A]	1.28	1.0	2.56
Total number of bunches		80	48	320
Energy spread σ_ε/E	[10^{-3}]	0.52	0.56	0.41
Bunch length σ_s	[m]	0.02	0.0048	0.0048
Number of cavities		4	40	8
Vacuum chamber loss factor	[V/pC]	2.2	12.5	8.9
Loss factor per cavity	[V/pC]	.12	0.5	0.5
HOM losses in vacuum chamber	[kW]	140	860	590
HOM losses per cavity	[kW]	8.2	36	34
Vacuum chamber contrib. to $(Z/n)_0$	[Ω]	0.24	0.78	0.55
Total cavity contrib. to $(Z/n)_0$	[Ω]	0.06	1.3	0.26
$(Z/n)_0$ expected	[Ω]	0.3	2.1	0.81
$(Z/n)_0$ threshold	[Ω]	0.33	2.3	1.1

dampers, have been studied in detail [17] and the corresponding growth rates calculated [16, 18]. The growth rate is in the worst case about 150 turns or 0.5 ms. In all cases the growth rate is faster than the natural damping rate, indicating the need for an active damping system described in the next section.

7.1 Feedback Systems

An active damping system has to be installed to control the multibunch instabilities. The voltage required per turn to stabilize the beam can be estimated from injection error considerations and the expected rise times of the instabilities. For errors of 1σ, in either the longitudinal or transverse directions, it is estimated that a longitudinal kick (expressed in volts) of 45 kV and a transverse kick (also expressed in volts) of 6 kV are needed.

The bandwidth of the damping system must be large enough to provide damping of all possible modes. For the case of 80 bunches in the machine a bandwidth of 12.5 MHz is required.

The longitudinal damping system will use mode detection electronics modelled after those developed for the CERN PS booster and subsequently employed in many other machines. RF cavities will be used in the longitudinal direction to provide the necessary damping voltage. Two cavities, with overlapping bandwidths of 8 MHz each, are sufficient to cover the entire 12.5 MHz range. These will be operated at approximately 800 MHz and will have loaded Q's on the order of 100 and shunt impedances of approximate 20 kΩ. The peak power required is then on the order of 50 kW. This is available using commercially available

Table 8.1: Characteristics of synchrotron radiation

Luminosity	[cm^{-2}s^{-1}]	10^{33}		10^{34}		6·10^{33}
Energy	[GeV]	3.5	8.0	3.5	8.0	5.3
Current	[A]	1.3	.56	2.6	1.1	1.0
Linear power density	[kW/m]	0.64	7.6	1.3	15.3	2.7
Photon flux	[10^{18} m^{-1}s^{-1}]	8.9	8.7	17.8	17.1	10.4
Critical energy	[keV]	1.5	18	1.5	18	5.1

standard UHF TV klystrons.

The transverse damping system will be similar in design to that presently in use in the NSLS booster. The transverse kicker plates will be approximately 1 m in length with a gap of approximately 40 mm. Proper design gives an input impedance of 50 Ω, thus the peak power required will be approximately 1 kW.

8 Vacuum System

The machine aperture of

$$a_H = \pm 50 \text{ mm}; \quad a_V = \pm 30 \text{ mm}.$$

has been defined by the ultimate symmetric machine requirements (Table 3.2).

The vacuum system design is based on LEP experience, with the exception of the vacuum chamber which follows HERA using copper. This is imposed by the large power deposited by the synchrotron radiation along the outer wall of the chamber. The characteristics of the synchrotron radiation are given in table 8.1 for the various machines considered. The chamber is designed for more than 20 KW/m power losses and, thanks to the low desorbtion coefficient of copper, the integrated dose required to store design currents with reasonable lifetime is only 10 A·h.

The two rings will have the same chamber for standardization and in order to avoid a special cooling circuit for aluminium. Moreover no lead shield is required with the copper chamber because of the large photon absorption coefficient [20].

The pumping system follows the LEP design. One ion pump (30 ℓ/s) is installed per drift chamber (or dipole) and two NEG ribbons are mounted in a separate pumping channel of the vacuum enclosure. This provides an average pumping speed of 500 ℓ/s/m. A model of cross section of the magnet-vacuum chamber assembly is presented in fig. 8.1.

At high luminosity the power deposited by parasitic mode losses in the vacuum chamber is considerable, 600 kW in the low energy ring of the asymmetric machines and 850 kW in each ring of the symmetric version. A special effort is required in order to develop bellows capable of coping with the high HOM power.

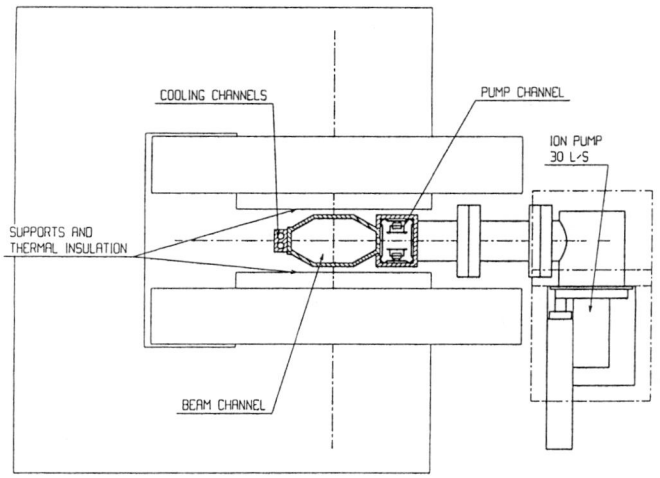

Figure 8.1: Copper vacuum chamber in dipole magnet.

This heat load must be added to the synchrotron radiation heat deposition. At full luminosity each vacuum chamber has to be cooled individually, at lower intensities they can be connected in series.

9 Background Considerations

An important criterion in the design of a high performance collider and detector for B-physics is a small diameter beam pipe around the interaction point and a low rate of synchrotron radiation photons and off momentum electrons hitting the beam pipe or absorber masks. We have investigated these two questions for the reference machine for a beam pipe of 25 mm radius and four absorber masks for synchrotron radiation at ±30 cm and at ±80 cm from the beam crossing point. The layout of the interaction region with beam pipe, masks and quadrupoles is shown in fig. 9.1.

Synchrotron Radiation
The main source of synchrotron radiation in the interaction region are the final focus quadrupoles $Q1 - Q4$. The β-function, and therefore the beam size, is largest in these quadrupoles and particles in the halo of the beam are bent strongly. The radiation due to the tilted detector solenoid can be neglected.

A strong source of photons scattered into the detector are the collimators. They are made from 8 mm of copper coated with 2 mm of tungsten, and should shield the direct radiation from the quadrupoles, but should not be hit by particles in the 10 σ tail of the bunch. A preliminary study using a modified version of

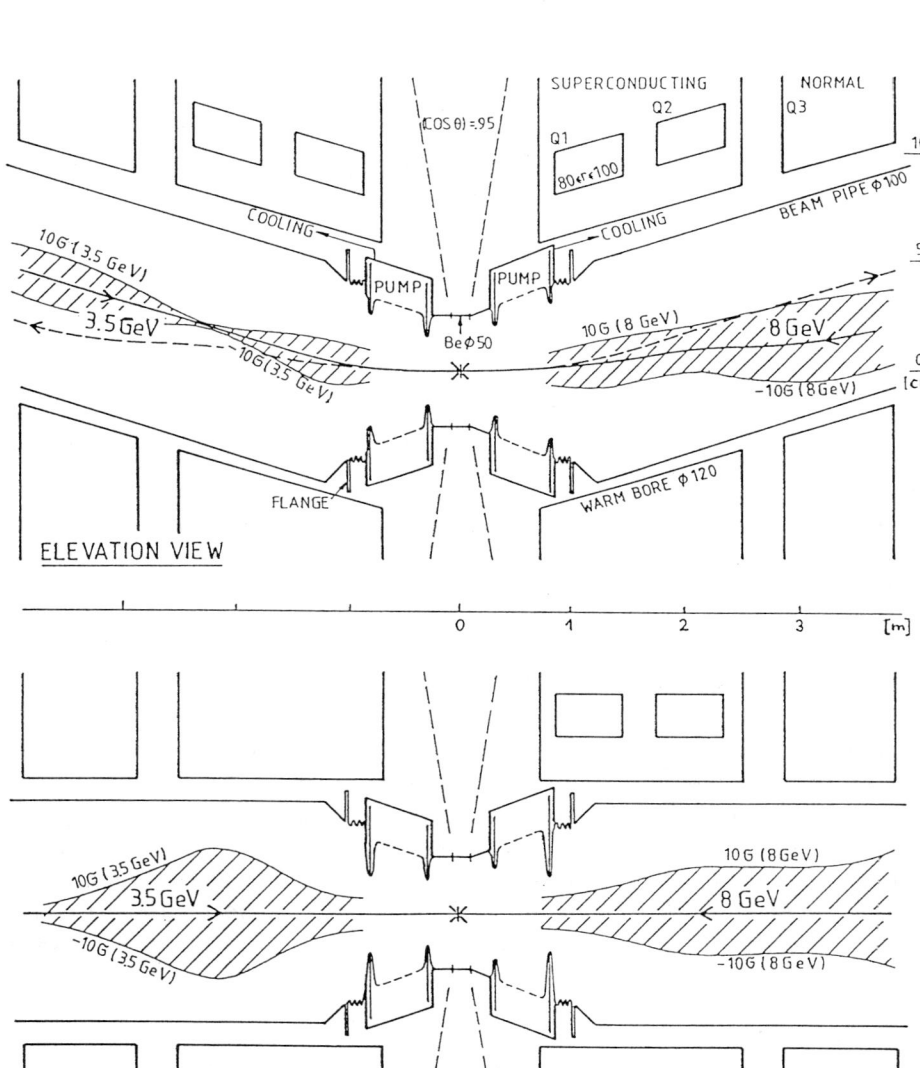

Figure 9.1: Layout of the interaction region.

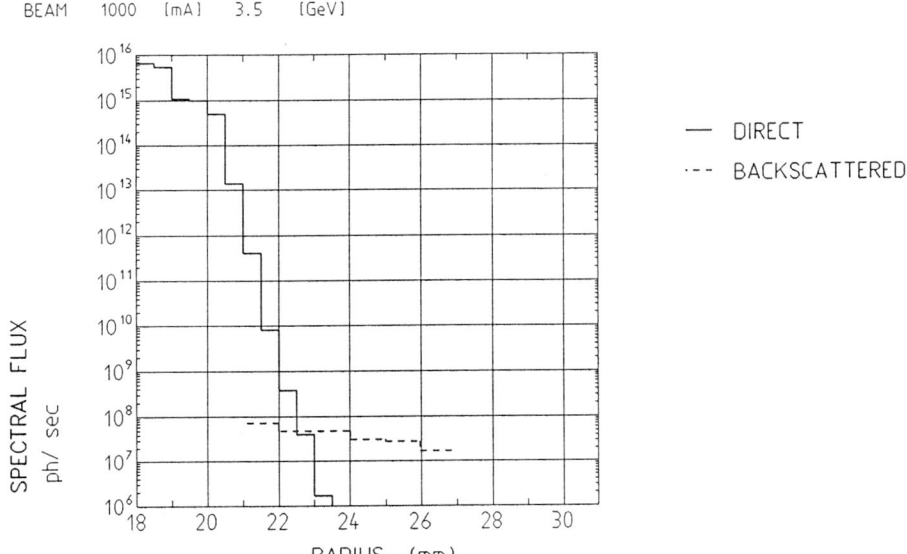

Figure 9.2: Number of photons hitting the beam pipe as a function of the beam pipe radius.

the program of ref. 21 has shown that circular collimators at $z = \pm 30$ cm with a vertical opening of ± 15 mm (3.5 GeV) and ± 20 mm (8 GeV) are necessary to shield radiation in the vertical plane. Radiation in the horizontal plane is shielded in addition by elliptical collimators at $z = \pm 80$ cm with horizontal openings of ± 16 mm. The number of direct and of backscattered photons from these masks into the detector ($z=\pm 8$ cm) as a function of the beam pipe radius is shown in fig. 9.2. A beam pipe radius of 25 mm is seen to be reasonable.

The spectrum of the photons scattered into the detector from the 8 GeV beam after passing through the beryllium beam pipe of 500 μm thickness is shown in fig. 9.3. The peak around 8 keV is from the Cu X-rays. The integral above E_γ=10 keV is of order 10^7 photons s^{-1}A^{-1}.

Another problem with the intense synchrotron radiation is the photodesorption in the beam pipe. For this, low energy photons ($E_\gamma \gtrsim 10$ eV) are important. The absorber masks are hit by 10^{16} photons/s with $E_\gamma > 10$ eV and in order to ensure a good vacuum (10^{-9} Torr) at the interaction point during storage ring operation, it is necessary to install pumps with a capacity of at least 300 ℓ/s close to the absorber masks. It has been estimated that a partial pressure of 10^{-9} Torr of CO or H_2 can be maintained after $10 - 100$ A·h of beam operation.

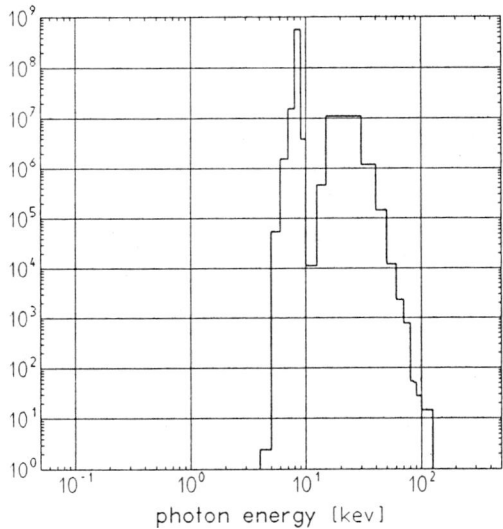

Figure 9.3: Photon spectrum of backscattered synchrotron radiation from the masks into the detector.

Beam Pipe Heating

Resistive heating in the 16 cm long berillium beam pipe (radius 2.5 cm) will produce 1.3 W of heat in case of the reference machine. This is increased by a factor of three if the number of bunches is reduced to 48 keeping the same average current. Smaller beam pipe diameters, shorter bunches, or more current would also increase the heating. For the ultimate asymmetric machine the heat load increases from 1.3 W to 11 W. If no care is taken to properly design the collimators geometry, the HOM losses could even be larger.

Off Momentum Particles

The numbers of off-momentum particles which hit the beam pipe wall in the vicinity of the interaction point have been estimated. The main locations along the circumference that contribute to the particle loss in the interaction region are located roughly 40 m and 70 m from the I.P., in the dispersion suppressors. Additional scrapers, installed 180° in betatron phase away from the I.P., would be effective in protecting the synchrotron radiation collimator, reducing the incident flux to less than 10^4 s^{-1}.

References

[1] The members of the machine study group of the CERN/PSI Collaboration were:
CERN: B. Autin, Y. Baconnier, P. Bossard, M. Boutheon, C. J. Brunet, P. Bryant, R. Cappi, J.P. Delahaye, B. De Raad, M. Fidecaro, R. Giannini, O. Gröbner, A. Hofmann, K. Hübner, J. Jowett, P. Lefèvre, D. Möhl, H. Moshammer, F. Pedersen, T. Pettersson, S. Pichler, W. Pirkl, J.P. Potier, T. Risselada, J.P. Riunaud, R. Schmidt, T. Taylor, R. Valbuena, T. Wikberg, B. Zotter
PSI:R. Abela, S. Adam, B. Berkes, H. Braun, J. Crawford, R. Eichler, K. Gabathuler, W. Joho, P. Marchand, S. Milton, T. Nakada, L. Rivkin, U. Schryber, K. Wille

[2] "Proposal for an Electron Positron Collider for Heavy Flavour Particle Physics and Synchrotron Radiation", PSI-PR-88-09 (July 1988)

[3] "Feasibility Study for a B-Meson Factory in the CERN ISR Tunnel", CERN 90-02 and PSI PR-90-08 (March 1990)

[4] A. Garren et al., "An asymmetric B-meson factory at PEP ", Proc. of the 1989 IEEE Particle Acc. Conf., pg 1847 (1989), and
S.Krishnagopal and R. Siemann, "Beam Energy Inequality in the Beam-Beam Interaction", Cornell CLNS 89/967, (1989)

[5] See the contributions on this subject in these Proceedings.

[6] Y. Baconnier and D. Möhl, "On the choice of parameters for a B-factory in the ISR tunnel", CERN -PS/AR Note 90-02.

[7] K. Wille, "Proposal for a High Luminosity Electron Positron Collider at PSI with an Option for Symmetric and Asymmetric Collision Mode", PSI Note PR-89-16 and Proc. of the Int.Conf. on High Energy Acc., Tsukuba, Japan, 1989.

[8] T. Risselada and T. Taylor, "Intersection Region Optics and Separation for an Asymmetric Energy B-factory in the ISR Tunnel", CERN-PS/PA Note 90-06.

[9] P. Marchand, "RF accelerating system for the BFI", PSI Technical Note, TM–12–90–01

[10] M. Billing, "Higher Mode Power Loss Limitations for Beam Currents in CESR", Cornell Internal Note CBN 84-15 (1984)

[11] L. I. Schiff, Rev. of Sci. Instr. 17 (1946) 6

[12] A. Hofmann, CERN LEP-TH Note 4 (1982)

[13] R. Cappi, R. Giannini, P. Marchand, and S. Milton, "Single Bunch Collective effects in BFI", CERN-PS/PA Note 90-01.

[14] A. Chao and J. Gareyte, SLAC Note SPEAR-197, PEP-224 (1976)

[15] R. Cappi, "Trapped Ion Effects in BFI", CERN-PS/PA Note 90-02.

[16] P. Marchand, "Design of the RF Accelerating System for the Storage Ring of the B–Meson Factory Proposed at PSI", PSI Technical Note TM–12–89–04

[17] P. Marchand, "Damping of the Parasitic Modes in the 500 MHz RF Cavities for the Storage Ring of the B–Meson Factory Proposed at PSI", PSI Technical Note, TM–12–89–06

[18] P. Marchand, "The Collective Phenomena in the B–Meson Factory Storage Ring Proposed at PSI", PSI Technical Note, TM–12–89–05

[19] S. Milton, "Multibunch Instability/Damping for the BFI", PSI-Note TM-12-90-02.

[20] O. Gröbner, H. Schuhbäck, R. Valbuena, T. Wikberg, "Design Considerations for the Vacuum System of a Beauty factory in the ISR", CERN Technical Note AT-VA/06/mpt-56.

[21] P. Münger and G.V. Holtey, Programme SYNBACK, CERN/LEP-B1/85/8.

B-FACTORY PLANS AT CORNELL UNIVERSITY

Maury Tigner
Cornell University, Ithaca, NY, 14853

OBJECTIVES

Ulitmate Goal: Asymmetric rings, 8 x 3.5 GeV, to be placed in CESR tunnel.

Immediate Goal: Symmetric, 2 ring machines capable of operation in CM energy range 9 to 14 GeV with Luminosity $10^{34} cm^{-2} s^{-1}$ at 11 GeV in CM to be converted t asymmetric operation as understanding warrants.

PARAMETERS

Following is a very preliminary parameter list and IR optics showing geometry and beam envelopes.

PARAMETER LIST

PRELIMINARY EXAMPLE #1
FLAT CROSSING BEAMS, EQUAL ENERGIES

E[GeV](beam energy)	5.3
$\mathcal{L}[10^{34}/cm^2 s]$(luminosity)	1.0
$(\xi/\beta^*)(1+r)[m^{-1}]$(lum. coefficient)	3.0
n_b(number bunches)	640
r (aspect ratio)	0
$N[10^{11}]$ (e/bunch)	0.73
I_{tot}[A](current in one beam)	2.9
Circumference [m]	768
β_h^*[m](IP focusing param.)	0.65
β_v^*[cm]	1.0
θ_c[mr](crossing angle)	±8.0
σ_l[cm](bunch length)	1.0
$\varepsilon_h[10^{-7}m]$(emittance)	1.05
$\alpha\ [10^{-2}]$(momentum compaction)	0.87
η^*[m](dispersion at IP)	0
Q_s(synchrotron tune)	0.06
Q_h(betatron tune)	10.7
$\sigma_E/E[10^{-4}]$(energy spread)	5.6
U_0[MeV/rev](SR loss)	1.04
P_{SR}[MW](per beam)	3.0
V_c[MV](cavity voltage)	10.6
$<Z/n>$[ohm](effective long. imped.)	0.65
P_{hom}[MW](per beam)	0.1
n_c(number of cavities per ring)	10
λ_{rf}[cm](rf wavelength)	60

WHILE SOME OF THESE PARAMETERS PRESENT TECHNICAL CHALLENGES, EACH OF THEM HAS BEEN, OR CAN BE, INVESTIGATED IN AN EXISTING FACILITY

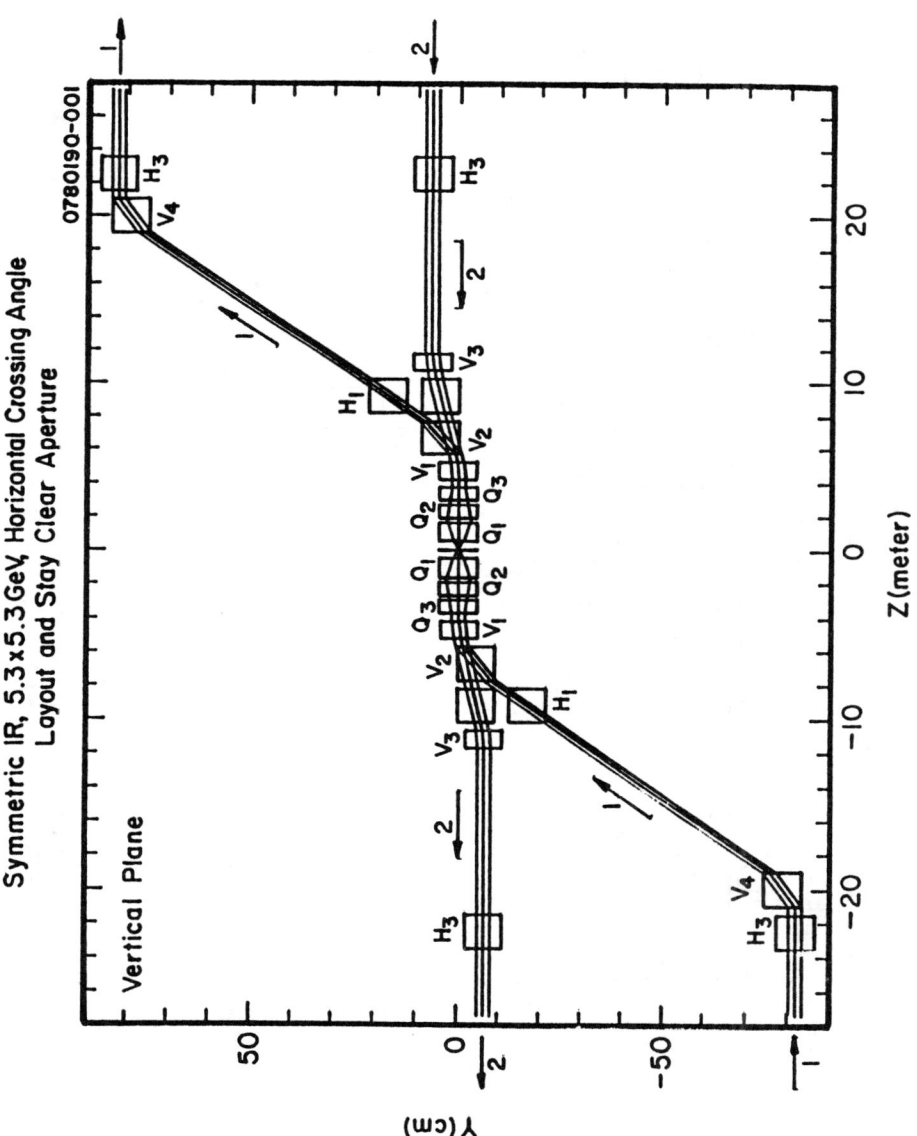

564 B-Factory Plans at Cornell University

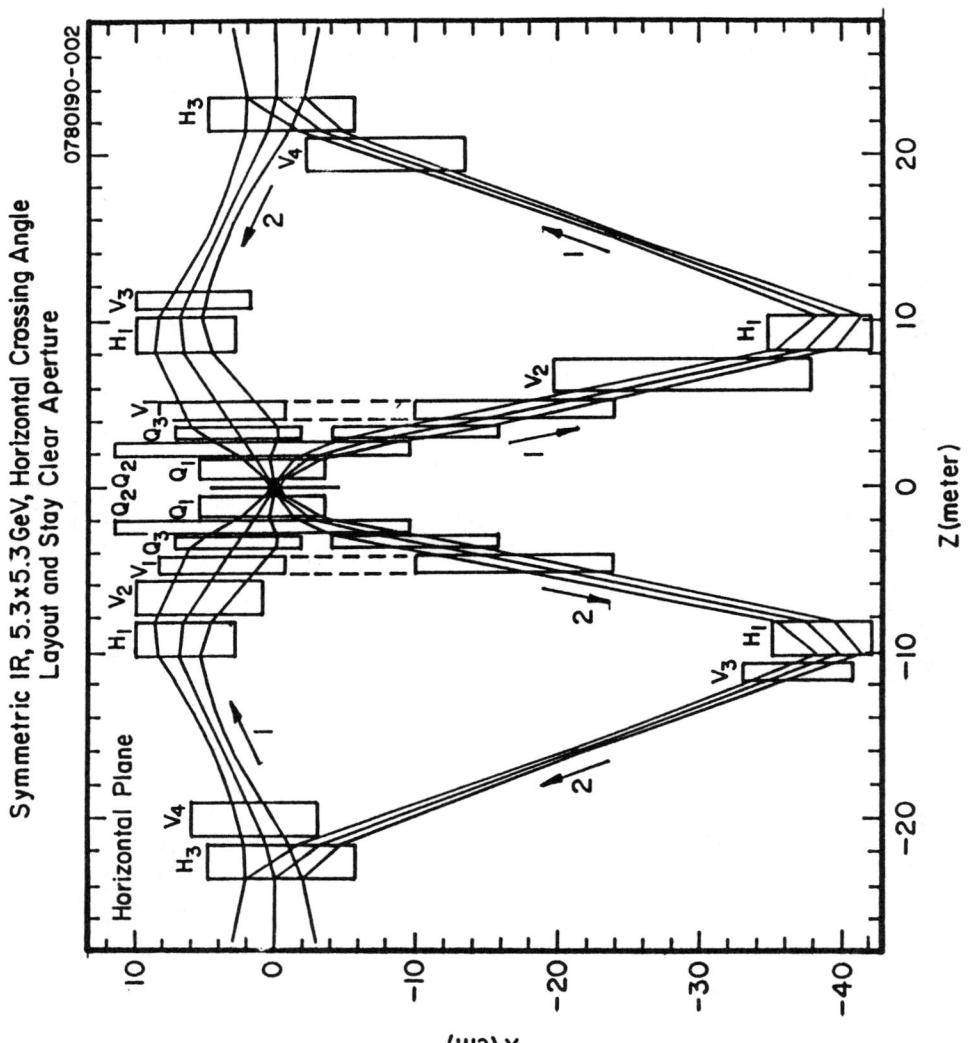

STUDY FOR AN ASYMMETRIC B-FACTORY

DESY B-Factory Study Group [1]
Deutsches Elektronensynchrotron DESY, Notkestrasse 85,
D-2000 Hamburg 52, West Germany.

Abstract

We report upon a feasibility study for an asymmetric $e^+ - e^-$-collider with beam energies of 10 GeV and 2.8 GeV. The PETRA storage ring is to be used to store the high energy beam. For the low energy beam, a new storage ring of 288 m circumference is to be built. The facility is designed for a peak luminosity of $\mathcal{L} = 3 \cdot 10^{33} cm^{-2} s^{-1}$.

1. INTRODUCTION

The study work at DESY on an asymmetric double ring $e^+ - e^-$ collider concentrates on designs with a high asymmetry of the beam energies E_1, E_2 (E_1/E_2 between 3.6 and 5.1). Using the existing PETRA ring at DESY for the high energy (HE) 10 GeV electrons and a new small ring for low energy (LE) positrons at an energy of 2.8 GeV allows to operate the asymmetric collider near the optimum Lorentz-boost of $\beta \cdot \gamma = 0.6$ for the investigation of CP-violation in the b-quark system [1].

Furthermore key experiments such as $B_s - \bar{B}_s$ mixing which require a large Lorentz-boost, are within the range of such a high asymmetry machine.

The special requirements and consequences for the accelerator design that arise from this high energy asymmetry are being studied and will be discussed below.

The luminosity $\mathcal{L}$ is given in equation (1). The current (I_1), the maximum tolerable tune shift $(\Delta \nu_1)$, the beam energy in units of the restmass γ_1 and the vertical beta function at the interaction point β_{y1}^* refer to the high energy beam. κ is the aspect ratio of both beams, r_0 the classical electron radius and e the elementary charge.

$$\mathcal{L} = \frac{I_1 \cdot \Delta\nu_1 \cdot \gamma_1 \cdot (1+\kappa)}{2er_0\beta_{y1}^*} \qquad (1)$$

The luminosity is proportional to the product of the beam energy, beam current and tune shift.

According to measurements in PETRA [2] the beam-beam tune shift limit becomes larger for higher energies. Experimental data fit well the law $\Delta\nu \sim \gamma^{3/2}$.

---

[1] K. Balewski, C. Geyer, B. Holzer, E. Jaeschke, D. Krämer, H. Nesemann, D. Proch, J. Sekutowicz F. Willeke and S.G. Wipf

Thus the required current in the HE ring decreases considerably as the energy asymmetry of the collider increases. The design of the LE ring is to be optimized so that the corresponding increase in the LE beam current does not limit the performance.

The total beam currents in both rings will be limited by coupled bunch instabilities unless they are damped by a broadband feedback system. Any B-factory with large luminosities relies on the technical feasibility of such very powerful damping systems.

At DESY we considered carefully two designs with different energy asymmetries, E_1/E_2 being 5.1 (investigated in [3]) and 3.6. Since in the lower asymmetry a higher current is needed to maintain the luminosity the reduction of the rf power to balance the synchrotron radiation losses is much lower than naively expected from the law $P \sim \gamma^4$. Moreover for low asymmetry the internal damping of the beam is reduced in the high energy beam. This drastically reduces the instability threshold for higher order coupled bunch modes for which there will be no feedback available. Therefore one is forced to reduce the cavity impedance but this increases the rf power dissipated in the cavities. Higher order coupled bunch instabilities are expected to limit the maximum beam current in low asymmetry solutions whereas the beam currents in high asymmetry solution will be limited only by the amount of available rf power. Therefore in the designs studied at DESY the reduction of rf power due to reduction of asymmetry from 5.1 to 3.6 amounted to only 30%. We draw the conclusion that the dependence of luminosity on the energy asymmetry is very weak.

The two beams have to be separated quickly after head-on collision. This is necessary to avoid parasitic beam-beam crossings and to focus the HE beam close to the interaction point IP without damaging the low energy beam. Since for low asymmetry the beams differ less in rigidity, separation becomes more difficult and requires more space assuming the synchrotron radiation power generated in the separator magnets remains constant. Furthermore the low beta quadrupoles for the low energy beam become less effective with lower asymmetry which has to be balanced by making them longer. As a result, the β- function at the IP , β^*, has to become larger which requires another increase in beam current to maintain the luminosity. All these problems are solvable if a magnetic separation at a high asymmetry is used.

The synchrotron radiation background problem is reduced considerably in a high asymmetry situation. For constant beam separation the magnetic fields can then be weaker which reduces the sum of the radiated power from both high and low energy beams. This eases the shielding of the detector for scattered photons from the high energy beam. The small magnetic rigidity allows to use permanent magnet low beta quadrupoles for the LE beam. They are very compact so that it is possible to place them close to the IP. For low asymmetry designs one is forced to use superconducting quadrupoles which imposes a

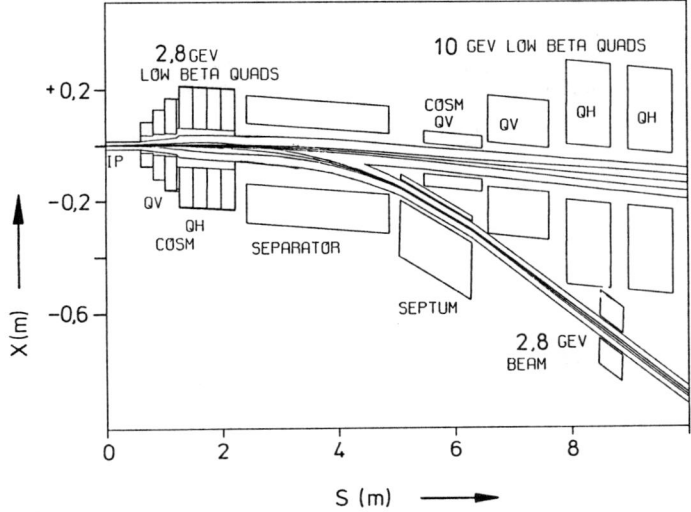

Figure 1: Layout of the interaction region and beam separation

large technical problem arising from the need to shield the cold beam tube (at liquid He temperature) from the large amount of synchrotron radiation power generated in the interaction region.

MAIN PARAMETERS AND LAYOUT

The present design of the DESY B-meson factory study is based on the lattice of the PETRA ring as an electron storage ring and a new additional small storage ring of 288m circumference for the positrons. The beam energies are 10 GeV and 2.8 GeV respectively and - assuming that a beam-beam tune shift of $\Delta \nu = 0.04$ can be reached in both rings - the beam currents for a luminosity of $\mathcal{L} = 3 \cdot 10^{33}$ are 1.2 A in the HE and 2.2 A in the LE machine. The vertical β-function at the interaction point is $\beta^*_{y,HE} = 4cm$ and $\beta^*_{y,LE} = 2cm$ respectively, with an aspect ratio of the beam cross section of one to ten.

The high energy ring is identical with the PETRA storage ring. The low energy ring consists of two half rings which are joined by S-shaped insertions at the interaction region and its opposite part in the lattice where the rf system is located. This asymmetric arrangement allows for both head-on collisions with a magnetic beam separation and collisions at a finite crossing angle (fig 1). The layout of the interaction region is based on the use of two permanent magnetic quadrupoles. Because of their high fields and small overall dimensions they

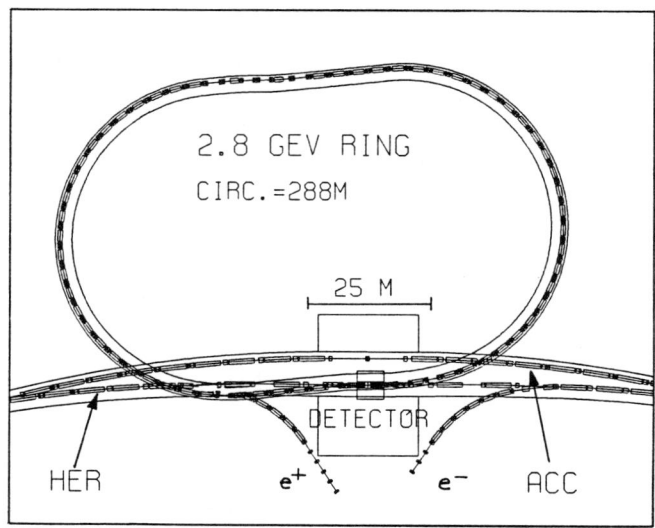

Figure 2: 2.8 GeV positron ring and interaction region located at the PETRA SE-hall

provide early focussing of the low energy beam. They are off-centre with respect to the head-on collision axis and act as combined function magnets. These two permanent magnets are followed by a conventional large aperture Panofsky-type quadrupole within which the LE beam is deflected to the high field region and quickly separated completely from the HE beam. With a minimum production of synchrotron radiation background this leads to a fast separation of more than 10 σ at a distance of 2.4 m from the interaction point and allows to distribute the total beam current over 480 (60) bunches in the HE (LE) ring. The last element of the separation scheme is a magnetic septum that affects only the LE beam and guides it to the normal FODO structure in the arcs.

The HE beam is focussed by conventional quadrupole magnets except for the half of the first low beta quadrupole which is a permanent magnet as well.

The horizontal emittances for both beams are adjusted to $1 \cdot 10^{-7} radm$ and $2 \cdot 10^{-7} radm$ respectively.

The layout of the S-shaped LE ring at its location at the **PETRA SE** hall is shown in fig 2. The main parameters of the B-meson factory are listed in table I.

Table 1: Main Parameters of the Asymmetric B-meson Factory		
	High Energy Ring	Low Energy Ring
particles	Electrons	Positrons
Beam Energy E/GeV	10	2.8
Circumference L/m	2304	288
Harmonic Number	3840	480
Beam Current I/A	1.2	2.2
Number of Bunches	480	60
Particles per Bunch	$1.2 \cdot 10^{11}$	$2.5 \cdot 10^{11}$
Hor. Emittance $\varepsilon_x/radm$	$1.2 \cdot 10^{-7}$	$2.3 \cdot 10^{-7}$
$\varepsilon_z/\varepsilon_x$	0.10	0.10
$\beta_x^*/m, \beta_z^*/m$	0.40, 0.04	0.20, 0.02
Tunes Q_x, Q_z, Q_s	25.70, 23.8, 0.047	9.19, 9.28, 0.033
Chromaticities ξ_x, ξ_z	$-51.6, -54.9$	$-18.8, -23.1$
Beam-Beam Tuneshift $\Delta\nu_x$	0.04	0.04
Beam-Beam Tuneshift $\Delta\nu_z$	0.04	0.04
Long. Damping Time τ_s/ms	16.5	9.1
Circumferential Voltage U/MV	9	2.7
Bunch Length σ_s/mm	16.5	14.1
Beam Power Loss P_{syn}/MV	5.6	0.66
Luminosity $\mathcal{L} = 3 \cdot 10^{33} cm^{-2} s^{-1}$		

RF-CONCEPT AND MAXIMUM BEAM CURRENT

The rf requirements of high input power and low parasitic impedance can be fullfilled by normal conducting 500 Mhz single cell cavities. It has been shown that an effective damping of the parasitic resonances is possible by applying a passive damping system [4].

Possible current limitations are due to single bunch and coupled bunch instabilities.

Coupled bunch instabilities are driven by parasitic modes of the cavities. The cavities have only one important longitudinal and one transverse resonance. The 18 monocell cavities of the high energy ring set the threshold current of 120 mA in 480 bunches.

In the case of the low energy ring the two important modes are less strongly damped but tuned by special antennas to reduce their influence on the beam. This allows to store a total current of 500 mA in 60 bunches.

Single bunch instabilities are caused by the broadband impedance of the ring. The contribution of the cavities has been calculated. The impedance of the vacuum chamber was estimated by using existing measurements and calculations of the impedance of PETRA. Using these impedances neither transverse nor longitudinal single bunch instabilities are of importance up to the design currents of 1.2 A in 480 bunches of the high energy ring and up to 2.2 A in the 60 bunches of the low energy ring.

The current in both rings is limited by coupled bunch instabilities. An active damping has to be installed in both rings to achieve the design current.

BEAM-BEAM INTERACTION

We follow the HERA concept in assuming that each beam in an asymmetric collider has individual beam-beam tune shift limits according to the beam energy, damping time, energy spread and collision frequency. We however impose the condition that the beam cross section at the IP must be the same for both beams in order to balance the nonlinear effects of the beam-beam interaction. This is suggested by experience with hadron colliders [5].

In order to make an estimate of the HE ring beam-beam tune shift limitation, we make use of extensive studies which have been performed on beam-beam interaction at PETRA [2]. These experiments indicates, that within the energy range of 7 GeV to 17 GeV the maximal tolerable tune shift scales as $\Delta \nu = 0.5 \cdot \sqrt{\delta}$ as proposed by Talman and Keil [6]. In addition scaling the tune shift with the square root of the number of crossing points is considered to be an experimentally well-established scaling law [7].

For a beam energy of 10 GeV and one beam beam crossing per revolution we are therefore confident that a beam-beam tuneshift in PETRA of $\Delta \nu = 0.04$ can be achieved.

Using the same scaling law for the low energy beam we arrive at the same number $\Delta \nu = 0.04$ for the maximum tolerable beam beam tune shift.

Recently it has been demonstrated that in a double ring collider with different circumferences of the two rings, the coherent beam beam interaction is enhanced by the fact that one bunch in one ring interacts with several bunches in the other one [8]. A coupled system with many degrees of freedom is formed which is - in the model of rigid bunches with a linear beam-beam force - subject to many additional linear sum-resonances. The width of the resonances increases with the strength of the beam-beam tuneshift parameters and the resonances overlap covering the whole tune range for a certain tune shift limit. We have investigated this effect for the present design where each bunch of the LE ring is coupled to eight bunches in the HE ring respectively. Fig 3 and 4 show a comparision of this effect for the cases of one to one and one to eight bunches. Plotted is the tune shift for which the motion becomes linearly unstable versus the tune of the LE beam. The number of resonances is multiplied but as their widths have decreased the total accessible range of tunes is only reduced by a factor

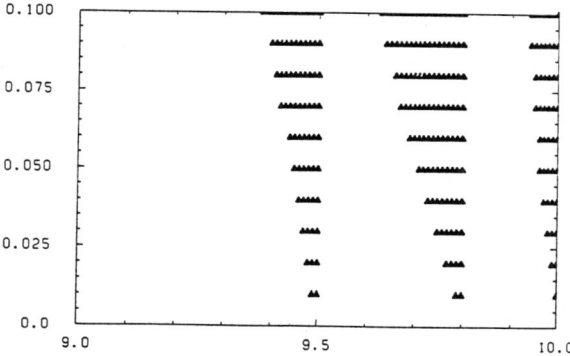

Figure 3: Coherent Tune Shift limit versus Tune for Equal Circumferences

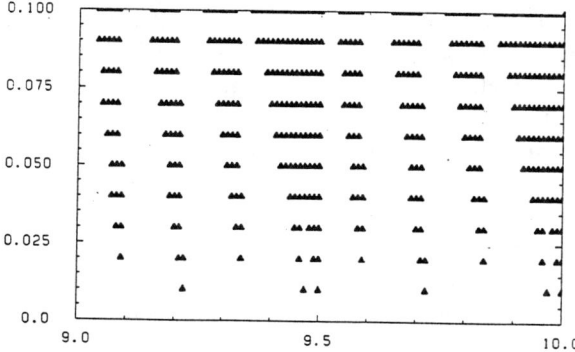

Figure 4: Coherent Tune Shift Limit versus Small Ring Tune in the Case of a Ratio of 1 to 8 in the Circumferences of the Two Rings

of two for the anticipated tune shift values of $\Delta\nu = 0.04$. The space between the resonances appears to be sufficient for operation. The "eight to one case" offers in some respect even more freedom to adjust the tune of the LE ring.

Concerning the HE ring, the coherent beam-beam interaction did not impose a performance limitation for PETRA as an $e^+ - e^-$ collider, though the achieved tune shift exceeded the level for which linear instability was expected by a factor of two for certain operation points [9].

We therefore - according to our present knowledge- do not consider coherent beam-beam interaction to impose a performance limitation for an asymmetric collider.

SYNCHROTRON RADIATION BACKGROUND

The fast separation in combination with the large beam currents generates high synchrotron radiation in the interaction region. In order to keep this radiation background as small as possible the separation is done by combined function magnets. The HE beam stays in the low field region which leads to a drastic reduction of the radiated power. The total amount of synchrotron radiation power produced in the separation magnets upstream from the detector is 3.7 kW for the LE beam and 15 kw for the HE beam. The maximum critical energies of the emitted spectra are 2.4 keV and 27 keV respectively.

With the present arrangement of the magnets and the synchrotron radiation masks, half of the synchrotron radiation power of the HE beam (6.9 kW) will travel through the interaction region without hitting any collimator. Even for this reduced synchrotron radiation power, one needs a very efficient collimator system to shield the detector. Fig 5 shows the collimator system and the total synchrotron radiation power on the different collimators. The detector beam pipe can be shielded against primary synchrotron radiation. The amount of scattered photons coming from the edges of the masks will have to be reduced by optimizing the position and material of the collimator system to reach a synchrotron radiation background as low as 10^9 photons per second with energies above 1 keV as required by the experiment[10].

We have calculated the power deposited at the interaction region caused by higher order modes and transient fields to be 14 kW.

SUMMARY

The layout of an asymmetric collider for an electron beam with a beam energy of E = 10 GeV and a positron energy of 2.8 GeV designed for a luminosity of $\mathcal{L} = 3 \cdot 10^{33} cm^{-2} s^{-1}$ has been presented. The most crucial aspects of the design are the high total beam currents of 1.2 A and 2.2 A for the high and low energy beams respectively and the synchrotron radiation produced in the interaction region. In order to store the high total beam currents an active feedback system is necessary to damp the lowest order coupled bunch instabilities. Other intensity limiting effects such as single bunch instabilities are not expected to be a problem for the anticipated parameters. The most serious problem at present is the synchrotron radiation masking systems which protect the detector. Although we are able to shield the detector from first order synchrotron radiation, there is still a considerable second and higher order photon flux into the detector.

Due to the larger asymmetry the layout of the interaction region can be achieved with room temperature magnets. The overall lattice design is fairly conservative.

Summerizing we conclude that the concept of high asymmetric design provides not only favorable solutions of many aspects like interaction region layout, beam separation and high beam intensity but also gives unique experimental possibilities due to the high Lorentz boost.

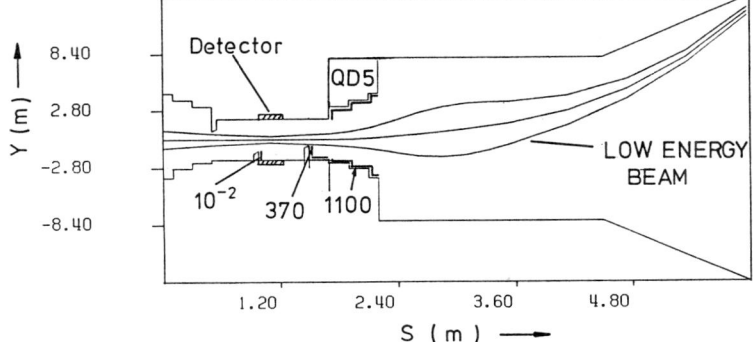

Figure 5: Synchrotron radiation masking system. Plottet are the collimators, the first permanent LE quadrupole, QD5 and the 10 σ envelopes of the LE beam. The radiation power hitting the masks is given in Watts

References

[1] H. Nesemann, M. Reidenbach. W. Schmidt-Parzefall, H.-D. Schulz and F. Willeke, "Ideas for Future B-Physics at DESY", DESY 89/80 (1989) W. Schmidt-Parzefall, these proceedings

[2] A. Piwinski, "Review of Mini Beta Luminosities in PETRA at Different Energies", DESY 81-066 (1981)

[3] K. Balewski, C. Geyer, G. Hemmie, B. Holzer, E. Jaeschke, R.-D. Kohaupt, D. Kraemer, H. Nesemann, D. Proch, J. Sekutowics, and F. Willeke, 14th Int. Conf. on High Energy Acc. Tsukuba, Aug. 1989

[4] C. Yinghua, D. Proch, J. Sekutowicz, "The Slotted Cavity-A Method of Broadband HOM Damping" Internat. Conf on High Energy Accelerators, Tsukuba (1989)

[5] L. Evans, private communication

[6] E. Keil and R. Talman, CERN-ISR-TH/81-33 (1981)

[7] D. Rice, these proceedings

[8] K. Hirata and E. Keil, CERN/LEP-TH/89-76 (1989)

[9] A. Piwinski priv. com.

[10] W. Schmidt-Parzefall priv. com.

ASYMMETRIC B-FACTORY PROJECT AT KEK

Y. Funakoshi, M. Anami, A. Asami, S. Enomoto, K. Hanaoka, T. Kageyama,
K. Kanazawa, E. Kikutani, J. Kobayashi, H. Koiso, S. Kurokawa, T. Ohsawa,
K. Oide, S. Sakanaka, K. Satoh, T. Shidara, M. Suetake, F.Takasaki,
Y.Yamazaki

KEK, National Laboratory for High Energy Physics, 1-1 Oho, Tsukuba-shi, Ibaraki-ken, 305 Japan

ABSTRACT

The present machine design of the future asymmetric B-factory at KEK is reviewed. Beam energies and a design luminosity are chosen as $8 \times 3.5 GeV^2$ and $1 \times 10^{34}/cm^2/sec$ in due consideration of requirements from physics experiments. In the design, it is aimed to squeeze the vertical beta-function at IP (β_y^*) to $1cm$. The beam-beam limit is assumed to be 0.05 with the requirement that the bunch length should be half of β_y^*; i.e. $0.5cm$. Flat beams are chosen. In order to attain the design luminosity, beam currents of $2.6A$ for the $3.5GeV$ ring and $1.1A$ for the $8GeV$ ring are necessary. An all RF buckets filling scheme and a crab crossing scheme are adopted to realize the short bunches and to decrease RF voltages. The damped cavity is adopted to suppress the coupled bunch instability. Growth rates of the instability are estimated in the case that the damped cavities are used. Further extensive studies and R&D's are necessary to realize the design luminosity.

1 INTRODUCTION

A working group of the future asymmetric B-factory machine was organized in June of 1989 at KEK. The group has laid out several machine designs and has studied several problems to realize design parameters. At first, the group investigated the possibility to construct a small ring as a low energy ring of the B-factory attached to the TRISTAN Main Ring which was also considered to be converted to a high energy ring of the factory. However, this option was abandoned after K. Hirata and E. Keil offered the so-called Hirata-Keil effect[1] on coherent beam-beam effects in the case that two rings have different circumferences. On the other hand, parallel to the study of the machine, physics issues of the future asymmetric B-factory have been studied actively by a working group of the physics division at KEK. The optimum energy asymmetry and the necessary luminosity which the future B-factory machine should satisfy were studied by this group and the results were offered to the working group of the accelerator. This report summarizes the present machine design for the future asymmetric B-factory at KEK, which has been made considering the above restrictions. The members of the working group are shown in the following with the items of works. In the workshop this status report was presented by one of the authors, Y. Funakoshi.

Coordinators	S. Kurokawa, K. Satoh
Lattice	K. Oide, K. Satoh, Y. Funakoshi, H. Koiso
Instability	Y. Funakoshi, K. Hanaoka, S. Sakanaka
Beam-Beam Effect	H. Koiso, K. Oide
RF System	E. Kikutani, M. Suetake, T. Kageyama, Y. Yamazaki
Vacuum System	K. Kanazawa
Beam Background	F. Takasaki
Injector Upgrade	T. Ohsawa, T. Shidara, S. Enomoto, J. Kobayashi, M. Anami, A. Asami
Operation Scheme	Y. Funakoshi, K. Satoh, H. Fukuma
Cost Estimation	K. Satoh
Group Secretary	Y. Funakoshi

2 REQUIREMENTS FROM PHYSICS EXPERIMENT

In the working group of the physics division at KEK, requirements to future asymmetric B-factory machines have been studied. The results of the study are summarized in Table 1. As seen in the table, physics people hope to carry out two different kinds of experiments. The first one is on CP violations. The optimum energy is around $8 \times 3.5 GeV^2$ and the required luminosity is about $1 \times 10^{34}/cm^2/sec$. In this case, the experiments will be done on the $\Upsilon(4S)$ resonance. The second experiment is on $B_s - \bar{B}_s$ mixing. The optimum energy is about $12 \times 2.47 GeV^2$ and the required luminosity is about $1 \times 10^{33}/cm^2/sec$ in this case. The experiments will be done on the $\Upsilon(5S)$ resonance. In the estimation of the required luminosities, the machine running time of $10^7 sec/year$ was assumed. Of the two options of the machines, the first one is much more difficult to be realized than the second. However, to detect CP violating effects in the B meson system is the main physics motivation to construct asymmetric B-factory machines. Therefore in the stage of the basic design, we have considered the first option as the design goal as shown in the following section.

Table 1 Requirements from physics experiments

Energy(GeV^2)	Luminosity($/cm^2/sec$)	Physics
$8 \times 3.5(\Upsilon(4S))$	1×10^{34}	CP violation
$12 \times 2.47(\Upsilon(5S))$	1×10^{33}	$B_s - \bar{B}_s$ mixing

3 DESIGN PRINCIPLE

In due consideration of the physics motivations described above, we have adopted the following as the basic principle for machine designs.

Basic principle: As the goal of the machine design, we aim to design the machine with the peak luminosity of $1 \times 10^{34}/cm^2/sec$ and the energies of $8 \times 3.5 GeV^2$.

4 GENERAL CONSIDERATION FOR LATTICE DESIGN

On some reasonable assumptions, the luminosity of ring colliders is expressed as:

$$L = \frac{\gamma}{2er_e} \frac{I\xi(1+r)}{\beta_y^*} \quad (1)$$

where γ is the beam energy in a unit of the electron rest mass, r_e the electron classical radius, e the elementary charge, I the total beam current, ξ the beam-beam parameter, r the beam aspect ratio and β_y^* the vertical beta-function at the collision point. We have fixed the beam energies and the luminosity at $8 \times 3.5 GeV^2$ and $1 \times 10^{34}/cm^2/sec$, respectively. The next problem is with which combination of four parameters ξ, I, r and β_y^* we should aim at the luminosity. The world record of the luminosity about $1 \times 10^{32}/cm^2/sec$ has been attained by CESR at Cornell University, which is working in the same energy region as ours. We must increase the luminosity by 2 order. Through many experiences of $e^+ - e^-$ colliders, it cannot be expected that the factor $\xi(1+r)/\beta_y^*$ would be increased drastically. Moreover the bunch current of the beam is strongly restricted due to the beam-beam effects. Therefore the major hope to attain the design luminosity is to increase the total beam current by increasing the number of bunches in the ring. On the other hand, beam currents more than ampere are necessary in order to obtain the design luminosity. Several estimations show that the problem of the coupled bunch instability is very severe with such high currents. In addition to this, other problems such as a rise of the vacuum pressure, heavy loads to injectors, a high beam background and so on may occur in the high current operation. So we have adopted the following strategies to aim at the luminosity of $1 \times 10^{34}/cm^2/sec$.

At first we make efforts to increase the factor $\xi(1+r)/\beta_y^*$ in order to obtain the luminosity of $1 \times 10^{34}/cm^2/sec$ with as small currents as possible.

After such efforts to decrease required currents, problems originated in the high current operation should be considered.

In this section, the basic ideas for a lattice design to maximize the factor $\xi(1+r)/\beta_y^*$ are described. And the problem related to the coupled bunch instability is mentioned in the next section together with a model design based on the basic ideas presented in this section.

4.1 Choice of basic parameters

Now the problem we should solve is how to maximize the factor $\xi(1+r)/\beta_y^*$. Firstly we have set up the goal of β_y^* at $1cm$. One of the main difficulties in squeezing beta-functions to smaller values is to decrease dynamic aperture. This rather aggressive value could be achieved if we have only one collision point. (The difficulty of the final focus system design originated in the energy asymmetry could be solved by adopting a finite angle crossing scheme discussed later.)

Another problem in squeezing β_y^* to a smaller value is that the beam-beam limit becomes lower with a small beta value unless the bunch length is not sufficiently short. A empirical rule suggests that the bunch length should be less than half of β_y^*, if one wishes to obtain beam-beam parameters more than 0.05. The origin of this rule is usually explained in terms of a synchro-beta resonance due to the phase modulation caused by the beam-beam force. According to the empirical rule, the beam-beam limit has been assumed to be 0.05 with the requirement that the bunch length should be half of β_y^*; $i.e.$ $0.5cm$.

We have chosen the more conventional flat beams rather than round beams, although Eq. (1) shows that the luminosity can be doubled if the round beams are used instead of the flat beams.

With the above rather aggressive parameters, still rather high currents are necessary to get the luminosity of $1 \times 10^{34}/cm^2/sec$; $1.1A$ for the $8GeV$ ring and $2.6A$ for the $3.5GeV$ ring.

4.2 All RF buckets filling scheme

The use of short bunches brings two additional problems. Firstly in order to keep the bunch length shorter, higher RF voltages or more number of RF cavities are required. This seems to induce the coupled bunch instability more easily. The second difficulty is that the use of short bunches lowers the threshold impedance for the turblent bunch lengthening. In order to overcome these two problems derived from the use of short bunches, we have adopted an all RF buckets filling scheme. With this scheme, the number of bunches increases extremely up to the same number as a harmonic number. Since the bunch currents become smaller, the threshold impedance for the bunch lengthening becomes higher. On the other hand, small bunch currents require low emittance so as to keep the beam-beam parameters same. The low emittance lattice results in a small momentum compaction factor. And to use a small momentum compaction factor is helpful to decrease necessary RF voltages. In the followings, somewhat quantitative discussions will be given.

If the total current, I is given, the number of particles per bunch is expressed as:

$$N = \frac{IS_B}{ec}, \qquad (2)$$

where S_B is a bunch spacing, e the elementary charge and c the speed of light. If a beam-beam parameter ξ and I are given, horizontal emittance is also given by:

$$\varepsilon_x = \frac{r_e N}{2\pi\gamma\xi(1+r)} \qquad (3)$$

or

$$\varepsilon_x = \frac{r_e I S_B}{2\pi\gamma ec\xi(1+r)}. \qquad (4)$$

This shows the horizontal emittance must be proportional to the bunch spacing, provided that ξ and I are conserved. In the case of a normal cell ring, ε_x and a momentum compaction factor α are approximated as:

$$\varepsilon_x \cong 2\frac{R}{\nu_x^3}\sigma_e^2 \qquad (5)$$

$$\alpha \cong \frac{1}{\nu_x^2}, \qquad (6)$$

where R is a ring average radius, σ_e the energy spread and ν_x the horizontal betatron tune. And so α is expressed as:

$$\alpha \cong (\frac{\varepsilon_x}{2R\sigma_e^2})^{\frac{2}{3}}. \qquad (7)$$

On the other hand, if the RF voltage is determined by the requirement to the bunch length, it is given by:

$$V_c = \frac{T_0 E}{eh\omega_0}\frac{\alpha c^2\sigma_e^2}{\sigma_z^2}, \qquad (8)$$

where T_0 is a revolution period, E the beam energy, h a harmonic number, ω_0 a revolution frequency and σ_z the bunch length. Combining the above equations, V_c can be expressed as:

$$V_c = \frac{E}{e}\frac{T_0}{h\omega_0}\frac{c^2\sigma_e^{\frac{2}{3}}}{(2R)^{\frac{2}{3}}}\frac{\varepsilon_x^{\frac{2}{3}}}{\sigma_z^2}, \qquad (9)$$

or

$$V_c \propto \frac{S_B^{\frac{2}{3}}}{\sigma_z^2}. \qquad (10)$$

With shorter bunches, V_c becomes higher. But that can be compensated to some extent by decreasing the bunch spacing.

The threshold impedance of the turbulent bunch lengthening is given by:

$$\frac{Z_\parallel}{n} = \frac{(2\pi)^{\frac{3}{2}}E/e\alpha\sigma_z\sigma_e^2}{I_b R}, \qquad (11)$$

where I_b is a bunch current. Since α is proportional to $\varepsilon_x^{\frac{2}{3}}$ or $S_B^{\frac{2}{3}}$ and I_b to S_B,

$$\frac{Z_\parallel}{n} \propto \frac{\sigma_Z}{S_B^{\frac{1}{3}}}. \tag{12}$$

With shorter bunches, $\frac{Z_\parallel}{n}$ becomes lower. But that can be compensated to some extent by decreasing the bunch spacing.

4.3 CRAB CROSSING SCHEME

The all buckets filling scheme carries another problem of beam separation. For the purpose of avoiding additional undesirable collisions or the effect of the long-range beam-beam force, a finite angle crossing scheme must be challenged. If the finite angle crossing scheme is adopted, a crab crossing scheme[2,3] is also necessary in order to avoid a geometrical loss of the luminosity and to prevent the beam-beam limit from lowering due to the synchro-beta resonance. A finite angle crossing scheme is also desirable to simplify the design of IR and to squeeze the β-function easily. In addition, the finite angle crossing scheme seems to be helpful to suppress the beam background, since no bending magnet for the beam separation is needed near IP.

5 DESIGN EXAMPLE

On the basis of the basic ideas presented in the previous section, a model design has been made. Fig. 1 shows a schematic view of our factory. Both rings have the same circumference of 805m. The magnet configuration of the $8GeV$ ring is drawn in the sketch of Fig. 2. The 3.5 GeV ring has almost the same configuration except that vertical translation magnets in the figure are used for horizontal translations in the $3.5GeV$ ring and that the length of dipole magnets is somewhat shorter than those of the $8GeV$ ring. The $8GeV$ ring has been chosen for positron in consideration of ion trapping effects. The collider has only one interaction point (IP). The two rings cross horizontally at IP with a angle of $50mrad$. The $8GeV$ ring will be mounted above the $3.5GeV$ ring except in IR. Both rings will be housed in the same tunnel which will be newly constructed. The existing two injectors for the TRISTAN main ring are also used for the injectors of the new B-factory rings. The TRISTAN accumulation ring (AR) is suitable for the injector of the $8GeV$ ring, since the maximum energy of AR is $8GeV$. The injection to the $3.5GeV$ ring is performed directly from the linac (Linac) through the AR tunnel. Linac should be upgraded in its energy from the present one of $2.5GeV$ to $3.5GeV$. The optics of the rings are shown in Fig. 3 and 4. The normal cells are made of FODO cells and the phase advances per cell are 90 deg in both directions. Machine parameters of the present design are listed in Table 2.

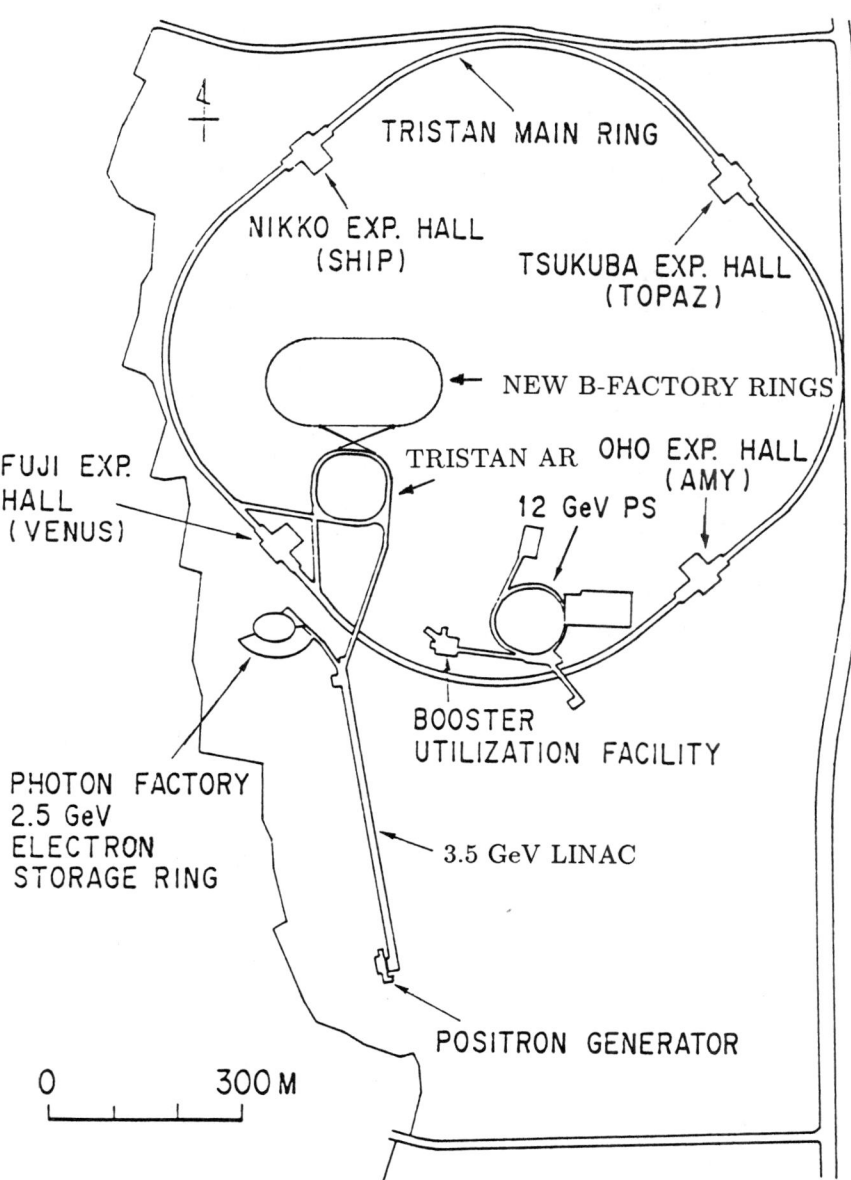

Fig. 1. Schematic view of the future B-factory machine

Table 2 Machine parameters of asymmetric B-factory

Energy	E	3.5	8	GeV
Circumference	C	805		m
Luminosity	L	1×10^{34}		$/cm^2/sec$
Beam-Beam limit	ξ_x/ξ_y	0.05/0.05		
Beta functions at IP	β_x^*/β_y^*	1.0/0.01		m
Total current	I	2.6	1.1	A
Natural bunch length	σ_z	0.50		cm
Energy spread	σ_ε	8.9×10^{-4}		
Bunch spacing	S_B	0.6		m
Particles/bunch	N	3.3×10^{10}	1.4×10^{10}	
Emittance	$\varepsilon_x/\varepsilon_y$	$1.9 \times 10^{-8}/1.9 \times 10^{-10}$		m
Synchrotron tune	ν_s	0.044	0.042	
Betatron tune	ν_x/ν_y	27/25	25/27	
Momentum compaction	α	1.9×10^{-3}	1.7×10^{-3}	
Energy loss/turn	U_0	1.2	6.2	MeV
RF voltage	V_c	16.8	36.0	MV
RF frequency	f_{RF}	500		MHz
longitudinal threshold	Z/n	0.27	1.3	Ω
Energy damping decrement	T_0/τ_E	3.4×10^{-4}	7.7×10^{-4}	
Crab angle	θ_x	25		$mrad$
Crab cavity frequency	f_x	500		MHz
Crab cavity voltage	V_x	0.93	2.1	MV

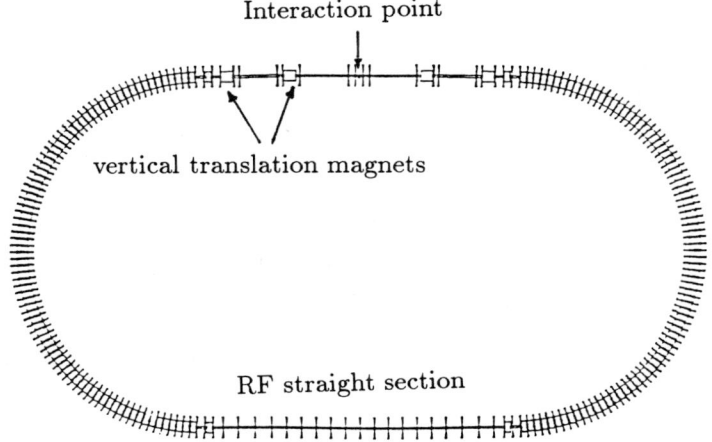

Fig. 2. Magnet configuration of $8GeV$ ring

Y. Funakoshi *et al.* 583

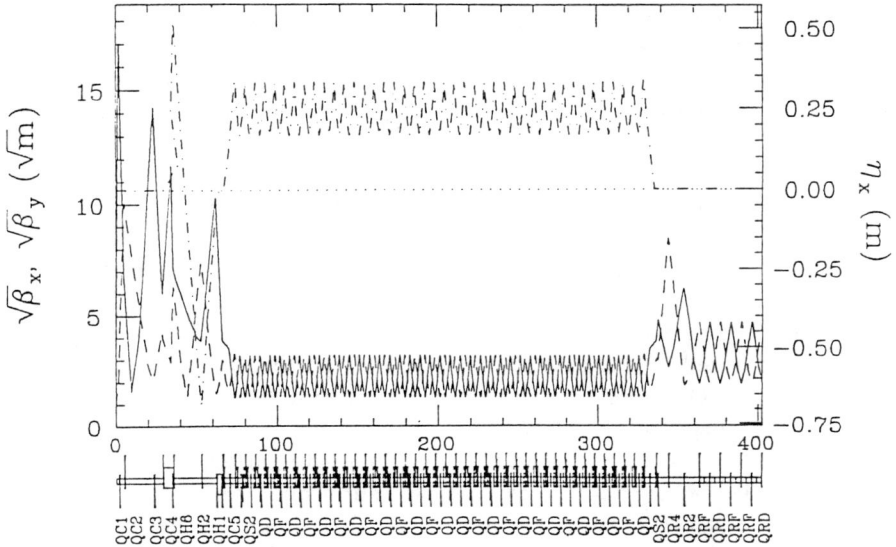

Fig. 3. Optics of $3.5 GeV$ ring

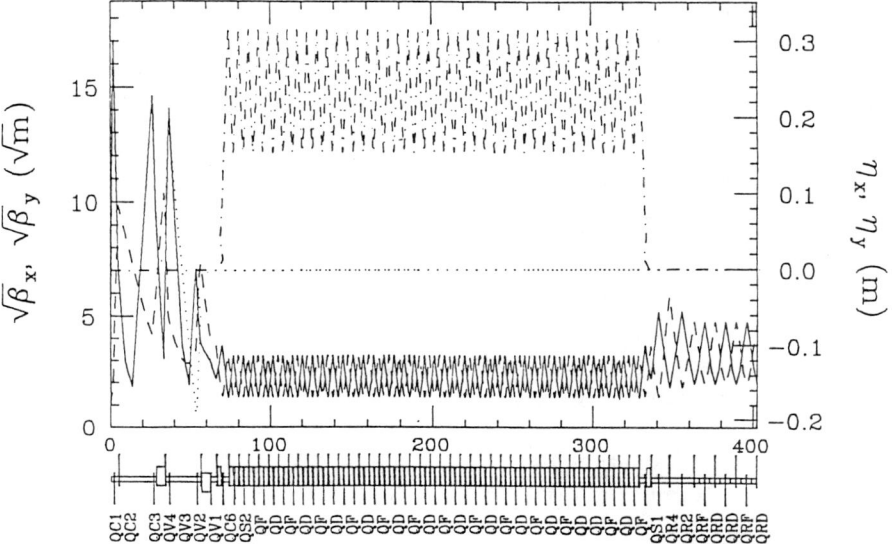

Fig. 4. Optics of $8 GeV$ ring

5.1 RF System and Coupled Bunch Instability

There are some methods to cure the coupled bunch instability such as to use a feedback system, to make missing buckets, to make a tune spread in a bunch or to make tune differences among bunches and so on. However, the essential and basic method to suppress the instability is to decrease the higher order mode (HOM) impedances of RF cavities, which are the main source of the instability. We have already made a effort to decrease the accelerating voltages in order to decrease the number of the cavities within the limit of the use of short bunches. Now, the RF system itself must be examined. Generally speaking, the HOM impedances are determined by three factors; the number of cavities, R/Q values and Q values of HOM's, although resonant frequencies of HOM's are also important parameters in considering the instability. Most of the efforts are usually devoted to decrease the Q values of the three factors, since drastic decreases of the Q values can be expected. Several methods to decrease the Q values have been proposed such as by using conventional HOM couplers, transverse slits or beam pipes with a large diameter attached to RF cavities to remove HOM fields from the cavities. Recently another possibility to damp the Q values of HOM's were proposed by R. B. Palmer for the linear collider (damped cavity)[4]. The basic idea of the damped cavity is to remove the HOM fields through slots which are cut on disks of disk-loaded type cavities. This type of cavity has been studied actively at KEK[5] both by calculations using the 3D code MAFIA[6] and by experiments using a 2 cell test cavity. Both of the calculations and the experiments showed that the Q values of the most dangerous a few HOM's can be damped to 10~30 without affecting the fundamental mode at all. Therefore the damped cavity is considered to be the most promising candidate for the RF system of the future B-factory machine at KEK.

Table 3 RF related parameters

Energy	E	3.5	8	GeV
RF frequency	f_{RF}	500		MHz
Number of cells/cavity	N_{ce}	3		
Operating mode		π-mode		
Shunt impedance	R_s	22.8		$M\Omega$
Maximum voltage/cavity	V_{max}	1.85		MV
Maximum cavity loss/cavity	P_c	150		kW
Accelerating Voltage	V_c	16.8	36.0	MV
Radiation loss	P_R	3.1	6.8	MW
Number of cavities	N_{ca}	9	20	
Window power	P_W	494	490	kW

Table 3 shows parameters related to the RF system of the B-factory using 3-cell damped cavities. Normal conducting copper cavities have been adopted.

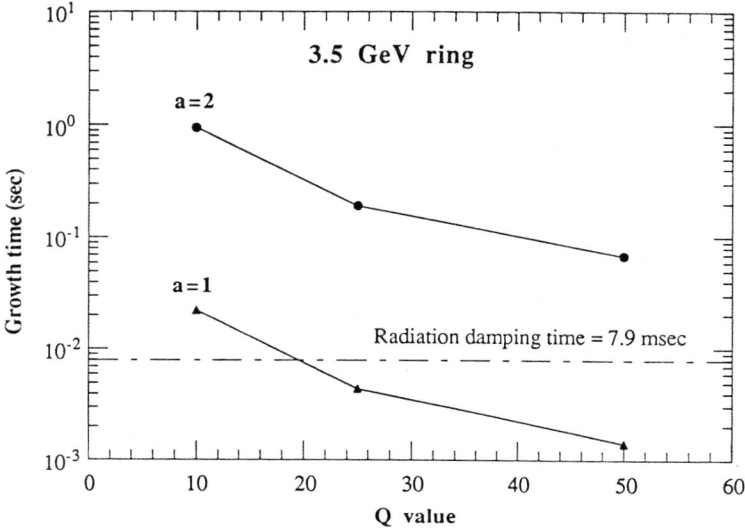

Fig. 5. Growth time of longitudinal coupled bunch instability

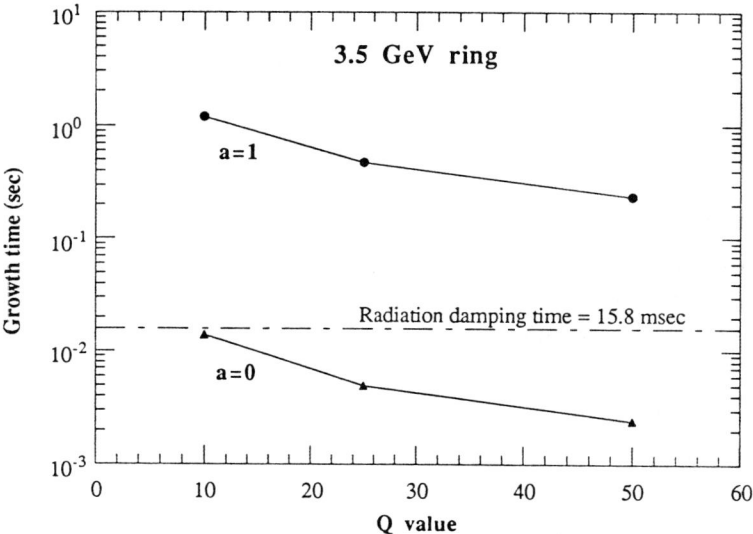

Fig. 6. Growth time of transverse coupled bunch instability

The shunt impedance of the fundamental mode was calculated by the URMEL code[7] to be $22.8 M\Omega$. Generally speaking, a field gradient of the cavity is restricted by the discharge in the cavity, the cooling power of the cavity and the permitted power of the input coupler window. The present value of the field gradient ($1.85MV$/cavity) is somewhat conservative for the discharge limit or the cooling power limit. Therefore if the higher power windows are available, somewhat higher field gradient could be attained, which results in a smaller number of RF cavities. In the calculation of the window powers HOM losses were ignored. At present the number of the cavities are 9 and 20 for the $3.5GeV$ ring and the $8GeV$ ring, respectively. And the total lengths of the RF systems are $10.8m$ for the $3.5GeV$ ring and $24m$ for the $8GeV$ ring. Both are well fitted in the rings, since about $70m$ of the RF straight sections are reserved in our design.

Using the damped cavity, there is a possibility that the coupled bunch instability could be suppressed without any other methods for cures. In order to investigate this possibility the growth rates of the instability were calculated by the ZAP code[8] using the parameters of our present design. The HOM impedances of the damped cavity were calculated by the URMEL code and in the growth rate calculations 7 longitudinal and 5 transverse HOM's were taken into account. Only the case of the $3.5GeV$ ring was examined. The instability of $8GeV$ ring is less serious, since the beam of $8GeV$ ring is more rigid, the beam current smaller and the damping time shorter, although the HOM impedances are twice higher than those of $3.5GeV$. Fig. 5 and 6 show the results of the calculations. As seen in the figures, only a=1 (dipole) mode of the longitudinal modes and a=0 (rigid dipole) mode of the transverse modes are problems. For those two modes, the effects of damping of the Q values by the use of damped cavity are still marginal, even if the Q values of all HOM's are decreased down to $10 \sim 30$. However, if other methods for cures are used together with the damped cavity, it seems easy to overcome the coupled bunch instability.

5.2 Dynamic Aperture

In our design the distance between IP and the edge of the first quadrupole magnet is about 1.6m. The natural chromaticities are about -35 and -100 in horizontal and vertical planes, respectively. The chromaticities are compensated by four families of sextupoles located close to the quadrupoles in the FODO cells. In order to investigate dynamic aperture of the rings, particle tracking studies have been performed about the $8GeV$ ring. The studies showed that the dynamic aperture of the $8GeV$ ring is wider than $10\sigma_x$, $10\sigma_y$ and $10\sigma_\varepsilon$. No tracking study about $3.5GeV$ ring has been performed yet.

5.3 Effect of Emission of Bremsstrahlung

In our luminosity region it is a important process that electrons or positrons emit Bremsstrahlung due to the electro-magnetic field of the counter-rotating beam at IP. If a particle loses sufficiently large energy, it can be lost from the beam. In our luminosity region this process may determine the beam life. Under the approximation that a particle is lost if it loses more energy than the energy aperture (ΔE_a) by this process, the cross section of particle loss is given by the following formula[9]:

$$\sigma = \frac{8\alpha r_e^2}{3}[(ln\varepsilon)^2 - (2ln\varepsilon + \frac{5}{4})(ln\frac{4E_1 E_2}{m_e^2} - \frac{1}{2}) - (\frac{\pi^2}{3} + \frac{3}{4}) + O(\varepsilon)] \quad (13)$$

where α is the fine structure constant, E_1, E_2 energies of beams, ε $\Delta E_a/E_1$ or $\Delta E_a/E_2$ and m_e the rest mass of electron. The results of the lifetime calculations using this formula are shown in Table 4. In the calculations the energy aperture of 1% or 3% was assumed. Since the cross section depends on $ln\varepsilon$, the dependence of the lifetime on the energy aperture is not so strong as is seen in the table. The calculations show that the luminosity lifetime is considerably short. Therefore we need powerful injectors in order to avoid a severe decrease of the integrated luminosity. The existing injectors should be upgraded in its intensity with regard to the particle loss rate shown here.

Table 4   Effect of emission of Bremsstrahlung

Energy	E	3.5	8	GeV
Beam current	I	2.6	1.1	A
Particles/beam	Np	4.35×10^{13}	1.84×10^{13}	
Energy aperture	$\Delta E_a/E$	0.01		
		(0.03)		
Cross section	σ	2.65×10^{-25}		cm^2
		(1.85×10^{-25})		cm^2
Luminosity	L	1×10^{34}		$/cm^2/sec$
Particle loss rate	$L\sigma$	2.65×10^9		$particles/sec$
		(1.85×10^9)		$particles/sec$
Beam life	τ_{Br}	4.6	1.9	$hour$
		(6.5)	(2.7)	$hour$

6   Remaining Problems and Necessary R & D's

6.1   Development of New RF cavity System

R & D of the RF cavity system is one of the most important points for the future B-factory project. For our present design, we need two different types of RF cavity; i.e. the damped cavity and the crab cavity.

In any design of the B-factory the high current operation with many bunches cannot be avoided, if one wants a high luminosity. And one of the most serious obstacles for the high current operation is the problem of the coupled bunch instability. Therefore, the conquest of the instability is a essential problem independent of factory designs. Some preliminary study about the damped cavity has shown the possibility to overcome the coupled bunch instability in our current region. However, in order to change the possibility into the certainty more extensive study has to be pursued for further developments of the damped cavity. Since in the study so far, Q-values of only a few HOM's were examined, Q-values of the rest modes should be investigated. In addition, another problem of the RF cavity system of the B-factory rings is that it has to be operated in the very high beam currents. The cavities must be tolerable for the heavy beam load. Especially for input couplers of cavities it is required that higher power than $500kW$ can pass through them stably in order to decrease both the broad band and the narrow band (HOM) impedances. The study of the damped cavity system is now in progress at KEK including a preparation for the beam test of 3-cell damped cavities in AR.

The possibility of the crab cavity is one of the key points of our present design. In other word, it could be said that in our design considerable part of the difficulty of the lattice design is squeezed in the crab scheme. Therefore, a high priority should be given to the developments of the crab cavity.

6.2 Development of New Vacuum System

A development of the vacuum system is also of importance for the future B-factory project. Requirements of the vacuum system mainly come from the vacuum pressure and the longitudinal broad band impedance. And the most severe requirements of the vacuum pressure seem to come from the suppression of the ion trapping effects and the beam background. Further studies must be done to fix the required vacuum pressure depending on the sort of the residual gas and the position of the rings. The possibility of the vacuum system to realize the tentative goal pressure of $10^{-10} \sim 10^{-9} Torr$ has been being studied in our situation of the extremely high beam currents.

6.3 Beam background

The beam background to the physics detector, particularly to a vertex detector, is one of the most severe difficulties for the future B-factory machine. The adoption of the finite angle crossing scheme is helpful to decrease the background in the sense that bending magnets for beam separation, in which beams emit a lot of synchrotron radiation, are not necessary. However, the effects of other sources of the beam background must be studied such as synchrotron radiation from quadrupoles in IR or from crab cavities and particles lost due to emission of Bremesstrahleng in colliding with the residual gas. Those studies should be done actively by both methods of simulations and experiments. Such studies are also important to offer a criterion for the design of the vacuum system.

6.4 Estimation of Ring Broad Band Impedance

The problem of bunch lengthening is one of the key points of our present design. Therefore, the longitudinal broad band impedance of the rings must be estimated as carefully as possibly. The design of ring elements such as the vacuum system or the devices for the cure of the ion trapping effect has to be done in due consideration of their impedances.

6.5 Beam-Beam Effects

In our design, it is an important supposition that the beam-beam parameter of 0.05 can be obtained if the bunch length is half of β_y^*. The effectiveness of this empirical rule should be investigated by the studies of the theory, simulations and experiments. The simulation studies on this problem are in progress at KEK. A preliminary result shows that the use of short bunches is effective to suppress the beam blow-up due to the beam-beam effect.

In principle the synchro-beta resonance, which comes from the beam-beam effect with a finite angle crossing, can be suppressed almost completely by using a crab crossing scheme. However, if there are some errors in the crab cavity system, the synchro-beta resonance might be induced owing to some slippage from the head-on collision. The tolerance for the errors of the crab scheme should be estimated by beam-beam simulations. Such problems are now under study.

6.6 IR design

A design of a physics detector, final focus quadrupole magnets, a crab cavity system, a vacuum system including a radiation masking system and a coupling compensation system must be done consistently. Of course this design study should be done parallel to the study of the beam background mentioned above.

6.7 Ion Trapping

The problem related to the ion trapping phenomena is now being studied using the present design parameters of the $8GeV$ ring mainly on the basis of the experiences at the KEK PF-ring. A preliminary and very rough estimation so far shows that the vacuum pressure of $10^{-11} \sim 10^{-10} Torr$ is necessary in order to suppress the two beam instability completely caused by trapped ions even if missing buckets , which are assumed to amount to $10 \sim 20\%$ of the full ring, are made and that assuming the vacuum pressure of $1 \times 10^{-9} Torr$, other methods, for instance a installation of ion cleaning electrodes, octupole magnets to make tune spread, a feedback system or RF quadrupoles to decouple the motion of each bunch, are also necessary in addition to the missing buckets methods. Further precise estimations about the effect of ion trapping should be done, together with the study of the methods for cures of the instability.

6.8 Coupled Bunch Effect

Together with the study of the damped cavity, the effectiveness of other methods to cure the instability should be estimated.

Recently importance of a transient effect for long bunch trains at injection was pointed out by K. A. Thompson and R. D. Ruth[10]. They found out that strong transverse wakes due to injection errors could be large enough to cause beam loss, even if all normal modes were stable. This transient effect should be studied also in our machine, since we use the very short bunch spacing.

6.9 Dynamic Aperture

Tracking studies for the $3.5 GeV$ ring should be done to examine the dynamic aperture of the ring. In addition, efforts to widen the dynamic apertures of both rings should be continued in order to enable the beam injections in the luminosity optics.

6.10 Injector upgrade

Linac and AR should be upgraded to be used as injectors for the B-factory rings. The energy of Linac should be increased from $2.5 GeV$ to $3.5 GeV$. A design for the energy upgrade of Linac is now in progress. As shown in the previous section, the particle loss rate of B-factory rings is rather high. Therefore the intensity of Linac should be increased so that it can compensate the particle loss at such high rate. So far as positron, a construction of a new damping ring might be necessary. AR should be also upgraded so as to enable the high current and the extremely multibunch operation.

7 FUTURE PROSPECTS

At KEK the operation of the TRISTAN has stepped into a new phase (phase II) with the start of the LEP. The operation in this phase (the luminosity run) is to be continued until the integrated luminosity of $300 pb^{-1}$ is accumulated. And the asymmetric B-factory is considered to be the phase III of the TRISTAN project.

An asymmetric B-factory project at KEK is now at a critical turning point. In the beginning of this year, the working group of the B-factory was reorganized and a new task-force has been established so that more intensive studies should be done. The task-force will make a more realistic and consistent design of the future B-factory by the end of this year or the beginning of the next year.

Some people at KEK have a plan to start the experiment of the B-meson physics with somewhat lower luminosity, say $3 \times 10^{33}/cm^2/sec$ and to aim at the luminosity of $1 \times 10^{34}/cm^2/sec$ after the accumulation of experiences and R&D's, although the design of only the machine with the luminosity of $1 \times 10^{34}/cm^2/sec$ is described in this report. It is now under discussion at KEK in what strategy we should aim at the machine with the luminosity of $1 \times 10^{34}/cm^2/sec$.

REFERENCES

1. K. Hirata, E. Keil, CERN/LEP-TH/89-54.
2. R. B. Palmer, SLAC-PUB 4707 (1988).
3. K. Oide, K. Yokoya, SLAC-PUB 4832 (1989).
4. R. B. Palmer, SLAC-PUB 4542 (1989).
5. M. Suetake, T. Higo, K. Takata, 7th symp. on accelerator Sci. Tech, Osaka, Japan (1989)p.103.
6. K. Klatt et al., Proc. of the 1986 Linear Accelerator Conf., SLAC 1986, SLAC report-303.
7. T. Weiland, Nucl. Instr. Meth. $\underline{216}$ 329 (1983).
8. M. S. Zisman, S. Chattopadhyay, J. J. Bisognano, LBL-21270 (1986).
9. K. Yokoya, private communication.
10. K. A. Thompson, R. Ruth, SLAC-PUB 4872 (1989).

NOVOSIBIRSK B - FACTORY

A.A. Zholents

Institute of Nuclear Physics, Novosibirsk, 630090, USSR

Abstract

Brief information about the B-factory, which is under the active study in the Institute of Nuclear Physics in Novosibirsk is presented.

Project Description

There are two major parts in the B-factory project in the Institute of Nuclear Physics in Novosibirsk. They are the injector for such a collider and the B-factory itself.

According to our plans the injector should consist of three main units. The first is a set of conventional linear pre-accelerators, which we will use for positron production and for the initial acceleration of electrons and positrons to the 510 MeV. The second is the damping ring with the nominal beam energy of 510 MeV. The third is the high gradient main linear accelerator, capable to accelerate the particles up to 8.5 GeV. The latter accelerator will based on new technology being developed for the linear collider VLEPP at the Institute of Nuclear Physics in Novosibirsk.

The design goals for the B-factory collider are high luminosity and small centre-of-mass energy spread of e^+e^- collisions. We consider a facility, consisting of two equal circumference storage rings with nominal beam energies of 6.5 GeV and 4.3 GeV. The main parameters of this machine are presented in the Table 1.

Table 1: Parameters of Novosibirsk B-factory

	Low-energy ring	High-energy ring
Energy, E[GeV]	4.3	6.5
Circumference, C[m]	655	655
Number of bunches, k_B	156	156
Particles per bunch, $N_b[10^{10}]$	9	6
Total current, I[A]	1	0.7
Longitudinal threshold impedance, $\left|\frac{Z}{n}\right|$ [Ω]	0.2	0.5
Emittance		
horizontal, ϵ_x [nm·rad]	8	6.5
vertical, ϵ_z [nm·rad]	0.25	0.25
Bunch length, σ_l [mm]	7.5	7.5
Energy spread, σ_ϵ [10^{-3}]	1	1
Energy loss / turn, [Mev]	1.2	2.7
Damping time, [msec]	17	13
Momentum compaction, α	0.002	0.002
Betatron tunes		
horizontal, Q_x	29	26
vertical, Q_z	20	13
Synchrotron tune, Q_s	0.028	0.028
RF parameters		
frequency, f_{rf} [MHz]	500	500
voltage, V_{rf} [MV]	8.8	15
Beta functions at IP		
horizontal, β_x [cm]	60	60
vertical, β_z [cm]	1	1
Horizontal dispersion function at IP, ψ [cm]	-40	40
Beam sizes at IP		
horizontal, σ_x [mm]	0.4	0.4
vertical, σ_z [mm]	0.0016	0.0016

Table 1: Continuation

	Low-energy ring	High-energy ring
Nominal beam-beam tune shift		
horizontal, ξ_x	0.012	0.012
vertical, ξ_z	0.05	0.05
Centre-of-mass energy spread, σ_W [MeV]	1.2	
Luminosity, L [cm^{-2}s^{-1}]	5×10^{33}	

Our approach to the realization of this parameter list was already presented in June, 1989 at the B-Factory Workshop in Blois, France [1]. After the Workshop no principal changes in the design were made with the exception of the synchrotron radiation masking scheme.

Our new masking scheme is based on the idea of vertical separation of flat synchrotron light beams in the vicinity of the interaction point (IP). Two masks with slits at different vertical position are placed symmetrically with respect to the IP (see Fig.1). A light beam approaching the IP is primary absorbed by the mask before the IP (whose slit lies above or below the beam). The remaining light which passes through the aperture of this mask can pass through the slit of the second mask. Thus, this radiation is absorbed far from the IP. With two such masks we can shield the vertex vacuum pipe from the direct synchrotron light as well as from the secondary scattering photons. In this scheme, photons have the possibility of hitting the vertex vacuum pipe only after the second scattering. This helps greatly reduce the background from the synchrotron radiation.

To produce the vertical separation for synchrotron light beams we plan to employ an S-bend in the vertical plane of the beam orbits. We can do this with a small tilt in the opposite directions of the first bending magnets in our separation scheme.

Acknowledgements

The author thanks all the members of B-factory group in Novosibirsk for their contributions to this work. He also wish to thank A. Dubrovin, V. Lebedev, A. Skrinsky and A. Vlasov for many useful discussions and G. Travish for his help in the preparation of this article.

References

[1] A.N. Dubrovin, A.M. Vlasov, A.A. Zholents, Interaction Region of 4x7 GeV Asymmetric B-Factory, Preprint INP 89-97, Novosibirsk 1989.

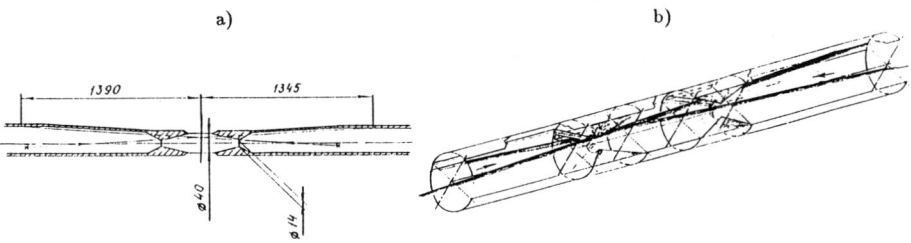

Fig.1 The masking scheme: a) the horizontal cross-section, b) an artistic view.
(all dimensions in mm)

AN ASYMMETRIC B FACTORY IN THE PEP TUNNEL*

A. Hutton

Stanford Linear Accelerator Center, Stanford University, Stanford, CA 94309, USA

M. S. Zisman

Lawrence Berkeley Laboratory, Berkeley, CA 94720, USA

ABSTRACT

This report addresses the feasibility of designing and constructing an asymmetric B-Factory—based on the PEP storage ring at SLAC—that can begin operation at a luminosity of $3 \times 10^{33} \mathrm{cm}^{-2}\mathrm{s}^{-1}$ and could ultimately reach even higher luminosity. Such a facility, operating at the $\Upsilon(4S)$ resonance, could be used to study mixing, rare decays, and CP violation in the BB system, and could also study tau and charm physics. The essential accelerator physics, engineering, and technology issues that must be addressed to successfully build this exciting and challenging facility are identified, and possible solutions, or R&D activities that will reasonably lead to such solutions, are described.

INTRODUCTION

It is now generally accepted that the best facility to study rare CP-violating B meson decays is a high-luminosity electron-positron collider in which the particles have different energies.[1] Such a B Factory based on the PEP storage ring at SLAC has been under study for about two years, originally as a collaboration among LBL, SLAC, Caltech, and many university groups, and recently joined by LLNL. The names of the participants in this study can be found in two internal reports,[2,3] which cover the design activities in more detail than can be contained in this paper.

The PEP storage ring is an ideal platform from which to launch an asymmetric B Factory facility, having a well-designed, flexible lattice with suitably long straight sections, a tunnel that permits the siting of a new low-energy storage ring without requiring extensive conventional facilities construction, and access to a powerful injector. These advantages, coupled with the existence—in close proximity—of a highly qualified and enthusiastic team of physicists from SLAC, LBL, LLNL, Caltech, and various University of California campuses, make this and ideal project for the SLAC site.

*Work supported by Department of Energy contracts DE-AC03-76SF00515 (SLAC) and DE-AC03-76SF00098 (LBL).

GENERAL

The project consists of two rings of equal circumference, both housed in the PEP tunnel. The High Energy Ring (HER) has a nominal energy of 9.0 GeV while the Low Energy Ring (LER) has a nominal energy of 3.1 GeV. An initial luminosity of 3×10^{33} cm^{-2}s^{-1}, with possible enhancement up to 10^{34} cm^{-2}s^{-1}, is required for an asymmetric storage ring at the $\Upsilon(4S)$ resonance. The design must be 'conservative' so that the initial luminosity can be achieved soon after turn-on. The injector will be the SLC, which has filling rates achieved today that meet the requirements for the B Factory. (This scenario does not preclude use of SLC for R&D for future linear colliders.) The infrastructure of PEP will be used wherever possible. The HER will use the PEP magnets while the LER will use some parts of the PEP vacuum system. The facility must be designed and built as a "factory," which means either adopting design parameters that are tried and tested or providing flexibility when we are obliged to operate in an unknown regime.

PARAMETER CHOICES

The general expression for luminosity in an asymmetric collider is cumbersome, involving various parameters of both beams at the IP. To simplify the choices, and to elucidate the general issues of luminosity for all B Factories, it is helpful to express the luminosity in an energy-transparent way. Here we express luminosity in terms of a single, common beam-beam tune shift parameter, ξ, along with a combination of other parameters taken from *either* the high-energy (e$^-$) or the low-energy (e$^+$) ring, irrespective of energy.

With a few plausible assumptions (e.g., complete beam overlap at the IP and equal beam-beam tune shifts for both beams in both transverse planes) such parameters as energy, intensity, emittance, and the values of the beta functions at the IP may be constrained to satisfy certain scaling relationships. It then becomes possible to express luminosity in a simple, energy-transparent form:[8]

$$\mathcal{L} = 2.17 \times 10^{34} \xi (1+r) \left(\frac{I \cdot E}{\beta_y^*} \right)_{+,-} \quad [\text{cm}^{-2}\text{s}-1] \quad (2.1-1) \qquad (1)$$

where

ξ is the maximum saturated dimensionless beam-beam interaction parameter (the same for both beams and for both the horizontal and the vertical transverse planes)

r is the aspect ratio characterizing the beam shape (1 for round, 0 for flat)

I is the average circulating current in amperes

E s the energy in GeV

β_y^* is the value of the vertical beta function at the IP in cm

The subscript on the combination $(I \cdot E/\beta_y^*)_{+,-}$ means that it may be taken from either ring.

A conservative value of 0.03 has been adopted for the beam-beam tune shift for both beams in both planes; practically every electron collider has reached this value. Flat beams ($r \approx 0$) are preferred, because the synchrotron radiation produced in the Interaction Region (IR) is about a factor of ten less for flat beams than for round beams.[4] This is because the angular divergence at the IP is smaller, reducing the beam sizes in the IR quadrupoles (which are also weaker), and thus minimizing the quadrupole synchrotron radiation. The smaller divergence also reduces the angle needed to separate the two beams. Finally an additional benefit is that the angular divergence of the synchrotron radiation is reduced, thereby easing the masking problems.

Given the distance to the nearest quadrupole, the values of β_y^* were chosen to ensure that the chromatic aberrations can be corrected. The smallest value must also be greater than the bunch length of 1 cm. The ratios of horizontal to vertical emittances and beta functions at the IP were chosen to be 25, comfortably less than the emittance ratio achieved routinely in PEP. A larger ratio would be easier for beam separation and for backgrounds, so this is a conservative choice.

These considerations lead to the parameter list for the collider shown in table 1. The single bunch current is less than that already achieved in PEP at the same energy. The new challenges are the high total current and the large number of bunches. The choice of positrons for the LER was to minimize the possible deleterious effects of ion trapping on collider performance by avoiding its occurrence in the lower energy of the two rings.

INTERACTION REGION LAYOUT

An interaction region with a crossing angle and crab crossing inherently creates the least synchrotron radiation but is totally untried. We believe that the beam dynamics must be studied in detail, the collisions simulated, and a test carried out on an existing collider before we could conservatively base our design on it. The head-on crossing scheme is based on the horizontal S-bend layout, which separates the masking of the two beams and is easily convertible to a crossing angle design if beam dynamics studies permit us to adopt this solution. We are continuing to study both options and are designing the facility to be compatible with either insertion. The global parameter list is already the same for both insertions and we are proceeding towards a unified layout that can be converted with a minimum of difficulty. This capability will be available when the machine is built and provides insurance against unforeseen problems.

It is now clear that the interaction region will be designed around special magnets in the same way that the injection and extraction regions require special

Table 1. Machine Parameters

	HER	LER	
Particle	Electron	Positron	
Energy	9.0	3.1	GeV
Luminosity	3×10^{33}		$cm^{-2}s^{-1}$
Tune shift	.03	.03	
RF frequency	476		MHz
Number of bunches	1746	1746	
Bunch spacing	1.26	1.26	m
β_y	3.0	1.5	cm
β_x	75.0	37.5	cm
Separation	Horizontal		
Beam current	1.48	2.14	
Particles per bunch	3.9×10^{10}	5.6×10^{10}	
Vert. emittance	1.8	3.6	nm·rad
Horiz. emittance	46	92	nm·rad
Vert. sigma at IP	7.4	7.4	μm
Horiz. sigma at IP	186	186	mm

magnets. Care in the design of the interaction makes the whole machine easier to design and eventually commission.

Table 2. Comparison of Interaction Regions

	Crab Crossing	Head-On Flat	Head-On Round[3]
Total synchrotron radiation power	0.42 kW	85.7 kW	715 kW
Total no. of > 4keV photons/crossing reaching beampipe	0.34	0.152	94.0

BACKGROUNDS

The total amount of synchrotron radiation produced in the interaction region is the sum of the quadrupole radiation and the bend radiation for both beams. Table 2 compares the present schemes with previous IR layouts. The crab crossing layout (fig. 1) produces the least amount of radiation and the head-on flat beam case (fig. 2) is the next best; both are considerably better than the round-beam cases.

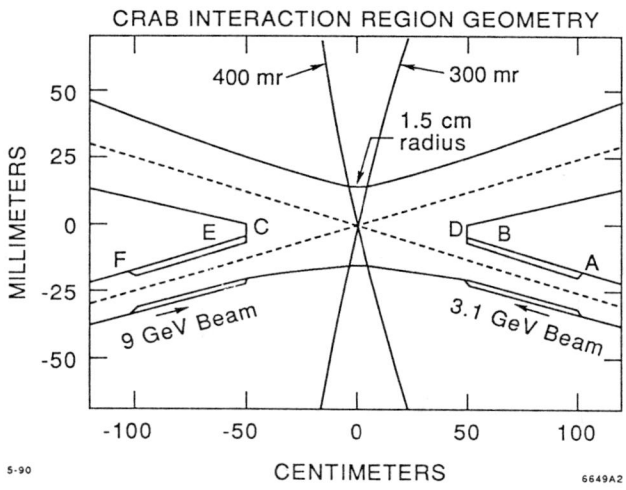

Figure 1. Crab interaction region geometry.

The numbers of photons hitting the beampipe in these two insertions were calculated assuming that the particle distribution is the sum of two gaussians— the core with the nominal sigmas and the tail with larger sigmas:

$$dN dx\, dy = \exp\left\{x^2 2\sigma_x^2 y^2 2\sigma_y^2\right\} + A\,\exp\left\{B_x^2 x^2 2\sigma_x^2 - B_y^2 y^2 2\sigma_y^2\right\} \qquad (2)$$

where $A = 10^{-4}$, $B_x = 0.65$ and $B_y = 0.1$. The vertical tail distribution is considerably larger than the core, corresponding to experience in existing storage rings. The total relative amount of particles in the tail is given by $A/B_x B_y = 1.5 \times 10^{-3}$. These values reproduce the background conditions at the MAC detector in PEP.[5] The backgrounds are presently dominated by bend radiation from the offset beams in the quadrupoles and hence are insensitive to the exact details of the tail distribution.

The synchrotron radiation background is shown in Table 2 and is extremely low in the preferred cases. It should be noted, however, that photons scattered

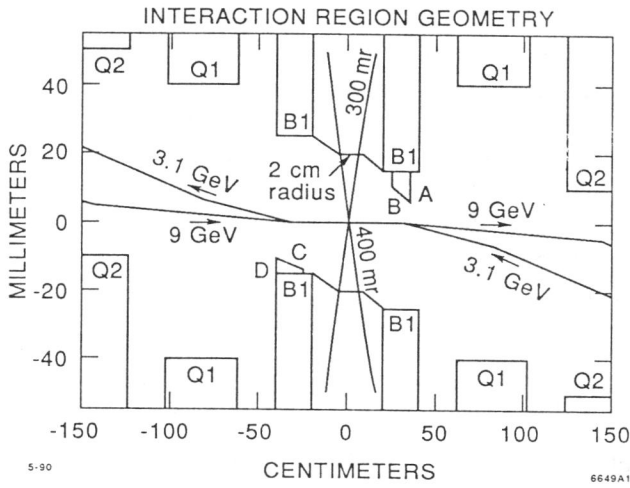

Figure 2. Flat beam interaction region geometry.

back from the beampipe beyond 3 meters have not yet been included; this contribution is being calculated now.

The power requirements of the relevant masks and collimators also need evaluating as do the 'crotches' which see high synchrotron radiation power.

The other heating mechanisms, ohmic and primarily higher order RF losses are also being evaluated to obtain a safe design for the vertex vacuum chamber.

The computer tools to evaluate particle backgrounds have now been prepared and the complete evaluation of the detector environment will soon be available.

THE LATTICE ARCS

Examining all of the possible scenarios leads to the conclusion that the component layout of the present PEP ring provides a range of emittances compatible with all requirements of the high energy ring (HER). The required beam stay-clear is somewhat smaller than PEP (±38mm by ±20mm compared with ±45mm by ±27mm). We are therefore confident in proposing to reuse all of the PEP magnetic components and their power supplies for the HER. The present PEP vacuum chamber will have to be replaced for the HER. The new chamber is being designed for 3 A to permit future upgrades. It is our present intention to adopt a copper vacuum chamber, profiting from the experience at HERA and from the BFI Proposal.[6] Copper seems better for thermal conductivity and outgassing, and detailed calculations show that such a chamber will be self-shielding for the synchrotron radiation. Assuming a photon desorption coefficient for copper of 2×10^{-6} molecules/photon (one tenth that of aluminium), the gas load in the HER at 3 A is very similar to that of PEP. The pumping speed required imposes a solution with distributed pumps in addition to lumped pumps. The distributed pumps will be ion pumps in the bends with the possible addition of NEG pumps in the quadrupoles. Removal of the synchrotron radiation power is limited by the heat transfer to the water (film transfer coefficient), so fins or multiple water passages will be used.

The LER will most likely have the same period length as PEP, a choice that still permits sufficient flexibility in the optics. The choice of bending radius will be determined by injection conditions and engineering considerations. We are currently devoting considerable effort to ensuring that the LER has sufficient flexibility to conform with "energy transparency" conditions if they prove important.[7,8] In particular, room will be left in the lattice for wigglers to control the damping time of the ring.

RF AND FEEDBACK

The RF system must provide 18.5 MV in the HER and 8 MV in the LER to maintain the bunch length at 1 cm. In addition, the synchrotron radiation power of 5.5 MW in the HER and 2.7 MW in the LER must be replaced by the RF systems. The solution adopted has 1 MW klystrons each feeding 2 or 4 single-cell cavities. The 4-cavity solution is easier from an RF power point of view as it only requires an RF window capable of handling 250 kW near the cavity. The 2-cavity solution is better for impedance and cost reasons so a research program is being initiated to develop 500 kW windows. Engineering has begun on laying out an RF module having 1 klystron and 4 cavities but that can easily have an additional klystron added. There is enough space in existing buildings to house these klystrons and the existing transformers have sufficient capacity to power them.

An RF frequency of 476 MHz is our preferred choice (one-sixth of the Linac frequency) and the detailed cavity shape is being designed. The cavities will be made of copper and the higher-order trapped modes must be heavily damped to reduce the multi-bunch instabilities to manageable levels. We are currently evaluating waveguide couplers on the sides of the single-cell cavities, both by computer simulation and using a model cavity. First results are encouraging. The preliminary design of a feedback system exists that could stabilize multi-bunch instabilities with growth rates of about 3 ms. The system uses a series array of 10 striplines at a central frequency of 1 GHz, each $-\lambda$ long with connecting transmission lines $\frac{1}{2}$-λ long. The total power required is estimated to be 3 kW with a bandwidth of 120 MHz.[3]

SUMMARY

The present study will lead to a Conceptual Design Report to be published in January 1991 that will also contain a detailed breakdown of cost and schedule. It is intended that the facility be built jointly by LBL, LLNL and SLAC. We believe that we are close to demonstrating that we have a totally viable solution for an asymmetric B Factory and that PEP is an ideal base for this machine.

REFERENCES

1. 'The Physics Program of a High-Luminosity Asymmetric B Factory at SLAC' SLAC-353, LBL-27856, CALT 68-1588 (October 1989).
2. 'Feasibility Study for an Asymmetric B Factory Based on PEP' LBL PUB-5244, SLAC-352, CALT68-1589 (October 1989).
3. 'Investigation of an Asymmetric B Factory in the PEP Tunnel' LBL PUB-5263, SLAC-359, CALT-68-1622 (March 1990).
4. A. Hutton, 'The Advantages of Flat Beams in Head-on Collisions.' Paper presented at this conference.
5. M. Sullivan, UC-IIRPA-88-01 (May 1988)
6. 'Feasibility Study for a B-Meson Factory in the CERN-ISR Tunnel' CERN-90-YY, PSI-PR-90-08 (March 1990).
7. Y. Chin 'Beam-Beam Interaction in an Asymmetric Collider for B-Physics', Proc IV International Conference on High Energy Accelerators, Tsukuba, Japan (1989).
8. S. Krishnagopal & R. H. Siemann, Phys. Rev. D, 41, 2312 (1990).

A Linac-on-Ring Collider B-Factory Study

P. Grosse Wiesmann, C. Johnson, D. Möhl, R. Schmidt
W. Weingarten, L. Wood[2]
CERN, 1211 Geneva 23, Switzerland
G. Coignet
LAPP, Annecy-le-Vieux, France

1 Abstract

A preliminary survey of the machine parameters required to achieve a luminosity of $10^{31} cm^{-2} s^{-1}$ at the $\Upsilon(4S)$ resonance in a linac-on-ring collider has been made. The low emittance electron source and recirculating superconducting linac based on LEP cavities appears to be within the scope of present technologies. The high-current, low-emittance positron storage ring with it's low-beta collision point can be broadly specified but a more detailed feasibility study is needed. Simulation of the beam-beam effect indicates that the beam-beam limit may be higher than in equivalent ring-on-ring colliders. The heavily disrupted electron beam poses no obvious problem.

2 Introduction

An e^+e^- collider at the $\Upsilon(4S)$ resonance with asymmetric beam energies and a luminosity of $10^{31} cm^{-2} s^{-1}$ is now actively sought for detailed studies of B-meson decays, and in particular CP-violation. The experimental requirement of very high integrated luminosity implies continuous operation at or near peak performance over months or even years and this sets an extraordinarily challenging goal for accelerator designers - a goal that entices new designs since it lies beyond the readily available performance of conventional colliders. Several feasibility studies of ring-on-ring colliders are under way or completed[1]. The very high performance requirements justify the study of alternative schemes such as the linac-on-ring collider

The main thrust of linac-on-ring collider studies has been towards higher collision energies[3], but additionaly, the different nature of the beam-beam effect and the smaller beam currents could be key elements in extending the luminosity limit beyond that presently attainable in low-energy colliders. The obstacle in the linac-linac B-Factories of achieving sufficiently high positron production rates[2] is avoided by storing the positron beam. Some preliminary studies have been made with encouraging results[4] and this contribution summarises the findings of an informal study at CERN on the long-term prospects for a linac-on-ring B-Factory.

---
[2] Present address: Exploration Consultants, Henley-on-Thames, U.K.

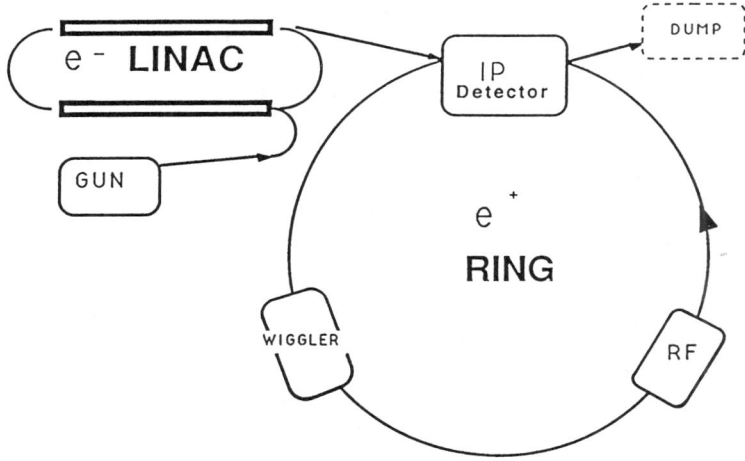

Figure 1: Linac-Ring-Collider overview

3 Design Constraints and Parameter

Figure 1 shows a layout of a linac-ring-collider. A positron beam has to be stored over long periods without significant emittance blow up and an electron beam has to be continuously renewed from a linac and dumped after the collision.

Linac-ring-colliders are a new concept and recipies for parameter lists based on experience do not yet exist. Both beams have different constraints and one expects the intensity of the two beams to be very different. Possible sets of parameters are given in Table 1. In the following we discuss how a few important parameters limit the luminosity; the most important goal for a B-factory. For a given power of the electron beam (P_{e^-}) the luminosity is determined by the transverse density of the stored positron bunch.

$$L = 10^{31} cm^{-2} s^{-1} \cdot \frac{P_e^-}{2.0 MW} \cdot \frac{N_{e^+}}{10^{11}} \cdot \frac{(\mu m)^2}{\sigma_x^{e^+} \sigma_y^{e^+}} \cdot H_D \cdot \frac{GeV}{E_{e^-}} \quad (1)$$

Where H_D is an enhancement factor due to the pinch effect in the beam-beam interaction. If one wants to achieve a luminosity of $10^{31} cm^{-2} s^{-1}$, an electron beam power of a few MW and a low emittance high peak current positron beam are needed. The electromagnetic forces, that accompany the positron bunch, act like a strong focusing lens on the electrons.

The disruption parameter(D), which relates the bunch length (σ_z) to the effective focal length of the beam force, is used to quantify the pinch effect of the beam-beam interaction in linear colliders.

Table 1: Linac Ring Collider Parameters.

Case	A	B	C	D	E
$E_{e-}(GeV)$	3.1	3.1	3.1	8.0	8.0
$E_{e+}(GeV)$	9.0	9.0	9.0	3.5	3.5
$I_{e-}(mA)$	1	2.6	2.6	2	2
$I_{e+}(A)$	1	0.5	0.9	0.9	1.6
$P_{e-}(MW)$	3.1	8.0	8.0	16.0	16.0
$\rho_{e+}(m)$	100	100	100	60	60
$P_{e+}(MW)$	5.8	2.9	5.2	0.2	0.4
$f_c(MHz)$	30	10	18	28	50
$N_{e-}(10^9)$	0.2	1.6	0.9	0.4	0.3
$N_{e+}(10^{11})$	2	3	3	2	2
$\sigma_x(\mu m)$	1.0	2.0	5.0	1.4	4.0
$\sigma_y(\mu m)$	1.0	2.0	0.8	1.4	0.5
σ_z (mm)	7	10	10	7	7
β_y^{e+} (mm)	7	10	10	7	7
D_{e-}^y	660	350	610	130	230
D_{e+}	0.22	0.6	0.6	0.6	.6
ξ_{e+}^y	0.018	0.05	0.05	0.05	0.05
$\delta_{bstr.}(10^{-4})$	1.9	0.7	0.4	2.6	1.0
H_D	1.0	1.0	1.0	1.0	1.0
$L(10^{34}cm^{-2}s^{-1})$	1.0	1.0	1.0	1.0	1.0

$$D_{e-}^y = 2.8 \cdot 10^3 \, \frac{MW}{P_{e-}} \cdot \frac{2}{1+\frac{\sigma_y}{\sigma_x}} \cdot \frac{\sigma_z^{e+}}{cm} \cdot \frac{L}{10^{31}cm^{-2}s^{-1}} \qquad (2)$$

The disruption parameter is inversely proportional to the electron beam power. For a luminosity of $10^{34}cm^{-2}s^{-1}$, a beam power of a few MW corresponds to a disruption parameter of several hundred. For such large values the electrons are strongly overfocused and undergo several oscillations through the high density positron bunch.

During the disruption process the electrons emit synchrotron radiation, so called beamstrahlung. At energies relevant for a B-factory the beamstrahlung losses ($\delta_{bstr.}$) are at the level of 10^{-4} and the energy smearing of the electron beam is of no concern.

The destabilizing effect that such a highly disrupted beam has on the storage ring beam cannot be simply quantified by the magnitude of D; in particular it is not obvious if fewer oscillations are less harmful to the stability of the ring beam (see chapter about beam-beam effect for further discussion).

Another important factor is obviously the intensity of the electron bunch, in Table 1 we therefore quantify the beam force by the linear tune shift(ξ_{e+}) that the nominal linac beam causes to the positron beam. In ring-ring-colliders ξ is used to characterize the strength of the non-linear forces. A lower limit for ξ_{e+} requires an increase of the positron current (I_{e+}) and of the collision frequency (f_c) proportional to the beta function (β^y_{e+}) at the collision point.

$$I_{e+} = 0.5 A \cdot \frac{2}{1 + \frac{\sigma_y}{\sigma_x}} \cdot \frac{\beta^y_{e+}}{cm} \cdot \frac{0.05}{\xi_{e+} H_D} \cdot \frac{9 GeV}{E_{e+}} \cdot \frac{L}{10^{34} cm^{-2} s^{-1}} \quad (3)$$

As a consequence of equation (3) higher positron beam energies imply smaller positron currents, but from synchrotron radiation power losses ($P_{e+} \propto L \cdot E^3/\rho$) a lower E_{e+} is preferred.

In Table 1 five sets of parameters each resulting in a luminosity of $10^{34} cm^{-2} s^{-1}$ are given. The first three cases(A,B,C) are for a 3.1 GeV electron beam and a 9 GeV storage ring beam; cases D and E are for an 8 GeV electron linac and a 3.5 GeV positron ring. For both energy choices round and flat beam examples are given.

In case A a one Ampere positron beam and a one mA electron beam are collided with a nominal collision spot size of 1 μm. This case corresponds to the beam-beam simulations described later. In cases B and C the essential input constraints are the electron beam power ($P_{e-} = 8 MW$), the betafunction of the positron beam at the collision point ($\beta^y_{e+} = 10 mm$), and the linear tuneshift caused by the nominal electron beam onto the positron beam ($\xi_{e+} = 0.05$).

In cases D and E, where the linac beam is the high energy beam, we allow for a larger linac beam power ($P_{e-} = 16 MW$), reduce the betafunction of the lower energy ring ($\beta_{e+} = 7 mm$) and keep the same tune shift parameter($\xi_{e+} = 0.05$).

4 Superconducting Electron Linac

Superconducting radiofrequency cavities offer the possibilty for a high current and high frequency electron beam with very efficient conversion of wall plug power into beam power. Those cavity designs have matured over the last years and are now applied in several projects[5] involving electron storage rings, nuclear physics linear accelerators and free electron laser applications. Figure 2 shows a standard LEP unit of four 350 MHz cavities with four cells each put into a common cryostat. The total length of such a subunit is about 10m; electrons are accelerated by 50 MeV with a gradient of 7 MV/m and a packing factor of 2/3. Such a unit has been tested sucessfully in LEP this year and for LEP200 it is forseen to install up to 64 units[6].

Based on the LEP units we outline in Figure 3 a recirculating linac. Assuming four recirculations and a gradient of 7MV/m, 16 of those units(64 cavities) and 8 standard LEP klystrons could accelerate a 2.6 mA beam up to 3GeV. The total amount of cavities and klystrons would be comparable to about a quarter

Figure 2: Four LEP cavities in one cryostat

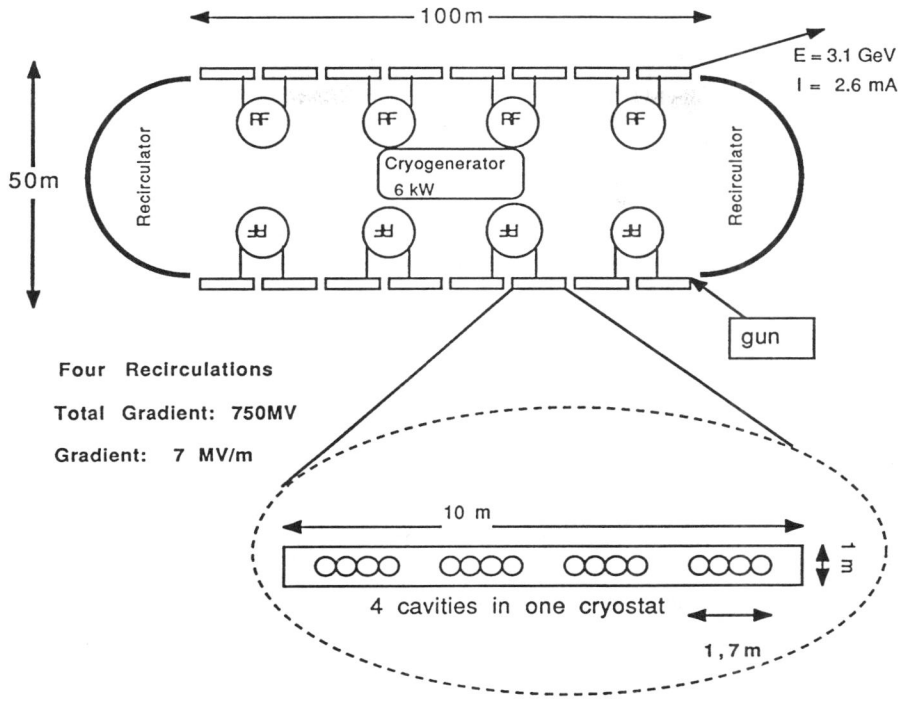

Figure 3: Recirculating linac based on LEP cavities

of what is planned for LEP200. The use of an existing production line allows a relatively reliable estimate of the complexity of such a linac and its industrial production costs.

In superconducting cavities a degradation of the beam quality due to transverse wakefields (emittance growth) or due to beam loading (energy resolution) is not expected, because of the possibility of large RF wavelength and large iris holes.

The higher order mode (HOM) couplers developed for the LEP cavities are adequate for the average and the peak currents envisaged in the recirculating linac. In Table 2 the requirements for a B-Factory are compared to some basic parameters of superconducting RF projects at LEP, HERA, TRISTAN and CEBAF. None of the quantities like total gradient, peak or average current are more demanding than those typically required in these projects.

In order to estimate the cryogenic losses we assume a $Q_0 = 5 \cdot 10^9$ at 4.2 Kelvin, which is not beyond reach at a gradient of 7MV/m. Cryogenic losses due to residual RF resistance of less than 4kW at 4.2 Kelvin are expected. In additon static heat losses of about 1.5 kW have to be envisaged. The expected cryogenic

Table 2: Comparison of Superconducting RF Projects.

Project	LEP	HERA	KEK	CEBAF	B-Fact.
Energy (GeV)	3.	0.3	0.2	0.8	0.8
rep. freq.(MHz)	0.04	10.	0.2	1500.	30
$N_{bunch}(10^{10})$	41.	2.	32.	.0003	.16
σ_z (mm)	16	8	12	1	7
$I_{peak}(KA)$	1.2	0.1	1.2	.0005	.01
$I_{average}(mA)$	6.	30.	20.	1.	10.
$f_{RF}(MHz)$	350	500	500	1500	350
$P_{RF}(MW)$	16.	9.	5.	.8	8.

load of less than 6 KW at 4.2Kelvin corresponds to about one third of the cryogenic power installed for the superconducting magnets at HERA.

An electron linac with 8 GeV beam energy and a beam power of 16 MW, as demanded for case D and E in Table 1, could be realized with twice the number of cavities and klystrons shown in Figure 3 and five recirculations.

Optics for the recirculators have been worked out for CEBAF [7] and for a similar project under study in SACLAY[8]. At the energies considered the required low level of energy smearing and emittance can be conserved in the recirculating arcs with sufficient bending radius and a low dispersion optics.

5 Electron Gun

A low emittance short electron bunch with a high repetition rate is demanded. The high repetition rates exclude the use of damping rings and low emittance electron beams directly from a cathode have to be used. For Free-Electron-Laser applications[9] electron guns with the required specifications have been developed. To avoid emittance blow up by the large space-charge forces in nonrelativistic dense electron bunches, high acceleration gradients at the photocathode are essential; photocathodes irradiated by a laser are directly placed into a high gradient RF cavity to overcome the space charge forces. In several laboratories[10] those electron guns are developed. Figure 4 shows the design for a gun with a superconducting RF cavity as proposed by Wuppertal,CEBAF and DESY(WCD) [11]. In Table 3 projects in Los Alamos, Brookhaven (BNL) and the design from Figure 4 are compared to typical requirements for a B-Factory gun. In these projects the peak currents and emittances are comparable to the requirements of a B-Factory. The BNL and the Los Alamos projects have repetition rate of a few Hz, because they are essentially developed for study reasons. As discussed in the Wuppertal-CEBAF-DESY RF gun design, a high repetition rate could be achieved with a commercially available mode-locked laser.

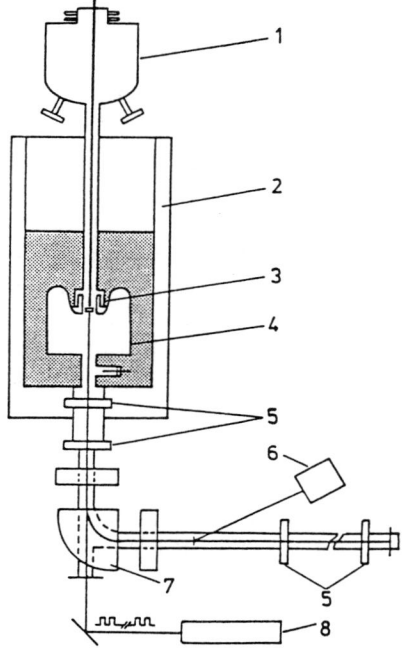

1: photocathode preparation chamber,
2: bath cryostat, 3: photocathode, 4: reentrant cavity,
5: wire scanner monitor, 6: streak camera,
7: spectrometer, 8: Nd:YAG laser

Figure 4: Electron gun; from ref. 11

Table 3: Comparison of Laser RF Guns.

Project	Los A.	BNL	WCD	B-Fact.
$N_{e^-}(10^9/bunch)$	60.	6.	1.	1
$\sigma_z (mm)$	9	.6	2.6	7
$I_{peak}(A)$	130	160	7.3	2.7
$\epsilon_n (mm-mrad)$	18	7.3	45	10
$f_{RF}(MHz)$	1300	2850	1300	—
rep. freq. (MHz)	few Hz	few Hz	125	30

6 Low Emittance High Current Positron Ring

The emittance requirements for the storage ring are comparable to those of damping rings for future linear colliders or advanced synchrotron light sources.

Synchrotron radiation provides a fast cooling mechanism, but to avoid beam heating, radiation losses in regions with large dispersion have to be avoided. Various lattice types have been considered including a high tune FODO lattice with wigglers in dispersion free zones. This type of ring has for example been studied for a CLIC damping ring in the SPS-tunnel[12] and a conversion of the PEP ring into a synchrotron radiation facility[13]. Tight alignment tolerances are required for those low emittance lattices.

Due to the low emittance, small aperture and high gradient quadrupoles can be used in the low β insertion and detector background problems from the ring are expected to be small. Since the electron beam is discarded after the interaction, chromaticity introduced by the low-beta insertion is of little concern for the electrons and is more easily minimized for the positron ring.

High peak and average currents are required like in the ring on ring scheme with the resulting challenge for beam stability. However ion trapping problems inherent to electron rings are avoided since only a positron ring is needed.

A more detailed feasibility study of the ring is indicated.

7 Beam-Beam Effect

The beam-beam limit in storage rings is caused largely by the tune-spread resulting from the non-linearity of the beam forces. A disrupted electron beam with a drastically reshaped charge distribution increases these non-linearities. On the other hand the fact that one beam is discarded after the collision opens up new possibilities: one can arrange for the discarded beam to have the lower energy thereby assigning the tune spread to the stiffer (less easily perturbed) beam, also coherent phenomena and flip-flop effects are of less concern.

Two different simulation programs have been used to study the set of beam parameters corresponding to case A in Table 1; in addition the size of the electron beam has been varied.

Figure 5a shows the 2σ contours of the positron beam with different electron trajectories oscillating through the positron bunch. At the center of the collision region the electron density is enhanced (pinch effect). The simulation presented in Figure 5b shows the trajectories of electrons within a beam with $\sigma_x = 1\mu m$; luminosity enhancement appears, but the maximum tuneshift and non-linearities increase. In Figure 5c a much broader electron beam distribution ($\sigma_x = 3\mu m$), but with nominal positron bunch ($\sigma_x = 1\mu m$) is represented; electrons are drawn in by the smaller e^+ bunch and the two bunches are better matched than in the former case (Fig. 5b). Due to the non-linear forces of the gaussian positron bunch there is no strong phase correlation between the electron trajectories.

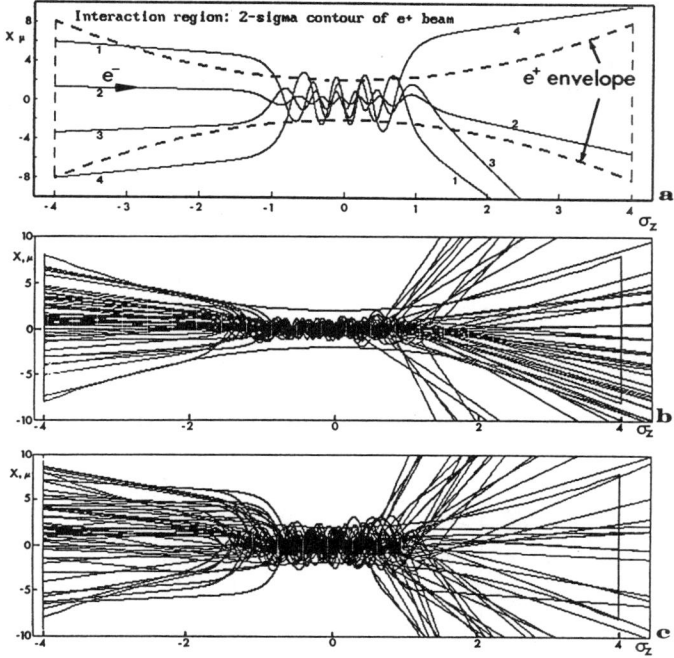

Figure 5: Electron trajectories in positron bunch.

In these simulations the positron field is generated for an unperturbed positron bunch ('weak-strong'). In a second program developed for the CLIC studies, both beam forces are simulated during the bunch crossing ('strong-strong').

Figure 6 shows how the envelopes of the two beams evolve during the collision for two different electron bunch widths. For the case of the broader electron distribution ($3\mu m$) the two bunches are well matched at the collision time (Fig. 6b,t=0.0). Table 4 gives the luminosity with and without the mutual pinch (forces on and off) for different e^- spot sizes normalised to an unperturbed collision size of $1\mu m$. The possibility to influence the nonlinear beam-beam forces by

Table 4: Luminosity Enhancement($H_D = \frac{L}{L^0}$)

$\sigma^{e^+}(\mu m)$	$\sigma^{e^-}(\mu m)$	$H_D(no force)$	$H_D(with force)$
1.	1.5	0.53	1.56
1.	2.0	0.40	1.27
1.	3.0	0.20	0.94

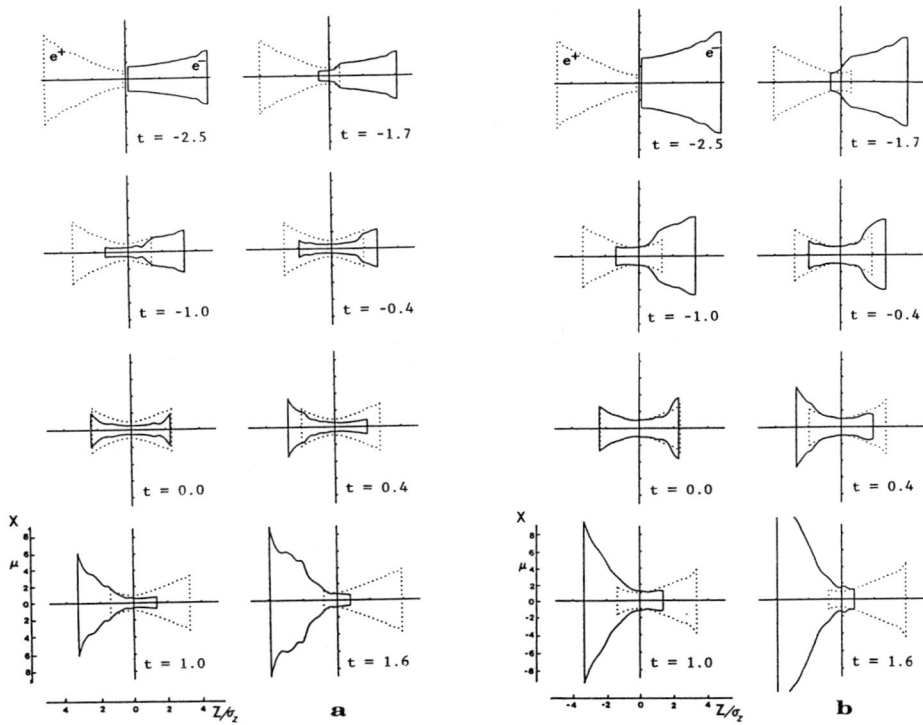

Figure 6: Beam envelope during collision.

adjusting the electron beam initial conditions is interesting and could give operational advantages; also the strong pinch renders the design relatively insensitive to linac beam quality.

A fairly extreme case with $\sigma_x = 1\mu m$ for both beams (other parameters as for case A in Table 1) results in an electron pinch down to $\sigma_x = 0.42\mu m$. The longitudinal form of the pinch is shown in Figure 7. The resulting distribution of the integral forces experienced by the opposing positrons (integrated over $\pm 3\sigma_z$) are illustrated in Figure 8. Because of the electron pinch the tuneshift of the positrons is correlated to their synchrotron amplitudes. Some preliminary studies of the consequences of this correlation and the behaviour with respect to bunch length have begun using a 'weak-strong' tracking program. This work is described briefly elsewhere in these proceedings[14]. To obtain a fully quantitative estimate of the acceptable limits of the linear tuneshift ξ_{e^+} and the bunch lengths, a storage ring with nonlinear elements and a low-β insertion must be

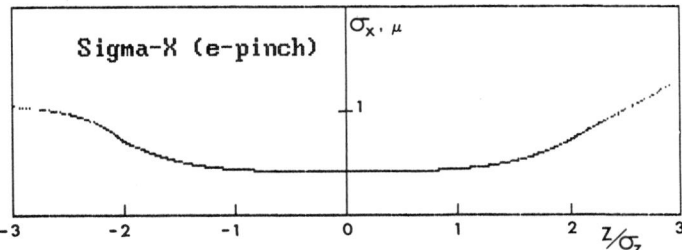

Figure 7: Longitudinal form of the pinched e^- beam.

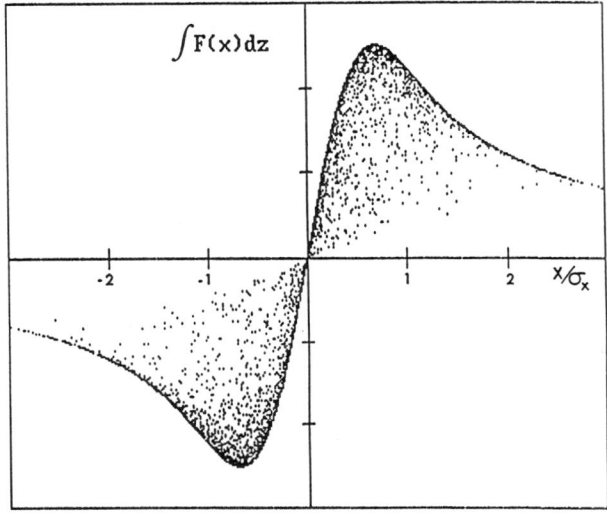

Figure 8: Integrated forces on the e^+ due to the pinched e^- beam

simulated. An experimental test of the linac-on-ring beam-beam interaction strategies would of course be invaluable.

8 Conclusions

A B-factory with a luminosity of $10^{34} cm^{-2} s^{-1}$ seems possible in the linac-ring-collider scheme. The superconducting radiofrequency electron linac together with a low emittance gun could be built based on existing technology. A design for the low emittance and high current positron ring looks possible along the lines of advanced synchrotron light sources or damping rings for future linear colliders. Compared to a ring-ring-collider the linac-ring-collider avoids a high current electron ring and allows lower emittance beams. New possibilities to improve the beam-beam limit might be given since the electron beam is discarded after the collision. Further studies have to be done, to see how much of the potential of a linac-ring collider could be realized to make it an attractive alternative for a low energy, high luminosity collider.

References

[1] A.M. Sessler, Particle World, **1**, 125 (1990).and
T.Nakada (Ed.), CERN 90-02, PSI PR-90-08 (1990).
and references therein

[2] U. Amaldi et al., EPAC Proc. , Rome, 754 (1988).

[3] P. Grosse-Wiesmann, Nucl. Instr. Methods, **A274**, 21 (1989).
C. Rubbia, EPAC Proc. , Rome, 290 (1988).

[4] J.J. Bisognano et al., CEBAF Tech. Note, November 1988.
U. Amaldi and G. Coignet, Moriond Proc. ,491 (1989).

[5] W. Weingarten, Particle World, **1**, 93 (1990).
D. Proch, EPAC Proc., Rome, 29 (1988).
S. Nogushi, EPAC Proc., Nice, (1990).

[6] C. Benvenuti et al., EPAC Proc., Nice, (1990).

[7] Conceptional Design Report, CEBAF, February 1986.

[8] J. P. Didelez et al., Projet d'un Accelerateur Supraconducteur d'Electrons de 4 GeV, Orsay/Saclay/Grenoble study, January 1990.

[9] R.L. Sheffield et al., Nucl. Instr. Methods, **A272**, 222 (1988).

[10] Y. Baconnier et al. EPAC Proc., Nice, (1990).
K. Batchelor et al., EPAC Proc., Rome, 954 (1988).

[11] H. Chaloupka et al., EPAC Proc., Rome, 1312 (1988).

[12] L. Evans and R. Schmidt, CLIC Note 58, Geneva, (1988).

[13] H. Wiedemann, US-CERN Accelerator Course, Texas, (1986).

[14] C.D Johnson and L. Wood, Incoherent Beam-Beam Effect - the Relationship Between Tune-shift, Bunch Length and Dynamic Aperture., These proceedings.

LIST OF PARTICIPANTS

WORKSHOP ON BEAM DYNAMICS ISSUES OF HIGH-LUMINOSITY ASYMMETRIC COLLIDER RINGS

February 12–16, 1990

Lawrence Berkeley Laboratory

PARTICIPANTS

Peter P. Bagley	Cornell University
William A. Barletta	Lawrence Livermore National Laboratory
Klaus H. Berkner	Lawrence Berkeley Laboratory
Elliott D. Bloom	Stanford Linear Accelerator Center
Karl L. Brown	Stanford Linear Accelerator Center
R. Cappi	CERN, Switzerland
Paul Channell	Los Alamos National Laboratory
Alex Chao	SSC Laboratory
Swapan Chattopadhyay	Lawrence Berkeley Laboratory
Pisin Chin	Stanford Linear Accelerator Center
Yu-Jiuan Chen	Lawrence Livermore National Laboratory
Yong Ho Chin	Lawrence Berkeley Laboratory
William Corbett	Stanford Linear Accelerator Center
Max Cornacchia	Stanford Linear Accelerator Center
Karel Cornelis	CERN, Switzerland
Ernest D. Courant	Brookhaven National Laboratory
George Craig	Lawrence Livermore National Laboratory
George F. Dell	Brookhaven National Laboratory
Domenico Dell'Orco	Lawrence Berkeley Laboratory
Herbert DeStaebler	Stanford Linear Accelerator Center
Martin H. R. Donald	Stanford Linear Accelerator Center
Jonathan Dorfan	Stanford Linear Accelerator Center
Thomas J. Fessenden	Lawrence Berkeley Laboratory
Hans R. Fitze	Stanford Linear Accelerator Center
Etienne Forest	Lawrence Berkeley Laboratory
Yoshihiro Funakoshi	KEK, Japan
Miguel A. Furman	Lawrence Berkeley Laboratory
Alper Garren	Lawrence Berkeley Laboratory
Terence Garvey	Lawrence Berkeley Laboratory
Andrey L. Gerasimov	Institute of Nuclear Physics, USSR
Christoph Geyer	DESY, West Germany
Samuel A. Heifets	CEBAF
Kohji Hirata	KEK, Japan
David G. Hitlin	California Institute of Technology
Bernard Holzer	DESY, West Germany
Ian Hsu	Stanford Linear Accelerator Center
Andrew Hutton	Stanford Linear Accelerator Center
Alan Jackson	Lawrence Berkeley Laboratory
Gerald P. Jackson	Fermi National Accelerator Laboratory

List of Participants

Ken Jacobs	MIT Bates Linac
Colin D. Johnson	CERN, Switzerland
Eberhard Keil	CERN, Switzerland
Roderich Keller	Lawrence Berkeley Laboratory
Semyon A. Kheifets	Stanford Linear Accelerator Center
Eiji Kikutani	KEK, Japan
Kwang Je Kim	Lawrence Berkeley Laboratory
Patrick Krejcik	Stanford Linear Accelerator Center
Srinivas Krishnagopal	Cornell University
Shin-ichi Kurokawa	KEK, Japan
Glen R. Lambertson	Lawrence Berkeley Laboratory
Valery Lebedev	Institute of Nuclear Physics, USSR
Edward P. Lee	Lawrence Berkeley Laboratory
Torsten Limberg	Stanford Linear Accelerator Center
Sig A. Martin	KFA Juelich, West Germany
Jay Marx	Lawrence Berkeley Laboratory
Masataka Mizobata	Mitsubishi Electric Corporation, Japan
Phil Morton	SAIC/SLAC
David V. Neuffer	Los Alamos National Laboratory
King Yuen Ng	Fermi National Accelerator Laboratory
Heinz-Dieter Nuhn	Stanford Synchrotron Radiation Laboratory
Piermaria Oddone	Lawrence Berkeley Laboratory
Yuri F. Orlov	Cornell University
Thaddeus Orzechowski	Lawrence Livermore National Laboratory
Flemming Pedersen	CERN, Switzerland
Rainer Pitthan	Stanford Linear Accelerator Center
Frank C. Porter	California Institute of Technology
Govindan Rangarajan	University of Maryland
John Rees	Stanford Linear Accelerator Center
David H. Rice	Cornell University
Thys F. Risselada	CERN, Switzerland
Leonid Z. Rivkin	Paul Scherrer Institute, Switzerland
David S. Robin	UCLA
Michael Ronan	Lawrence Berkeley Laboratory
David L. Rubin	Cornell University
Francesco Ruggiero	CERN, Switzerland
Rudiger Schmidt	CERN, Switzerland
Walter Schmidt-Parzefall	DESY, West Germany
John T. Seeman	Stanford Linear Accelerator Center
Roger V. Servranckx	TRIUMF, Canada
Andrew Sessler	Lawrence Berkeley Laboratory
Toshiya Tanabe	Columbia University
Clyde E. Taylor	Lawrence Berkeley Laboratory
Lee C. Teng	Argonne National Laboratory
Jeffrey L. Tennyson	California Institute of Technology
Maury Tigner	Cornell University
Gaetano Vignola	Laboratory Nazionali Di Frascati, Italy
Ferd Voelker	Lawrence Berkeley Laboratory
Nick Walker	Stanford Linear Accelerator Center
Tony Warwick	Lawrence Berkeley Laboratory
Ferdinand Willeke	DESY, West Germany

Edmund J. Wilson	CERN, Switzerland
Lindsay Wood	CERN, Switzerland
Simon S. Yu	Lawrence Livermore National Laboratory
Xiao-Tong Yu	Massachusetts Institute of Technology
Alexander Zholents	Institute of Nuclear Physics, USSR
Michael Zisman	Lawrence Berkeley Laboratory
Abbi Zolfaghari	Massachusetts Institute of Technology

APPENDIX

Reprinted from Particle World, Vol. 1, No. 5, p. 125-132, 1990[†]

Beam Dynamics Issues of High-Luminosity Asymmetric Collider Rings[*]

Andrew M. Sessler

● Abstract

The beam dynamics issues presented by a high-luminosity asymmetric electron collider ring (such as is required for a B meson factory) are described. Attention is focused on lattice aspects, on single-beam effects, and on beam-beam interaction effects. The over-all conclusion is that a facility with a beam of (about) 3 GeV in one ring and a beam of (about) 9 GeV in a second ring having a luminosity of between 10^{33} and 10^{34} cm^{-2}s^{-1} is a feasible concept.

1. Introduction

The desire to study, in great detail, the B-$\bar{\text{B}}$ system and, in particular, to study the CP-violation in that system, has motivated the development of very high-luminosity asymmetric collider rings.[1] The development of such a collider presents new challenges to accelerator physicists, and in order to explore and assess the beam dynamics issues that this quest raises, a Workshop on the subject was called by the Lawrence Berkeley Laboratory and the Stanford Linear Accelerator Center in February of this year.[2]

The physics to be done at a B Factory requires an integrated luminosity of more than 30 fb^{-1}/year.[1,3] This is equivalent to a collider delivering a luminosity of at least 3×10^{33} cm^{-2}s^{-1} for a third of each year (10^7 seconds). The required luminosity is larger than the present performance of colliders, in the same energy range, by a factor of at least 30. In addition, the collider must have a center of mass energy of 10-11 GeV with beam energy ratios of up to 5 to 1. (If the collider is symmetric in energy, then the luminosity required is larger than that given above, by an additional factor of about 5.) From the machine physicist's point of view the extrapolation in luminosity is much more of a challenge than the extrapolation from symmetric colliders to asymmetric ones.

In this article, we shall draw heavily upon the Workshop. On the very closing day of the Workshop, a small group of physicists gathered together and attempted to summarize the conclusions in a succinct form. Their Summary is presented in Part I of these Proceedings.

[†] © Gordon and Breach Science Publishers S.A.
[*] This work was supported by the Director, Office of Energy Research, Office of High Energy and Nuclear Physics, Division of High Energy Physics, of the US Department of Energy under contract DE-AC03-76SF00098.

Beam dynamics issues may conveniently be broken into three categories. The first is that having to do with "single particle" phenomena. Under this comes the design of a proper focusing lattice, RF acceleration, injection, extraction, radiation damping, quantum fluctuations, etc.

The second category consists of single-beam phenomena arising from the many-body aspects of a beam. Within this category are conventional "space charge phenomena" (negligibly small at relativistic energies), and also rather sophisticated phenomena such as intrabeam scattering, synchro-betatron mode coupling, and single- and multi-bunch coherent instabilities.

The third category consists of those phenomena that result from the interaction between beams where the non-linear forces are the primary source of concern.

In this paper we shall consider the beam dynamics of B Factories by discussing, in turn, single-particle phenomena, single-beam phenomena, and beam-beam phenomena. We shall not be concerned with various "practical" issues such as injection, e^+ production, vacuum, etc. They are, of course, important. We do note that the large luminosity implies a short beam lifetime and hence a dedicated injector and (probably) the ability to take data while "topping off" the beam. We start, first, with some general considerations.

2. General Considerations

Some of the elements that must be considered by the machine physicist are shown in Figure 1. Of course, we are not starting from scratch; circular colliders have been built, and carefully studied, for two decades. It is quite appropriate to ask in that context what must be done differently from that which has been done in existing colliders in order to achieve the performance specifications of a B Factory. In fact, this mode of reasoning is very simple, and almost unique in its results, so that all of the various proposals for B Factories (see Section 6) are quite similar in general nature.

The reasoning begins as follows: The beam-beam interaction puts a limit on the luminosity created by one bunch (meeting one bunch of the other beam) which we presently do not know how to exceed. We can make this limit as large as possible by focusing the beams to very small size at the crossing point ("low β^*"). But to get the required very large luminosity with a reasonable beam emittance will still require many bunches in each beam. Because of the many potential near crossings (even with the separation that can be achieved electrostatically) the collider needs to have two rings.

What are the consequences of this direction for the design? The first thing with which we must be concerned is multi-bunch effects, and we shall discuss this more in Section 4. Suffice it to note here that, due to the large current in each ring, there are severe multi-bunch instabilities and they must be handled by strong feedback systems. Even then it is critical to reduce their growth rates in the first place by proper design of RF cavities with reduced higher-order mode response. A second major consequence of the design is that the bunches must be separated rather close to the interaction point (because unwanted crossings must be avoided and the many bunches are close together). If the collisions are head-on—and experience suggests that the deleterious aspects of the beam-beam interaction are greatly enhanced if the crossing is not head-on—then powerful magnets are required near the crossing point and these produce synchrotron radiation background from which the detector must be shielded. We shall discuss this in Section 3. Alternatively, the crossing could be at an angle, but appear as if it is head-on; this approach would employ the suggestion by Bob Palmer of "crab crossing" (described below).[4] This scheme, which has not yet been studied very extensively, does not require the use of separation magnets and consequently is good from a masking point of view, but requires strong crab RF cavities near the interaction point. The technical feasibility of this scheme is unknown at this time. Being able to focus both a high energy beam (HEB) and a low energy beam (LEB) by a common set of magnets in the interaction region (IR) implies a novel and challenging feature of asymmetric machines. A third consequence of the design is that the very low β^* implies a concomitant need for very short bunches and, hence, a very powerful radio frequency system. It is clear from both the above that the RF system must be of special design that can deliver a large amount of power and voltage to the beam with a minimum number of cavities.

There are other consequences of our design direction, and some of them will be touched upon below. Much more can be found in the various design study documents being produced in the laboratories mentioned in Section 6, but the major consequences are as listed above and depicted in Figure 2. One cannot help but notice from Figure 2 and Figure 1, where all the issues seem to converge on the IR layout circle, that the design of the interaction region optics plays a central and crucial role in any high-luminosity collider design. No other aspect is as intricately connected to all others as is the interaction region.

3. Lattices

Perhaps one should start with consideration of the beam-beam interaction, for that is central to a B Factory design. Fortunately, however, the consequences of this subject can be summarized very succinctly, and that allows us to proceed in the logical order of designing a collider for single-particle effects and then, subsequently, concerning ourselves with many-particle phenomena. Of course, life is not that simple and there must be continual interchange between the experts in lattices and in many-body phenomena.

The physics of single-particle behavior in colliders has been set out in the classic work by Matt Sands.[5] Although that work is 20 years old, it includes just about everything one needs to know to design a collider. We shall not go through considerations that are well known, such as betatron tune, chromaticity, dispersion, radiation damping times and emittance, although all of these are needed to design a collider. (For example, we shall not comment upon the required beam emittance which is low, but in the range that has already been achieved.) Rather, we shall comment only in a very general way, upon the novel features that enter into B Factory design.

Perhaps the central complicating feature of the design is that there must be two rings. (Not completely new ground; think of the ISR, or HERA.) Thus the interaction region, with its separation of particles, and its production of a very low β^* at the crossing point is the most difficult part of the design. Of course, one must be concerned with chromaticity corrections, making straight sections in which wigglers can be inserted to produce and control low beam emittance and short damping times, and the myriad of other things that go into a lattice design. But the main complication comes about with designing the interaction region.

The difficulty is in the combined aspect of producing a low β^* and separating the beams, while at the same time not producing too much synchrotron radiation very near the interaction point. The low β^* can be produced by powerful focusing quadrupoles, but as the beams are separated, the one that is off-center in the quadrupole feels a large field and consequently bends and radiates. (The one on-center also radiates, but only because of its finite size.) In an obvious way, any dipole magnets that are employed to separate the unequal energy beams also produce synchrotron radiation from both beams. Rather strong magnets are needed to get prompt separation (as one moves away from the interaction point) because of the close bunch spacing.

A number of different suggestions have been made, and presently are being explored, for the interaction region geometry (see Section 6). In Figure 3 we have indicated the essential elements of two of these suggestions. As of this writing, no completely acceptable solution has been produced, although there is no reason to believe that one cannot be achieved. Of course, there needs to be considerable attention to the quality and nature of the required synchrotron radiation masks and the sensitivity of the detector to radiation. In addition to synchrotron radiation, there is the background from lost particles which is strongly affected by the beam-beam interaction that is the primary mechanism for putting particles into the tail of beam distributions.

One issue in the design is whether the beam is flat (aspect ratio of say 40 to 1) or round. It is unclear how much one gains in the beam-beam interaction with round beams (as discussed in Section 5) and it appears to be more difficult to design an interaction region with a round (but small cross section) beam rather than with a flat beam (very small vertically, but big horizontally), thus the obvious advantage, of a factor of two, in round beams versus flat beams is washed out. Also in favor of flat beams, there appears to be less synchrotron radiation in that case because the required focusing gradients are lower than those needed to produce round beams. Presently, there is no unanimity of thought on the subject of round vs. flat beams.

Still another aspect of the interaction region is whether or not the collisions are head-on or crossing at an angle. Certainly a non-zero crossing angle reduces the masking problem greatly, but crab crossing, which would be necessary, has not yet been tried. In Figure 4 we indicate the nature of crab crossing. The luminosity of a head-on configuration would be maintained in the crossing case but, much more importantly, the transverse beam-beam kick does not couple to the longitudinal degree of freedom of the particles, and hence the beam-beam interaction is no different in the crossing case than in the head-on case. Study and simulations of the effects of crab crossing, in synergism with beam-beam effects, is just starting. Most projects are not "counting on crab crossing," but are allowing for the possibility of incorporating this feature in the future (i.e., by having S-bends in the case of head-on collisions).

4. Single-Beam Phenomena

The subject of single beam instabilities has been well-studied through the years in connection with storage rings and colliders. The new synchrotron radiation sources are being built with very short bunches (so as to get good time resolution of the radiation) and with very small emittance bunches (so as to have very bright sources).

Their construction has been based upon our knowledge and experience with colliders, but the frontiers of research on single beam instabilities are now being pushed by the people concerned with synchrotron radiation sources.[6]

A comprehensive discussion of intense beam phenomena can be found in many laboratory design study reports and, in particular, in two recent papers.[7,8] One must consider the longitudinal microwave instability, transverse mode-coupling instability, and coupled-bunch instabilities. It is the last that are the most serious. They are driven by the impedance of the RF cavities and for the regime of total current under consideration for a B Factory, have growth times for the worst modes on the order of a millisecond. (Recall that synchrotron radiation demands RF cavities with power in the 10 MW range.) Such rapid instabilities must be controlled by very powerful feedback systems; that is, systems of wide bandwidth and having considerable amplifier power. It is not novel to employ such systems (they are presently used on a number of machines) but the present demands on power and bandwidth are in excess of current practice.

Because coupled-bunch instabilities need to be reduced as much as possible, there is the need to reduce the impedances of the higher modes in the RF cavities as much as possible. This can be accomplished by making the cavity bore large, damping the higher-order modes, and using as few cavities as possible (i.e., operate at a high gradient). The last demands the ability of "windows" to transmit great RF power, and that requires new technology. The issue of room temperature or superconducting cavities is not yet settled. Notice that the crab-cavities (which will give increased impedance, and therefore are a negative element in the crab crossing scheme) will likely be superconducting cavities as they demand voltage, but do not demand power.

Finally, we should mention other single-beam phenomena that are not limiting, but need to be considered in the design. These include radiation damping, quantum excitation, intra-beam scattering, Touschek scattering, and gas scattering. For example, consideration is being given to whether the vacuum chamber should be made of aluminum and have an antechamber to absorb the synchrotron heating, or whether it is allowable to have a single chamber made of copper.

Of more than passing interest is the collection of ions in the electron beam. This matter is well-known, but still not entirely understood. The clearing of unwanted ions (without introducing excessive impedance from the clearing electrodes, which will drive various instabilities) or without losing luminosity, as will be the case if one imposes a long gap in the train of bunches, is possible, but not easy.

5. Two-Beam Phenomena

The beam-beam interaction is the heart of any collider. But the beam-beam coherent electromagnetic interaction—a particle of one beam interacting with the total electric and magnetic fields of the other beam—is an unwanted component of the collision and, very importantly, puts a severe limit on the operation of the collider. The beam-beam interaction has been studied, both theoretically and experimentally, for decades.[9] This effect is often treated in the "weak-strong" approximation, which consists of one particle interacting with a prescribed intense beam. A proper analysis must, however, include strong-strong phenomena such as coherent beam-beam effects and instabilities.

The beam-beam effect is usually quantified in terms of the linear lens effect of one beam on the other. It is clear, of course, that any linear effect can be compensated and that it is really the *non-linear* part of the interaction (which is proportional to the linear lens effect) that is important. Luminosity, L, of a collider is given by

$$L = \frac{N^+ N^- f k}{4\pi \sigma_x \sigma_y} \quad (1)$$

where N^+, N^- are the bunch particle numbers, f is the frequency of rotation, k is the number of bunches in the collider, and σ_x, σ_y are the horizontal and vertical beam sizes, respectively, (assumed the same in the two rings). The vertical beam-beam strength parameter, ξ_y, is given by

$$\xi_y^\pm = \frac{N^\mp r_e \beta_y^{*\pm}}{2\pi \gamma^\pm (\sigma_x + \sigma_y) \sigma_y} \quad (2)$$

where r_e is the classical electron radius, γ is the energy of the beam in units of rest mass energy, and the β^* value is introduced explicitly.

Combining these formulas we arrive at

$$L = \frac{fk(1+r)}{2r_e} \left[\frac{N^+ \gamma^+ \xi_y^+}{\beta_y^{*+}}\right]^{1/2} \left[\frac{N^- \gamma^- \xi_y^-}{\beta_y^{*-}}\right]^{1/2} \quad (3)$$

where r is the aspect ratio of the beams (1 for round, 0 for flat). In deriving this formula it has been assumed that the beam-beam interaction in the vertical direction is the limiting phenomenon. The beam-beam strength parameter, ξ, both experimentally and theoretically, is within the range 0.03 to 0.06. Thus we see that high luminosity requires high beam current and low β^* (and that these two quantities can be varied arbitrarily provided the beam size is properly adjusted). There are, of course, other limits on the low β^* value and the beam current.

At first sight, it appears that round beams are better than flat beams (by a factor of two), and this effect may be even greater than is explicit if ξ depends on the beam aspect ratio. At present, the dependence of ξ on aspect ratio is moot. It appears to be more difficult to make a low-β^* lattice for round beams than for flat beams, by about a factor of two, which removes the obvious advantage of round beams. Thus it is unclear at this time whether round beams offer any advantage over flat beams.

The beam-beam interaction will be more severe if the bunch is comparable to, or long or longer than, the β^* at the crossing point. This is because β^* increases quickly (quadratically with distance) as one moves away from the crossing point. Thus it is necessary to have short bunches which requires lots of RF voltage. In fact, the necessary length of bunches precludes making β^* very small (and hence limits the amount of luminosity possible with a single pair of bunches).

The beam-beam interaction tends to throw particles out to large amplitudes and this results in short beam lifetime and aggravated detector background. Radiation damping has the opposite effect and it is true that a collider performs better when the radiation damping is large. Just how much damping is required for various operating conditions is not yet clear. It is a matter under study at this time.

The beam-beam interaction can also lead to motion of the beam as a whole (rather than the incoherent effect discussed above). It is important to avoid coherent instabilities, and that appears possible in practice. Finally, then, all projects are not considering moving into new ground with the beam-beam interaction (except in having β^* very small; i.e. of the order of the bunch length), but plan on obtaining the improved luminosity over present colliders by means of having many bunches.

6. Projects

There is great interest, throughout the world, in the development of a B Factory. Serious design studies are now under way at six different institutions; namely, Cornell in Ithaca[10], KEK in Tsukuba[11], INP in Novosibirsk[12], CERN in Geneva (in collaboration with the Paul Scherrer Institute)[13], DESY in Hamburg[14], and SLAC/LBL in Stanford[15]. Four of the projects are based on existing rings; namely PEP at SLAC, the ISR at CERN, CESR at Cornell, and PETRA at DESY. In addition, there are studies, at CERN[16] and at CEBAF[17], of a linac colliding with a ring.

The projects are still in a very preliminary state, with some of them hoping to have a reasonably firm parameter list before the end of the year. It appears at present that there is a convergence of design parameters (linac options aside) so that there is considerable similarity among the various projects. (A year ago, one could not have said that.) To illustrate the range of parameters under study, we show in Table 1 the present design parameters of three of these projects. It seems likely that many of the parameters will change before the projects become actual proposals. The SLAC/LBL parameters are for round beams, but that group is now developing a flat-beam case, which may be what it actually proposes. The Cornell group plans to start with a symmetric collider and then go to an asymmetric case. (The asymmetric case is the one listed.) The Novosibirsk beams have correlated dispersions at the interaction point that result in a "narrowing" of center-of-mass energy spread. This may be desirable from an experimental point of view, but may worsen the beam-beam effect. The last is being studied right now, with initial results looking encouraging.

In conclusion, the construction of a B Factory to study B Meson physics and CP-violation in that system seems, from the beam physics point of view, to be feasible, but challenging. Feasibility studies are now under way to quantify the challenge.

• Acknowledgments

   Thanks are given to the attendees at the Workshop on High-Luminosity Asymmetric Collider Rings, Lawrence Berkeley Laboratory, February 12-16, 1990. I wish to thank Gil Travish for making Figures 3 and 4. I also wish to thank S. Chattopadhyay, G. Lambertson, M. Tigner and M. Zisman for a careful reading of the paper and a number of useful suggestions. This work was supported by the Director, Office of Energy Research, Office of High Energy and Nuclear Physics, Division of High Energy Physics, of the US Department of Energy under contract DE-AC03-76SF00098.

• **References**

1. "Linear-Collider $B\bar{B}$ Factory Conceptual Design", editor D.H. Stock, World Scientific (1987); "The Physics Program of a High-Luminosity Asymmetric B-Factory", SLAC 353, LBL PUB-5245, CALT-68-1588, October, 1989; J. Dorfan, Particle World, to be published (1990).

2. Proceedings of the Workshop on Beam Dynamics Issues of High-Luminosity Asymmetric Collider Rings, American Institute of Physics Conference Proceedings, to be published (1990).

3. The Physics Program of a High-Luminosity Asymmetric B-Factory, SLAC-353, LBL PUB-5245, CALT-68-1588, October 1989.

4. R.B. Palmer, "Energy Scaling, Crab Crossing and the Pair Problem", SLAC-PUB-4707 (1988); K. Oide and K. Yokoya, "Crab Crossing Scheme for Storage Ring Colliders", SLAC-PUB-4832 (1989); K. Oide, "Asymmetric Collider with Crossing Angle", Workshop on High-Luminosity Asymmetric Rings for B-physics, Cal Tech, April 1989, CALT-68-1552 (1989).

5. M. Sands, "The Physics of Electron Storage Rings—An Introduction", Stanford Linear Accelerator Center Report No. SLAC-121. National Technical Information Service, Springfield, Virginia (1970).

6. M.S. Zisman, "Influence of Collective Effects on the Performance of High-Brightness Synchrotron Radiation Sources", Proceedings of JAERI-Riken Symposium on Accelerator Technology for the High-Brilliance Synchrotron Radiation Sources, Tokyo, September 1988, 311-346; S. Chattopadhyay, "Stability of High Brilliance Synchrotron Radiation Sources", 6th National Conference on Synchrotron Radiation Instrumentation, Berkeley, August 1989, to be published in Nucl. Instr. & Methods A; A. Jackson, "The Challenges of Third Generation Synchrotron Light Sources", Proceedings of the XIVth International Conference on High Energy Accelerators, Tsukuba, August 1989, to be published in Particle Accelerators.

7. R. H. Siemann, "B-Factories: A Prospective of B-Physics and Possible Accelerator Design Approaches", Proceedings of the 1988 Linear Accelerator Conference, Newport News, October 1988.

8. R.H. Siemann, "The Accelerator Challenges of B-Factories", Proceedings of the XIVth International Conference on High Energy Accelerators, Tsukuba, August 1989, to be published in Particle Accelerators.

9. E. Keil, "Beam-Beam Effects in Electron and Proton Colliders", Proceedings of the 1989 IEEE Particle Accelerator Conference, IEEE-89CH2669-0, 1731 (1990).

10. K. Berkelman, invited talk presented at the La Thuile Symposium, February 29-March 5, 1988; M. Tigner, Proceedings of the Workshop on Beam Dynamics Issues of High-Luminosity Asymmetric Collider Rings, American Institute of Physics Conference Proceedings, to be published (1990).

11. F. Abe et al., "Proposal for Study of B-Physics by a Detector with Particle Identification and High Resolution Calorimetry at Tristan Accumulator Ring", KEK-Report 1988.

12. A.N. Dubrovin, A.N. Skrinsky, G.N. Tumaiki and A.A. Zholents, "Conceptual Design of a Ring Beauty Factory", EPAC Accelerator Conference, Rome, June 1988, 1, p. 467; A.A. Zholents, Proceedings of the Workshop on Beam Dynamics Issues of High-Luminosity Asymmetric Collider Rings, American Institute of Physics Conference Proceedings, to be published (1990).

13. "Proposal for an Electron Positron Collider for Heavy Flavor Particle Physics and Synchrotron Radiation", PR-88-09, July 1988, report from Paul Scherrer Institute, CH-5234 Villigen, Switzerland.

14. H. Nesemann, W. Schmidt-Parzefall and F. Willeke, "The Use of Petra as a B-Factory", EPAC Accelerator Conference, Rome, June 1988, 1, p. 439.

15. Feasibility Study for an Asymmetric B Factory Based on PEP, Oct. 1989, LBL Pub-5244/SLAC-3521, CALT-68-1589; Investigation of an Asymmetric B Factory in the PEP Tunnel, March 1990, LBL Pub. 5263/SLAC-359/CALT-68-1622.

16. P. Grosse-Wiessmann, C.D. Johnson, D. Möhl, R. Schmidt, W. Weingarten, L. Wood and G. Coignet, "CERN Linac-on-Ring Option", in Proceedings of the Workshop on Beam Dynamics Issues of High-Luminosity Asymmetric Collider Rings, American Institute of Physics Conference Proceedings, to be published (1990).

17. S. Heifets, "The CEBAF B Factory Project", in Proceedings of the Workshop on Beam Dynamics Issues of High-Luminosity Asymmetric Collider Rings, American Institute of Physics Conference Proceedings, to be published (1990).

• *Received and reviewed by A. Poskanzer, May 1990.*

Figures

Fig 1. A diagram showing the various phenomena, and their major interconnections, that must be considered in designing a high-luminosity circular collider. Of course, at some level, every circle is connected to every other circle. Technical feasibility is a dominant consideration and is included, really, at all levels by "knowing what can, and cannot, be done". Notice that cost, which in the last analysis is the determining factor, is completely left out of the diagram. (Figure due to Maury Tigner.)

Fig 2. The logical steps that one takes in designing a high-luminosity collider. Some explanation, and further analysis, are given in Section 2 of the paper along with further details in Sections 3, 4, and 5.

Fig 3. The design of the interaction region of a collider is still in a state of flux, with a number of interesting ideas being considered, but with no consensus as to how best to proceed. One possibility is an S-bend, head-on configuration, which is shown in Fig. 3a. This appears to be good for masking of the detector, while allowing for subsequent modification so as to have crossing at an angle. Other ideas include a configuration where the high-energy beam goes through the centers of focusing quadrupoles, use of combined function magnets, and "tilting" of the detector solenoid so as to facilitate beam separation. In Fig. 3b we show a three dimensional bend (S-vertically and C-horizontally), a "propeller blade" crossing which might be quite advantageous as far as masking is concerned.

Fig 4. A diagram of "crab crossing" which shows how by tilting the bunches (by half the crossing angle, which typically is about 25 mrad) the crossing appears "head-on" in a moving frame (up in the diagram). Notice that one needs to tilt the bunches and then un-tilt them after the crossing. Powerful RF cavities are required to do the necessary gymnastics and they have to be reasonably close to the interaction point and carefully adjusted to avoid introducing synchro-betatron resonances.

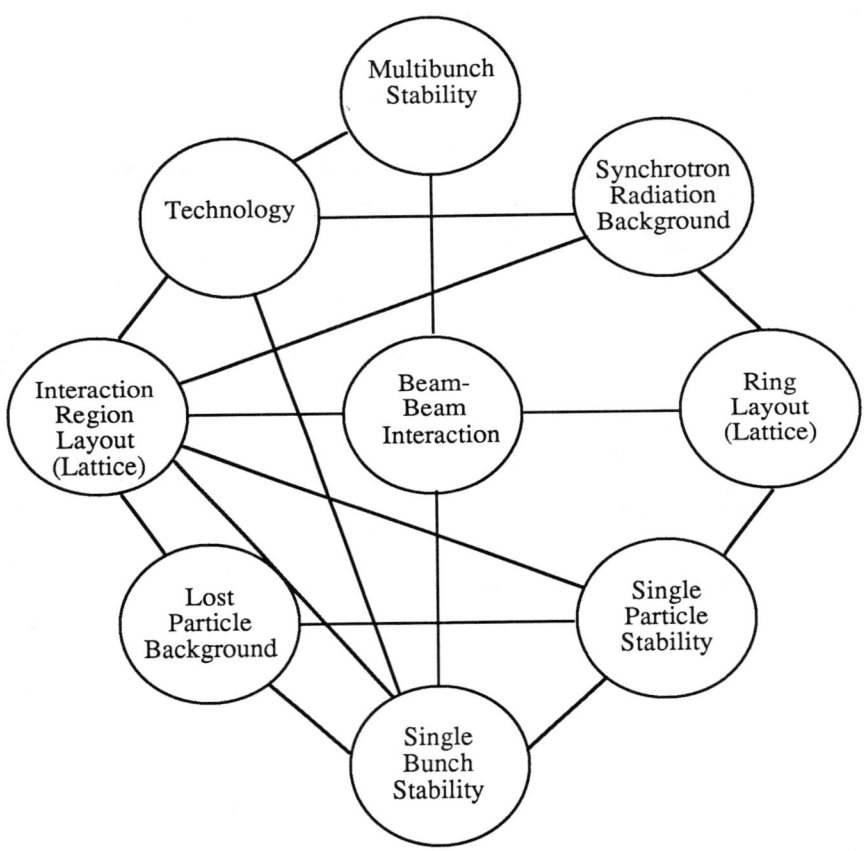

Fig. 1

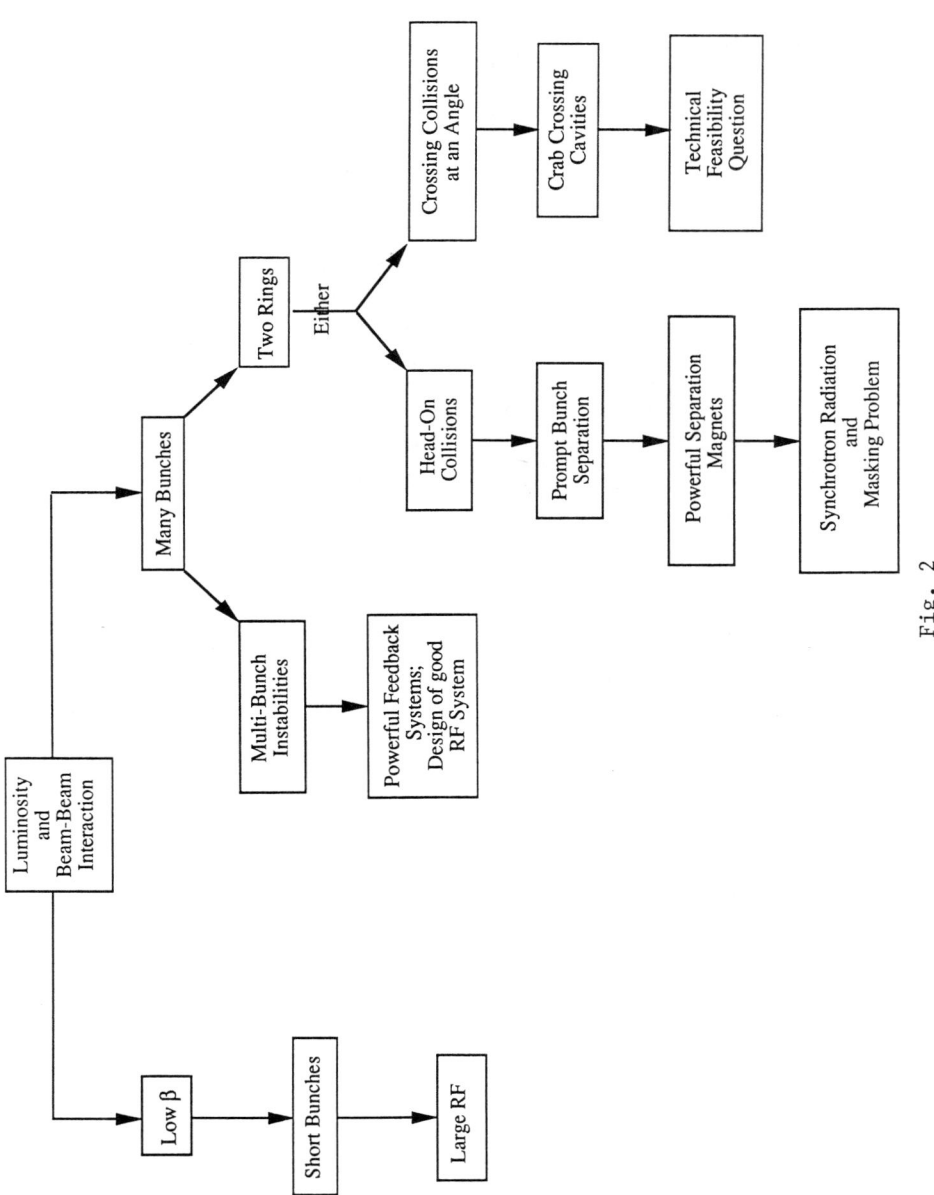

Fig. 2

634 Beam Dynamics Issues

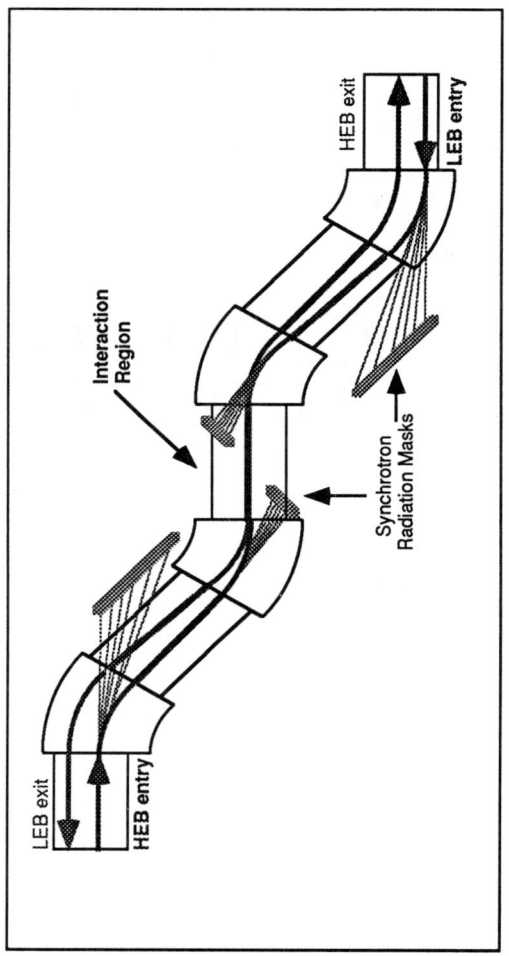

Fig. 3a

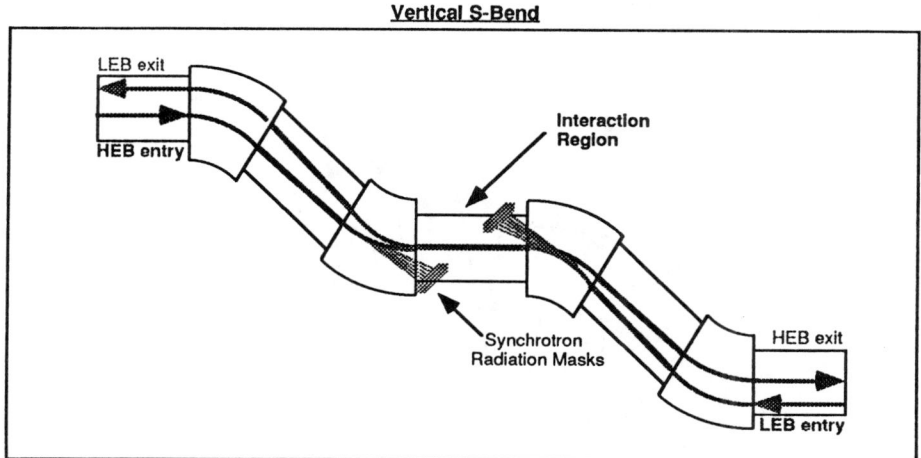

Fig. 3b

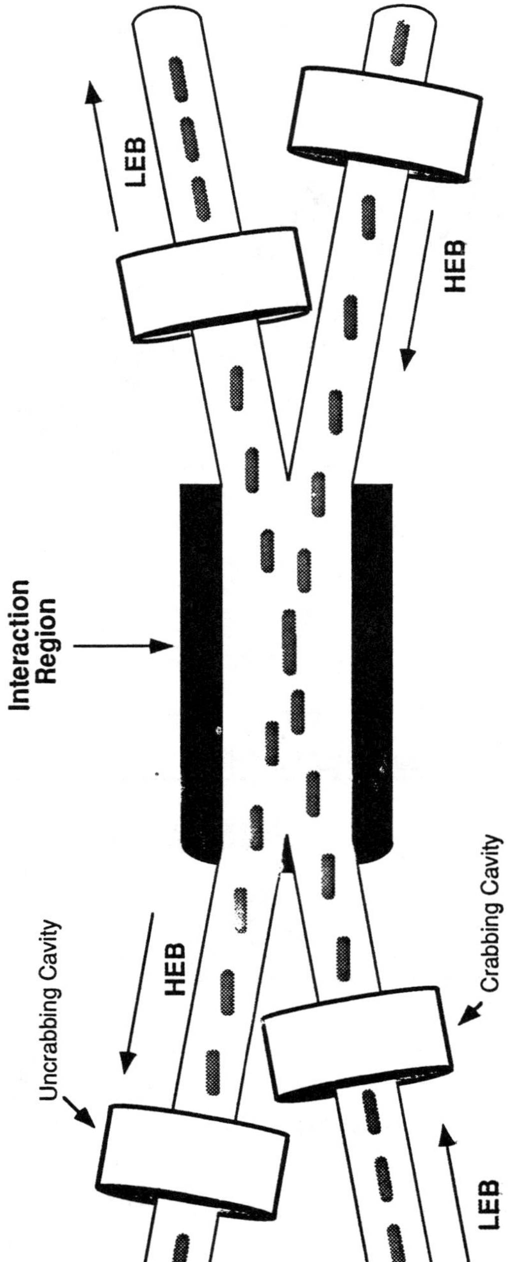

636 Beam Dynamics Issues

Fig. 4

Table I. Typical Parameters of B Factory Projects

	SLAC/LBL[f]		Novosibirsk[c]		Cornell[a]	
	Low-energy ring	High-energy ring	Low-energy ring	High-energy ring	Low-energy ring	High-energy ring
Energy, E [GeV]	3.1	9	4.3	6.5	3.5	8.0
Circumference, C [m]	2200	2200	655	655	768	768
Number of bunches, k_B	1296	1296	156	156	96	96
Particles per bunch, N_b [10^{10}]	7.88	5.44	9	6	37	16
Total current, I [A]	2.23	1.54	1	0.7	2.19	0.96
Emittance						
ε_x [nm·rad]	66	33	8	6.5	78	78
ε_y [nm·rad]	66	33	0.25	0.25	78	78
Bunch length, σ_l [mm]	10	10	7.5	7.5	18	18
Momentum spread, σ_p [10^{-4}]	9.5	6.1	10	10	3.6	8.4
Damping time						
$\tau_{x,y}$ [ms]	32.3	37	17	13	26	26
τ_E [ms]	17.3	18.5	–	–	13	13
Beta functions at IP						
β_x^* [cm]	3	6	60	60	3	3
β_y^* [cm]	3	6	1	1	3	3
Betatron tune						
horizontal, ν_x	37.76	21.28	29	26	7.04	10.7
vertical, ν_y	35.79	18.20	20	13	7.04	10.7
Synchrotron tune, Q_s	0.039	0.053	0.028	0.028	0.05	0.05
Momentum compaction, α	0.00115	0.00245	0.002	0.002	2.02	0.88
RF parameters						
frequency, f_{rf} [MHz]	353.2	353.2	500	500	500	500
voltage, V_{rf} [MV]	10	25	8.8	15	2.2	11.3
Nominal beam-beam tune shift						
ξ_{ox}	0.03	0.03	0.012	0.012	0.045	0.045
ξ_{oy}	0.03	0.03	0.05	0.05	0.045	0.045
Luminosity, L [cm^{-2}s^{-1}]	3×10^{33}		5×10^{33}		5×10^{33}	

Author Index

A

Anami, M., 575
Asami, A., 575
Autin, B., 320

B

Baconnier, Y., 320
Bagley, P., 298
Baigley, P., 6
Balewski, K., 565
Bassetti, M., 389

C

Cappi, R., 419
Channell, P. J., 464
Chao, A., 2
Chattopadhyay, S., 2
Chin, Y. H., 424
Coignet, G., 602
Courant, E., 2

D

DeStaebler, H., 59
Donald, M., 6
Dubrovin, A., 347

E

Enomoto, S., 575

F

Fripp, M., 484
Funakoshi, Y., 575

G

Garren, A., 6, 43, 347
Geyer, C., 6, 565
Grosse Wiesmann, P., 602

H

Hanaoka, K., 575
Hartung, W., 508
Heifets, S. A., 405, 484
Hirata, K., 175, 320, 389, 441
Hoffmann, A., 320
Holzer, B., 6, 565
Hutton, A., 2, 6, 284, 594

J

Jackson, G., 327
Jaeschke, E., 565
Johnson, C. D., 454, 602
Jowett, J., 320

K

Kageyama, T., 575
Kanazawa, K., 575
Keil, E., 2, 25
Kikutani, E., 381, 434, 575
Kirchgesner, J., 508
Kobayashi, J., 575
Koiso, H., 575
Krafft, G. A., 484
Krajcik, P., 6
Krämer, D., 565
Krishnagopal, S., 278
Kurokawa, S., 2, 575

L

Lambertson, G., 2, 35
Lebedev, V. A., 289
Lengeler, H., 320

M

McDowell, C., 484
Moffat, D., 508
Möhl, D., 320, 602
Morton, P., 19
Moshammer, H., 320, 389

N

Nesemann, H., 565
Ng, K.-Y., 411

O

Ohsawa, T., 575
Oide, K., 575
Orlov, Yu., 6, 336

P

Padamsee, H., 508
Pedersen, F., 2, 246
Pellegrini, C., 364
Porter, F., 6
Proch, D., 565

R

Rees, J., 2
Rice, D. H., 219
Risselada, T., 6, 309, 320
Rivkin, L., 536
Robin, D., 364
Rubin, D. L., 235, 508
Ruggiero, F., 389

S

Sakanaka, S., 575
Samed, Y., 508
Saraniti, D., 508
Satoh, K., 575
Schmidt, R., 6, 314, 602
Schmidt-Parzefall, W., 42
Sears, J., 508
Seeman, J., 2, 386
Sekutowics, J., 565
Sessler, A., 2, 622
Shidara, T., 575

Shu, Q. S., 508
Siemann, R. H., 278
Suetake, M., 575

T

Takasaki, F., 575
Tanabe, T., 6
Taylor, T., 320
Teng, L., 6
Tennyson, J. L., 130
Tigner, M., 2, 270, 561

U

Umstätter, H. H., 320

V

Verdier, A., 320
Voelker, F., 35

W

Walker, N., 6
Wang, T., 320
Weingarten, W., 602
Willeke, F., 2, 6, 565
Wilson, E., 6
Wipf, S. G., 565
Wood, L., 454, 602

Y

Yamazaki, Y., 575

Z

Zholents, A., 2, 6, 347, 592
Zisman, M., 2, 81, 594

AIP Conference Proceedings

		L.C. Number	ISBN
No. 83	The Galactic Center (Cal. Inst. of Tech., 1982)	82-71635	0-88318-182-7
No. 84	Physics in the Steel Industry (APS/AISI, Lehigh University, 1981)	82-72033	0-88318-183-5
No. 85	Proton-Antiproton Collider Physics –1981 (Madison, WI)	82-72141	0-88318-184-3
No. 86	Momentum Wave Functions – 1982 (Adelaide, Australia)	82-72375	0-88318-185-1
No. 87	Physics of High Energy Particle Accelerators (Fermilab Summer School, 1981)	82-72421	0-88318-186-X
No. 88	Mathematical Methods in Hydrodynamics and Integrability in Dynamical Systems (La Jolla Institute, 1981)	82-72462	0-88318-187-8
No. 89	Neutron Scattering – 1981 (Argonne National Laboratory)	82-73094	0-88318-188-6
No. 90	Laser Techniques for Extreme Ultraviolet Spectroscopy (Boulder, CO, 1982)	82-73205	0-88318-189-4
No. 91	Laser Acceleration of Particles (Los Alamos, NM, 1982)	82-73361	0-88318-190-8
No. 92	The State of Particle Accelerators and High Energy Physics (Fermilab, 1981)	82-73861	0-88318-191-6
No. 93	Novel Results in Particle Physics (Vanderbilt, 1982)	82-73954	0-88318-192-4
No. 94	X-Ray and Atomic Inner-Shell Physics – 1982 (International Conference, U. of Oregon)	82-74075	0-88318-193-2
No. 95	High Energy Spin Physics – 1982 (Brookhaven National Laboratory)	83-70154	0-88318-194-0
No. 96	Science Underground (Los Alamos, NM, 1982)	83-70377	0-88318-195-9
No. 97	The Interaction Between Medium Energy Nucleons in Nuclei – 1982 (Indiana University)	83-70649	0-88318-196-7
No. 98	Particles and Fields – 1982 (APS/DPF University of Maryland)	83-70807	0-88318-197-5
No. 99	Neutrino Mass and Gauge Structure of Weak Interactions (Telemark, 1982)	83-71072	0-88318-198-3
No. 100	Excimer Lasers – 1983 (OSA, Lake Tahoe, NV)	83-71437	0-88318-199-1
No. 101	Positron-Electron Pairs in Astrophysics (Goddard Space Flight Center, 1983)	83-71926	0-88318-200-9
No. 102	Intense Medium Energy Sources of Strangeness (UC-Santa Cruz, CA, 1983)	83-72261	0-88318-201-7
No. 103	Quantum Fluids and Solids – 1983 (Sanibel Island, FL)	83-72440	0-88318-202-5

No. 104	Physics, Technology and the Nuclear Arms Race (APS, Baltimore, MD, 1983)	83-72533	0-88318-203-3
No. 105	Physics of High Energy Particle Accelerators (SLAC Summer School, 1982)	83-72986	0-88318-304-8
No. 106	Predictability of Fluid Motions (La Jolla Institute, 1983)	83-73641	0-88318-305-6
No. 107	Physics and Chemistry of Porous Media (Schlumberger-Doll Research, 1983)	83-73640	0-88318-306-4
No. 108	The Time Projection Chamber (TRIUMF, Vancouver, 1983)	83-83445	0-88318-307-2
No. 109	Random Walks and Their Applications in the Physical and Biological Sciences (NBS/La Jolla Institute, 1982)	84-70208	0-88318-308-0
No. 110	Hadron Substructure in Nuclear Physics (Indiana University, 1983)	84-70165	0-88318-309-9
No. 111	Production and Neutralization of Negative Ions and Beams (3rd Int'l Symposium) (Brookhaven, NY, 1983)	84-70379	0-88318-310-2
No. 112	Particles and Fields – 1983 (APS/DPF, Blacksburg, VA)	84-70378	0-88318-311-0
No. 113	Experimental Meson Spectroscopy – 1983 (7th International Conference, Brookhaven, NY)	84-70910	0-88318-312-9
No. 114	Low Energy Tests of Conservation Laws in Particle Physics (Blacksburg, VA, 1983)	84-71157	0-88318-313-7
No. 115	High Energy Transients in Astrophysics (Santa Cruz, CA, 1983)	84-71205	0-88318-314-5
No. 116	Problems in Unification and Supergravity (La Jolla Institute, 1983)	84-71246	0-88318-315-3
No. 117	Polarized Proton Ion Sources (TRIUMF, Vancouver, 1983)	84-71235	0-88318-316-1
No. 118	Free Electron Generation of Extreme Ultraviolet Coherent Radiation (Brookhaven/OSA, 1983)	84-71539	0-88318-317-X
No. 119	Laser Techniques in the Extreme Ultraviolet (OSA, Boulder, CO, 1984)	84-72128	0-88318-318-8
No. 120	Optical Effects in Amorphous Semiconductors (Snowbird, UT, 1984)	84-72419	0-88318-319-6
No. 121	High Energy e^+e^- Interactions (Vanderbilt, 1984)	84-72632	0-88318-320-X
No. 122	The Physics of VLSI (Xerox, Palo Alto, CA, 1984)	84-72729	0-88318-321-8
No. 123	Intersections Between Particle and Nuclear Physics (Steamboat Springs, CO, 1984)	84-72790	0-88318-322-6
No. 124	Neutron-Nucleus Collisions: A Probe of Nuclear Structure (Burr Oak State Park, 1984)	84-73216	0-88318-323-4
No. 125	Capture Gamma-Ray Spectroscopy and Related Topics – 1984 (Int'l Symposium, Knoxville, TN)	84-73303	0-88318-324-2

No. 126	Solar Neutrinos and Neutrino Astronomy (Homestake, 1984)	84-63143	0-88318-325-0
No. 127	Physics of High Energy Particle Accelerators (BNL/SUNY Summer School, 1983)	85-70057	0-88318-326-9
No. 128	Nuclear Physics with Stored, Cooled Beams (McCormick's Creek State Park, IN, 1984)	85-71167	0-88318-327-7
No. 129	Radiofrequency Plasma Heating (Sixth Topical Conference) (Callaway Gardens, GA, 1985)	85-48027	0-88318-328-5
No. 130	Laser Acceleration of Particles (Malibu, CA, 1985)	85-48028	0-88318-329-3
No. 131	Workshop on Polarized ^{3}He Beams and Targets (Princeton, NJ, 1984)	85-48026	0-88318-330-7
No. 132	Hadron Spectroscopy–1985 (International Conference, Univ. of Maryland)	85-72537	0-88318-331-5
No. 133	Hadronic Probes and Nuclear Interactions (Arizona State University, 1985)	85-72638	0-88318-332-3
No. 134	The State of High Energy Physics (BNL/SUNY Summer School, 1983)	85-73170	0-88318-333-1
No. 135	Energy Sources: Conservation and Renewables (APS, Washington, DC, 1985)	85-73019	0-88318-334-X
No. 136	Atomic Theory Workshop on Relativistic and QED Effects in Heavy Atoms (Gaithersburg, MD, 1985)	85-73790	0-88318-335-8
No. 137	Polymer-Flow Interaction (La Jolla Institute, 1985)	85-73915	0-88318-336-6
No. 138	Frontiers in Electronic Materials and Processing (Houston, TX, 1985)	86-70108	0-88318-337-4
No. 139	High-Current, High-Brightness, and High-Duty Factor Ion Injectors (La Jolla Institute, 1985)	86-70245	0-88318-338-2
No. 140	Boron-Rich Solids (Albuquerque, NM, 1985)	86-70246	0-88318-339-0
No. 141	Gamma-Ray Bursts (Stanford, CA, 1984)	86-70761	0-88318-340-4
No. 142	Nuclear Structure at High Spin, Excitation, and Momentum Transfer (Indiana University, 1985)	86-70837	0-88318-341-2
No. 143	Mexican School of Particles and Fields (Oaxtepec, México, 1984)	86-81187	0-88318-342-0
No. 144	Magnetospheric Phenomena in Astrophysics (Los Alamos, NM, 1984)	86-71149	0-88318-343-9
No. 145	Polarized Beams at SSC & Polarized Antiprotons (Ann Arbor, MI & Bodega Bay, CA, 1985)	86-71343	0-88318-344-7
No. 146	Advances in Laser Science–I (Dallas, TX, 1985)	86-71536	0-88318-345-5
No. 147	Short Wavelength Coherent Radiation: Generation and Applications (Monterey, CA, 1986)	86-71674	0-88318-346-3
No. 148	Space Colonization: Technology and The Liberal Arts (Geneva, NY, 1985)	86-71675	0-88318-347-1

No. 149	Physics and Chemistry of Protective Coatings (Universal City, CA, 1985)	86-72019	0-88318-348-X
No. 150	Intersections Between Particle and Nuclear Physics (Lake Louise, Canada, 1986)	86-72018	0-88318-349-8
No. 151	Neural Networks for Computing (Snowbird, UT, 1986)	86-72481	0-88318-351-X
No. 152	Heavy Ion Inertial Fusion (Washington, DC, 1986)	86-73185	0-88318-352-8
No. 153	Physics of Particle Accelerators (SLAC Summer School, 1985) (Fermilab Summer School, 1984)	87-70103	0-88318-353-6
No. 154	Physics and Chemistry of Porous Media—II (Ridge Field, CT, 1986)	83-73640	0-88318-354-4
No. 155	The Galactic Center: Proceedings of the Symposium Honoring C. H. Townes (Berkeley, CA, 1986)	86-73186	0-88318-355-2
No. 156	Advanced Accelerator Concepts (Madison, WI, 1986)	87-70635	0-88318-358-0
No. 157	Stability of Amorphous Silicon Alloy Materials and Devices (Palo Alto, CA, 1987)	87-70990	0-88318-359-9
No. 158	Production and Neutralization of Negative Ions and Beams (Brookhaven, NY, 1986)	87-71695	0-88318-358-7
No. 159	Applications of Radio-Frequency Power to Plasma: Seventh Topical Conference (Kissimmee, FL, 1987)	87-71812	0-88318-359-5
No. 160	Advances in Laser Science–II (Seattle, WA, 1986)	87-71962	0-88318-360-9
No. 161	Electron Scattering in Nuclear and Particle Science: In Commemoration of the 35th Anniversary of the Lyman-Hanson-Scott Experiment (Urbana, IL, 1986)	87-72403	0-88318-361-7
No. 162	Few-Body Systems and Multiparticle Dynamics (Crystal City, VA, 1987)	87-72594	0-88318-362-5
No. 163	Pion–Nucleus Physics: Future Directions and New Facilities at LAMPF (Los Alamos, NM, 1987)	87-72961	0-88318-363-3
No. 164	Nuclei Far from Stability: Fifth International Conference (Rosseau Lake, ON, 1987)	87-73214	0-88318-364-1
No. 165	Thin Film Processing and Characterization of High-Temperature Superconductors (Anaheim, CA, 1987)	87-73420	0-88318-365-X
No. 166	Photovoltaic Safety (Denver, CO, 1988)	88-42854	0-88318-366-8
No. 167	Deposition and Growth: Limits for Microelectronics (Anaheim, CA, 1987)	88-71432	0-88318-367-6
No. 168	Atomic Processes in Plasmas (Santa Fe, NM, 1987)	88-71273	0-88318-368-4
No. 169	Modern Physics in America: A Michelson-Morley Centennial Symposium (Cleveland, OH, 1987)	88-71348	0-88318-369-2

No. 170	Nuclear Spectroscopy of Astrophysical Sources (Washington, DC, 1987)	88-71625	0-88318-370-6
No. 171	Vacuum Design of Advanced and Compact Synchrotron Light Sources (Upton, NY, 1988)	88-71824	0-88318-371-4
No. 172	Advances in Laser Science–III: Proceedings of the International Laser Science Conference (Atlantic City, NJ, 1987)	88-71879	0-88318-372-2
No. 173	Cooperative Networks in Physics Education (Oaxtepec, Mexico, 1987)	88-72091	0-88318-373-0
No. 174	Radio Wave Scattering in the Interstellar Medium (San Diego, CA, 1988)	88-72092	0-88318-374-9
No. 175	Non-neutral Plasma Physics (Washington, DC, 1988)	88-72275	0-88318-375-7
No. 176	Intersections Between Particle Land Nuclear Physics (Third International Conference) (Rockport, ME, 1988)	88-62535	0-88318-376-5
No. 177	Linear Accelerator and Beam Optics Codes (La Jolla, CA, 1988)	88-46074	0-88318-377-3
No. 178	Nuclear Arms Technologies in the 1990s (Washington, DC, 1988)	88-83262	0-88318-378-1
No. 179	The Michelson Era in American Science: 1870–1930 (Cleveland, OH, 1987)	88-83369	0-88318-379-X
No. 180	Frontiers in Science: International Symposium (Urbana, IL, 1987)	88-83526	0-88318-380-3
No. 181	Muon-Catalyzed Fusion (Sanibel Island, FL, 1988)	88-83636	0-88318-381-1
No. 182	High T_c Superconducting Thin Films, Devices, and Application (Atlanta, GA, 1988)	88-03947	0-88318-382-X
No. 183	Cosmic Abundances of Matter (Minneapolis, MN, 1988)	89-80147	0-88318-383-8
No. 184	Physics of Particle Accelerators (Ithaca, NY, 1988)	89-83575	0-88318-384-6
No. 185	Glueballs, Hybrids, and Exotic Hadrons (Upton, NY, 1988)	89-83513	0-88318-385-4
No. 186	High-Energy Radiation Background in Space (Sanibel Island, FL, 1987)	89-83833	0-88318-386-2
No. 187	High-Energy Spin Physics (Minneapolis, MN, 1988)	89-83948	0-88318-387-0
No. 188	International Symposium on Electron Beam Ion Sources and their Applications (Upton, NY, 1988)	89-84343	0-88318-388-9
No. 189	Relativistic, Quantum Electrodynamic, and Weak Interaction Effects in Atoms (Santa Barbara, CA, 1988)	89-84431	0-88318-389-7
No. 190	Radio-frequency Power in Plasmas (Irvine, CA, 1989)	89-45805	0-88318-397-8
No. 191	Advances in Laser Science–IV (Atlanta, GA, 1988)	89-85595	0-88318-391-9

No. 192	Vacuum Mechatronics (First International Workshop) (Santa Barbara, CA, 1989)	89-45905	0-88318-394-3
No. 193	Advanced Accelerator Concepts (Lake Arrowhead, CA, 1989)	89-45914	0-88318-393-5
No. 194	Quantum Fluids and Solids—1989 (Gainesville, FL, 1989)	89-81079	0-88318-395-1
No. 195	Dense Z-Pinches (Laguna Beach, CA, 1989)	89-46212	0-88318-396-X
No. 196	Heavy Quark Physics (Ithaca, NY, 1989)	89-81583	0-88318-644-6
No. 197	Drops and Bubbles (Monterey, CA, 1988)	89-46360	0-88318-392-7
No. 198	Astrophysics in Antarctica (Newark, DE, 1989)	89-46421	0-88318-398-6
No. 199	Surface Conditioning of Vacuum Systems (Los Angeles, CA, 1989)	89-82542	0-88318-756-6
No. 200	High T_c Superconducting Thin Films: Processing, Characterization, and Applications (Boston, MA, 1989)	90-80006	0-88318-759-0
No. 201	QED Stucture Functions (Ann Arbor, MI, 1989)	90-80229	0-88318-671-3
No. 202	NASA Workshop on Physics From a Lunar Base (Stanford, CA, 1989)	90-55073	0-88318-646-2
No. 203	Particle Astrophysics: The NASA Cosmic Ray Program for the 1990s and Beyond (Greenbelt, MD, 1989)	90-55077	0-88318-763-9
No. 204	Aspects of Electron–Molecule Scattering and Photoionization (New Haven, CT, 1989)	90-55175	0-88318-764-7
No. 205	The Physics of Electronic and Atomic Collisions (XVI International Conference) (New York, NY, 1989)	90-53183	0-88318-390-0
No. 206	Atomic Processes in Plasmas (Gaithersburg, MD, 1989)	90-55265	0-88318-769-8
No. 207	Astrophysics from the Moon (Annapolis, MD, 1990)	90-55582	0-88318-770-1
No. 208	Current Topics in Shock Waves (Bethlehem, PA, 1989)	90-55617	0-88318-776-0
No. 209	Computing for High Luminosity and High Intensity Facilities (Santa Fe, NM, 1990)	90-55634	0-88318-786-8
No. 210	Production and Neutralization of Negative Ions and Beams (Brookhaven, NY, 1990)	90-55316	0-88318-786-8
No. 211	High-Energy Astrophysics in the 21st Century (Taos, NM, 1989)	90-55644	0-88318-803-1
No. 212	Accelerator Instrumentation (Brookhaven, NY, 1989)	90-55838	0-88318-645-4
No. 213	Frontiers in Condensed Matter Theory (New York, NY, 1989)	90-6421	0-88318-771-X 0-88318-772-8 (pbk.)
No. 214	Beam Dynamics Issues of High-Luminosity Asymmetric Collider Rings (Berkeley, CA, 1990)	90-55857	0-88318-767-1